Ken Wilson Memorial Volume:

Renormalization, Lattice Gauge Theory, the Operator Product Expansion and Quantum Fields

Ken Wilson Memorial Volume:

Renormalization, Lattice Gauge Theory, the Operator Product Expansion and Quantum Fields

Editors

Belal E Baaquie
National University of Singapore, Singapore

Kerson Huang
Massachusetts Institute of Technology, USA

Michael E Peskin
Stanford University, USA

K K Phua
Nanyang Technological University, Singapore

World Scientific

NEW JERSEY • LONDON • SINGAPORE • BEIJING • SHANGHAI • HONG KONG • TAIPEI • CHENNAI

Published by

World Scientific Publishing Co. Pte. Ltd.
5 Toh Tuck Link, Singapore 596224
USA office: 27 Warren Street, Suite 401-402, Hackensack, NJ 07601
UK office: 57 Shelton Street, Covent Garden, London WC2H 9HE

British Library Cataloguing-in-Publication Data
A catalogue record for this book is available from the British Library.

KEN WILSON MEMORIAL VOLUME
Renormalization, Lattice Gauge Theory, the Operator Product Expansion and Quantum Fields

ISBN 978-981-4619-21-9
ISBN 978-981-4619-22-6 (pbk)

In-house Editor: Ng Kah Fee

Typeset by Stallion Press
Email: enquiries@stallionpress.com

Printed in Singapore

Foreword

Kenneth Geddes Wilson (1936–2013) was a legend in his own time, with his ideas having a wide ranging and lasting impact on many branches of Physics.

The unifying theme in Wilson's research was the study of quantum field theory, starting from his PhD degree, earned in 1961. The 1960's was a lean decade for quantum field theory, with questions raised on its mathematical consistency and soundness and with competing theories for strong interactions. Wilson was undeterred by these challenges being faced by quantum field theory and continued his study on its fundamental foundations.

It would be accurate to state that Wilson brought an unprecedented degree of clarity to our understanding of quantum fields. The vexing problem of infinities was shown to be the natural and inevitable consequence of quantum fields being constituted by infinitely many coupled degrees of freedom. In Wilson's view, to define a quantum field, one must first introduce a cutoff, something analogous to the concept of epsilon in calculus. One lowers the cutoff by integrating out — "thinning" — the high momentum degrees of freedom and which is achieved by the renormalization group transformation. The repeated application of the transformation gives rise to the divergences of quantum field theory. The existence of a fixed point of the renormalization group transformation, termed as Wilson's vision, ensures that a finite theory results from the procedure of renormalization, which entails taking the cutoff to infinity.

The physics of phase transitions was the first place where Wilson applied his concept of quantum fields. It is worth noting that the theory of phase transitions is based on classical statistical mechanics, and what it has in common with quantum fields is that both are driven by infinitely many coupled degrees of freedom. Wilson's work showed that the mathematics of quantum field theory is applicable to phenomena that lie outside quantum physics, and in doing so provided a leading exemplar of what can be termed as quantum mathematics.

Wilson's work on the operator product expansion opened the way for understanding high energy scattering experiments; his formulation of gauge fields and fermions on a lattice provided a rigorous foundation of the Standard Model, as well as a construction that is amenable to numerical studies.

The contributions to this memorial volume span the diverse subjects to which Wilson contributed to, as well as recollections by those who knew him personally. It is a worthwhile tribute to a great scholar and thinker, whose seminal and profound discoveries will surely continue to illuminate theoretical physics for many years to come.

Belal E. Baaquie
January, 2015

Foreword

"Thinking outside the box" has become a cliché; but it may be said that Ken Wilson truly did that.

Even theoretical physics has its "boxes". In Newton's time, for example, geometrical proofs were regarded as superior to analytical ones, and Newton broke out of it by inventing calculus.

What was Ken Wilson's box?

It was the notion that only "fundamental" theories and exact solutions are worthy goals. Outsiders, such as phenomenology and numerical computation, were relegated to second-class status.

People in the box were embarrassed by the "ultraviolet catastrophe", and the need of a high-momentum cutoff in quantum electrodynamics. They sighed in relief when the mathematical trick of renormalization buried the cutoff in observable quantities. They declared, falsely, that the cutoff had been "sent to infinity".

Ken has always had a fondness for computation. When Hewlett-Packard came out with a programmable calculator in the nineteen-seventies, he was among the first to own one, on which he promptly coded the Schrödinger equation. And he invented lattice gauge theory to put QCD on a computer, realizing that it requires a new way of thinking, and not merely replacing derivatives by finite differences. The lattice spacing here furnishes a built-in high-momentum cutoff, and it becomes clear that the cutoff cannot be "sent to infinity" by declaration, for it is a scale parameter, and does not explicitly appear in the lattice equations. The only way to change the scale parameter is to "tune" the coupling constants without changing the basic identity of the theory, through a modification of the effective Lagrangian. And this is renormalization.

The renormalization theory leads to novel concepts, such as renormalization-group trajectories, fixed points, and crossovers. It blurs the distinction between "fundamental" and phenomenological theories. It opens the door to a deeper understanding of scale transformations, which is yet to be achieved.

In this volume of essays and memoirs, we remember Ken and his landmark contributions to physics.

Kerson Huang
January, 2015

Contents

Chapter 1

Effective field theory, past and future

Steven Weinberg

Theory Group, Department of Physics, University of Texas
Austin, TX, 78712
weinberg@physics.utexas.edu

This is based on a talk I gave at the 6th International Conference on Chiral Dynamics at Bern in July, 2009. (It is available on line at http://pos.sissa.it/cgi-bin/reader/conf.cgi?confid=86.) In this talk I reminisced about the early development of effective field theories of the strong interactions, and acknowledged the important influence of the work of Kenneth Wilson. I then commented briefly on some other applications of effective field theories, and took up the idea that the Standard Model and General Relativity are the leading terms in an effective field theory that is not renormalizable in the usual sense. I noted the essential role of the Wilsonian renormalization group in formulating such theories. Finally, I cited recent calculations that suggest that the effective field theory of gravitation and matter is asymptotically safe, a possibility that had been suggested to me by the existence in various theories of a Wilson-Fisher fixed point.

I have been asked by the organizers of this meeting to "celebrate 30 years" of a paper[1] on effective field theories that I wrote in 1979. I am quoting this request at the outset, because in the first half of this talk I will be reminiscing about my own work on effective field theories, leading up to this 1979 paper. I think it is important to understand how confusing these things seemed back then, and no one knows better than I do how confused I was. But I am sure that many in this audience know more than I do about the applications of effective field theory to the strong interactions since 1979, so I will mention only some early applications to strong interactions and a few applications to other areas of physics. I will then describe how we have come to think that our most fundamental theories, the Standard Model and General Relativity, are the leading terms in an effective field theory. Finally, I will report on recent work of others that lends support to a suggestion that this effective field theory may actually be a fundamental theory, valid at all energies.

It all started with current algebra. As everyone knows, in 1960 Yoichiro Nambu had the idea that the axial vector current of beta decay could be considered to be conserved in the same limit that the pion, the lightest hadron, could be considered massless.[2] This assumption would follow if the axial vector current was associated with a spontaneously broken approximate symmetry, with the pion playing the role of a Goldstone boson.[3] Nambu used this idea to explain the success of the Goldberger-Treiman formula[4] for the pion decay amplitude, and with his collaborators he was able to derive formulas for the rate of emission of a single low energy pion in various collisions.[5] In this work it was not necessary to assume anything about the nature of the broken symmetry — only that there was some approximate symmetry responsible for the approximate conservation of the axial vector current and the approximate masslessness of the pion. But to deal with processes involving more than one pion, it was necessary to use not only the approximate conservation of the current but also the current commutation relations, which of course do depend on the underlying broken symmetry. The technology of using these properties of the currents, in which one does not use any specific Lagrangian for the strong interactions, became known as current algebra.[a] It scored a dramatic success in the derivation of the Adler-Weisberger sum rule[7] for the axial vector beta decay coupling constant g_A, which showed that the current commutation relations are those of $SU(2) \times SU(2)$.

When I started in the mid-1960s to work on current algebra, I had the feeling that, despite the success of the Goldberger-Treiman relation and the Adler-Weisberger sum rule, there was then rather too much emphasis on the role that the axial vector current plays in weak interactions. After all, even if there were no weak interactions, the fact that the strong interactions have an approximate but spontaneously broken $SU(2) \times SU(2)$ symmetry would be a pretty important piece of information about the strong interactions.[b] To demonstrate the point, I was able to use current algebra to derive successful formulas for the pion-pion and pion-nucleon scattering lengths.[9] When combined with a well-known dispersion relation[10] and the Goldberger-Treiman relation, these formulas for the pion–nucleon scattering lengths turned out to imply the Adler-Weisberger sum rule.

In 1966 I turned to the problem of calculating the rate of processes in which arbitrary numbers of low energy massless pions are emitted in the collision of other hadrons. This was not a problem that urgently needed to be solved. I was interested in it because a year earlier I had worked out simple formulas for the rate of emission of arbitrary numbers of soft gravitons or photons in any collision,[11] and I was curious whether anything equally simple could be said about soft pions. Calculating the amplitude for emission of several soft pions by use of the technique of current algebra turned out to be fearsomely complicated; the non-vanishing commutators of the currents associated with the soft pions prevented the derivation of anything

[a] The name may be due to Murray Gell-Mann. The current commutation relations were given in Ref. 6.

[b] I emphasized this point in my rapporteur's talk on current algebra at the 1968 "Rochester" conference; see Ref. 8.

as simple as the results for soft photons or gravitons, except in the special case in which all pions have the same charge.[12]

Then some time late in 1966 I was sitting at the counter of a café in Harvard Square, scribbling on a napkin the amplitudes I had found for emitting two or three soft pions in nucleon collisions, and it suddenly occurred to me that these results looked very much like what would be given by lowest order Feynman diagrams in a quantum field theory in which pion lines are emitted from the external nucleon lines, with a Lagrangian in which the nucleon interacts with one, two, and more pion fields. Why should this be? Remember, this was a time when theorists had pretty well given up the idea of applying specific quantum field theories to the strong interactions, because there was no reason to trust the lowest order of perturbation theory, and no way to sum the perturbation series. What was popular was to exploit tools such as current algebra and dispersion relations that did not rely on assumptions about particular Lagrangians.

The best explanation that I could give then for the field-theoretic appearance of the current algebra results was that these results for the emission of n soft pions in nucleon collisions are of the minimum order, G_π^n, in the pion-nucleon coupling constant G_π, except that one had to use the exact values for the collision amplitudes without soft pion emission, and divide by factors of the axial vector coupling constant $g_A \simeq 1.2$ in appropriate places. Therefore any Lagrangian that satisfied the axioms of current algebra would have to give the same answer as current algebra in lowest order perturbation theory, except that it would have to be a field theory in which soft pions were emitted only from external lines of the diagram for the nucleon collisions, for only then would one know how to put in the correct factors of g_A and the correct nucleon collision amplitude.

The time-honored renormalizable theory of nucleons and pions with conserved currents that satisfied the assumptions of current algebra was the "linear σ-model,"[c] with Lagrangian (in the limit of exact current conservation):

$$\begin{aligned}\mathcal{L} = &-\frac{1}{2}\left[\partial_\mu\vec{\pi}\cdot\partial^\mu\vec{\pi} + \partial_\mu\sigma\,\partial^\mu\sigma\right]\\ &-\frac{m^2}{2}\left(\sigma^2+\vec{\pi}^2\right) - \frac{\lambda}{4}\left(\sigma^2+\vec{\pi}^2\right)^2\\ &-\bar{N}\gamma^\mu\partial_\mu N - G_\pi\bar{N}\left(\sigma + 2i\gamma_5\vec{\pi}\cdot\vec{t}\right)N\,, \end{aligned}\tag{1.1}$$

where N, $\vec{\pi}$, and σ are the fields of the nucleon doublet, pion triplet, and a scalar singlet, and $\vec{t}$ is the nucleon isospin matrix (with $\vec{t}^2 = 3/4$). This Lagrangian has an $SU(2)\times SU(2)$ symmetry (equivalent as far as current commutation relations are concerned to an $SO(4)$ symmetry), that is spontaneously broken for $m^2 < 0$ by the expectation value of the σ field, given in lowest order by $<\sigma>= F/2 \equiv \sqrt{-m^2/\lambda}$,

[c]This theory, with the inclusion of a symmetry-breaking term proportional to the σ field, was intended to provide an illustration of a "partially conserved axial current," that is, one whose divergence is proportional to the pion field. See Ref. 13.

which also gives the nucleon a lowest order mass $2G_\pi F$. But with a Lagrangian of this form soft pions could be emitted from internal as well as external lines of the graphs for the nucleon collision itself, and there would be no way to evaluate the pion emission amplitude without having to sum over the infinite number of graphs for the nucleon collision amplitude.

To get around this obstacle, I used the chiral $SO(4)$ symmetry to rotate the chiral four-vector into the fourth direction

$$\left(\vec{\pi}, \sigma\right) \mapsto \left(0, \sigma'\right), \qquad \sigma' = \sqrt{\sigma^2 + \vec{\pi}^2}\,, \tag{1.2}$$

with a corresponding chiral transformation $N \mapsto N'$ of the nucleon doublet. The chiral symmetry of the Lagrangian would result in the pion disappearing from the Lagrangian, except that the matrix of the rotation (1.2) necessarily, like the fields, depends on spacetime position, while the theory is only invariant under spacetime-*independent* chiral rotations. The pion field thus reappears as a parameter in the $SO(4)$ rotation (1.2), which could conveniently be taken as

$$\vec{\pi}' \equiv F\vec{\pi}/[\sigma + \sigma']\,. \tag{1.3}$$

But the rotation parameter $\vec{\pi}'$ would not appear in the transformed Lagrangian if it were independent of the spacetime coordinates, so wherever it appears it must be accompanied with at least one derivative. This derivative produces a factor of pion four-momentum in the pion emission amplitude, which would suppress the amplitude for emitting soft pions, if this factor were not compensated by the pole in the nucleon propagator of an external nucleon line to which the pion emission vertex is attached. Thus, with the Lagrangian in this form, pions of small momenta can only be emitted from external lines of a nucleon collision amplitude. This is what I needed.

Since σ' is chiral-invariant, it plays no role in maintaining the chiral invariance of the theory, and could therefore be replaced with its lowest-order expectation value $F/2$. The transformed Lagrangian (now dropping primes) is then

$$\begin{aligned}\mathcal{L} = &-\frac{1}{2}\left[1 + \frac{\vec{\pi}^2}{F^2}\right]^{-2} \partial_\mu\vec{\pi}\cdot\partial^\mu\vec{\pi} \\ &-\bar{N}\Bigg[\gamma^\mu\partial_\mu + G_\pi F/2 \\ &\quad + i\gamma^\mu\left[1 + \frac{\vec{\pi}^2}{F^2}\right]^{-1}\left[\frac{2}{F}\gamma_5\vec{t}\cdot\partial_\mu\vec{\pi} + \frac{2}{F^2}\vec{t}\cdot(\vec{\pi}\times\partial_\mu\vec{\pi})\right]\Bigg]N\,.\end{aligned} \tag{1.4}$$

In order to reproduce the results of current algebra, it is only necessary to identify F as the pion decay amplitude $F_\pi \simeq 184$ MeV, replace the term $G_\pi F/2$ in the nucleon bilinear with the actual nucleon mass m_N (given by the Goldberger–Treiman relation as $G_\pi F_\pi/2g_A$), and replace the pseudovector pion-nucleon coupling $1/F$ with its actual value $G_\pi/2m_N = g_A/F_\pi$. This gives an effective Lagrangian

$$\begin{aligned}\mathcal{L}_{\text{eff}} = &-\frac{1}{2}\left[1+\frac{\vec{\pi}^2}{F_\pi^2}\right]^{-2}\partial_\mu\vec{\pi}\cdot\partial^\mu\vec{\pi} \\ &-\bar{N}\left[\gamma^\mu\partial_\mu + m_N\right. \\ &\left.+\,i\gamma^\mu\left[1+\frac{\vec{\pi}^2}{F_\pi^2}\right]^{-1}\left[\frac{G_\pi}{m_N}\gamma_5\vec{t}\cdot\partial_\mu\vec{\pi}+\frac{2}{F_\pi^2}\vec{t}\cdot(\vec{\pi}\times\partial_\mu\vec{\pi})\right]\right]N\,. \end{aligned} \tag{1.5}$$

To take account of the finite pion mass, the linear sigma model also includes a chiral-symmetry breaking perturbation proportional to σ. Making the chiral rotation (1.2), replacing σ' with the constant $F/2$, and adjusting the coefficient of this term to give the physical pion mass m_π gives a chiral symmetry breaking term

$$\Delta\mathcal{L}_{\text{eff}} = -\frac{1}{2}\left[1+\frac{\vec{\pi}^2}{F_\pi^2}\right]^{-1} m_\pi^2\,\vec{\pi}^2\,. \tag{1.6}$$

Using $\mathcal{L}_{\text{eff}} + \Delta\mathcal{L}_{\text{eff}}$ in lowest order perturbation theory, I found the same results for low-energy pion-pion and pion-nucleon scattering that I had obtained earlier with much greater difficulty by the methods of current algebra.[14]

A few months after this work, Julian Schwinger remarked to me that it should be possible to skip this complicated derivation, forget all about the linear σ-model, and instead infer the structure of the Lagrangian directly from the non-linear chiral transformation properties of the pion field appearing in (1.5).[d] It was a good idea. I spent the summer of 1967 working out these transformation properties, and what they imply for the structure of the Lagrangian.[16] It turns out that if we require that the pion field has the usual linear transformation under $SO(3)$ isospin rotations (because isospin symmetry is supposed to be not spontaneously broken), then there is a *unique* $SO(4)$ chiral transformation that takes the pion field into a function of itself — unique, that is, up to possible redefinition of the field. For an infinitesimal $SO(4)$ rotation by an angle ϵ in the $a4$ plane (where $a = 1, 2, 3$), the pion field π_b (labelled with a prime in Eq. (1.3)) changes by an amount

$$\delta_a\pi_b = -i\epsilon F_\pi\left[\frac{1}{2}\left(1-\frac{\vec{\pi}^2}{F_\pi^2}\right)\delta_{ab}+\frac{\pi_a\pi_b}{F_\pi^2}\right]\,. \tag{1.7}$$

Any other field ψ, on which isospin rotations act with a matrix $\vec{t}$, is changed by an infinitesimal chiral rotation in the $a4$ plane by an amount

$$\delta_a\psi = \frac{\epsilon}{F_\pi}\left(\vec{t}\times\vec{\pi}\right)_a\psi\,. \tag{1.8}$$

[d] For Schwinger's own development of this idea, see Ref. 15. It is interesting that in deriving the effective field theory of goldstinos in supersymmetry theories, it is much more transparent to start with a theory with linearly realized supersymmetry and impose constraints analogous to setting $\sigma' = F/2$, than to work from the beginning with supersymmetry realized non-linearly, in analogy to Eq. (1.7); see Z. Komargodski and N. Seiberg, to be published.

This is just an ordinary, though position-dependent, isospin rotation, so a non-derivative isospin-invariant term in the Lagrangian that does not involve pions, like the nucleon mass term $-m_N\bar{N}N$, is automatically chiral-invariant. The terms in Eq. (1.5):

$$-\bar{N}\left[\gamma^\mu\partial_\mu + \frac{2i}{F_\pi^2}\gamma^\mu\left[1+\frac{\vec{\pi}^2}{F_\pi^2}\right]^{-1}\vec{t}\cdot(\vec{\pi}\times\partial_\mu\vec{\pi})\right]N\ , \tag{1.9}$$

and

$$-i\frac{G_\pi}{m_N}\left[1+\frac{\vec{\pi}^2}{F_\pi^2}\right]^{-1}\bar{N}\gamma^\mu\gamma_5\vec{t}\cdot\partial_\mu\vec{\pi}N\ , \tag{1.10}$$

are simply proportional to the most general chiral-invariant nucleon–pion interactions with a single spacetime derivative. The coefficient of the term (1.9) is fixed by the condition that N should be canonically normalized, while the coefficient of (1.10) is chosen to agree with the conventional definition of the pion-nucleon coupling G_π, and is not directly constrained by chiral symmetry. The term

$$-\frac{1}{2}\left[1+\frac{\vec{\pi}^2}{F^2}\right]^{-2}\partial_\mu\vec{\pi}\cdot\partial^\mu\vec{\pi} \tag{1.11}$$

is proportional to the most general chiral invariant quantity involving the pion field and no more than two spacetime derivatives, with a coefficient fixed by the condition that $\vec{\pi}$ should be canonically normalized. The chiral symmetry breaking term (1.6) is the most general function of the pion field without derivatives that transforms as the fourth component of a chiral four-vector. None of this relies on the methods of current algebra, though one can use the Lagrangian (1.5) to calculate the Noether current corresponding to chiral transformations, and recover the Goldberger-Treiman relation in the form $g_A = G_\pi F_\pi/2m_N$.

This sort of direct analysis was subsequently extended by Callan, Coleman, Wess, and Zumino to the transformation and interactions of the Goldstone boson fields associated with the spontaneous breakdown of any Lie group G to any subgroup H.[17] Here, too, the transformation of the Goldstone boson fields is unique, up to a redefinition of the fields, and the transformation of other fields under G is uniquely determined by their transformation under the unbroken subgroup H. It is straightforward to work out the rules for using these ingredients to construct effective Lagrangians that are invariant under G as well as H.[e] Once again, the key point is that the invariance of the Lagrangian under G would eliminate all

[e]There is a complication. In some cases, such as $SU(3)\times SU(3)$ spontaneously broken to $SU(3)$, fermion loops produce G-invariant terms in the action that are not the integrals of G-invariant terms in the Lagrangian density; see Ref. 18. The most general such terms in the action, whether or not produced by fermion loops, have been cataloged.[19] It turns out that for $SU(N)\times SU(N)$ spontaneously broken to the diagonal $SU(N)$, there is just one such term for $N\geq 3$, and none for $N=1$ or 2. For $N=3$, this term is the one found by Wess and Zumino.

presence of the Goldstone boson field in the Lagrangian if the field were spacetime-independent, so wherever functions of this field appear in the Lagrangian they are always accompanied with at least one spacetime derivative.

In the following years, effective Lagrangians with spontaneously broken $SU(2) \times SU(2)$ or $SU(3) \times SU(3)$ symmetry were widely used in lowest-order perturbation theory to make predictions about low energy pion and kaon interactions.[20] But during this period, from the late 1960s to the late 1970s, like many other particle physicists I was chiefly concerned with developing and testing the Standard Model of elementary particles. As it happened, the Standard Model did much to clarify the basis for chiral symmetry. Quantum chromodynamics with N light quarks is automatically invariant under a $SU(N) \times SU(N)$ chiral symmetry,[f] broken in the Lagrangian only by quark masses, and the electroweak theory tells us that the currents of this symmetry (along with the quark number currents) are just those to which the $W^{\pm}$, Z^0, and photon are coupled.

During this whole period, effective field theories appeared as only a device for more easily reproducing the results of current algebra. It was difficult to take them seriously as dynamical theories, because the derivative couplings that made them useful in the lowest order of perturbation theory also made them nonrenormalizable, thus apparently closing off the possibility of using these theories in higher order.

My thinking about this began to change in 1976. I was invited to give a series of lectures at Erice that summer, and took the opportunity to learn the theory of critical phenomena by giving lectures about it.[g] In preparing these lectures, I was struck by Kenneth Wilson's device of "integrating out" short-distance degrees of freedom by introducing a variable ultraviolet cutoff, with the bare couplings given a cutoff dependence that guaranteed that physical quantities are cutoff independent. Even if the underlying theory is renormalizable, once a finite cutoff is introduced it becomes necessary to introduce every possible interaction, renormalizable or not, to keep physics strictly cutoff independent. From this point of view, it doesn't make much difference whether the underlying theory is renormalizable or not. Indeed, I realized that even without a cutoff, as long as every term allowed by symmetries is included in the Lagrangian, there will always be a counterterm available to absorb every possible ultraviolet divergence by renormalization of the corresponding coupling constant. Non-renormalizable theories, I realized, are just as renormalizable as renormalizable theories.

---

[f] For a while it was not clear why there was not also a chiral $U(1)$ symmetry, that would also be broken in the Lagrangian only by the quark masses, and would either lead to a parity doubling of observed hadrons, or to a new light pseudoscalar neutral meson, both of which possibilities were experimentally ruled out. It was not until 1976 that 't Hooft pointed out that the effect of triangle anomalies in the presence of instantons produced an intrinsic violation of this unwanted chiral $U(1)$ symmetry; see Ref. 21.

[g] My constant companion in the summer of 1976 was the review article by K. G. Wilson and J. B. Kogut, *Physics Reports* **12**, 25 (1974). For the published version of my Erice lectures, see Ref. 22.

This opened the door to the consideration of a Lagrangian containing terms like (1.5) as the basis for a legitimate dynamical theory, not limited to the tree approximation, provided one adds every one of the infinite number of other, higher-derivative, terms allowed by chiral symmetry.[h] But for this to be useful, it is necessary that in some sort of perturbative expansion, only a finite number of terms in the Lagrangian can appear in each order of perturbation theory.

In chiral dynamics, this perturbation theory is provided by an expansion in powers of small momenta and pion masses. At momenta of order m_π, the number ν of factors of momenta or m_π contributed by a diagram with L loops, E_N external nucleon lines, and V_i vertices of type i, for any reaction among pions and/or nucleons, is

$$\nu = \sum_i V_i \left(d_i + \frac{n_i}{2} + m_i - 2 \right) + 2L + 2 - \frac{E_N}{2} , \tag{1.12}$$

where d_i, n_i, and m_i are respectively the numbers of derivatives, factors of nucleon fields, and factors of pion mass (or more precisely, half the number of factors of u and d quark masses) associated with vertices of type i. As a consequence of chiral symmetry, the minimum possible value of $d_i + n_i/2 + m_i$ is 2, so the leading diagrams for small momenta are those with $L = 0$ and any number of interactions with $d_i + n_i/2 + m_i = 2$, which are the ones given in Eqs. (1.5) and (1.6). To next order in momenta, we may include tree graphs with any number of vertices with $d_i + n_i/2 + m_i = 2$ and just one vertex with $d_i + n_i/2 + m_i = 3$ (such as the so-called σ-term). To next order, we include any number of vertices with $d_i + n_i/2 + m_i = 2$, plus either a single loop, or a single vertex with $d_i + n_i/2 + m_i = 4$ which provides a counterterm for the infinity in the loop graph, or two vertices with $d_i + n_i/2 + m_i = 3$. And so on. Thus one can generate a power series in momenta and m_π, in which only a few new constants need to be introduced at each new order. As an explicit example of this procedure, I calculated the one-loop corrections to pion–pion scattering in the limit of zero pion mass, and of course I found the sort of corrections required to this order by unitarity.[i]

But even if this procedure gives well-defined finite results, how do we know they are true? It would be extraordinarily difficult to justify any calculation involving loop graphs using current algebra. For me in 1979, the answer involved a radical reconsideration of the nature of quantum field theory. From its beginning in the late 1920s, quantum field theory had been regarded as the application of quantum mechanics to fields that are among the fundamental constituents of the universe — first the electromagnetic field, and later the electron field and fields for other known "elementary" particles. In fact, this became a working definition of an elementary particle — it is a particle with its own field. But for years in teaching courses on

[h]I thought it appropriate to publish this in a festschrift for Julian Schwinger; see Ref. 1.

[i]Unitarity corrections to soft-pion results of current algebra had been considered earlier, for example see Ref. 23.

quantum field theory I had emphasized that the description of nature by quantum field theories is inevitable, at least in theories with a finite number of particle types, once one assumes the principles of relativity and quantum mechanics, plus the cluster decomposition principle, which requires that distant experiments have uncorrelated results. So I began to think that although specific quantum field theories may have a content that goes beyond these general principles, quantum field theory itself does not. I offered this in my 1979 paper as what Arthur Wightman would call a folk theorem: "if one writes down the most general possible Lagrangian, including *all* terms consistent with assumed symmetry principles, and then calculates matrix elements with this Lagrangian to any given order of perturbation theory, the result will simply be the most general possible S-matrix consistent with perturbative unitarity, analyticity, cluster decomposition, and the assumed symmetry properties." So current algebra wasn't needed.

There was an interesting irony in this. I had been at Berkeley from 1959 to 1966, when Geoffrey Chew and his collaborators were elaborating a program for calculating S-matrix elements for strong interaction processes by the use of unitarity, analyticity, and Lorentz invariance, without reference to quantum field theory. I found it an attractive philosophy, because it relied only on a minimum of principles, all well established. Unfortunately, the S-matrix theorists were never able to develop a reliable method of calculation, so I worked instead on other things, including current algebra. Now in 1979 I realized that the assumptions of S-matrix theory, supplemented by chiral invariance, were indeed all that are needed at low energy, but the most convenient way of implementing these assumptions in actual calculations was by good old quantum field theory, which the S-matrix theorists had hoped to supplant.

After 1979, effective field theories were applied to strong interactions in work by Gasser, Leutwyler, Meissner, and many other theorists. My own contributions to this work were limited to two areas — isospin violation, and nuclear forces.

At first in the development of chiral dynamics there had been a tacit assumption that isotopic spin symmetry was a better approximate symmetry than chiral $SU(2) \times SU(2)$, and that the Gell-Mann–Ne'eman $SU(3)$ symmetry was a better approximate symmetry than chiral $SU(3) \times SU(3)$. This assumption became untenable with the calculation of quark mass ratios from the measured pseudoscalar meson masses.[24] It turns out that the d quark mass is almost twice the u quark mass, and the s quark mass is very much larger than either. As a consequence of the inequality of d and u quark masses, chiral $SU(2) \times SU(2)$ is broken in the Lagrangian of quantum chromodynamics not only by the fourth component of a chiral four-vector, as in (1.6), but also by the third component of a different chiral four-vector proportional to $m_u - m_d$ (whose fourth component is a pseudoscalar). There is no function of the pion field alone, without derivatives, with the latter transformation property, which is why pion–pion scattering and the pion masses are described by (1.6) and the first term in (1.5) in leading order, with no isospin breaking aside of course from that due to electromagnetism. But there are non-derivative corrections to pion–nucleon interactions,[25] which at momenta of order m_π are suppressed relative to the derivative coupling terms in (1.5) by just one factor of m_π or momenta:

$$\begin{aligned}\Delta'\mathcal{L}_{\text{eff}} = & -\frac{A}{2}\left(\frac{1-\pi^2/F_\pi^2}{1+\pi^2/F_\pi^2}\right)\bar{N}N \\ & -B\left[\bar{N}t_3N - \frac{2}{F_\pi^2}\left(\frac{\pi_3}{1+\pi^2/F_\pi^2}\right)\bar{N}\vec{t}\cdot\vec{\pi}N\right] \\ & -\frac{iC}{1+\vec{\pi}^2/F_\pi^2}\bar{N}\gamma_5\vec{\pi}\cdot\vec{t}N \\ & -\frac{iD\pi_3}{1+\vec{\pi}^2/F_\pi^2}\bar{N}\gamma_5 N\,, \end{aligned} \tag{1.13}$$

where A and C are proportional to $m_u + m_d$, and B and D are proportional to $m_u - m_d$, with $B \simeq -2.5$ MeV. The A and B terms contribute isospin conserving and violating terms to the so-called σ-term in pion nucleon scattering.

My work on nuclear forces began one day in 1990 while I was lecturing to a graduate class at Texas. I derived Eq. (1.12) for the class, and showed how the interactions in the leading tree graphs with $d_i + n_i/2 + m_i = 2$ were just those given here in Eqs. (1.5) and (1.6). Then, while I was standing at the blackboard, it suddenly occurred to me that there was one other term with $d_i + n_i/2 + m_i = 2$ that I had never previously considered: an interaction with no factors of pion mass and no derivatives (and hence, according to chiral symmetry, no pions), but *four* nucleon fields — that is, a sum of Fermi interactions $(\bar{N}\Gamma N)(\bar{N}\Gamma' N)$, with any matrices Γ and Γ' allowed by Lorentz invariance, parity conservation, and isospin conservation. This is just the sort of "hard core" nucleon–nucleon interaction that nuclear theorists had long known has to be added to the pion-exchange term in theories of nuclear force. But there is a complication — in graphs for nucleon–nucleon scattering at low energy, two-nucleon intermediate states make a large contribution that invalidates the sort of power-counting that justifies the use of the effective Lagrangian (1.5), (1.6) in processes involving only pions, or one low-energy nucleon plus pions. So it is necessary to apply the effective Lagrangian, including the terms $(\bar{N}\Gamma N)(\bar{N}\Gamma' N)$ along with the terms (1.5) and (1.6), to the two-nucleon irreducible nucleon–nucleon potential, rather than directly to the scattering amplitude.[26] This program was initially carried further by Ordoñez, van Kolck, Friar, and their collaborators,[27] and eventually by several others.

The advent of effective field theories generated changes in point of view and suggested new techniques of calculation that propagated out to numerous areas of physics, some quite far removed from particle physics. Notable here is the use of the power-counting arguments of effective field theory to justify the approximations made in the BCS theory of superconductivity.[28] Instead of counting powers of small momenta, one must count powers of the departures of momenta from the Fermi surface. Also, general features of theories of inflation have been clarified by recasting these theories as effective field theories of the inflaton and gravitational fields.[29]

Perhaps the most important lesson from chiral dynamics was that we should keep an open mind about renormalizability. The renormalizable Standard Model of elementary particles may itself be just the first term in an effective field theory that contains every possible interaction allowed by Lorentz invariance and the $SU(3) \times SU(2) \times U(1)$ gauge symmetry, only with the non-renormalizable terms suppressed by negative powers of some very large mass M, just as the terms in chiral dynamics with more derivatives than in Eq. (1.5) are suppressed by negative powers of $2\pi F_\pi \approx m_N$. One indication that there is a large mass scale in some theory underlying the Standard Model is the well-known fact that the three (suitably normalized) running gauge couplings of $SU(3) \times SU(2) \times U(1)$ become equal at an energy of the order of 10^{15} GeV (or, if supersymmetry is assumed, 2×10^{16} GeV, with better convergence of the couplings.)

In 1979 papers by Frank Wilczek and Tony Zee[30] and me[31] independently pointed out that, while the renormalizable terms of the Standard Model cannot violate baryon or lepton conservation,[j] this is not true of the higher non-renormalizable terms. In particular, four-fermion terms can generate a proton decay into antileptons, though not into leptons, with an amplitude suppressed on dimensional grounds by a factor M^{-2}. The conservation of baryon and lepton number in observed physical processes thus may be an accident, an artifact of the necessary simplicity of the leading renormalizable $SU(3) \times SU(2) \times U(1)$-invariant interactions. I also noted at the same time that interactions between a pair of lepton doublets and a pair of scalar doublets can generate a neutrino mass, which is suppressed only by a factor M^{-1}, and that therefore with a reasonable estimate of M could produce observable neutrino oscillations. The subsequent confirmation of neutrino oscillations lends support to the view of the Standard Model as an effective field theory, with M somewhere in the neighborhood of 10^{16} GeV.

Of course, these non-renormalizable terms can be (and in fact, had been) generated in various renormalizable grand-unified theories by integrating out the heavy particles in these theories. Some calculations in the resulting theories can be assisted by treating them as effective field theories.[k] But the important point is that the existence of suppressed baryon- and lepton-nonconserving terms, and some of their detailed properties, should be expected on much more general grounds, *even if the*

[j]This is not true if the effective theory contains fields for the squarks and sleptons of supersymmetry. However, there are no renormalizable baryon or lepton violating terms in "split supersymmetry" theories, in which the squarks and sleptons are superheavy, and only the gauginos and perhaps higgsinos survive to ordinary energies; see Ref. 32.

[k]The effective field theories derived by integrating out heavy particles had been considered.[33] In 1980, in a paper titled "Effective Gauge Theories," I used the techniques of effective field theory to evaluate the effects of integrating out the heavy gauge bosons in grand unified theories on the initial conditions for the running of the gauge couplings down to accessible energies.[34]

underlying theory is not a quantum field theory at all. Indeed, from the 1980s on, it has been increasingly popular to suppose that the theory underlying the Standard Model as well as general relativity is a string theory.

Which brings me to gravitation. Just as we have learned to live with the fact that there is no renormalizable theory of pion fields that is invariant under the chiral transformation (1.7), so also we should not despair of applying quantum field theory to gravitation just because there is no renormalizable theory of the metric tensor that is invariant under general coordinate transformations. It increasingly seems apparent that the Einstein–Hilbert Lagrangian $\sqrt{g}R$ is just the least suppressed term in the Lagrangian of an effective field theory containing every possible generally covariant function of the metric and its derivatives. The application of this point of view to long range properties of gravitation has been most thoroughly developed by John Donoghue and his collaborators.[35] One consequence of viewing the Einstein–Hilbert Lagrangian as one term in an effective field theory is that any theorem based on conventional general relativity, which declares that under certain initial conditions future singularities are inevitable, must be reinterpreted to mean that under these conditions higher terms in the effective action become important.

Of course, there is a problem — the effective theory of gravitation cannot be used at very high energies, say of the order of the Planck mass, no more than chiral dynamics can be used above a momentum of order $2\pi F_\pi \approx 1$ GeV. For purposes of the subsequent discussion, it is useful to express this problem in terms of the Wilsonian renormalization group. The effective action for gravitation takes the form

$$\begin{aligned} I_{\text{eff}} = -\int d^4x\,\sqrt{-\text{Det}g}\Big[& f_0(\Lambda) + f_1(\Lambda)R \\ & + f_{2a}(\Lambda)R^2 + f_{2b}(\Lambda)R^{\mu\nu}R_{\mu\nu} \\ & + f_{3a}(\Lambda)R^3 + \dots\Big]\,, \end{aligned} \tag{1.14}$$

where here Λ is the ultraviolet cutoff, and the $f_n(\Lambda)$ are coupling parameters with a cutoff dependence chosen so that physical quantities are cutoff-independent. We can replace these coupling parameters with dimensionless parameters $g_n(\Lambda)$:

$$\begin{aligned} & g_0 \equiv \Lambda^{-4} f_0\,;\; g_1 \equiv \Lambda^{-2} f_1\,;\; g_{2a} \equiv f_{2a}\,; \\ & g_{2b} \equiv f_{2b}\,;\; g_{3a} \equiv \Lambda^2 f_{3a}\,;\;\dots\quad. \end{aligned} \tag{1.15}$$

Because dimensionless, these parameters must satisfy a renormalization group equation of the form

$$\Lambda \frac{d}{d\Lambda} g_n(\Lambda) = \beta_n\Big(g(\Lambda)\Big)\,. \tag{1.16}$$

In perturbation theory, all but a finite number of the $g_n(\Lambda)$ go to infinity as $\Lambda \to \infty$, which if true would rule out the use of this theory to calculate anything at very high energy. There are even examples, like the Landau pole in quantum electrodynamics and the phenomenon of "triviality" in scalar field theories, in which the couplings blow up at a *finite* value of Λ.

It is usually assumed that this explosion of the dimensionless couplings at high energy is irrelevant in the theory of gravitation, just as it is irrelevant in chiral dynamics. In chiral dynamics, it is understood that at energies of order $2\pi F_\pi \approx m_N$, the appropriate degrees of freedom are no longer pion and nucleon fields, but rather quark and gluon fields. In the same way, it is usually assumed that in the quantum theory of gravitation, when Λ reaches some very high energy, of the order of 10^{15} to 10^{18} GeV, the appropriate degrees of freedom are no longer the metric and the Standard Model fields, but something very different, perhaps strings.

But maybe not. It is just possible that the appropriate degrees of freedom at all energies are the metric and matter fields, including those of the Standard Model. The dimensionless couplings can be protected from blowing up if they are attracted to a finite value g_{n*}. This is known as *asymptotic safety.*[1]

Quantum chromodynamics provides an example of asymptotic safety, but one in which the theory at high energies is not only safe from exploding couplings, but also free. In the more general case of asymptotic safety, the high energy limit g_{n*} is finite, but not commonly zero.

For asymptotic safety to be possible, it is necessary that all the beta functions should vanish at g_{n*}:

$$\beta_n(g_*) = 0 \,. \tag{1.17}$$

It is also necessary that the physical couplings should be on a trajectory that is attracted to g_{n*}. The number of independent parameters in such a theory equals the dimensionality of the surface, known as the *ultraviolet critical surface*, formed by all the trajectories that are attracted to the fixed point. This dimensionality had better be finite, if the theory is to have any predictive power at high energy. For an asymptotically safe theory with a finite-dimensional ultraviolet critical surface, the requirement that couplings lie on this surface plays much the same role as the requirment of renormalizability in quantum chromodynamics — it provides a rational basis for limiting the complexity of the theory.

This dimensionality of the ultraviolet critical surface can be expressed in terms of the behavior of $\beta_n(g)$ for g near the fixed point g_*. Barring unexpected singularities, in this case we have

$$\beta_n(g) \to \sum_m B_{nm}(g_m - g_{*m}) \,, \quad B_{nm} \equiv \left(\frac{\partial \beta_n(g)}{\partial g_m}\right)_* \,. \tag{1.18}$$

The solution of Eq. (1.16) for g near g_* is then

$$g_n(\Lambda) \to g_{n*} + \sum_i u_{in}\, \Lambda^{\lambda_i} \,, \tag{1.19}$$

[1]This was first proposed in my 1976 Erice lectures; see Ref. 22.

where λ_i and u_{in} are the eigenvalues and suitably normalized eigenvectors of B_{nm}:

$$\sum_m B_{nm}\, u_{im} = \lambda_i\, u_{in}\ . \tag{1.20}$$

Because B_{nm} is real but not symmetric, the eigenvalues are either real, or come in pairs of complex conjugates. The dimensionality of the ultraviolet critical surface is therefore equal to the number of eigenvalues of B_{nm} with negative real part. The condition that the couplings lie on this surface can be regarded as a generalization of the condition that quantum chromodynamics, if it were a fundamental and not merely an effective field theory, would have to involve only renormalizable couplings.

It may seem unlikely that an infinite matrix like B_{nm} should have only a finite number of eigenvalues with negative real part, but in fact examples of this are quite common. As we learned from the Wilson–Fisher theory of critical phenomena, when a substance undergoes a second-order phase transition, its parameters are subject to a renormalization group equation that has a fixed point, with a single infrared-repulsive direction, so that adjustment of a single parameter such as the temperature or the pressure can put the parameters of the theory on an infrared attractive surface of co-dimension one, leading to long-range correlations. The single infrared-repulsive direction is at the same time a unique ultraviolet-attractive direction, so the ultraviolet critical surface in such a theory is a one-dimensional curve. Of course, the parameters of the substance on this curve do not really approach a fixed point at very short distances, because at a distance of the order of the interparticle spacing the effective field theory describing the phase transition breaks down.

What about gravitation? There are indications that here too there is a fixed point, with an ultraviolet critical surface of finite dimensionality. Fixed points have been found (of course with $g_{n*} \neq 0$) using dimensional continuation from $2+\epsilon$ to 4 spacetime dimensions,[36] by a $1/N$ approximation (where N is the number of added matter fields),[37] by lattice methods,[38] and by use of the truncated exact renormalization group equation,[39] initiated in 1998 by Martin Reuter. In the last method, which had earlier been introduced in condensed matter physics[40] and then carried over to particle theory,[41] one derives an exact renormalization group equation for the total vacuum amplitude $\Gamma[g,\Lambda]$ in the presence of a background metric $g_{\mu\nu}$ with an *infrared* cutoff Λ. This is the action to be used in calculations of the true vacuum amplitude in calculations of graphs with an *ultraviolet* cutoff Λ. To have equations that can be solved, it is necessary to truncate these renormalization group equations, writing $\Gamma[g,\Lambda]$ as a sum of just a finite number of terms like those shown explicitly in Eq. (1.14), and ignoring the fact that the beta function inevitably does not vanish for the couplings of other terms in $\Gamma[g,\Lambda]$ that in the given truncation are assumed to vanish.

Initially only two terms were included in the truncation of $\Gamma[g,\Lambda]$ (a cosmological constant and the Einstein–Hilbert term $\sqrt{g}R$), and a fixed point was found with two eigenvalues λ_i, a pair of complex conjugates with negative real part. Then a third operator ($R_{\mu\nu}R^{\mu\nu}$ or the equivalent) was added, and a third eigenvalue was found, with λ_i real and negative. This was not encouraging. If each time that new terms were included in the truncation, new eigenvalues appeared with negative real part, then the ultraviolet critical surface would be infinite dimensional, and the theory, though free of couplings that exploded at high energy, would lose all predictive value at high energy.

In just the last few years calculations have been done that allow more optimism. Codello, Percacci, and Rahmede[42] have considered a Lagrangian containing all terms $\sqrt{g}R^n$ with n running from zero to a maximum value $n_{\rm max}$, and find that the ultraviolet critical surface has dimensionality 3 even when $n_{\rm max}$ exceeds 2, up to the highest value $n_{\rm max} = 6$ that they considered, for which the space of coupling constants is 7-dimensional. Furthermore, the three eigenvalues they find with negative real part seem to converge as $n_{\rm max}$ increases, as shown in the following table of ultraviolet-attractive eigenvalues:

$n_{\rm max} = 2:$	$-1.38 \pm 2.32i$	-26.8
$n_{\rm max} = 3:$	$-2.71 \pm 2.27i$	-2.07
$n_{\rm max} = 4:$	$-2.86 \pm 2.45i$	-1.55
$n_{\rm max} = 5:$	$-2.53 \pm 2.69i$	-1.78
$n_{\rm max} = 6:$	$-2.41 \pm 2.42i$	-1.50

In a subsequent paper[43] they added matter fields, and again found just three ultraviolet-attractive eigenvalues. Further, this year Benedetti, Machado, and Sauer-essig[44] considered a truncation with a different four terms, terms proportional to $\sqrt{g}R^n$ with $n = 0, 1$ and 2 and also $\sqrt{g}C_{\mu\nu\rho\sigma}C^{\mu\nu\rho\sigma}$ (where $C_{\mu\nu\rho\sigma}$ is the Weyl tensor) and they too find just three ultraviolet-attractive eigenvalues, also when matter is added. If this pattern of eigenvalues continues to hold in future calculations, it will begin to look as if there is a quantum field theory of gravitation that is well-defined at all energies, and that has just three free parameters.

The natural arena for application of these ideas is in the physics of gravitation at small distance scales and high energy — specifically, in the early universe. A start in this direction has been made by Reuter and his collaborators,[45] but much remains to be done.

Acknowledgments

I am grateful for correspondence about recent work on asymptotic safety with D. Benedetti, D. Litim, R. Percacci, and M. Reuter, and to G. Colangelo and J. Gasser for inviting me to give this talk. This material is based in part on work supported by the National Science Foundation under Grant NO. PHY-0455649 and with support from The Robert A. Welch Foundation, Grant No. F-0014.

References

1. S. Weinberg, *Physica* **A96**, 327 (1979).
2. Y. Nambu, *Phys. Rev. Lett.* **4**, 380 (1960).
3. J. Goldstone, *Nuovo Cimento* **9**, 154 (1961); Y. Nambu and G. Jona-Lasinio, *Phys. Rev.* **122**, 345 (1961); J. Goldstone, A. Salam, and S. Weinberg, *Phys. Rev.* **127**, 965 (1962).
4. M. L. Goldberger and S. B. Treiman, *Phys. Rev.* **111**, 354 (1956).
5. Y. Nambu and D. Lurie, *Phys. Rev.* **125**, 1429 (1962); Y. Nambu and E. Shrauner, *Phys. Rev.* **128**, 862 (1962).
6. M. Gell-Mann, *Physics* **1**, 63 (1964).
7. S. L. Adler, *Phys. Rev. Lett.* **14**, 1051 (1965); *Phys. Rev.* **140**, B736 (1965); W. I. Weisberger, *Phys. Rev. Lett.* **14**, 1047 (1965).
8. S. Weinberg, in *Proceedings of the 14th International Conference on High-Energy Physics*, p. 253.
9. S. Weinberg, *Phys. Rev. Lett.* **17**, 616 (1966). The pion-nucleon scattering lengths were calculated independently by Y. Tomozawa, *Nuovo Cimento* **46A**, 707 (1966).
10. M. L. Goldberger, Y. Miyazawa, and R. Oehme, *Phys. Rev.* **99**, 986 (1955).
11. S. Weinberg, *Phys. Rev.* **140**, B516 (1965).
12. S. Weinberg, *Phys. Rev. Lett.* **16**, 879 (1966).
13. J. Bernstein, S. Fubini, M. Gell-Mann, and W. Thirring, *Nuovo Cimento* **17**, 757 (1960); M. Gell-Mann and M. Lévy, *Nuovo Cimento* **16**, 705 (1960); K. C. Chou, *Soviet Physics JETP* **12**, 492 (1961).
14. S. Weinberg, *Phys. Rev. Lett.* **18**, 188 (1967).
15. J. Schwinger, *Phys. Lett.* **24B**, 473 (1967).
16. S. Weinberg, *Phys. Rev.* **166**, 1568 (1968).
17. S. Coleman, J. Wess, and B. Zumino, *Phys. Rev.* **177**, 2239 (1969); C. G. Callan, S. Coleman, J. Wess, and B. Zumino, *Phys. Rev.* **177**, 2247(1969).
18. J. Wess and B. Zumino, *Phys. Lett.* **37B**, 95 (1971); E. Witten, *Nucl. Phys.* **B223**, 422 (1983).
19. E. D'Hoker and S. Weinberg, *Phys. Rev.* **D50**, R6050 (1994).
20. For reviews, see S. Weinberg, in *Lectures on Elementary Particles and Quantum Field Theory — 1970 Brandeis University Summer Institute in Theoretical Physics, Vol. 1*, ed. S. Deser, M. Grisaru, and H. Pendleton (The M.I.T. Press, Cambridge, MA, 1970); B. W. Lee, *Chiral Dynamics* (Gordon and Breach, New York, 1972).
21. G. 't Hooft, *Phys. Rev.* **D14**, 3432 (1976).
22. S. Weinberg, "Critical Phenomena for Field Theorists," in *Understanding the Fundamental Constituents of Matter*, ed. A. Zichichi (Plenum Press, New York, 1977).
23. H. Schnitzer, *Phys. Rev. Lett.* **24**, 1384 (1970); *Phys. Rev.* **D2**, 1621 (1970); L.-F. Li and H. Pagels, *Phys. Rev. Lett.* **26**, 1204 (1971); *Phys. Rev.* **D5**, 1509 (1972); P. Langacker and H. Pagels, *Phys. Rev.* **D8**, 4595 (1973).
24. S. Weinberg, contribution to a festschrift for I. I. Rabi, *Trans. N. Y. Acad. Sci.* **38**, 185 (1977).
25. S. Weinberg, in *Chiral Dynamics: Theory and Experiment — Proceedings of the Workshop Held at MIT, July 1994* (Springer-Verlag, Berlin, 1995). The terms in Eq. (13) that are odd in the pion field are given in Section 19.5 of S. Weinberg, *The Quantum Theory of Fields*, Vol. II (Cambridge University Press, 1996).

26. S. Weinberg, *Phys. Lett.* **B251**, 288 (1990); *Nucl. Phys.* **B363**, 3 (1991); *Phys. Lett.* **B295**, 114 (1992).
27. C. Ordoñez and U. van Kolck, *Phys. Lett.* **B291**, 459 (1992); C. Ordoñez, L. Ray, and U. van Kolck, *Phys. Rev. Lett.* **72**, 1982 (1994); U. van Kolck, *Phys. Rev.* **C49**, 2932 (1994); U. van Kolck, J. Friar, and T. Goldman, *Phys. Lett. B* **371**, 169 (1996); C. Ordoñez, L. Ray, and U. van Kolck, *Phys. Rev. C* **53**, 2086 (1996); C. J. Friar, *Few-Body Systems Suppl.* **99**, 1 (1996).
28. G. Benfatto and G. Gallavotti, *J. Stat. Phys.* **59**, 541 (1990); *Phys. Rev.* **42**, 9967 (1990); J. Feldman and E. Trubowitz, *Helv. Phys. Acta* **63**, 157 (1990); **64**, 213 (1991); **65**, 679 (1992); R. Shankar, *Physica* **A177**, 530 (1991); *Rev. Mod. Phys.* **66**, 129 (1993); J. Polchinski, in *Recent Developments in Particle Theory, Proceedings of the 1992 TASI*, eds. J. Harvey and J. Polchinski (World Scientific, Singapore, 1993); S. Weinberg, *Nucl. Phys.* **B413**, 567 (1994).
29. C. Cheung, P. Creminilli, A. L. Fitzpatrick, J. Kaplan, and L. Senatore, *J. High Energy Physics* **0803**, 014 (2008); S. Weinberg, *Phys. Rev. D* **73**, 123541 (2008).
30. F. Wilczek and A. Zee, *Phys. Rev. Lett.* **43**, 1571 (1979).
31. S. Weinberg, *Phys. Rev. Lett.* **43**, 1566 (1979).
32. N. Arkani-Hamed and S. Dimopoulos, *JHEP* **0506**, 073 (2005); G. F. Giudice and A. Romanino, *Nucl. Phys. B* **699**, 65 (2004); N. Arkani-Hamed, S. Dimopoulos, G. F. Giudice, and A. Romanino, *Nucl. Phys. B* **709**, 3 (2005); A. Delgado and G. F. Giudice, *Phys. Lett.* **B627**, 155 (2005).
33. T. Appelquist and J. Carrazone, *Phys. Rev.* **D11**, 2856 (1975).
34. S. Weinberg, *Phys. Lett.* **91B**, 51 (1980).
35. J. F. Donoghue, *Phys. Rev.* **D50**, 3874 (1884); *Phys. Lett.* **72**, 2996 (1994); lectures presented at the Advanced School on Effective Field Theories (Almunecar, Spain, June 1995), gr-qc/9512024; J. F. Donoghue, B. R. Holstein, B. Garbrecth, and T. Konstandin, *Phys. Lett.* **B529**, 132 (2002); N. E. J. Bjerrum-Bohr, J. F. Donoghue, and B. R. Holstein, *Phys. Rev.* **D68**, 084005 (2003).
36. S. Weinberg, in *General Relativity*, ed. S. W. Hawking and W. Israel (Cambridge University Press, 1979) p. 700; H. Kawai, Y. Kitazawa, and M. Ninomiya, *Nucl. Phys. B* **404**, 684 (1993); *Nucl. Phys. B* **467**, 313 (1996); T. Aida and Y. Kitazawa, *Nucl. Phys. B* **401**, 427 (1997); M. Niedermaier, *Nucl. Phys. B* **673**, 131 (2003).
37. L. Smolin, *Nucl. Phys.* **B208**, 439 (1982); R. Percacci, *Phys. Rev. D* **73**, 041501 (2006).
38. J. Ambjørn, J. Jurkewicz, and R. Loll, *Phys. Rev. Lett.* **93**, 131301 (2004); *Phys. Rev. Lett.* **95**, 171301 (2005); *Phys. Rev.* **D72**, 064014 (2005); *Phys. Rev.* **D78**, 063544 (2008); and in *Approaches to Quantum Gravity*, ed. D. Oríti (Cambridge University Press).
39. M. Reuter, *Phys. Rev. D* **57**, 971 (1998); D. Dou and R. Percacci, *Class. Quant. Grav.* **15**, 3449 (1998); W. Souma, *Prog. Theor. Phys.* **102**, 181 (1999); O. Lauscher and M. Reuter, *Phys. Rev. D* **65**, 025013 (2001); *Class. Quant. Grav.* **19**, 483 (2002); M. Reuter and F. Saueressig, *Phys Rev. D* **65**, 065016 (2002); O. Lauscher and M. Reuter, *Int. J. Mod. Phys. A* **17**, 993 (2002); *Phys. Rev. D* **66**, 025026 (2002); M. Reuter and F. Saueressig, *Phys. Rev. D* **66**, 125001 (2002); R. Percacci and D. Perini, *Phys. Rev. D* **67**, 081503 (2002); *Phys. Rev. D* **68**, 044018 (2003); D. Perini, *Nucl. Phys. Proc. Suppl. C* **127**, 185 (2004); D. F. Litim, *Phys. Rev. Lett.* **92**, 201301 (2004); A. Codello and R. Percacci, *Phys. Rev. Lett.* **97**, 221301 (2006); A. Codello, R. Percacci, and C. Rahmede, *Int. J. Mod. Phys.* **A23**, 143 (2008); M. Reuter and F. Saueressig, 0708.1317; P. F. Machado and F. Saueressig, *Phys. Rev.* **D77**, 124045 (2008); A. Codello, R. Percacci, and C. Rahmede, *Ann. Phys.* **324**, 414 (2009); A. Codello and R. Percacci, 0810.0715; D. F. Litim 0810.3675; H. Gies and M. M. Scherer, 0901.2459; D. Benedetti,

P. F. Machado, and F. Saueressig, 0901.2984, 0902.4630; M. Reuter and H. Weyer, 0903.2971.
40. F. J. Wegner and A. Houghton, *Phys. Rev.* **A8**, 401 (1973).
41. J. Polchinski, *Nucl. Phys.* **B231**, 269 (1984); C. Wetterich, *Phys. Lett. B* **301**, 90 (1993).
42. A. Codello, R. Percacci, and C. Rahmede, *Int. J. Mod. Phys.* **A23**, 143 (2008).
43. A. Codello, R. Percacci, and C. Rahmede, *Ann. Phys.* **324**, 414 (2009).
44. D. Benedetti, P. F. Machado, and F. Saueressig, 0901.2984, 0902.4630.
45. A. Bonanno and M. Reuter, *Phys. Rev. D* **65**, 043508 (2002); *Phys. Lett.* **B527**, 9 (2002); M. Reuter and F. Saueressig, *J. Cosm. and Astropart. Phys.* **09**, 012 (2005).

Chapter 2

A critical history of renormalization*

Kerson Huang
Massachusetts Institute of Technology, Cambridge, MA 02139, USA
Institute of Advanced Studies, Nanyang Technological University,
Singapore 639673
kersonh@mit.edu

The history of renormalization is reviewed with a critical eye, starting with Lorentz's theory of radiation damping, through perturbative QED with Dyson, Gell-Mann and Low, and others, to Wilson's formulation and Polchinski's functional equation, and applications to "triviality," and dark energy in cosmology.

Keywords: Renormalization; quantum electrodynamics; quantum field theory.

2.1. Dedication

Renormalization, that astounding mathematical trick that enabled one to tame divergences in Feynman diagrams, led to the triumph of quantum electrodynamics. Ken Wilson made it physics, by uncovering its deep connection with scale transformations. The idea that scale determines the perception of the world seems obvious. When one examines an oil painting, for example, what one sees depends on the resolution of the instrument one uses for the examination. At resolutions of the naked eye, one sees art, perhaps, but upon greater and greater magnifications, one sees pigments, then molecules and atoms, and so forth. What is nontrivial is to formulate this mathematically, as a physical theory, and this is what Ken Wilson had achieved. To remember him, I recall some events at the beginning of his physics career.

I first met Ken around 1962, when I was a fresh assistant professor at M.I.T., and Ken a Junior Fellow at Harvard's Society of Fellows. He had just gotten his Ph.D. from Cal. Tech. under Gell-Mann's supervision. In his thesis, he obtained exact solutions of the Low equation, which describes π-meson scattering from a

*To the memory of Kenneth G. Wilson (1936–2013).
This article was originally published in *Int. J. Mod. Phys. A* **28**, 1330050 (2013).

fixed-source nucleus. (He described himself as an "aficionado" of the equation.) I had occasion to refer to this thesis years later, when Francis Low and I proved that the equation does not possess the kind of "bootstrap solution" that Geoffrey Chew advocated.[2,3]

While at the Society of Fellows, Ken spent most of his time at M.I.T. using the computing facilities. He was frequently seen dashing about with stacks of IBM punched cards used then for Fortran programming.

He used to play the oboe in those days, and I played the violin, and we had talked about getting together to play the Bach concerto for oboe and violin with a pianist (for we dare not contemplate an orchestra), but we never got around to that. I had him over for dinner at our apartment on Wendell Street in Cambridge, and received a thank-you postcard a few days later, with an itemized list of the dishes he liked.

At the time, my M.I.T. colleague Ken Johnson was working on nonperturbative QED, on which Ken Wilson had strong opinions. One day, when Francis Low and I went by Johnson's office to pick him up for lunch, we found the two of them in violent argument at the blackboard. So Francis said, "We'll go to lunch, and leave you two scorpions to fight it out."

That was quite a while ago, and Ken went on to do great things, including the theory of renormalization that earned him the Nobel Prize of 1982. In this article, I attempt to put myself in the role of a "physics critic" on this subject. I will concentrate on ideas, and refer technical details to Refs. 4 and 5.

While Ken's work has a strong impact on the theory of critical phenomena, I concentrate here on particle physics.

2.2. Lorentz: electron self-force and radiation damping

After J. J. Thomson discovered the first elementary particle, the electron,[6] the question naturally arose about what it was made of. Lorentz ventured into the subject by regarding the electron as a uniform charge distribution of radius a, held together by unknown forces. As indicated in Fig. 2.1, the charge elements of this distribution exert Coulomb forces on each other, but they do not cancel out, due to retardation. Thus, there is a net "self-force," and Lorentz obtained it in the limit $a \to 0$[7]:

$$F_{\text{self}} = -m_{\text{self}}\,\ddot{x} + \frac{2e^2}{3c^3}\dddot{x} + O(a). \tag{2.1}$$

Internal Coulomb interactions give rise to a "self-mass":

$$m_{\text{self}}\,c^2 = \iint \frac{dq\,dq'}{|r - r'|} \underset{a \to 0}{\longrightarrow} O\left(\frac{1}{a}\right), \tag{2.2}$$

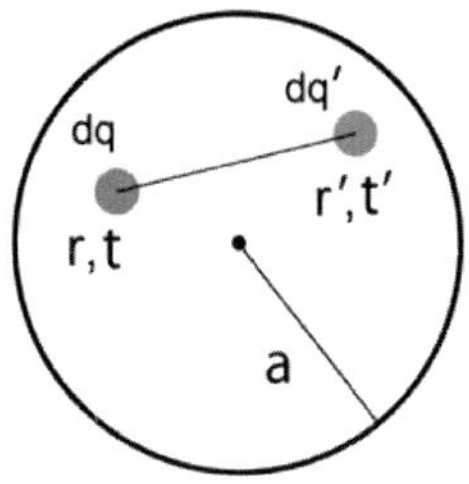

Fig. 2.1. Modeling the classical electron as charge distribution of radius a. The Coulomb forces between charge elements do not add up to zero, because of retardation: $|t' - t| = |r' - r|/c$. Consequently, there is a "self-force," featuring a "self-mass" that diverges in the limit $a \to 0$, but can be absorbed into the physical mass. The finite remainder gives the force of radiation damping.

which diverges linearly when $a \to 0$. This was the first occurrence of the "ultraviolet catastrophe," which befalls anyone toying with the inner structure of elementary particles.

One notices, with great relief, that the self-mass can be absorbed into the physical mass in the equation of motion:

$$\begin{aligned} m_0 \ddot{x} &= F_{\text{self}} + F_{\text{ext}}, \\ \left(m_0 + m_{\text{self}}\right)\ddot{x} &= \frac{2e^2}{3c^3}\dddot{x} + F_{\text{ext}} + O(a), \end{aligned} \tag{2.3}$$

where m_0 is the bare mass. One can takes the physical mass from experiments, and write

$$m\ddot{x} = \frac{2e^2}{3c^3}\dddot{x} + F_{\text{ext}} \tag{2.4}$$

with $m = m_0 + m_{\text{self}}$. One imagines that the divergence of m_{self} is canceled by m_0, which comes from the unknown forces that hold the electron together. This is the earliest example of "mass renormalization." Thus, the $\dddot{x}$ term, the famous radiation damping, is exact in the limit $a \to 0$ within the classical theory. Of course, when a approaches the electron Compton wavelength, this model must be replaced by a quantum-mechanical one, and this leads us to QED (quantum electrodynamics).

2.3. The triumph of QED

Modern QED took shape soon after the advent of the Dirac equation in 1928,[8] and the hole theory in 1930.[9] These theories make the vacuum a dynamical medium containing virtual electron–positron pairs. Weisskopf[10,11] was the first to investigate the electron self-energy in this light, and found that screening by induced pairs reduces the linear divergence in the Lorentz theory to a logarithmic one.[12,13,a]

[a]Weisskopf was then Pauli's assistant. According to his recollection (private communication), he made an error in his first paper and got a quadratic divergence. On day, he got a letter from "an

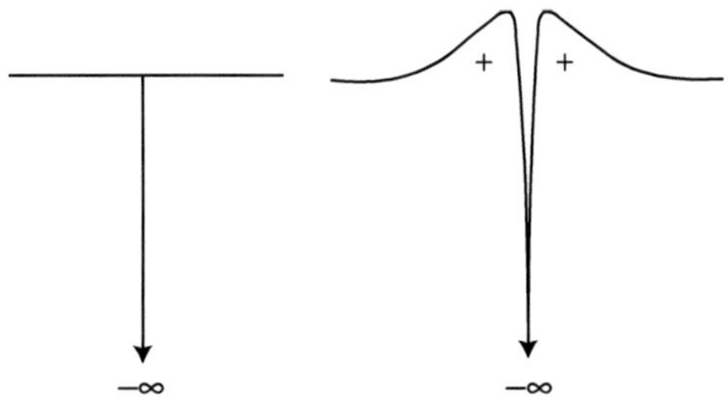

Fig. 2.2. Charge density of the bare electron (left) and that of the physical electron, which is "dressed" by vacuum polarization (virtual pairs in the Dirac vacuum). The bare charge is logarithmically divergent.

Heisenberg, Dirac, and others[14–17] studied the electron's charge distribution due to "vacuum polarization," i.e. momentary charge separation in the Dirac vacuum. The unscreened "bare charge" was found to be divergent, again logarithmically. A sketch of the charge distribution of the electron is shown in Fig. 2.2. The mildness of the logarithmic divergence played an important role in the subsequent renormalization of QED. But it was delayed for a decade because of World War II.

The breakthrough in QED came in 1947, with the measurements of the Lamb shift[18] and the electron anomalous moment.[19] In the first post-war physics conference at Shelter Island, L.I., N.Y., June 2–4, 1947, participants thrashed out QED issues (Fig. 2.3 shows a group picture). Bethe[20] made an estimate of the Lamb shift immediately after the conference (reportedly on the train back to Ithaca, N.Y.) by implementing charge renormalization, in addition to Lorentz's mass renormalization. This pointed the way to the successful calculation of the Lamb shift[18–20] in lowest-order perturbation theory.

As for the electron anomalous moment, Schwinger[24] calculated it to lowest order as $\alpha/2\pi$, where α is the fine-structure constant, without encountering divergences.

2.4. Dyson: subtraction = multiplication, or the magic of perturbative renormalization

Dyson made a systematic study of renormalization in QED in perturbation theory.[25] The dynamics in QED can be described in terms of scattering processes. In perturbation theory, one expands the scattering amplitude as power series in the electron bare charge e_0 (the charge that appears in the Lagrangian). Terms in this expansion

obscure physicist at Harvard" by the name of Wendell Furry, who pointed out that the divergence should have been logarithmic. Greatly distressed, Weisskopf showed Pauli the letter, and asked whether he should "quit physics." The usually acerbic Pauli became quite restrained at moments like this, and merely huffed, "I never make mistakes!"

1 I. Rabi, 2 L. Pauling, 3 J. Van Vleck , 4 W. Lamb, 5 G. Breit, 6 Local Sponsor, 7 K. Darrow, 8 G. Uhlenbeck, 9 J. Schwinger, 10 E. Teller, 11 B. Rossi, 12 A. Nordsiek, 13 J. v. Neumann, 14 J. Wheeler, 15 H. Bethe, 16 R. Serber, 17 R. Marshak, 18 A. Pais, 19 J. Oppenheimer, 20 D. Bohm, 21 R. Feynman, 22 V. Weisskopf, 23 H. Feshbach.

Fig. 2.3. Shelter Island conference on QED, June 2–4, 1947.

are associated with Feynman graphs, which involve momentum-space integrals that diverge at the upper limit. To work with them, one introduces a high-momentum cutoff Λ. Dyson shows that mass and charge renormalization remove all divergences, to all orders of perturbation theory.

The divergences can be traced to one of three basic divergent elements in Feynman graphs, contained in the full electron propagator S', the full photon propagator D', and the full vertex Γ. They can be reduced to the following forms:

$$\begin{aligned} S'(p) &= \left[(p\cdot\gamma) - m_0 + \Sigma(p)\right]^{-1}, \\ D'(k^2) &= -k^{-2}\left[1 - e_0^2\Pi(k^2)\right]^{-1}, \\ \Gamma(p_1,p_2) &= \gamma + \Lambda^*(p_1,p_2), \end{aligned} \tag{2.5}$$

which must be all regarded as power series expansions in e_0^2. The divergent elements are Σ, Π, Λ^*, called respectively the self-energy, the vacuum polarization, and the proper vertex part. The Feynman graphs for these quantities are shown in Fig. 2.4, and they are all logarithmically divergent. Thus, one subtraction will suffice to render them finite.

The divergent part of Σ can absorbed into the bare mass m_0, as in the Lorentz theory. What is new is that the subtracted divergent part of Π can be converted into multiplicative charge renormalization, whereby e_0 is replaced by the renormalized charge $e = Ze_0$. The divergence in Λ^* can be similarly disposed of. We illustrate how this happen to lowest order.

The electron charge can be defined via the electron–electron scattering amplitude, which is given in QED by the Feynman graphs in Fig. 2.5. The two electrons

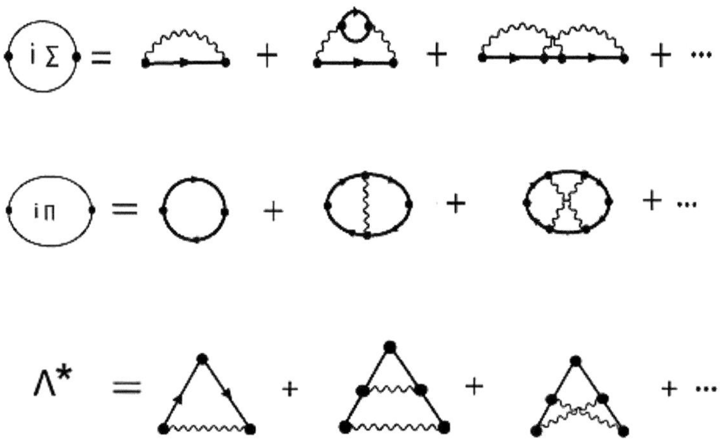

Fig. 2.4. The basic divergent elements in Feynman graphs. They are all logarithmically divergent.

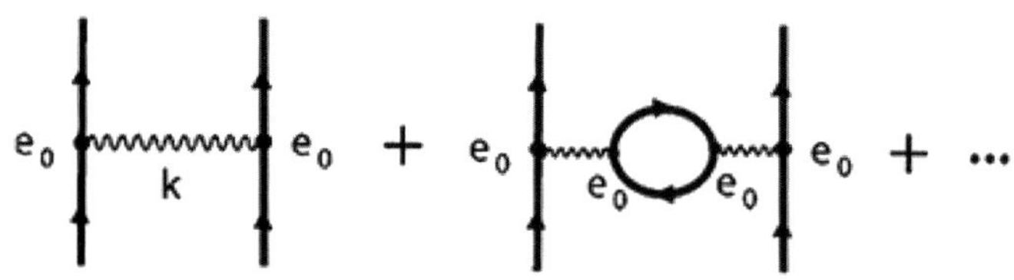

Fig. 2.5. Electron–electron scattering. The limit of zero four-momentum transfer $k \to 0$ defines the electron charge.

exchange a photon. We write the propagator has the form

$$\begin{aligned} D'(k^2) &= -\frac{d'(k^2)}{k^2}, \\ d'(k^2) &= \frac{1}{1 - e_0^2 \Pi(k^2)}, \end{aligned} \tag{2.6}$$

where k is the four-momentum transfer.

To lowest order, the vacuum polarization is given by

$$\Pi(k^2) = -\frac{1}{12\pi^2} \ln \frac{\Lambda}{m} + R(k^2) + O(e_0^2), \tag{2.7}$$

where Λ is the high-momentum cutoff, and m is the electron mass. (To this order it does not matter whether it is the bare mass or renormalized mass.) The first term is logarithmically divergent when $\Lambda \to \infty$, and the term R is convergent. One subtraction at some momentum μ makes Π convergent:

$$\Pi_c(k^2) \equiv \Pi(k^2) - \Pi(\mu^2). \tag{2.8}$$

We now write

$$e_0^2\, d'(k^2) = \frac{e_0^2}{Z^{-1}(\mu^2) - e_0^2 \Pi_c(k^2)} = \frac{e_0^2 Z(\mu^2)}{1 - e_0^2 Z(\mu^2) \Pi_c(k^2)}, \tag{2.9}$$

where

$$Z^{-1}(\mu^2) \equiv 1 - e_0^2\Pi(\mu^2). \tag{2.10}$$

Both Z and Z^{-1} are power series with divergent coefficients, and both diverge when $\Lambda \to \infty$. The combination $e_0^2 Z(\mu^2)$ gives a renormalized fine-structure constant

$$\alpha(\mu^2) = e_0^2 Z(\mu^2). \tag{2.11}$$

The physical fine-structure constant corresponds to zero momentum transfer:

$$\alpha \equiv \alpha(0) \approx (137.036)^{-1}. \tag{2.12}$$

We see that the subtraction of $\Pi(\mu^2)$ in (2.8) has been turned into a multiplication by $Z(\mu^2)$ in (2.11); but only to order e_0^4 in perturbation theory. Dyson proves the seeming miracle, that this is valid order by order, to all orders of perturbation theory.

2.5. Gell-Mann and low: it's all a matter of scale

Gell-Mann and Low[26] reformulates Dyson's renormalization program, using a functional approach, in which the divergent elements Σ, Π, Λ^* are regarded as functionals of one other, and functional equations for them can be derived from general properties of Feynman graphs. The divergent parts of these functionals can be isolated via subtractions, and the subtracted parts can be absorbed into multiplicative renormalization constant, by virtue of the behaviors of the functionals under scale transformations.

One sees the cutoff Λ in a new light, as a scale parameter. In fact, it is the only scale parameter in a self-contained theory. When one performs a subtraction at momentum μ, and absorbs the Λ-dependent part into renormalization constants, one effectively lowers the scale from Λ to μ. The degrees of freedom between Λ and μ are not discarded, but hidden in the renormalization constants; the identity of the theory is preserved. The situation is illustrated in Fig. 2.6.

The renormalized charge to order α^2, and for $|k|^2 \gg m^2$, is given by

$$\alpha(k^2) = \alpha + \frac{\alpha^2}{3\pi} \ln \frac{|k|^2}{m^2}. \tag{2.13}$$

This is called a "running coupling constant," because it depends on the momentum scale k. It has been measured at a high momentum,[27]

$$\alpha(k_0^2) \approx (127.944)^{-1}, \tag{2.14}$$

where $k_0 \approx 91.2$ GeV. The Fourier transform of $\alpha(k^2)$ gives the electrostatic potential of an electron.[28] As expected, it approaches the Coulomb potential er^{-1} as $r \to \infty$, where e is the physical charge. For $r \ll \hbar/mc$, it is given by

$$V(r) \approx \frac{e}{r}\left[1 + \frac{2\alpha}{3\pi} \ln \frac{r_0}{r} + O(\alpha^2)\right], \tag{2.15}$$

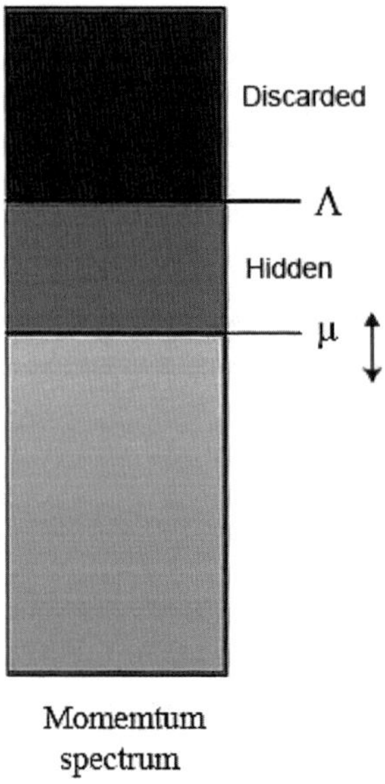

Fig. 2.6. The degrees of freedom of the system at higher momenta than the cutoff Λ are omitted from the theory, by definition. The degrees of freedom between Λ and the sliding renormalization point μ are "hidden" in the renormalization constants. Thus, μ is an effective cutoff, the scale at which one is observing the system.

where $r_0 = (\hbar/mc)(e^{5/6}\gamma)^{-1}$, $\gamma \approx 1.781$. We see that the bare charge e_0 of the electron, namely that residing at the center, diverges like $\ln(1/r)$.

Gell-Mann and Low[26] give the following physical interpretation of charge renormalization:

> A test body of "bare charge" q_0 polarizes the vacuum, surrounding itself by a neutral cloud of electrons and positrons; some of these, with a net charge δq, of the same sign as q_0, escape to infinity, leaving a net charge $-\delta q$ in the part of the cloud which is closely bound to the test body (within a distance of $\hbar/mc$). If we observe the body from a distance much greater than $\hbar/mc$, we see an effective charge $q = q_0 - \delta q$, the renormalized charge. However, as we inspect more closely and penetrate through the cloud to the core of the test charge, the charge that we see inside approaches the bare charge q_0 concentrated at a point at the center.

2.6. Asymptotic freedom

The running coupling constant "runs" at a rate described by the β-function (introduced as ψ by Gell-Mann and Low):

$$\beta(\alpha(\mu^2)) = \mu^2 \frac{\partial \alpha(\mu^2)}{\partial \mu^2}. \tag{2.16}$$

For QED we can calculate this from (2.13) to lowest order in α:

$$\beta_{\text{QED}}(\alpha) = \frac{\alpha^2}{3\pi}. \tag{2.17}$$

That this is positive means that α increases with the momentum scale. But it has the opposite sign in QCD (quantum chromodynamics)[29,30]:

$$\beta_{\mathrm{QCD}}(\alpha) = -\frac{\alpha^2}{6\pi}\left(\frac{33}{2} - N_f\right), \tag{2.18}$$

where α here is the analog of the fine-structure constant, and $N_f = 6$ is the number of quark flavors. Thus, QCD approaches a free theory in the high-momentum limit. This is called "asymptotic freedom."

QCD is a gauge theory like QED, but there are eight "color" charges, and eight gauge photons, called gluons. Unlike the photon, which is neutral, the gluons carry color charge. When a bare electron emits or absorbs a photon, its charge distribution does not change, because the photon is neutral. In contrast, when a quark emits or absorbs a gluon, its charge center is shifted, since the gluon is charged. Consequently, the "dressing" of a bare quark smears out its charge to a distribution without a central singularity. As one penetrates the cloud of vacuum polarization of a dressed quark, one see less and less charge inside, and finally nothing at the center. This is the physical origin of asymptotic freedom. Figure 2.7 shows a comparison between the dressed electron and the dressed quark, with relevant Feynman graphs that contribute to the dressing.

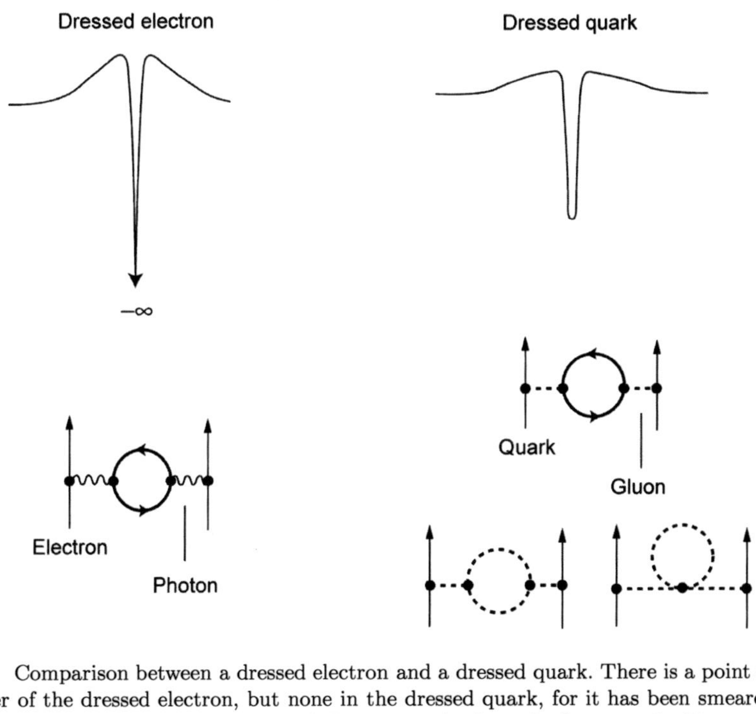

Fig. 2.7. Comparison between a dressed electron and a dressed quark. There is a point charge at the center of the dressed electron, but none in the dressed quark, for it has been smeared out by the gluons, which are themselves charged. Lower panels show the lowest-order Feynman graphs for charge renormalization. For the quark, there are two extra graphs arising from gluon–gluon interactions.

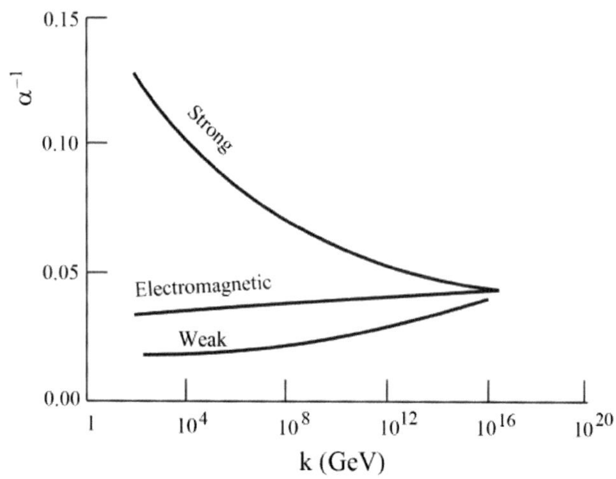

Fig. 2.8. Extrapolation of the running coupling constants for the strong, electromagnetic, and weak interactions indicate that they would meet at a momentum $k \approx 10^{17}$ GeV, giving rise to speculations of a "grand unification."

In the standard model of particle physics, there are three forces: strong, electromagnetic and weak, whose strengths can be characterized respectively by $\alpha_{\rm QCD}$, $\alpha_{\rm QED}$, $\alpha_{\rm Weak}$, which stand in the approximate ratios $10{:}10^{-2}{:}10^{-5}$ at low momenta. While $\alpha_{\rm QCD}$ is asymptotically free, the other two are not. Consequently $\alpha_{\rm QCD}$ will decrease with momentum scale, whereas the other two increase. Extrapolation of present trend indicates they would all meet at about 10^{17} GeV, as indicated in Fig. 2.8. This underlies the search for a "grand unified theory" at that scale.

2.7. The renormalization group (RG)

The transformations of the scale μ form a group, and the running coupling constant $\alpha(\mu^2)$ gives a representation of this group, which was named RG (the renormalization group) by Bogoliubov.[31] The β-function is a "tangent vector" to the group. By integrating (2.16), we obtain the relation

$$\ln \frac{\mu^2}{\mu_0^2} = \int_{\alpha(\mu_0^2)}^{\alpha(\mu^2)} \frac{d\alpha}{\beta(\alpha)}. \tag{2.19}$$

As $\mu \to \infty$, the left side diverges, and therefore $\alpha(\mu^2)$ must either diverge, or approach a zero of $\beta(\alpha)$. The latter is a fixed point of RG, at which the system is scale-invariant. This shows that the scale μ is determined by the value of $\alpha(\mu^2)$.

Figure 2.9 shows plots of the β-function for QED and QCD. As the momentum scale k increases, $\alpha(k^2)$ runs along the direction of the arrows determined by the sign of β. For QED, α increases with k, and since perturbation theory becomes invalid at high k, we lose control over high-energy QED. For QCD, on the other hand, α

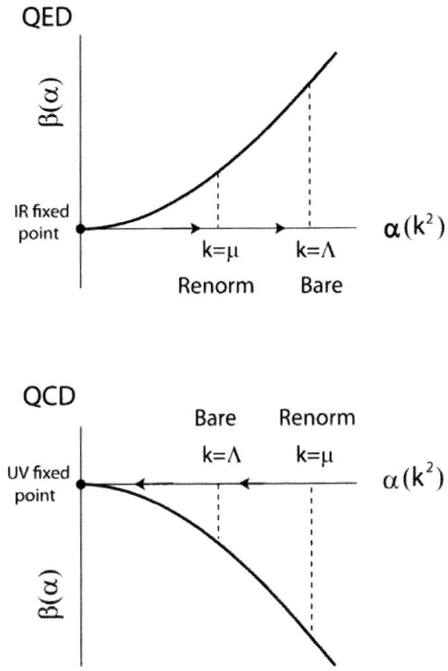

Fig. 2.9. The sign of $\beta(\alpha)$ determines the direction of arrows on the α axis, indicating the direction of running as the momentum scale k increases. Fixed points are where $\beta(\alpha) = 0$, at which the system is scale-invariant. The origin is an IR (infrared) fixed point for QED, for the system goes towards it when $k \to 0$. It is an UV (ultraviolet) fixed point for QCD, which approaches it when $k \to \infty$.

runs towards the UV fixed point at zero, perturbation theory becomes increasingly accurate, and we have a good understanding in this regime. The other side of the coin is that QCD becomes a hard problem at low energies, where it exhibits quark confinement.

The plots clarify the relation between the cutoff scale Λ that defines the bare system, and the effective scale μ, which defines the renormalized system.

We now have a better understanding of what can be done with the original cutoff Λ. Being a scale parameter, Λ is determined by (2.19) and the limit $\Lambda \to \infty$ can be achieved only by moving $\alpha(\Lambda^2)$ to a fixed point, and in QCD this means $\alpha(\Lambda^2) = 0$. And there is no problem with this.

In QED, on the other hand, there is no known fixed point except the one at the origin. In practice, one keeps Λ finite, whose value is not important. In this way, one can perform calculations that agree with experiments to one part in 10^{12}, in the case of the electron anomalous moment.[30,31] If one insists on making $\Lambda \to \infty$, one must make $\alpha(\Lambda^2) = 0$, but that makes $\alpha(\mu^2) = 0$ for all $\mu < \Lambda$, and one has a trivial free theory. We will expand on this "triviality problem" later.

Particle theorists have a peculiar sensitivity to the cutoff, because they regard it as a stigma that exposes an imperfect theory. In the early days of renormalization, when the cutoff was put out of sight by renormalization, some leaped to declare that the cutoff has been "sent to infinity." That, of course, cannot be done by fiat. Only in QCD can one achieve that, owing to asymptotic freedom.

A more general statement of renormalization refers to any correlation function G',

$$G'(p;\Lambda,e_0^2) = Z^*\left(\frac{\Lambda}{\mu, e_0^2}\right) G(p;\mu,\alpha(\mu^2)), \tag{2.20}$$

where p collectively denotes all the external momenta, Z^* is a dimensionless renormalization constant that diverges when $\Lambda \to \infty$, μ is an arbitrary momentum scale less than Λ, $\alpha(\mu^2)$ is given by (2.11), and G is a convergent correlation function. Since the left side is independent of μ, we have

$$\frac{d}{d\mu}\left[Z^*\left(\frac{\Lambda}{\mu}, e_0^2\right) G(p;\mu,\alpha)\right] = 0 \tag{2.21}$$

which leads to the Callan–Symanzik equation[33,34]

$$\left[\mu\frac{\partial}{\partial\mu} + \beta(\alpha)\frac{\partial}{\partial\alpha} + \gamma(\mu)\right] G(p;\mu,\alpha) = 0, \tag{2.22}$$

where β is defined by (2.16), and $\gamma(\mu) = \mu\frac{\partial}{\partial\mu}\ln Z^*(\Lambda/\mu,\alpha_0)$ is called the "anomalous dimension." This shows how renormalization accompanies a scale transformation, so as to preserve the basic identity of the theory.

2.8. The Landau ghost

Between the great triumph of quantum field theory in QED in 1947, and the emergence of the standard model of particle physics around 1975, particle theorists wandered like Moses in some desert, for nearly three decades. During that time they get disenchanted with quantum field theory, because the great hope they had pinned on the theory to explain the strong interactions did not materialize.[b] The was a feeling that something crazy was called for, like quantum mechanics,[c] or maybe the "bootstrap."[2,3] Landau thought he has at least disposed of quantum field theory by exposing a fatal flaw.

[b] I recall that, in the late fifties, E. Fermi and F. J. Dyson separately gave the Morris Loeb Lecture at Harvard University. Fermi talked about a newly discovered pio-nucleon "33 resonance," with spin-3/2 and isospin-3/2 (now called the delta baryon), and said, "I will not understand this in my lifetime." Dyson talked about the so-called "Tamm–Dancoff approximation" for pion-nucleon scattering, and said, "We will not understand this problem in a hundred years."

[c] In 1958, Heisenberg and Pauli proposed a "unified field theory." Pauli gave a seminar at Columbia University with Niels Bohr in attendance. When the seminar began, Bohr said, "To be right, the theory had better be crazy." Pauli said, "It's crazy! You will see. It's crazy!" The theory turns out be a version of the four-fermion interaction.

Substituting (2.16) in (2.19) and performing the integration, one obtains

$$\alpha(k^2) = \frac{\alpha}{1 - (\alpha/3\pi)\ln(k^2/m^2)}. \tag{2.23}$$

This is supposed to be an improvement on (2.13), equivalent to summing a certain class of Feynman graphs — the so-called "leading logs" with terms of the form $(e_0^2 \ln \Lambda)^n$. Landau[35] pointed out that there is a pole with negative residue:

$$\alpha(k^2) \approx \frac{-3\pi k_{\text{ghost}}^2}{k^2 - k_{\text{ghost}}^2}, \quad \frac{k_{\text{ghost}}^2}{m^2} = \exp\frac{3\pi}{\alpha}. \tag{2.24}$$

This represents a photon excited state, whose wave function has negative squared modulus, and is called a "ghost state." Its mass is of order 10^{280} m. It can be shown that $\Lambda < k_{\text{ghost}}$, and thus the ghost occurs only if we continue the theory to beyond the preset cutoff. However, if one insists on making $\Lambda \to \infty$, one must push the ghost to infinity, and this means $\alpha \to 0$, leading to a trivial theory. Landau said that this possibility exposes a fundamental flaw in quantum field theory,[d] which "should be buried with honors."

The triviality problem also occurs in other theories, for example the scalar ϕ^4 Higgs field in the standard model. Earlier, it was found in the Lee model,[36] an exactly soluble model of meson scattering. Källén and W. Pauli[37] showed that the ghost state renders the S-matrix nonunitary, and this pathology cannot be cured by redefining Hilbert space to admit negative norms.[e]

We shall see that the triviality problem is a general property of IR fixed points. The moral is: to get infinite cutoff, get yourself a UV fixed point!

Quantum field theory did not die, but bounced back with a vengeance, in the form of Yang–Mills gauge theory in the standard model.

2.9. Renormalizability

Renormalization in perturbation theory hinges on the degree of divergence K of Feynman graphs, which is determined via a power-counting procedure. It depends on the form of coupling — how many lines meet at a vertex, etc. Renormalization in QED relies on the fact the interaction $\bar{\psi}A\psi$ gives $K = 0$ (logarithmic divergence). One can imagine interactions that would give $K > 0$, and that would be nonrenormalizable. An example is the four-fermion interaction $(\bar{\psi}\psi)^2$. There is thus a criterion of renormalizability: under a scale change, while the existing coupling constants undergo renormalization, no new coupling should arise. In other words, the system should be self-similar.

[d] Apparently, Landau regarded the ghost state as a hallmark of quantum field theories. He reportedly calculated the β-function of Yang–Mills theory (on which QCD is based), but made a sign error, and missed asymptotic freedom.

[e] There are other ghost states in quantum field theory, arising from gauge-fixing, such as the Faddeev–Popov ghost.[5] But these are mathematical devices that have no physical consequences.

Such considerations are based on the presumption that each new coupling brings in its own scale. In a self-contained system, however, the cutoff Λ sets the only scale in the theory, and all coupling constants must be proportional to an appropriate power of Λ. When this is taken into account in power counting, what was considered a nonrenormalizable interaction can become renormalizable. If all coupling constants are made dimensionless in this manner, then they could freely arise under scale transformations, and the system need not be self-similar to be renormalizable.

As illustration, consider scalar field theory with a Lagrangian density of the form (with $\hbar = c = 1$)

$$\mathcal{L} = \frac{1}{2}(\partial\phi)^2 - V(\phi),$$
$$V(\phi) = g_2\phi^2 + g_4\phi^4 + g_6\phi^6 + \cdots . \tag{2.25}$$

The theory is called ϕ^M theory, where M is the highest power that occurs. Each coupling g_n corresponds to a vertex in a Feynman graph, at which n lines meet, and each line carries momentum. The momenta of the internal lines are integrated over, and produce divergences. Thus, each Feynman graph is proportional to Λ^K, with a degree of divergence K that can be found by a counting procedure. The relation between K and topological properties of Feynman graphs, such as the number of vertices and internal lines, determines renormalizability.

It was said conventionally that only the ϕ^4 theory is renormalizable. This determination, however, assumes that the g_n are arbitrary parameters. The dimensionality of g_n in d-dimensional space–time is

$$[g_n] = (\text{Length})^{nd/2-n-d}. \tag{2.26}$$

Treating them as independent will means that each g_n bring into the system an independent scale. But the only intrinsic length scale in a self-contained system is the inverse cutoff Λ^{-1}. Thus each g_n should be scaled with appropriate powers of Λ:

$$g_n = u_n\Lambda^{n+d-nd/2} \tag{2.27}$$

so that u_n is dimensionless. When this is done, the cutoff dependence of g_n enters into power counting, and all ϕ^K theories become renormalizable.[38]

With the scaling (2.27), one can construct an asymptotically free scalar field, one that is free from the triviality problem. For an N-component scalar field in $d = 4$, $V(\phi)$ is uniquely given by the Halpern–Huang potential[39]

$$V(\phi) = c\Lambda^{4-b}[M(-2+b/2, N/2, z) - 1], \quad z = \frac{8\pi^2}{\Lambda^2}\sum_{n=1}^{N}\phi_n^2, \tag{2.28}$$

where c, b are arbitrary constants, and $M(a, b; z)$ is the Kummer function, which has exponential behavior for large fields:

$$M(p, q, z) \approx \Gamma(q)\Gamma^{-1}(p)z^{p-q}\exp z. \tag{2.29}$$

The theory is asymptotically free for $b > 0$. This has applications in the Higgs sector of the standard model and in cosmology, to be discussed later.

Not all theories are renormalizable, even with the scaling of coupling constants. There is a true spoiler, namely, the "axial anomaly" in fermionic theories. It arises from the fact that the classically conserved axial vector current becomes nonconserved in quantum theory, due to the existence of topological charges (see Refs. 2 and 4). This leads to Feynman graphs with the "wrong" scaling behavior, and the only way to get rid of divergences arising from such Feynman graphs is to cancel them with similar graphs. The practical consequence is that quarks and leptons in the standard model must occur in a family, such that their anomalies cancel. We know of three families: $\{u, d, e, \nu_e\}$, $\{s, c, \mu, \nu_\mu\}$, $\{t, b, \tau, \nu_\tau\}$. If a new quark or lepton is discovered, it should bring with it an entire family.

2.10. Wilson's renormalization theory

Wilson reformulates renormalization independent of perturbation theory, and puts scale transformations at the forefront. He was concerned with critical phenomena in matter, where there is a natural cutoff, the atomic lattice spacing a. When one writes down a Hamiltonian, a does not explicitly appear, because it only supplies the length scale. The scaling (2.27) of coupling constants is natural and automatic. This is an important psychological factor in one's approach to the subject.

The first hint of how to do renormalization on a spatial lattice space comes from Kadanoff's "block spin" transformations.[40] This is a coarse-graining process, as illustrated in Fig. 2.10. Spins with only up-down states are represented by the black dots, with nearest-neighbor (nn) interactions. In the first level of coarse-graining, spins are grouped into blocks, indicated by the solid enclosures. The original spins are replaced by a single averaged spin at the center. The lattice spacing becomes 2, but is rescaled back to 1. The block–block interactions now have renormalized coupling constants; however, new couplings arise, for the blocking process generates nnn and longer-ranged interactions. Kadanoff concentrates on the fixed points of iterative blocking, and ignores the new couplings for this purpose. Wilson takes the new couplings into account, by providing "hooks" for them from the beginning. That is, the coupling-constant space is enlarged to include all possible couplings: nnn, nnnn, etc. In the beginning, when there were only the nn couplings, one regarded the rest as potentially present, but negligible. The couplings can grow or decrease in successive blocking transformations.

Wilson implements renormalization using the Feynman path integral, as follows. A quantum field theory can be described through its correlation functions. For a scalar field, for example, these are the functional averages $\langle\phi\phi\rangle$, $\langle\phi\phi\phi\rangle$, $\langle\phi\phi\phi\phi\rangle, \ldots,$ and they can be obtained from the generating functional

$$W[J] = \mathcal{N} \int D\phi \exp i(S[\phi, \Lambda] - (J, \phi)) \tag{2.30}$$

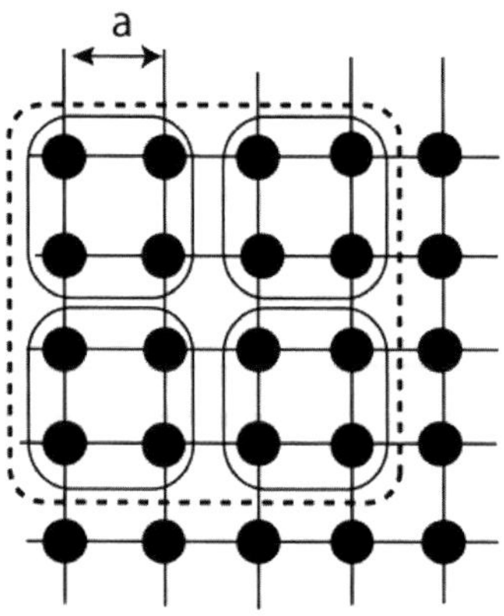

Fig. 2.10. Block–spin transformations. In the spin lattice, the up-down spins, represented by the black dots, interact with each other via nearest-neighbor (nn) interactions. In the first level of coarse-graining, they are grouped into blocks of 4, indicated by the solid enclosures, and replaced by a single averaged spin at the center. The original lattice spacing a now becomes $2a$, but is rescaled back to a. In the next level, these blocks are grouped into higher blocks indicated by the dotted enclosure, and so forth. However, the block–block interactions will include nnn, nnnn interactions, and so forth.

by repeated functional differentiation with respect to the external current $J(x)$. Here $S[\phi] = \int d^dx\,\mathcal{L}$ is the classical action, where $\mathcal{L}$ is the classical Lagrangian density (2.25), and $\int D\phi$ denotes functional integration. There is a short-distance cutoff Λ^{-1}, which is the only scale in $S[\phi]$. Of course, J introduces an scale, but that is external rather than intrinsic. For simplicity we set $J \equiv 0$ in this discussion.

By making the time pure imaginary (Euclidean time, in the language of relativistic quantum field theory) one can regard $W[J]$ as the partition function for a thermal system described by an order parameter $\phi(x)$, and the imaginary time corresponds to the inverse temperature. In this way, a result from quantum field theory can be translated into that in statistical mechanics, and vice versa.

The functional integration $\int D\phi$ extends over all possible functional forms of $\phi(x)$. It may be carried out by discretizing x as a spatial lattice, and integrating over the field at each site. Alternatively one can integrate over all Fourier transforms in momentum space, made discrete by enclosing the system in a large spatial box. Here we choose the latter route:

$$\int D\phi = \prod_{|k|<\Lambda} \int_{-\infty}^{\infty} d\phi_k, \tag{2.31}$$

where ϕ_k denotes a Fourier component of the field, and Λ is the high-momentum cutoff. We lower the effective cutoff to μ by "hiding" the degrees of freedom between Λ and μ, as indicated in Fig. 2.6. To do this, we integrate over the momenta in this interval, and put the result in the form of a new effective action. That is, we write

$$\begin{aligned}
&\mathcal{N} \prod_{|k|<\Lambda} \int_{-\infty}^{\infty} d\phi_k \exp iS[\phi] \\
&= \mathcal{N} \prod_{|k|<\mu} \int_{-\infty}^{\infty} d\phi_k \left\{ \prod_{\mu<|k|<\Lambda} \int_{-\infty}^{\infty} d\phi_{k'} \exp iS[\phi] \right\} \\
&= \mathcal{N}' \prod_{|k|<\mu} \int_{-\infty}^{\infty} d\phi_k \exp iS'[\phi]. \qquad (2.32)
\end{aligned}$$

The integrations in the brackets { } define the new action $S'[\phi]$, which contains only degrees of freedom below momentum μ.[f] From this, we can obtain a new Lagrangian density $\mathcal{L}'$, which contains new couplings $\{u'_n\}$ that are functions of the old ones $\{u_n\}$.[g] This, in a nutshell, is Wilson's renormalization transformation.

Successive renormalization transformations give a series of effective Lagrangians:

$$\mathcal{L} \to \mathcal{L}' \to \mathcal{L}'' \to \mathcal{L}''' \to \cdots \qquad (2.33)$$

which describe how the appearance of the system changes under coarse-graining. The identity of the system is preserved, because the generating functional W is not changed. We allow for all possible couplings u_n, and thus the parameter space is that of all possible Lagrangians. Renormalization generates a trajectory in that space — the RG trajectory. Couplings that were originally negligible can grow, and so the trajectory can break out into new dimensions, as illustrated in Fig. 2.11.

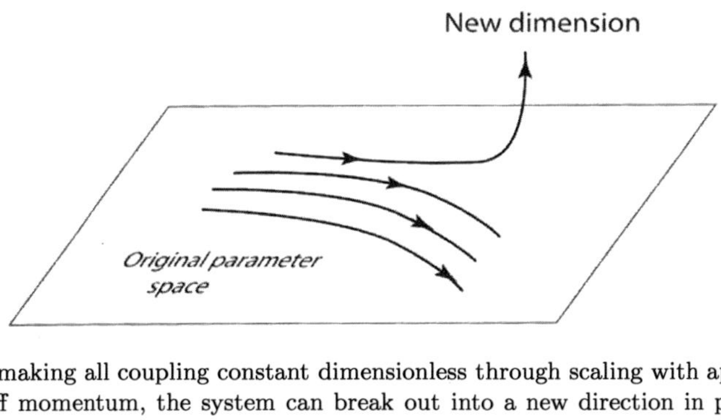

Fig. 2.11. By making all coupling constant dimensionless through scaling with appropriate powers of the cutoff momentum, the system can break out into a new direction in parameter space under renormalization. The trajectories sketched here represent RG trajectories with various initial conditions.

[f] The cutoff Λ actually does not appear in any of the formulas, because it merely supplies a scale. Lowering the cutoff from Λ to μ actually means lowering it from 1 to μ/Λ (see Ref. 4 for details).

[g] Some rescaling need to be done to put the new Lagrangian in a standard form. These operations can affect the normalization constant $\mathcal{N}$.

There is no requirement that the theory be self-similar, and thus it appears that all theories are renormalizable.[h]

That this method of renormalization reduces to that in perturbation theory can be proven by deriving (2.20) with this approach.[4]

2.11. In the space of all possible Lagrangians

Under the coarse-graining steps, the effective Lagrangian traces out a trajectory in parameter space, the RG trajectory.[i] With different initial conditions, one goes on different trajectories, and the whole parameter space is filled with them, like stream lines in a hydrodynamic flow. There are sources and sinks in the flow, and these are fixed points, where the system remains invariant under scale changes. The correlation length becomes infinite at these fixed points. This means that the lattice approaches a continuum: $a \to 0$, or $\Lambda \to \infty$.

Let us define the direction of flow along an RG trajectory to be the coarse-graining direction, or towards low momentum. If it flows out of a fixed point, then the fixed point appears to be a UV fixed point, for it is to be reached by going opposite to the flow, towards the high-momentum limit. Such a trajectory is called a UV trajectory. If it flows into a fixed point, it is called an IR trajectory, along which the fixed point appears to be an IR fixed point. This is illustrated in Fig. 2.12.

Actually, Λ is infinite along the entire IR trajectory, because this is so at the fixed point, and Λ can only decrease upon coarse-graining. Thus, one cannot place a system on an IR trajectory, but only on an adjacent trajectory. When we get closer and closer to the IR trajectory, $\Lambda \to \infty$, and the system more closely resembles

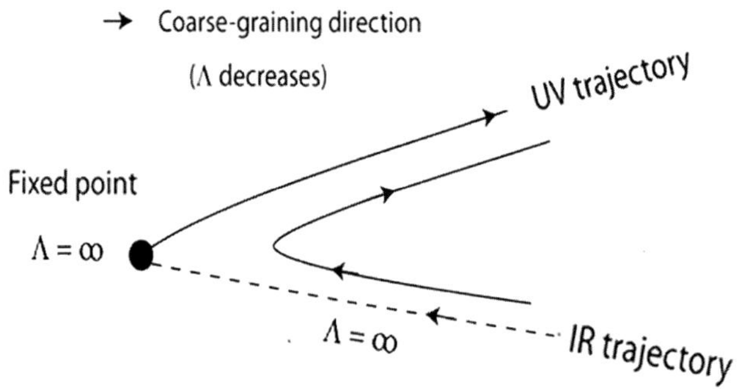

Fig. 2.12. Fix points are sources and sinks of RG trajectories, at which the cutoff is infinite.

---

[h]Notable exceptions, as mentioned previously, are fermionic theories exhibiting the axial anomaly. In terms of the Feynman path integral, certain scaling operations fail, owing to noninvariance of the integration measure (see Refs. 2 and 4).

[i]It might appear that the coarse-graining proceeds only in one direction; but that is a matter of defining the RG trajectory. Once defined, one can travel back and forth along the trajectory.

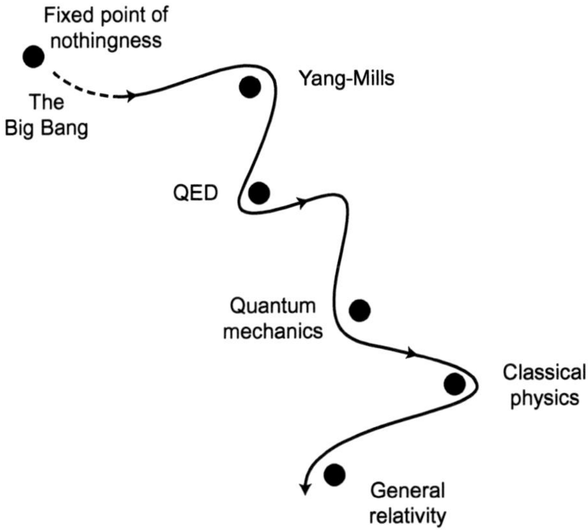

Fig. 2.13. Different physical theories govern different length scales. Each theory can be represented by a fixed point in the space of Lagrangians. The world is like a ship navigating this space, and the fixed points are the ports of call. As the scale changes, the world sails from port to port, and lingers for a while at each port.

that at the IR fixed point. It is most common to have a free theory at the fixed point, since it is scale-invariant, and this gives a more physical understanding of the triviality problem.

The flow velocity along an RG trajectory can be measured by the arc length covered in a coarse-graining step. It slows down in the neighborhood of a fixed point, and speeds up between fixed points. Thus, it darts from one fixed point to the next, like a ship sailing between ports of call. Some couplings grow as it approaches a fixed point, and these are called "relevant" interactions. The one that die out are called "irrelevant," and may be neglected. Thus, each port corresponds to a characteristic set of interactions, and the system puts on a certain face at that port. This is illustrated in Fig. 2.13.

2.12. Polchinski's functional equation

We have used a sharp momentum cutoff in (2.31). In a significant improvement, Polchinski[42] generalizes the method to an arbitrary cutoff, and derives a functional equation for the renormalized action. The cutoff is introduced by modifying the free propagator k^{-2} of the field theory, by replacing it by

$$\Delta(k^2) = \frac{f(k^2/\Lambda^2)}{k^2}, \quad f(z) \underset{z\to\infty}{\longrightarrow} 0, \tag{2.34}$$

where the detailed form of the cutoff function $f(k^2/\Lambda^2)$ is not important. What is important is that Λ is the only scale in the theory. The regulated propagator in configurational space will be denoted by $K(x,\Lambda)$.

The action is written as

$$S[\phi,\Lambda] = S_0[\phi,\Lambda] + S'[\phi,\Lambda], \tag{2.35}$$

where the first term corresponds to the free field, and the second term represents the interaction. We have

$$S_0[\phi,\Lambda] = \frac{1}{2}\int d^dx\, d^dy\, \phi(x)K^{-1}(x-y,\Lambda)\phi(y), \tag{2.36}$$

where $K^{-1}(x-y,\Lambda)$ is the inverse of the propagator $K(x-y,\Lambda)$, in an operator sense. It differs from the Laplacian operator significantly only in a neighborhood of $|x-y|=0$, of radius Λ^{-1}. The generating functional is written as

$$W[J,\Lambda] = \mathcal{N}\int D\phi\, e^{-S[\phi,\Lambda]-(J,\phi)}, \tag{2.37}$$

where the normalization constant $\mathcal{N}$ may depend on Λ.

In the Wilson method, one integrates out mode between Λ and μ to lower the effective cutoff to μ. A more general point of view is that any change in Λ is compensated by a change in $S'[\phi,\Lambda]$, in order to preserve the basic identity of the theory:

$$\frac{dW[J,\Lambda]}{d\Lambda} = 0. \tag{2.38}$$

This is the generalization of (2.21) in perturbative renormalization.

The remarkable thing is that Polchinski solves (2.38) by finding a functional integro-differential equation for $S'[\phi,\Lambda]$. For $J\equiv 0$, it reads[j]

$$\frac{dS'}{d\Lambda} = -\frac{1}{2}\int dx\, dy \frac{\partial K(x-y,\Lambda)}{\partial\Lambda}\left[\frac{\delta^2 S'}{\delta\phi(x)\delta\phi(y)} - \frac{\delta S'}{\delta\phi(x)}\frac{\delta S'}{\delta\phi(y)}\right]. \tag{2.39}$$

Periwal[43] shows how one can use this to derive the Halpern–Huang potential in "two lines." (The original derivation involves summing one-loop Feynman graphs[39]). Assuming that there are no derivative couplings,[k] we can write S' as the integral of a local potential:

$$\begin{aligned} S'[\phi,\Lambda] &= \Lambda^d\int d^dx\, U(\varphi(x),\Lambda),\\ \varphi(x) &= \Lambda^{1-d/2}\phi(x), \end{aligned} \tag{2.40}$$

where U is a dimensionless function and φ is a dimensionless field. The scalar potential is given by $V=\Lambda^d U$.

---

[j] See Ref. 4 for a proof.

[k] The original derivation did not assume this, but showed that no derivative couplings arise from renormalization, if none were present originally.

Near the Gaussian fixed point, where $S' \equiv 0$, one can linearize (2.39), and obtain a linear differential equation for $U(\varphi, \Lambda)$:

$$\Lambda\frac{\partial U}{\partial \Lambda} + \frac{\kappa}{2}U'' + \left(1 - \frac{d}{2}\right)\varphi U' + Ud = 0, \tag{2.41}$$

where a prime denote partial derivative with respect to φ, and $\kappa = \Lambda^{3-d}\partial K(0, \Lambda)/\partial\Lambda$. Now we seek eigenpotentials $U_b(\varphi, \Lambda)$ with the property

$$\Lambda\frac{\partial U_b}{\partial \Lambda} = -bU_b. \tag{2.42}$$

In the language of perturbative renormalization theory, the right side is the linear approximation to the β-function, and the solution is asymptotically free for $b > 0$. Substitution into the previous equation leads to the differential equation

$$\left[\frac{\kappa}{2}\frac{d^2}{d\varphi^2} - \frac{1}{2}(d-2)\varphi\frac{d}{d\varphi} + (d-b)\right]U_b = 0. \tag{2.43}$$

Since this equation does not depend on Λ, the Λ-dependence of the potential is contained in a multiplicative factor. In view of (2.42), the factor is Λ^{-b}.

For $d \neq 2$, (2.43) can be transformed into Kummer's equation:

$$\left[z\frac{d^2}{dz^2} + (q-z)\frac{d}{dz} - p\right]U_b = 0, \tag{2.44}$$

where

$$q = \frac{1}{2}, \quad p = \frac{b-d}{d-2}, \quad z = (2\kappa)^{-1}(d-2)\varphi^2. \tag{2.45}$$

The solution is

$$U_b(z) = c\Lambda^{-b}[M(p, q, z) - 1], \tag{2.46}$$

where c is an arbitrary constant, and M is the Kummer function. We have subtracted 1 to make $U_b(0) = 0$. This is permissible, since it merely changes the normalization of the generating functional. For $d = 2$, (2.43) leads to the so-called sine-Gordon theory.

2.13. Why triviality is not a problem

The massless free scalar is scale-invariant, and corresponds to the Gaussian fixed point. When the length scale increases from zero, and we imagine the system being displaced infinitesimally from this fixed point, it will sail along some RG trajectory, along some direction in parameter space, the function space spanned by possible forms of $V(\phi)$. Equation (2.42) describes the properties associated with various directions. Along directions with $b > 0$, the system will be on a UV trajectory.

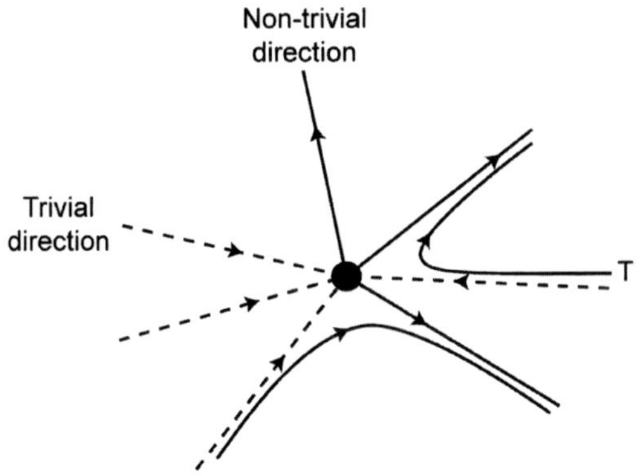

Fig. 2.14. The Gaussian fixed point. RJ trajectories emanate along all possible directions in parameter space. Arrows denote direction of increasing length scale. Nontrivial directions correspond to a theory with asymptotic freedom. Trivial direction signifies that the theory remains at the fixed point.

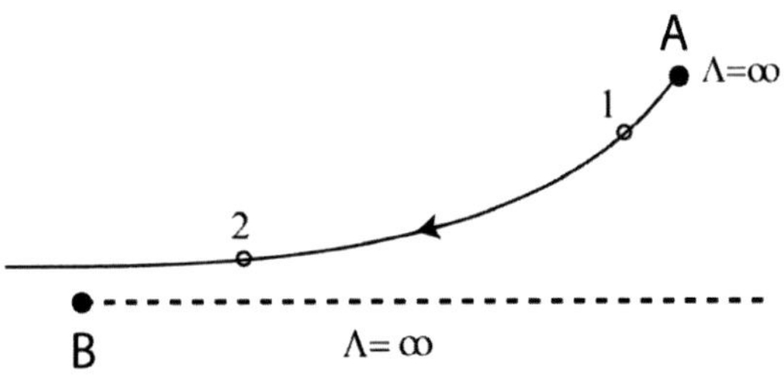

Fig. 2.15. How "triviality" may arise, and why it is not a problem. Here, A and B represent two Gaussian fixed points. The system at point 1 is on a UV trajectory and asymptotically free. It crosses over to a neighborhood of B, skirting an IR trajectory. At point 2 it resembles a trivial ϕ^4 theory, because higher couplings have become irrelevant.

With $b < 0$, the system is on an IR trajectory, and behave as if it had never left the fixed point. This is illustrated in Fig. 2.14.

The existence of UV directions suggests a possible solution to the triviality problem, as illustrated in Fig. 2.15. Consider two Gaussian fixed points A and B. A scalar field leaves A along a UV trajectory, and crosses over to a neighborhood of B, skirting an IR trajectory of B. At point 1, the potential is Halpern–Huang, but at point 2 it becomes ϕ^4, with all higher coupling becoming irrelevant. The original cutoff Λ is infinite, being pushed into A. The effective cutoff at 2 is the renormalization scale μ. It is the scale in which we observe the system, and it should not be made infinite.

In the case of QED, the fixed point B would correspond to our QED Lagrangian, and A could represent some asymptotically free Yang–Mills gauge theory, such as one with internal symmetry group corresponding to a grand unified theory.

2.14. Asymptotic freedom and the big bang

The vacuum carries complex scalar fields. We know of at least the Higgs field of the standard model, which generates mass for gauge bosons in the weak sector. Grand unified theories call for more scalar fields. A complex scalar field serves as order parameter for superfluidity, and from this point of view the universe is filled with superfluids. In a recent theory, a generic cosmic superfluid is used to explain dark energy and dark matter. Briefly, dark energy is the energy density of the superfluid, and dark matter is the manifestation of density fluctuations of the superfluid from its equilibrium vacuum value.[44–46]

At the big bang, the scalar field is assumed to emerge from the Gaussian fixed point along some direction in parameter space, as indicated in Fig. 2.14. If the chosen direction corresponds to an IR trajectory, then the system never left the fixed point, and nothing happens. If it is a UV trajectory, however, the potential will be a Halpern–Huang potential, and, as the scale expands will spawn a possible universe. There should be only one scale at the big bang, the radius of the universe $a(t)$ in the Robertson–Walker metric, and it should be identified with the cutoff Λ of the scalar field:

$$\Lambda = \frac{\hbar}{a(t)}.$$

There is thus a dynamical feedback: the scalar field generates gravity, which supplies the cutoff to the field. Einstein's equation then leads to a power-law expansion of the form

$$a(t) \sim \exp t^{1-p}, \tag{2.47}$$

where $p < 1$. This describes a universe with accelerated expansion, i.e. there is dark energy. The equivalent cosmological constant decays in time like t^{-2p}, and this could circumvent the usual "fine-tuning problem." Vortex activities in the superfluid create quantum turbulence, in which all matter was created during a initial "inflation era." Many observed phenomena, such as dark mass halos around galaxies, can be explained in terms of the cosmic superfluid.

References

1. K. G. Wilson, An investigation of the low equation and the Chew–Mandelstam equations, Ph.D. thesis, Cal. Tech. (1961).
2. K. Huang and F. E. Low, *J. Math. Phys.* **6**, 795 (1965).
3. K. Huang and A. H. Mueller, *Phys. Rev. Lett.* **14**, 396 (1965).
4. K. Huang, *Quantum Field Theory, from Operators to Path Integrals*, 2nd edn. (Wiley-VCH, Weinheim, Germany, 2010).
5. K. Huang, *Quarks, Leptons, and Gauge Fields*, 2nd edn. (World Scientific, Singapore, 1992).
6. J. J. Thomson, *The Electrician* **39**, 104 (1897).

7. H. A. Lorentz, *The Theory of the Electron*, 2nd edn. (Dover Publications, New York, 1952).
8. P. A. M. Dirac, *Proc. R. Soc. A* **117**, 610 (1928).
9. P. A. M. Dirac, *Proc. R. Soc. A* **126**, 360 (1930).
10. V. F. Weisskopf, *Z. Phys.* **89**, 27 (1934).
11. V. F. Weisskopf, *Z. Phys.* **90**, 817 (1934).
12. V. F. Weisskopf, *Phys. Rev.* **56**, 72 (1939).
13. K. Huang, *Phys. Rev.* **101**, 1173 (1956).
14. W. Heisenberg, *Z. Phys.* **90**, 209 (1934).
15. P. A. M. Dirac, *Proc. Camb. Phil. Soc.* **30**, 150 (1934).
16. R. Serber, *Phys. Rev.* **48**, 49 (1935).
17. A. E. Uehling, *Phys. Rev.* **48**, 55 (1935).
18. W. E. Lamb and R. C. Retherford, *Phys. Rev.* **72**, 241 (1947).
19. P. Kusch and H. M. Foley, *Phys. Rev.* **74**, 250 (1948).
20. H. A. Bethe, *Phys. Rev.* **72**, 339 (1947).
21. J. Schwinger, *Phys. Rev.* **73**, 1439 (1948).
22. R. P. Feynman, *Phys. Rev.* **74**, 1430 (1948).
23. B. French and V. F. Weisskopf, *Phys. Rev.* **75**, 1240 (1949).
24. J. Schwinger, *Phys. Rev.* **73**, 416L (1948).
25. F. J. Dyson, *Phys. Rev.* **75**, 486 (1949).
26. M. Gell-Mann and F. E. Low, *Phys. Rev.* **95**, 1300 (1954).
27. Particle Data Group (J. Beringer *et al.*), *Phys. Rev. D* **86**, 016001 (2012).
28. J. Schwinger, *Phys. Rev.* **75**, 651 (1949).
29. D. J. Gross and F. Wilczek, *Phys. Rev. Lett.* **30**, 1343 (1973).
30. H. D. Politzer, *Phys. Rev. Lett.* **30**, 1346 (1973).
31. N. N. Bogoliubov and D. V. Shirkov, *Introduction to the Theory of Quantum Fields* (Wiley-Interscience, New York, 1959).
32. G. Gabrielse *et al.*, *Phys. Rev. Lett.* **97**, 030802 (2006).
33. C. G. Callan, *Phys. Rev. D* **12**, 1541 (1970).
34. K. Symanzik, *Commun. Math. Phys.* **18**, 27 (1970).
35. L. D. Landau, *Niels Bohr and the Development of Physics*, ed. W. Pauli (McGraw-Hill, New York, 1955).
36. T. D. Lee, *Phys. Rev.* **95**, 1329 (1954).
37. G. Källén and W. Pauli, *Dan. Mat. Fys. Medd.* **30**, No. 7 (1955).
38. K. Halpern and K. Huang, *Phys. Rev. Lett.* **74**, 3526 (1995).
39. K. Halpern and K. Huang, *Phys. Rev.* **53**, 3252 (1996).
40. L. P. Kadanoff, *Physics* (N.Y.) **2**, 263 (1966).
41. K. G. Wilson, *Rev. Mod. Phys.* **55**, 583 (1983).
42. J. Polchinski, *Nucl. Phys. B* **231**, 269 (1984).
43. V. Periwal, *Mod. Phys. Lett. A* **11**, 2915 (1996).
44. K. Huang, H. B. Low and R. S. Tung, *Class. Quantum Grav.* **29**, 155014 (2012), arXiv:1106.5282.
45. K. Huang, H. B. Low and R. S. Tung, *Int. J. Mod. Phys. A* **27**, 1250154 (2012), arXiv:1106.5283.
46. K. Huang, C. Xiong and X. Zhao, *Int. J. Mod. Phys. A* **29**, 1450074 (2014), arXiv:1304.1595.

Chapter 3

Renormalization group theory, the epsilon expansion, and Ken Wilson as I knew him*

Michael E. Fisher
Institute for Physical Science and Technology,
University of Maryland
College Park, MD 20742-8510, USA.
xpectnil@umd.edu

The tasks posed for renormalization group theory within statistical physics by critical phenomena theory in the 1960's are set out briefly in contradistinction to quantum field theory, which was the origin for Ken Wilson's concerns. Kadanoff's 1966 block spin scaling picture and its difficulties are presented; Wilson's early vision of flows is described from the author's perspective. How Wilson's subsequent breakthrough ideas, published in 1971, led to the epsilon expansion and the resulting clarity is related. Concluding sections complete the general picture of flows in a space of Hamiltonians, universality, and scaling. The article represents a 40% condensation (but with added items) of an earlier account: *Rev. Mod. Phys.* **70** (1998) 653–681.

Foreword

In March 1996 the Departments of Philosophy and of Physics at Boston University cosponsored a two-day Colloquium "On the Foundations of Quantum Field Theory." But in the full title, this was preceded by the phrase "A Historical Examination and Philosophical Reflections," which set the aims of the meeting. The participants were mainly high energy physicists, experts in field theories, and interested philosophers of science.[1] I was called on to speak, essentially in a service role, presumably because I had witnessed and had some hand in the development of renormalization group

*This article was originally published in *Int. J. Mod. Phys. B*.
[1]The proceedings of the conference were published under the title *Conceptual Foundations of Quantum Field Theory* (Cao, 1999): for details see the references collected in the Selected Bibliography.

concepts and because I had played a role in applications where these ideas really mattered. A version of my talk was published in *Reviews of Modern Physics* (Fisher, 1998)[2] and, with some revisions, in the Proceedings of the Colloquium (Cao, 1999) edited by the prime organizer Tian Yu Cao.

On 15 June 2013 my former Cornell colleague and friend, Kenneth Geddes Wilson, who had played such a major role in the conception, formulation, and initial applications of renormalization group theory died unexpectedly a week after his 77th birthday. A Memorial Symposium was held in his honor at Cornell University on 16 November that year; it was my sad but rewarding duty to speak on the occasion to a general and varied audience. For this memorial volume, however, I felt it would be more appropriate to edit and, in places, to expand my earlier talk to especially stress the role that I, myself, saw Ken Wilson play in conceiving and formulating the renormalization group approach for critical phenomena and in its effective application *via* the so-called *epsilon expansion.* This is what I hope readers will appreciate in the following text and figures.

3.1. Introduction

It is held by some that the "Renormalization Group" (RG) — or, better, renormalization group*s* or, let us say, *Renormalization Group Theory* (or RGT) is "one of the underlying ideas in the theoretical structure of Quantum Field Theory." That belief suggests the potential value of a historical and conceptual account of RG theory and the ideas and sources from which it grew, as viewed from the perspective of statistical mechanics and condensed matter physics. Especially pertinent are the roots in the theory of critical phenomena.

The proposition just stated regarding the significance of RG theory for Quantum Field Theory (or QFT, for short) is open to debate even though experts in QFT have certainly invoked RG ideas. Indeed, one may ask: How far is some concept only instrumental? How far is it crucial? It is surely true in physics that when we have ideas and pictures that are extremely useful, they acquire elements of reality in and of themselves. But, philosophically, it is instructive to look at the degree to which such objects are purely instrumental — merely useful tools — and the extent to which physicists seriously suppose they embody an essence of reality. Certainly, many parts of physics are well established and long precede RG ideas. Among these is statistical mechanics itself, a theory *not* reduced and, in a deep sense, *not* directly reducible to lower, more fundamental levels without the introduction of specific, new postulates.

Furthermore, statistical mechanics has reached a stage where it is well posed mathematically; many of the basic theorems (although by no means all)[3] have

[2] References in this form or as a name in the text followed by a date in parentheses will be found below in the Selected Bibliography.

[3] See, *e.g.*, Fisher on p. 165 of (Cao, 1999).

been proved with full rigor. In that context, I believe it is possible to view the renormalization group as merely an instrument or a computational device. On the other hand, at one extreme, one might say: "Well, the partition function itself is really just a combinatorial device." But most practitioners tend to think of it (and especially its logarithm, the free energy) as rather more basic!

Now my aim here is not to instruct those who understand these matters well.[4] Rather, I hope to convey to nonexperts and, in particular, to any with a philosophical interest, a little more about what Renormalization Group Theory is[5] — at least in the eyes of some of those who have earned a living by using it! One hopes such information may be useful to those who might want to discuss its implications and significance or assess how it fits into QFT in particular or into physics more broadly.

Whence came Renormalization Group Theory? This is a good question to start with.[6] I will try to respond, sketching the foundations of RG theory in the *critical exponent relations* and crucial *scaling concepts* of Leo P. Kadanoff, Benjamin Widom, and myself developed in 1963–66[7] — among, of course, other important workers, particularly Cyril Domb[8] and his group at King's College London, of which, originally, I was a member, George A. Baker, Jr., whose introduction of Padé approximant techniques proved so fruitful in gaining quantitative knowledge,[9] and Valery L. Pokrovskii and A.Z. Patashinskii in the Soviet Union who were, perhaps, the first to bring field-theoretic perspectives to bear.[10] Especially I will say something of the genesis of the full RG concept — the systematic integrating out of appropriate degrees of freedom and the resulting RG flows — in the inspired work of Ken Wilson[11] when he was a colleague of Ben Widom and myself at Cornell University in 1965–1972. One must point also to the general, clarifying formulation of RG theory by Franz J. Wegner (1972a) when he was associated with Leo Kadanoff at

---

[4]Such as the field theorists D. Gross and R. Shankar (see Cao, 1999, and Shankar, 1994). Note also Bagnuls and Bervillier (1997).

[5]It is worthwhile to stress, at the outset, what a "renormalization group" is *not*! Although in many applications the particular renormalization group employed may be invertible, and so constitute a continuous or discrete, group of transformations, it is, in general, only a *semigroup*. In other words a renormalization group is not necessarily invertible and, hence, cannot be 'run backwards' without ambiguity: in short it is *not* a "group." This misuse of mathematical terminology may be tolerated since these aspects play, at best, a small role in RG theory. The point will be returned to later.

[6]Five influential reviews antedating renormalization-group concepts are Domb (1960), Fisher (1965, 1967b), Kadanoff *et al.* (1967) and Stanley (1971). Early reviews of renormalization group developments are provided by Wilson and Kogut (1974) and Fisher (1974): see also Wilson (1983) and Fisher (1983).

[7]See Essam and Fisher (1963), Widom (1965a,b), Kadanoff (1966), and Fisher (1967a).

[8]Note Domb (1960), Domb and Hunter (1965), and the account in Domb (1996).

[9]See Baker (1961) and the overview in Baker (1990).

[10]The original paper is Patashinskii and Pokrovskii (1966); their text (1979), which includes a chapter on RG theory, appeared in Russian around 1975 but did not then discuss RG theory.

[11]Wilson (1971a,b) which he later described within a QFT context in Wilson (1983).

Brown University: their focus on *relevant*, *irrelevant* and *marginal* 'operators' (or perturbations) played a central role.[12]

3.2. The task for renormalization group theory

Let us, at this point, step back briefly by highlighting, from the viewpoint of statistical physics, what it is one would wish RG theory to accomplish. First and foremost, (A) it should explain the ubiquity of power laws at and near critical points.

To be more explicit, consider gas-liquid criticality in single-component fluids and the closely anologous situation of an anisotropic, single-axis ferromagnet. As the temperature, T, approaches the critical point value, T_c, it is convenient to introduce the reduced temperature variable

$$t \equiv (T - T_c)/T_c \to 0\pm\,. \tag{3.1}$$

Then, measuring the specific heat (at constant volume for a fluid or in zero magnetic field, $H = 0$, for a ferromagnet, *etc.*) one finds a power-law divergence to infinity of the form

$$C(T) \approx A^{\pm}/|t|^{\alpha} \quad \text{as} \quad t \to 0\pm\,. \tag{3.2}$$

The amplitudes A^+ and A^- are *non*universal depending on the system (as, of course, is T_c); but their dimensionless ratio, A^+/A^-, and the characteristic *critical exponent* (or index), α are both found to be *universal.* (In fact, for bulk ($d = 3$)-dimensional systems one has $A^+/A^- \simeq 0.52$ and $\alpha \simeq 0.11$; but for $d = 2$, as predicted by Onsager (1944), $|t|^{-\alpha}$ is replaced by $\log|t|$ while $A^+/A^- = 1$. The power law (3.2) is markedly different from the 'classical' (or 'traditional' or 'mean-field') prediction of a mere jump in specific heat, $\Delta C = C_c^- - C_c^+ > 0$, for all d.

Next, one should recognize, following Landau[13] and Ginzburg,[14] the existence, at points $\mathbf{r}$ in d-dimensional Euclidean space, of a locally defined *order parameter*, say $\Psi(\mathbf{r})$. In a fluid this might well be the fluctuating density $\rho(\mathbf{r})$; in a ferromagnet it might be the local magnetization, $M(\mathbf{r})$, or spin variable, $S(\mathbf{r})$.

The order parameter and its exponent are directly revealed in the critical region of a fluid via the shape of the coexistence curve. This is described by

$$\Delta\rho \equiv \tfrac{1}{2}\left[\rho_{\text{liq}}(T) - \rho_{\text{gas}}(T)\right] \approx B|t|^{\beta} \quad \text{as} \quad t \to 0-, \tag{3.3}$$

in which the amplitude B is *non*universal but the *critical exponent* β, takes a *universal* value close to $\beta \simeq 0.325$ for $d = 3$: see, *e.g.*, Heller and Benedek (1962). For

[12]Note the reviews by Kadanoff (1976) and Wegner (1976).

[13]See Landau and Lifshitz (1958), especially Sec. 135.

[14]In particular for the theory of superconductivity: see V.L. Ginzburg and L.D. Landau, 1959, "On the Theory of Superconductivity," *Zh. Eksp. Teor. Fiz.* **20**, 1064; and, for a personal historical account, V.L. Ginzburg, 1997, "Superconductivity and Superfluidity (What was done and what was not)," *Phys.-Uspekhi* **40**, 407–432.

$d = 2$ the result is $\beta = \frac{1}{8}$ (Onsager, 1949; Yang, 1952; Kim and Chan, 1984); but both values differ significantly from the classical value $\beta = \frac{1}{2}$ predicted, *e.g.*, by the van der Waals equation.

By the same token, the spontaneous magnetization, $M_0(T)$, of an anisotropic ferromagnet with $M = \pm M_0(T)$, varies as $B|t|^\beta$ (the + or − depending on whether the field, H, below T_c, approaches 0 from above or below).

In terms of the order parameter one may also define the basic two-point or *pair correlation function*, $G(\mathbf{r})$, via

$$G(\mathbf{r}; T, \cdots) \equiv \langle \Delta\Psi(\mathbf{r})\Delta\Psi(\mathbf{0})\rangle, \tag{3.4}$$

where $\Delta\Psi(\mathbf{r}) = \Psi(\mathbf{r}) - \langle\Psi\rangle$ while the angular brackets $\langle\cdot\rangle$ denote a statistical average over the thermal (and, if relevant, quantal) fluctuations. Physically, $G(\mathbf{r})$ is important because it provides a direct measure of the development of order as criticality is approached. Indeed, the correlation length, $\xi(T)$, specifies the scale (or range) of the decay of $G(\mathbf{r})$ — typically exponential in character[15] relative to its long-range limit. The power law

$$\xi(T) \approx \xi_0^\pm/|t|^\nu \quad \text{as} \quad t \to 0\pm, \tag{3.5}$$

then describes the divergence of the correlation length; the exponent ν is close to 0.63 for fluids, *etc.*, when $d = 3$ (while for $d = 2$ one has[16] $\nu = 1$). It is worth remarking that in quantum field theory, the inverse correlation length ξ^{-1} is basically equivalent to the *renormalized mass* of the field $\psi(\mathbf{r})$.

Precisely at the critical point itself, however, an exponential decay is replaced, rather generally, by a power-law decay, namely,

$$G_c(\mathbf{r}) \approx D/r^{d-2+\eta} \qquad r \to \infty. \tag{3.6}$$

The critical exponent η, sometimes referred to as the *dimensional anomaly*, vanishes identically in all classical theories! However, it takes the value $\eta = \frac{1}{4}$ in $d = 2$ dimensions[17] while being nonzero and close to 0.036 for ($d = 3$)-dimensional fluids and anisotropic ferromagnets.[18]

Another central quantity is the divergent isothermal compressibility $\chi(T)$ (for a fluid) or isothermal susceptibility, $\chi(T) \propto (\partial M/\partial H)_T$ (for a ferromagnet). For this function, we write

$$\chi(T) \approx C^\pm/|t|^\gamma \quad \text{as} \quad t \to 0\pm, \tag{3.7}$$

and find the universal value $\gamma \simeq 1.24$ for $d = 3$ fluid systems and anisotropic ferromagnets (while $\gamma = 1\frac{3}{4}$ for $d = 2$).[19] The classical value is simply $\gamma = 1$.

---

[15] See Onsager (1944) and, *e.g.*, Fisher (1962, 1964).

[16] As shown by Onsager (1944).

[17] As can be shown from the work of Kaufman and Onsager (1949); see Fisher (1959) and also Fisher (1965, Sec. 29; 1967b, Sec. 6.2) and Fisher and Burford (1967).

[18] See, *e.g.*, Fisher and Burford (1967), Fisher (1983), Baker (1990) and Domb (1996).

[19] First advanced for $d = 2$ by Fisher (1959).

It is important to realize that there are other *universality classes* known theoretically although only a few are found experimentally.[20] Nevertheless, distinct universality classes yield distinct, observably different exponent values. Indeed, one of the early successes of RG theory was delineating and sharpening our grasp of various important universality classes: this will be illustrated briefly below. Furthermore, if and when a control variable, such as the pressure or a magnetic field, can change the universality class, it becomes important to know the value of the corresponding *crossover exponent* ϕ.[21] Its value, positive or negative, speaks to the *relevance* or *irrelevance* of the controlled variable.

In demanding, as we agreed, that RG theory should explain, in the context of critical phenomena, "the ubiquity of universal power laws," it may be helpful to compare the issue with the challenge to atomic physics of explaining the ubiquity of sharp spectral lines. Quantum mechanics responds, crudely speaking, by saying: "Well, (a) there is some wave — or a *wave function* ψ — needed to describe electrons in atoms, and (b) to fit a wave into a confined space the wave length must be quantized: hence (c) only certain definite energy levels are allowed and, thence, (d) there are sharp, spectral transitions between them!"

Of course, that is far from being the whole story in quantum mechanics; but I believe it captures an important essence. Neither is the first RG response the whole story: but, *to anticipate*, in Wilson's conception RG theory crudely says: "Well, (a) there is a *flow* in some *space*, $\mathbb{H}$, *of Hamiltonians* (or "coupling constants"); (b) the critical point of a system is associated with a *fixed point* (or stationary point) of that flow; (c) the flow operator — technically the *RG transformation*,[22] $\mathbb{R}$ — can be *linearized* about that fixed point; and (d) typically, such a linear operator (as in quantum mechanics) has a spectrum of discrete, but nontrivial eigenvalues, say λ_k; then (e) each (asymptotically independent) exponential term in the flow varies as $e^{\lambda_k l}$, where l is the *flow* (or renormalization) *parameter*, and corresponds to a physical power law, say $|t|^{\phi_k}$, with critical exponent ϕ_k proportional to the eigenvalue λ_k." How one may find suitable transformations $\mathbb{R}$ and why the flows matter, are the subjects for the following chapters of our story.

Within this picture, distinct fixed points may correspond to distinct universality classes of distinct character (which are often blessed with names such as: Ising, XY, Heisenberg, ...).

---

[20] See, *e.g.*, the survey in Fisher (1974a) and Aharony (1976).

[21] Fisher and Pfeuty (1972), Wegner (1972b).

[22] As explained in more detail in Secs. 3.5 and 3.6 below, a specific renormalization transformation, say $\mathbb{R}_b$, acts on some 'initial' Hamiltonian $\mathcal{H}^{(0)}$ in the space $\mathbb{H}$ to transform it into a new Hamiltonian, $\mathcal{H}^{(1)}$. Under repeated operation of $\mathbb{R}_b$ the initial Hamiltonian "flows" into a sequence $\mathcal{H}^{(l)}$ $(l = 1, 2, \ldots)$ corresponding to the iterated RG transformation $\mathbb{R}_b \ldots \mathbb{R}_b$ (l times) which, in turn, specifies a new transformation $\mathbb{R}_{b^l}$. These "products" of repeated RG operations serve to define a *semigroup* of transformations that, in general, does *not* actually give rise to a group: see Footnote 5 above and the discussion below in Sec. 3.5 associated with Eq. (3.23).

Just as quantum mechanics does much more than explain sharp spectral lines, so RG theory should also explain, at least in principle, (B) the values of the leading thermodynamic and correlation exponents, α, β, γ, δ, ν, η, and ω (most already mentioned above) and (C) clarify why and how the classical values are in error, including the existence of borderline dimensionalities, like $d_\times = 4$, above which classical theories become valid. Beyond the leading exponents, one wants (D) the correction-to-scaling exponent θ (and, ideally, the higher-order correction exponents) and, especially, (E) one needs a method to compute crossover exponents, ϕ, to check for the relevance or irrelevance of a multitude of possible perturbations. Two central issues, of course, are (F) the understanding of *universality* with non-trivial exponents and (G) a derivation of *scaling*. The establishment of scaling leads directly to exponent relations such as

$$\alpha + 2\beta + \gamma = 2 \quad \text{and} \quad (2-\eta)\nu = \gamma, \quad \textit{etc.} \tag{3.8}$$

But much more is implied in terms of *data collapse*, the existence of *scaling functions*, *etc.*[23]

And, more subtly, one wants (H) to understand the *breakdown* of universality and scaling in certain circumstances — one might recall continuous spectra in quantum mechanics — and (J) to handle effectively logarithmic and more exotic dependences on temperature, *etc.*[24]

An important further requirement as regards condensed matter physics is that RG theory should be firmly related to the science of statistical mechanics as perfected by Gibbs. Certainly, there is no need and should be no desire, to replace standard statistical mechanics as a basis for describing equilibrium phenomena in pure, homogeneous systems.[25] Accordingly, it is appropriate to summarize briefly the demands of statistical mechanics in a way suitable for describing the formulation of RG transformations.

We may start by supposing that one has a set of microscopic, fluctuating, mechanical variables: in QFT these would be the various quantum fields, $\psi(\mathbf{r})$, defined — one supposes — at all points in a Euclidean (or Minkowski) space. In

[23]Historically, scaling concepts grew slowly, starting with exponent relations (Fisher, 1959, 1962, 1964; Buckinham and Gunton, 1969; Essam and Fisher, 1963; Rushbrooke, 1963; and Griffiths, 1965, 1972). Most transparant was Widom (1965a,b) but note also Domb and Hunter (1965); Kadanoff (1966); Kadanoff *et al.* (1967); Fisher (1967, 1971, 1974a); and Stanley (1971, Chaps. 11, 12). For *corrections-to-scaling* and the related concepts of *irrelevance*, *marginality* and *relevance* one must cite Wegner (1972, 1976); see also Kadanoff (1976) and Fisher (1974, 1983).

[24]See, *e.g.*, Ahlers *et al.* (1975), Aharony (1973), and Kadanoff and Wegner (1971).

[25]One may, however, raise legitimate concerns about the adequacy of customary statistical mechanics when it comes to the analysis of random or impure systems, such as, *e.g.*, "vortex-glass superconductors" (Koch *et al.*, 1984) — or in applications to systems far from equilibrium or in metastable or steady states — *e.g.*, in fluid turbulence, in sandpiles and earthquakes, *etc.* And the use of RG ideas in chaotic mechanics and various other topics listed by Benfatto and Gallavotti (1995) Chap. 1, clearly does *not* require a statistical mechanical basis.

statistical physics we will, rather, suppose that in a physical system of volume V there are N discrete "degrees of freedom." For classical fluid systems one would normally use the coordinates $\mathbf{r}_1, \mathbf{r}_2, \cdots, \mathbf{r}_N$ of the constituent particles. However, it is simpler mathematically — and the analogies with QFT are closer — if we consider here a set of "*spins*" $s_{\mathbf{x}}$ (which could be vectors, tensors, operators, *etc.*) associated with discrete lattice sites located at uniformly spaced points $\mathbf{x}$. If the lattice spacing is a, one can take $V = Na^d$ and the density of degrees of freedom in d spatial dimensions is $N/V = a^{-d}$.

In terms of the basic variables $s_{\mathbf{x}}$, one can form various "local operators" (or "physical densities" or "observables") like the local magnetization and energy densities,

$$M_{\mathbf{x}} = \mu_B s_{\mathbf{x}}, \quad \mathcal{E}_{\mathbf{x}} = -\tfrac{1}{2}J\sum_{\boldsymbol{\delta}} s_{\mathbf{x}} s_{\mathbf{x}+\boldsymbol{\delta}}, \quad \cdots, \tag{3.9}$$

(where μ_B and J are fixed coefficients while $\boldsymbol{\delta}$ runs over the nearest-neighbor lattice vectors). A physical system of interest is then specified by its *Hamiltonian* $\mathcal{H}[\{s_{\mathbf{x}}\}]$ — or energy function, as in mechanics — which is usually just a spatially uniform sum of local operators. The crucial function is the *reduced Hamiltonian*

$$\bar{\mathcal{H}}[s; t, h, \cdots, h_j, \cdots] = -\mathcal{H}[\{s_{\mathbf{x}}\}; \cdots, h_j, \cdots]/k_B T, \tag{3.10}$$

where s denotes the set of all the microscopic spins $s_{\mathbf{x}}$ while $t, h, \ldots, h_j, \ldots$ are various "*thermodynamic fields*" (in QFT — the coupling constants). We may suppose that one or more of the thermodynamic fields, in particular the temperature, can be controlled directly by the experimenter [see Eq. (3.1)]; but others may be "given" since they will, for example, embody details of the physical system that are "fixed by nature."

Normally in condensed matter physics one thus focuses on some specific form of $\bar{\mathcal{H}}$ with at most two or three variable parameters — the Ising model is one such particularly simple form with just two variables, t, the reduced temperature, and $h = \mu_B H/k_B T$, the reduced field. An important feature of Wilson's approach, however, is to regard any such "physical Hamiltonian" as merely specifying a subspace (spanned, say, by "coordinates" t and h) in a very large space of possible (reduced) Hamiltonians, $\mathbb{H}$: see the schematic illustration in Fig. 3.1. This change in perspective proves crucial to the proper formulation of a renormalization group: in principle, it enters also in QFT although in practice, it is usually given little attention.

Granted a microscopic Hamiltonian, statistical mechanics promises to tell one the thermodynamic properties of the corresponding macroscopic system! First one must compute the partition function

$$Z_N[\bar{\mathcal{H}}] = \mathrm{Tr}_N^s\left\{e^{\bar{\mathcal{H}}[s]}\right\}, \tag{3.11}$$

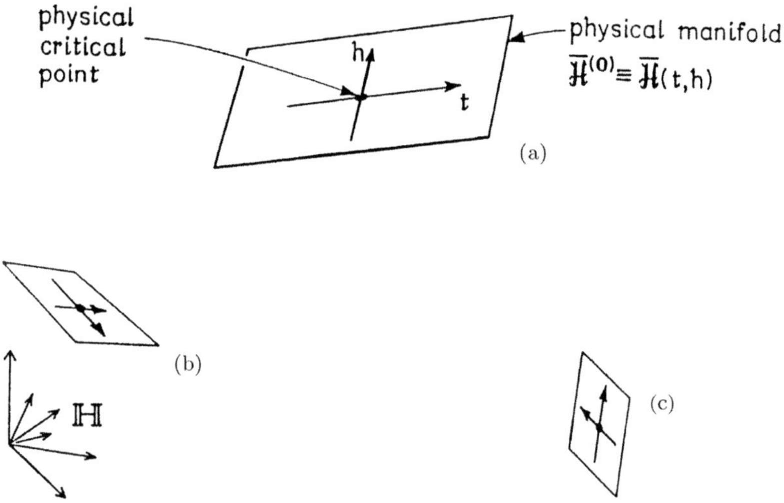

Fig. 3.1. Schematic illustration of the space of Hamiltonians, $\mathbb{H}$, having, in general, infinitely many dimensions (or coordinate axes). A particular physical system or model representing, say, the ferromagnet, iron, is specified by its reduced Hamiltonian $\bar{\mathcal{H}}(t,h)$, with $t = (T - T_c)/T_c$ and $h = \mu_B H/k_B T$ defined for *that* system: but in $\mathbb{H}$ this Hamiltonian specifies only a submanifold — the physical manifold, labelled (a), that is parametrized by the 'local coordinates' t and h. Other submanifolds, (b), (c), located elsewhere in $\mathbb{H}$, depict the physical manifolds for Hamiltonians corresponding to other particular physical systems, say, the ferromagnets nickel and gadolinium, *etc.*

where the *trace operation*, $\mathrm{Tr}_N^s\{\cdot\}$, denotes a summation or integration[26] over all the possible values of all the N spin variables $s_{\mathbf{x}}$ in the system of volume V. The *Boltzmann factor*, $\exp(\bar{\mathcal{H}}[s])$, measures, of course, the probability of observing the microstate specified by the set of values $\{s_{\mathbf{x}}\}$ in an equilibrium ensemble at temperature T. Then the thermodynamics follow from the total free energy density, which is given by[27]

$$f\left[\bar{\mathcal{H}}\right] \equiv f(t, h, \cdots, h_j, \cdots) = \lim_{N,V\to\infty} V^{-1} \log Z_N\left[\bar{\mathcal{H}}\right]; \tag{3.12}$$

this includes a *singular part* $f_s\left[\bar{\mathcal{H}}\right]$ near a critical point of interest as well as smooth *background terms* which are analytic (in t, h, $\cdots$) through the critical point. Correlation functions are defined similarly in stardard manner.

To the degree that one can actually perform the trace operation in Eq. (3.11) for a particular model system and take the "thermodynamic limit" in Eq. (3.12) one will obtain the precise critical exponents, scaling functions, and so on. This was

---

[26]Here, for simplicity, we suppose the $s_{\mathbf{x}}$ are classical, commuting variables. If they are operator-valued then, in the standard way, the trace must be defined as a sum or integral over diagonal matrix elements computed with a complete basis set of N-variable states.

[27]In Eq. (3.12) we have explicitly indicated the thermodynamic limit in which N and V become infinite maintaining the ratio $V/N = a^d$ fixed: in QFT this corresponds to an infinite system with an ultraviolet lattice cutoff.

Onsager's (1944) route in solving the $d = 2$, spin-$\frac{1}{2}$ Ising models in zero magnetic field. At first sight one then has no need of RG theory. That surmise, however, turns out to be far from the truth. The issue is "simply" one of understanding! (Should one ever achieve truly high precision in simulating critical systems on a computer — a prospect which now seems closer than in the past — the same problem would remain.) In short, while one knows for sure that $\alpha = 0$ (log), $\beta = \frac{1}{8}$, $\gamma = 1\frac{3}{4}$, $\nu = 1$, $\eta = \frac{1}{4}$, $\cdots$ for the planar Ising models one does not know *why* the exponents have these values or *why* they satisfy the exponent relations Eq. (3.8) or why the scaling laws are obeyed. Indeed, the seemingly inevitable mathematical complexities of solving even such physically oversimplified models exactly[28] serve to conceal almost all traces of general, underlying mechanisms and principles that might "explain" the results. Thus it comes to pass that even a rather crude and approximate solution of a two-dimensional Ising model by a real-space RG method can be truly instructive.[29]

3.3. Kadanoff's scaling picture

The twelve months from late-1965 through 1966 saw the clear formulation of scaling for the thermodynamic properties in the critical region and the fuller appreciation of scaling for the correlation functions.[30] One may highlight Widom's (1965) approach since it was the most direct and phenomenological — a bold, new thermodynamic hypothesis was advanced by generalizing a particular feature of the classical theories. But Domb and Hunter (1965) reached essentially the same conclusion for the thermodynamics based on analytic and series-expansion considerations, as did Patashinskii and Pokrovskii (1966) using a more microscopic formulation that brought out the relations to the full set of correlation functions (of all orders).[31]

Kadanoff (1966), however, derived scaling by introducing a completely new concept, namely, the *mapping* of a critical or near-critical system onto itself by a reduction in the effective number of degrees of freedom.[32] This paper attracted

[28]As expounded in the monograph by Rodney Baxter (1982).

[29]See Niemeijer and van Leeuwen (1976), Burkhardt and van Leeuwen (1982), and Wilson (1975, 1983) for discussion of real-space RG methods.

[30]Note the remarks made above in Footnote 23. One might also recall, in this respect, earlier work (Fisher, 1959, 1962, 1964) restricted (in the application to ferromagnets) to zero magnetic field.

[31]It was later seen, furthermore, (Kiang and Stauffer, 1970; Fisher, 1971, Sec. 4.4) that thermodynamic scaling with general exponents (but particular forms of scaling function) was embodied in the "droplet model" partition function advanced by Essam and Fisher (1963) from which the exponent relations $\alpha' + 2\beta + \gamma' = 2$, *etc.*, were originally derived.

[32]Novelty is always relative! From a historical perspective one should recall a suggestive contribution by M.J. Buckingham, presented in April 1965, in which he proposed a division of a lattice system into cells of geometrically increasing size, $L_n = b^n L_0$, with controlled intercell couplings. This led him to propose "the *existence* of an asymptotic 'lattice problem' such that the description of the nth order in terms of the $(n-1)$th is the same as that of the $(n+1)$th in terms of the nth." This is practically a description of "scaling" or "self similarity" as we recognize

much favorable notice since, beyond obtaining all the scaling properties, it seemed to lay out a direct route to the actual *calculation* of critical properties. On closer examination, however, the implied program seemed — as I will explain briefly — to run rapidly into insuperable difficulties and interest faded. In retrospect, however, Kadanoff's scaling picture embodied important features eventually seen to be basic to Wilson's conception of the full renormalization group. Accordingly, it is appropriate to present a sketch of Kadanoff's seminal ideas.

For simplicity, consider with Kadanoff (1966), a lattice of spacing a (and dimensionality $d > 1$) with $S = \frac{1}{2}$ Ising spins $s_{\mathbf{x}}$ which, by definition, take only the values $+1$ or -1. Spins on nearest-neighbor sites are coupled by an energy parameter or coupling constant, $J > 0$, which favors parallel alignment [see, *e.g.*, Eq. (3.9) above]. Thus at low temperatures the majority of the spins point "up" ($s_{\mathbf{x}} = +1$) or, alternatively, "down" ($s_{\mathbf{x}} = -1$); in other words, there will be a spontaneous magnetization, $M_0(T)$, which decreases when T rises until it vanishes at the critical temperature $T_c > 0$: recall paragraph after Eq. (3.3).

Now divide the lattice up into (disjoint) blocks, of dimensions $L \times L \times \cdots \times L$ with $L = ba$ so that each block contains b^d spins. Then associate with each block, say $\mathcal{B}_{\mathbf{x}'}$ centered at point $\mathbf{x}'$, a new, effective *block spin*, $s'_{\mathbf{x}'}$: see Fig. 3.2. If, finally, we *rescale* all spatial coordinates according to

$$\mathbf{x} \Rightarrow \mathbf{x}' = \mathbf{x}/b, \tag{3.13}$$

the new lattice of block spins $s'_{\mathbf{x}'}$ looks just like the original lattice of spins $s_{\mathbf{x}}$. Note, in particular, the density of degrees of freedom is unchanged: see Fig. 3.2.

But if this appearance is to be more than superficial one must be able to relate the new or "renormalized" coupling J' between the block spins to the original coupling J, or, equivalently, the renormalized temperature deviation t' to the original value t. Likewise one must relate the new, renormalized magnetic field h' to the original field h.

To this end, Kadanoff supposes that b is large but less than the ratio, ξ/a, of the *correlation length*, $\xi(t, h)$, to the lattice spacing a; since ξ diverges at criticality — see Eq. (3.5) — this allows, asymptotically, for b to be chosen *arbitrarily*. Then Kadanoff argues that the total coupling of the magnetic field h to a block of b^d

---

it today. Unfortunately, however, Buckingham failed to draw any significant, correct conclusions from his conception and his paper seemed to have little influence despite its presentation at the notable international conference on *Phenomena in the Neighborhood of Critical Points* organized by M.S. Green (with G.B. Benedek, E.W. Montroll, C.J. Pings, and the author) and held at the National Bureau of Standards, then in Washington, D.C. The Proceedings, complete with discussion remarks, were published, in December 1966, under the editorship of Green and J.V. Sengers (1966). Nearly all the presentations addressed the rapidly accumulating experimental evidence, but many well known theorists from a range of disciplines attended including P.W. Anderson, P. Debye, C. de Dominicis, C. Domb, S.F. Edwards, P.C. Hohenberg, K. Kawasaki, J.S. Langer, E. Lieb, W. Marshall, P.C. Martin, T. Matsubara, E.W. Montroll, O.K. Rice, J.S. Rowlinson, G.S. Rushbrooke, L. Tisza, G.E. Uhlenbeck, and C.N. Yang; but B. Widom, L.P. Kadanoff, and K.G. Wilson are *not* listed among the participants.

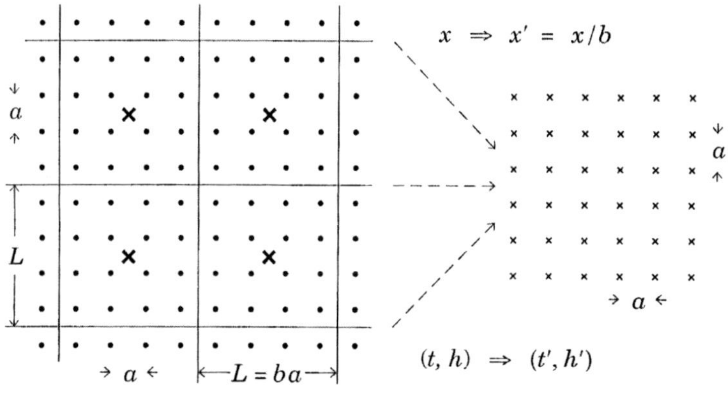

Fig. 3.2. A lattice of spacing a of Ising spins $s_{\mathbf{x}} = \pm 1$ (in $d = 2$ dimensions) marked by solid dots, divided up into Kadanoff blocks or cells of dimensions $(L = ba) \times (L = ba)$ each containing a block spin $s'_{\mathbf{x}'} = \pm 1$, indicated by a cross. After a rescaling, $\mathbf{x} \Rightarrow \mathbf{x}' = \mathbf{x}/b$, the lattice of block spins appears identical with the original lattice. However, one supposes that the temperature t, and magnetic field h, of the original lattice can be renormalized to yield appropriate values, t' and h', for the rescaled, block-spin lattice: see text. In this illustration the spatial rescaling factor is $b = 4$.

spins is equivalent to a coupling to b^d average spins

$$\bar{s}_{\mathbf{x}'} \equiv b^{-d} \sum_{\mathbf{x} \in \mathcal{B}_{\mathbf{x}'}} s_{\mathbf{x}} \cong \zeta(b) s'_{\mathbf{x}'}, \tag{3.14}$$

where the sum runs over all the sites $\mathbf{x}$ in the block $\mathcal{B}_{\mathbf{x}'}$, while the "asymptotic equivalence" to the new, Ising block spin $s'_{\mathbf{x}'}$ entails, Kadanoff proposes, some "spin rescaling or renormalization factor" $\zeta(b)$. Introducing a similar thermal renormalization factor, $\theta(b)$, leads to the *recursion relations*

$$t' \approx \theta(b) t \quad \text{and} \quad h' \approx b^d \zeta(b) h. \tag{3.15}$$

Correspondingly, the basic correlation function — compare with Eq. (3.4) — should renormalize as

$$G(\mathbf{x}; t, h) \equiv \langle s_{\mathbf{0}} s_{\mathbf{x}} \rangle \approx \zeta^2(b) G(\mathbf{x}'; t', h'). \tag{3.16}$$

In summary, under a spatial scale transformation and the integration out of all but a fraction b^{-d} of the original spins, the system asymptotically *maps back into itself* although at a renormalized temperature and field! However, the map is *complete* in the sense that *all* the statistical properties should be related by similarity.

But how should one choose — or, better, determine — the renormalization factors ζ and θ? Let us consider the basic relation Eq. (3.16) *at* criticality, so that $t = h = 0$ and, by Eq. (3.15), $t' = h' = 0$. Then, if we accept the observation/expectation Eq. (3.6) of a power law decay, *i.e.*, $G_c(\mathbf{x}) \sim 1/|\mathbf{x}|^{d-2+\eta}$, one soon finds that $\zeta(b)$

must be just a power of b. It is natural, following Kadanoff (1966), then to propose the forms

$$\zeta(b) = b^{-\omega} \quad \text{and} \quad \theta(b) = b^{\lambda}, \tag{3.17}$$

where the two exponents ω and λ characterize the critical point under study while b is an essentially unrestricted *scaling parameter.*

By capitalizing on the freedom to choose b as t, $h \to 0$, or, more-or-less equivalently, by *iterating* the recursion relations Eqs. (3.15) and (3.16), one can, with some further work, show that the previous exponent relations, Eqs. (3.8), must hold. But, beyond that, the full scalling laws also follow. Thus one may write the correlation function in zero field ($h = 0$) as

$$G(\mathbf{r}; T) \approx \frac{D}{r^{d-2+\eta}} \mathcal{G}\left(\frac{r}{\xi(T)}\right), \tag{3.18}$$

where for consistency with Eq. (3.6), the *scaling function*, $\mathcal{G}(x)$, satisfies the normalization condition $\mathcal{G}(0) = 1$. Integrating $\mathbf{r}$ over all space yields the compressibility/susceptibility $\chi(T)$ and, thence, one may derive the relation $\gamma = (2-\eta)\nu$, stated in Eqs. (3.8).

Similarly, one can derive scaling for the singular part of the free energy (introduced in Eq. (3.12) above). If, for the sake of some extra generality we introduce another field, say, $g = P/k_BT$, where *e.g.*, P denotes the pressure, the scaling hypothesis may be written

$$f_s(t, h, g) \approx |t|^{2-\alpha} \mathcal{F}\left(\frac{h}{|t|^{\Delta}}, \frac{g}{|t|^{\phi}}\right), \tag{3.19}$$

where α is the specific heat exponent, introduced in Eq. (3.2), while the new exponent Δ, which determines "how h scales with t," is simply given by $\Delta = \beta + \gamma = \beta\delta$ where the exponent δ describes the order parameter variation *at* $t = 0$ via $\langle\Psi\rangle \sim h^{1/\delta}$. The crossover exponent, ϕ, was mentioned in Sec. 3.2, shortly before Eqs. (3.8), but is not, at first sight, directly relevant for Kadanoff's analysis.

Of course, all the exponents are now determined by ω and λ in Eqs. (3.17); thus one finds $\nu = 1/\lambda$ and $\beta = \omega\nu$. Furthermore, the analysis leads to new exponent relations, namely, the so-called *hyperscaling laws*[33] which explicitly involve the spatial dimensionality d: most notable is[34]

$$d\nu = 2 - \alpha. \tag{3.20}$$

Although this relation must be questioned for $d > 4$, Kadanoff's scaling picture is greatly strengthened by the fact that it holds *exactly* for the $d = 2$ Ising model! And likewise for all other exactly soluble models when $d < 4$.[35]

[33] See (Fisher, 1974a) where the special character of the hyperscaling relations is stressed.
[34] See Kadanoff (1966), Widom (1965a), and Stell (1965, unpublished, quoted in Fisher, 1969). The relation $\delta = (d + 2 - \eta)/(d - 2 + \eta)$ is also worthy of note.).
[35] See, *e.g.*, Fisher (1983) and, for the details of the exactly solved models, Baxter (1982).

Historically, the careful numerical studies of the $d = 3$ Ising models by series expansions[36] for many years suggested a small but significant deviation from Eq. (3.20) as allowed by pure scaling phenomenolgy.[37] But, in later years, the accumulating weight of evidence, when critically reviewed, convinced even the most cautious skeptics of the validity of Eq. (3.20) in three dimensions![38]

Nevertheless, all is not roses! Unlike the previous exponent relations (all being independent of d) hyperscaling fails for the classical theories unless $d = 4$. And since one knows (rigorously for certain models) that the classical exponent values are valid for $d > 4$, it follows that hyperscaling cannot be generally valid. Thus something is certainly missing from Kadanoff's picture. Now, thanks to RG insights, we know that the breakdown of hyperscaling is to be understood via the second argument in the "fuller" scaling form Eq. (3.19): when d exceeds the appropriate borderline dimension, $d_\times$, a "dangerous irrelevant variable" appears and must be allowed for.[39] In essence one finds that the scaling function limit $\mathcal{F}(y, z \to 0)$, previously accepted without question, is no longer well defined but, rather, diverges as a power of z: asymptotic scaling survives but $d^* \equiv (2 - \alpha)/\nu$ sticks at the value 4 for $d > d_\times = 4$.

However, the issue of hyperscaling was *not* the main road block to the analytic development of Kadanoff's picture. The principal difficulties arose in explaining the *power-law* nature of the rescaling factors in Eqs. (3.15)–(3.17) and, in particular, in justifying the idea of a *single*, effective, renormalized coupling J' between adjacent block spins, say $s'_{\mathbf{x}'}$ and $s'_{\mathbf{x}'+\delta'}$. Thus the interface between two adjacent $L \times L \times L$ blocks (taking $d = 3$ as an example) separates two block faces each containing b^2, strongly interacting, original lattice spins $s_{\mathbf{x}}$. Well below T_c all these spins are frozen, "up" or "down," and a single effective coupling could well suffice; but at and above T_c these spins must fluctuate on many scales and a single effective-spin coupling seems inadequate to represent the inherent complexities.[40]

One may note, also that Kadanoff's picture, like the scaling hypothesis itself, provides no real hints as to the origins of universality: the rescaling exponents ω and λ in Eqs. (3.17) might well change from one system to another. Wilson's (1971a) conception of the renormalization group answered *both* the problem of the "lost microscopic details" of the original spin lattice *and* provided a natural explanation of universality.

---

[36] For accounts of series expansion techniques and their important role see: Domb (1960, 1996), Baker (1961, 1990), Essam and Fisher (1963), Fisher (1965, 1967b), and Stanley (1971).

[37] As expounded systematically in (Fisher, 1974a) with hindsight enlightened by RG theory.

[38] See Fisher and Chen (1985) and Baker and Kawashima (1995, 1996).

[39] See Fisher in Gunton and Green (1974, p. 66) where a "dangerous irrelevant variable" is characterized as a "hidden relevant variable;" and Fisher (1983, App. D).

[40] In hindsight, we know this difficulty is profound: in general, it is *impossible* to find an adequate single coupling. However, for certain special models it does prove possible and Kadanoff's picture goes through: see Nelson and Fisher (1975) and Fisher (1983). Further, in defense of Kadanoff, the condition $b \ll \xi/a$ was supposed to "freeze" the original spins in each block sufficiently well to justify their replacement by a simple block spin.

3.4. Wilson's quest

Now because this account has a historical perspective, and since I was Ken Wilson's colleague at Cornell for some twenty years, I will say something about how his search for a deeper understanding of quantum field theory led him to formulate renormalization group theory as we know it today. The first remark to make is that Ken Wilson was a markedly independent and original thinker as well as being a rather private and reserved person. Secondly, in his 1975 article, in *Reviews of Modern Physics*, from which I have already quoted, Ken Wilson gave his own considered overview of RG theory which, in my judgement, still stands well today. In 1982 he received the Nobel Prize and in his Nobel lecture, published in 1983, he devotes a section to "Some History Prior to 1971" in which he recounts his personal scientific odyssey.

He explains that as a student at Caltech in 1956–60, he failed to avoid "the default for the most promising graduate students [which] was to enter elementary-particle theory." There he learned of the 1954 paper by Gell-Mann and Low "which was the principal inspiration for [his] own work prior to Kadanoff's (1966) formulation of the scaling hypothesis." By 1963 Ken Wilson had resolved to pursue quantum field theories as applied to the strong interactions. Prior to summer 1966 he heard Ben Widom present his scaling equation of state in a seminar at Cornell "but was puzzled by the absence of any theoretical basis for the form Widom wrote down." Later, in summer 1966, on studying Onsager's solution of the Ising model in the reformulation of Lieb, Schultz, and Mattis,[41] Wilson became aware of analogies with field theory and realized the applicability of his own earlier RG ideas (developed for a truncated version of fixed-source meson theory[42]) to critical phenomena. This gave him a scaling picture but he discovered that he "had been scooped by Leo Kadanoff." Thereafter Ken Wilson amalgamated his thinking about field theories on a lattice and critical phenomena learning, in particular, about Euclidean QFT[43] and its close relation to the transfer matrix method in statistical mechanics — the basis of Onsager's (1944) solution.

That same summer of 1966 I joined Ben Widom at Cornell and we jointly ran an open and rather wide-ranging seminar loosely centered on statistical mechanics. Needless to say, the understanding of critical phenomena and of the then new scaling theories was a topic of much interest. Ken Wilson frequently attended and, perhaps partially through that route, soon learned a lot about critical phenomena. He was, in

---

[41] See Schultz *et al.* (1964).
[42] See Wilson (1983).
[43] As stressed by Symanzik (1966) the Euclidean formulation of quantum field theory makes more transparent the connections to statistical mechanics. Note, however, that in his 1966 article Symanzik did not delineate the special connections to critical phenomena *per se* that were gaining increasingly wide recognition; see, *e.g.*, Patashinskii and Pokrovskii (1966), Fisher (1969, Sec. 12) and the remarks below concerning Fisher and Burford (1967).

particular, interested in the series expansion and extrapolation methods for estimating critical temperatures, exponents, amplitudes, *etc.*, for lattice models that had been pioneered by Cyril Domb and the King's College, London group. This approach is, incidentally, still one of the most reliable and precise routes available for estimating critical parameters.[44] At that time I, myself, was completing a paper on work with a London University student, Robert J. Burford, using high-temperature series expansions to study in detail the correlation functions and scattering behavior of the two- and three-dimensional Ising models.[45] Our theoretical analysis had already brought out some of the analogies with field theory revealed by the transfer matrix approach. Ken himself undertook large-scale series expansion calculations in order to learn and understand the techniques. Indeed, relying on the powerful computer programs Ken Wilson developed and kindly made available to us, another one of my students, Howard B. Tarko, extended the series analysis of the Ising correlation functions to temperatures below T_c and to all values of the magnetic field.[46] Our results lasted fairly well and many of them were only later revised and improved.[47]

Typically, then, Ken Wilson's approach was always "hands on" and his great expertise with computers was ever at hand to check his ideas and focus his thinking.[48] From time to time Ken would intimate to Ben Widom or myself that he might be ready to tell us where his thinking about the central problem of explaining scaling had got to. Of course, we were eager to hear him speak at our seminar although his talks were frequently hard to grasp. From one of his earlier talks and the discussion afterwards, however, I carried away a powerful and vivid picture of *flows* — flows in a large space. And the point was that at the initiation of the flow, when the "time" or "flow parameter" l, was small, two nearby points would travel close together; see Fig. 3.3. But as the flow developed a point could be reached — a bifurcation point (and hence, as one later realized, a stationary or fixed point of the flow) — beyond which the two originally close points could separate and, as l increased, diverge to vastly different destinations: see Fig. 3.3. At the time, I vaguely understood this as indicative of how a sharp, nonanalytic phase transition could grow from smooth analytic initial data.[49]

But it was a long time before I understood the nature of the space — the space $\mathbb{H}$ of Hamiltonians — and the *mechanism* generating the flow, that is, a renormalization group transformation. Nowadays, when one looks at Fig. 3.3, one sees the locus of initial points, $l = 0$, as identifying the manifold corresponding to the original or 'bare' Hamiltonian (see Fig. 3.1) while the trajectory leading to the

[44] See the reviews Domb (1960), Fisher (1965, 1967b), Stanley (1971).
[45] Fisher and Burford (1967).
[46] Tarko and Fisher (1975).
[47] See, *e.g.*, Zinn and Fisher (1996), Zinn, Lai, and Fisher (1996), Butera and Comi (1997), Guida and Zinn-Justin (1998), Campostrini *et al.* (2002), and El-Showk *et al.* (2014a) who exploit conformal symmetry, and references therein.
[48] See his remarks in Wilson (1983) on page 591, column 1.
[49] See the (later) introductory remarks in Wilson (1971a) related to Fig. 6.1 there.

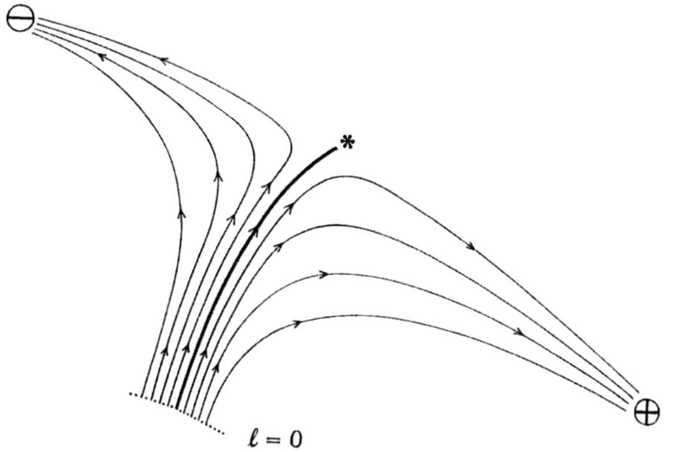

Fig. 3.3. A "vision" of flows in some large space inspired by a seminar of K. G. Wilson in the period 1967–1970. The idea conveyed is that initially close, smoothly connected points at the start of the flow — the locus $l = 0$ — can eventually separate and run to far distant regions representing very different "final" physical states: the essence of a phase transition. In modern terms the flow is in the space $\mathbb{H}$ of Hamiltonians; the intersection of the separatrix, shown bolder, with the initial locus ($l = 0$) represents the physical critical point; $*$ denotes the controlling fixed point, while $\oplus$ and $\ominus$, represent asymptotic high-T, disordered states, and low-T, ordered states, respectively.

bifurcation point represents a locus of critical points; the two distinct destinations for $l \to \infty$ then typically, correspond to a high-temperature, fully disordered system and to a low-temperature fully ordered system: see Fig. 3.3.

In 1969 word reached Cornell that two Italian theorists, C. Di Castro and G. Jona-Lasinio, were claiming[50] that the "multiplicative renormalization group," as expounded in the field-theory text by Bogoliubov and Shirkov (1959), could provide "a microscopic foundation" for the scaling laws (which, by then, were well established phenomenologically). The formalism and content of the field-theoretic renormalization group was totally unfamiliar to most critical-phenomena theorists: but the prospect of a microscopic derivation was clearly exciting! However, the articles[51] proved hard to interpret as regards concrete progress and results. Nevertheless, the impression is sometimes conveyed that Wilson's final breakthrough was somehow anticipated by Di Castro and Jona-Lasinio.[52]

Such an impression would, I believe, be quite misleading. Indeed, Di Castro was invited to visit Cornell where he presented his ideas in a seminar that was listened to attentively. Again I have a vivid memory: walking to lunch at the Statler Inn after the seminar I checked my own impressions with Ken Wilson by asking:

[50]The first published article was Di Castro and Jona-Lasinio (1969).
[51]See the later review by Di Castro and Jona-Lasinio (1976) for references to their writings in the period 1969–1972 prior to Wilson's 1971 papers and the ϵ-expansion in 1972.
[52]See, for example, Benfatto and Gallavotti (1995) on page 96 in *A Brief Historical Note*, which is claimed to represent only the authors' personal "cultural evolution through the subject."

"Well, did he really say anything new?" (By "new" I meant some fresh insight or technique that carried the field forward.) The conclusion of our conversation was "No." The point was simply that none of the problems then outstanding — see the "tasks" outlined above (in Sec. 3.2) — had been solved or come under effective attack. In fairness, I must point out that the retrospective review by Di Castro and Jona-Lasinio themselves (1976) is reasonably well balanced: One accepted a scaling hypothesis and injected that as an ansatz into a general formalism; then certain insights and interesting features emerged; but, in reality, only scaling theory had been performed; and, in the end, as Di Castro and Jona-Lasinio say: "Still one did not see how to perform explicit calculations." Incidentally, it is also interesting to note Wilson's sharp criticism[53] of the account presented by Bogoliubov and Shirkov (1959) of the original RG ideas of Stueckelberg and Petermann (who, in 1953, coined the phrase "groupes de normalization") and of Gell-Mann and Low (1954).

One more personal anecdote may be permissible here. In August 1973 I was invited to present a tutorial seminar on renormalization group theory while visiting the Aspen Center for Physics. Ken Wilson's thesis advisor, Murray Gell-Mann, was in the audience. In the discussion period after the seminar Gell-Mann expressed his appreciation for the theoretical structure created by his famous student that I had set out in its generality, and he asked: "But tell me, what has all that got to do with the work Francis Low and I did so many years ago?"[54] In response, I explained the connecting thread and the far-reaching intellectual inspiration: certainly there is a thread but — to echo my previous comments — I believe that its length is comparable to that reaching from Maxwell, Boltzmann, and ideal gases to Gibbs' general conception of ensembles, partition functions, and their manifold inter-relations.

3.5. The construction of renormalization group transformations: the epsilon expansion

In telling my story I have purposefully incorporated a large dose of hindsight by emphasizing the importance of viewing a particular physical system — or its reduced Hamiltonian, $\bar{\mathcal{H}}(t, h, \cdots)$: see Eq. (3.10) — as specifying only a relatively small manifold in a large space, $\mathbb{H}$, of possible Hamiltonians. But why is that more than a mere formality? One learns the answer as soon as, following Wilson (1975, 1983), one attempts to implement Kadanoff's scaling description in some concrete, computational way. In Kadanoff's picture (in common with the Gell-Mann–Low, Callan–Symanzik, and general QFT viewpoints) one *assumes* that after a "rescaling" or "renormalization" the new, renormalized Hamiltonian (or, in QFT, the Lagrangean) has the *identical form* except for the renormalization of a single parameter (or coupling constant) or — as in Kadanoff's picture — of at most a small *fixed* number, like the temperature t and field h. That assumption is the dangerous and, unless one

[53]See, especially, Wilson (1975) on page 796, column 1, and Footnote 10 in Wilson (1971a).
[54]That is, in Gell-Mann and Low (1954).

is especially lucky,[55] the *generally false* step! Wilson (1975, p. 592) has described his "liberation" from this straight jacket and how the freedom gained opened the door to the systematic design of RG transformations.

To explain, we may state matters as follows: Gibbs' prescription for calculating the partition function — see Eq. (3.11) — tells us to sum (or to integrate) over the allowed values of *all* the N spin variables $s_{\mathbf{x}}$. But this is very difficult! Let us, instead, adopt a strategy of "divide and conquer," by separating the set $\{s_{\mathbf{x}}\}$ of N spins into two groups: first, $\{s_{\mathbf{x}}^{<}\}$, consisting of $N' = N/b^d$ spins which we will leave as untouched fluctuating variables; and, second, $\{s_{\mathbf{x}}^{>}\}$ consisting of the remaining $N - N'$ spin variables over which we will integrate (or sum) so that they drop out of the problem. If we draw inspiration from Kadanoff's (or Buckingham's[56]) block picture we might reasonably choose to integrate over all but one central spin in each block of b^d spins. This process, which Kadanoff has dubbed "decimation" (after the Roman military practice), preserves translational invariance and clearly represents a concrete form of "coarse graining" (which, in earlier days, was typically cited as a way to derive, "in principle," mesoscopic or Landau–Ginzburg descriptions).

Now, after taking our partial trace we must be left with some new, *effective Hamiltonian*, say, $\bar{\mathcal{H}}_{\text{eff}}[s^{<}]$, involving only the preserved, unintegrated spins. On reflection one realizes that, in order to be faithful to the original physics, such an effective Hamiltonian must be defined via its Boltzmann factor: recalling our brief outline of statistical mechanics, that leads directly to the explicit formula

$$e^{\bar{\mathcal{H}}_{\text{eff}}[s^{<}]} = \text{Tr}_{N-N'}^{s^{>}}\left\{e^{\bar{\mathcal{H}}_{\text{eff}}[s^{<}\cup s^{>}]}\right\}, \tag{3.21}$$

where the 'union,' $s^{<}\cup s^{>}$, simply stands for the full set of original spins $s \equiv \{s_{\mathbf{x}}\}$. By a spatial rescaling, as in Eq. (3.13), and a relabelling, namely, $s_{\mathbf{x}}^{<} \Rightarrow s'_{\mathbf{x}'}$, we obtain the "renormalized Hamiltonian," $\bar{\mathcal{H}}'[s'] \equiv \bar{\mathcal{H}}_{\text{eff}}[s^{<}]$. Formally, then, we have succeeded in defining an *explicit renormalization transformation*. We will write

$$\bar{\mathcal{H}}'[s'] = \mathbb{R}_b\{\bar{\mathcal{H}}[s]\}, \tag{3.22}$$

where we have elected to keep track of the spatial rescaling factor, b, as a subscript on the RG operator $\mathbb{R}$.

Note that if we complete the Gibbsian prescription by taking the trace over the renormalized spins we simply get back to the desired partition function, $Z_N[\bar{\mathcal{H}}]$. (The formal derivation for those who might be interested is set out in the footnote below.[57]) Thus nothing has been lost: the renormalized Hamiltonian retains all the

[55] See Footnote 40 above and Nelson and Fisher (1975) and Fisher (1983).

[56] Recall Footnote 32 above.

[57] We start with the definition Eq. (3.21) and recall Eq. (3.11) to obtain

$$\begin{aligned} Z_{N'}[\bar{\mathcal{H}}'] &\equiv \text{Tr}_{N'}^{s'}\left\{e^{\bar{\mathcal{H}}[s']}\right\} = \text{Tr}_{N'}^{s^{<}}\left\{e^{\bar{\mathcal{H}}_{\text{eff}}[s^{<}]}\right\} = \text{Tr}_{N'}^{s^{<}}\text{Tr}_{N-N'}^{s^{>}}\left\{e^{\bar{\mathcal{H}}[s^{<}\cup s^{>}]}\right\} \\ &= \text{Tr}_{N}^{s}\left\{e^{\bar{\mathcal{H}}[s]}\right\} = Z_N[\bar{\mathcal{H}}], \end{aligned}$$

from which the free energy $f[\bar{\mathcal{H}}]$ follows via Eq. (3.12).

thermodynamic information. On the other hand, experience suggests that, rather than try to compute Z_N directly from $\bar{\mathcal{H}}'$, it will prove more fruitful to *iterate* the transformation so obtaining a sequence, $\bar{\mathcal{H}}^{(l)}$, of renormalized Hamiltonians, namely,

$$\bar{\mathcal{H}}^{(l)} = \mathbb{R}_b\left[\bar{\mathcal{H}}^{(l-1)}\right] = \mathbb{R}_{b^l}\left[\bar{\mathcal{H}}\right], \tag{3.23}$$

with $\bar{\mathcal{H}}^{(0)} \equiv \bar{\mathcal{H}}$, $\bar{\mathcal{H}}^{(1)} = \bar{\mathcal{H}}'$. It is these iterations that give rise to the *semigroup* character of the RG transformation.[58]

But now comes the crux: thanks to the rescaling and relabelling, the microscopic variables $\{s'_{\mathbf{x}'}\}$ are, indeed, completely equivalent to the original spins $\{s_{\mathbf{x}}\}$. However, when one proceeds to determine the nature of $\bar{\mathcal{H}}_{\text{eff}}$, and thence of $\bar{\mathcal{H}}'$, by using the formula (3.21), one soon discovers that one *cannot* expect the original form of $\bar{\mathcal{H}}$ to be reproduced in $\bar{\mathcal{H}}_{\text{eff}}$. Consider, for concreteness, an initial Hamiltonian, $\bar{\mathcal{H}}$, that describes Ising spins ($s_{\mathbf{x}} = \pm 1$) on a square lattice in zero magnetic field with just nearest-neighbor interactions of coupling strength $K_1 = J_1/k_BT$: in the most conservative Kadanoff picture there must be *some* definite recursion relation for the renormalized coupling, say, $K_1' = \mathcal{T}_1(K_1)$, embodied in a definite function $\mathcal{T}_1(\cdot)$. But, in fact, one finds that $\bar{\mathcal{H}}_{\text{eff}}$ must actually contain *further* nonvanishing spin couplings, K_2, between second-neighbor spins, K_3, between third-neighbors, and so on up to *indefinitely* high orders. Worse still, four-spin coupling terms like $K_{\square 1}s_{\mathbf{x}_1}s_{\mathbf{x}_2}s_{\mathbf{x}_3}s_{\mathbf{x}_4}$ appear in $\bar{\mathcal{H}}_{\text{eff}}$, again for *all* possible arrangements of the four spins! And also six-spin couplings, eight-spin couplings, $\cdots$. Indeed, for any given set Q of $2m$ Ising spins on the lattice (and its translational equivalents), a nonvanishing coupling constant, K_Q', is generated and appears in $\bar{\mathcal{H}}'$!

The only saving grace is that further iteration of the decimation transformation Eq. (3.21) cannot (in zero field) lead to anything worse! In other words the space $\mathbb{H}_{\text{Is}}$ of Ising spin Hamiltonians in zero field may be specified by the infinite set $\{K_Q\}$, of all possible spin couplings, and is *closed* under the decimation transformation Eq. (3.21). Formally, one can thus describe $\mathbb{R}_b$ by the full set of *recursion relations*

$$K_P' = \mathcal{T}_P(\{K_Q\}) \quad (\text{all}\, P). \tag{3.24}$$

Clearly, this answers our previous questions as to what becomes of the complicated across-the-faces-of-the-block interactions in the original Kadanoff picture:

[58]Thus successive decimations with scaling factors b_1 and b_2 yield the quite general relation

$$\mathbb{R}_{b_2}\mathbb{R}_{b_1} = \mathbb{R}_{b_2b_1},$$

which essentially defines a unitary *semigroup* of transformations. See Footnotes 5 and 22 above, and the formal algebraic definition in MacLane and Birkhoff (1967): a unitary semigroup (or 'monoid') is a set M of elements, u, v, w, x, $\cdots$ with a binary operation, $xy = w \in M$, which is associative, so $v(wx) = (vw)x$, and has a unit u, obeying $ux = xu = x$ (for all $x \in M$) — in RG theory, the unit transformation corresponds simply to $b = 1$. Hille (1948) and Riesz and Sz.-Nagy (1955) describe semigroups within a continuum, functional analysis context and discuss the existence of an infinitesimal generator when the flow parameter l is defined for continuous values $l \geq 0$: see Eq. (3.33) below and Wilson's (1971a) introductory discussion.

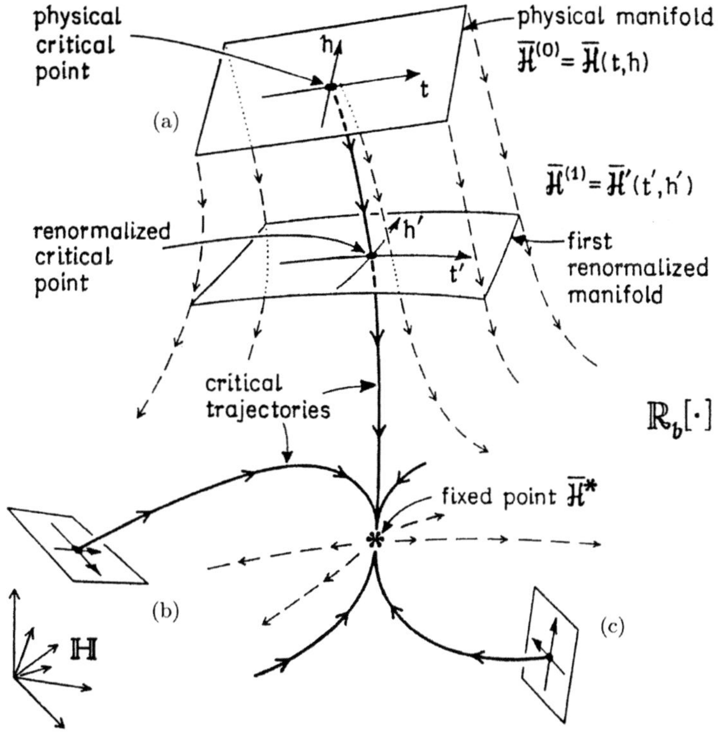

Fig. 3.4. A depiction of the space of Hamiltonians $\mathbb{H}$ — compare with Fig. 3.1 — showing initial or physical manifolds [labelled (a), (b), ..., as in Fig. 3.1] and the flows induced by repeated application of a discrete RG transformation $\mathbb{R}_b$ with a spatial rescaling factor b (or induced by a corresponding continuous or differential RG). Critical trajectories are shown bold: they all terminate, in the region of $\mathbb{H}$ shown here, at a fixed point $\bar{\mathcal{H}}^*$. The full space contains, in general, other nontrivial, critical fixed points, describing multicritical points and distinct critical-point universality classes; in addition, trivial fixed points, including high-temperature "sinks" with no outflowing or relevant trajectories, typically appear. *Lines of fixed points* and other more complex structures may arise and, indeed, play a crucial role in certain problems. [After Fisher (1983).]

They actually carry the renormalized Hamiltonian *out* of the (too small) manifold of nearest-neighbor Ising models and introduce (infinitely many) further couplings. The resulting situation is portrayed schematically in Fig. 3.4: the renormalized manifold for $\bar{\mathcal{H}}'(t', h')$ generally has no overlap with the original manifold. Further iterations, and *continuous* [see Eq. (3.33) below] as against discrete RG transformations, are suggested by the flow lines or "trajectories" also shown in Fig. 3.4. We will return to some of the details of these below.

In practice, the naive decimation transformation specified by Eq. (3.21) generally fails as a foundation for useful calculations.[59] Indeed, the design of effective

[59] See Kadanoff and Niemeijer in Gunton and Green (1974), Niemeijer and van Leeuwen (1976), Fisher (1983).

RG transformations turns out to be an art more than a science: there is no standard recipe! Nevertheless, there are guidelines: the general philosophy enunciated by Wilson and expounded well, for example, in a lecture by Shankar treating fermionic systems,[60] is to attempt to *eliminate* first those microscopic variables or degrees of freedom of "least direct importance" to the macroscopic phenomenon under study, while *retaining* those of most importance. In the case of ferromagnetic or gas-liquid critical points, the phenomena of most significance take place on long length scales — the correlation length, ξ, diverges; the critical point correlations, $G_c(\mathbf{r})$, decay slowly at long-distances; long-range order sets in below T_c.

Thus in his first, breakthrough articles in 1971, Wilson used an ingenious "phase-space cell" decomposition for *continuously variable scalar spins* (as against ± 1 Ising spins) to treat a lattice Landau–Ginzburg model with a general, single-spin or 'on-site' potential $V(s_{\mathbf{x}})$ acting on each spin, $-\infty < s_{\mathbf{x}} < \infty$. Blocks of cells of the smallest spatial extent were averaged over to obtain a single, renormalized cell of twice the linear size (so that $b = 2$). By making sufficiently many simplifying approximations (that, incidentally, imposed $\eta = 0$), Wilson obtained an explicit *nonlinear, integral recursion relation* that transformed the l-times renormalized potential, $V^{(l)}(\cdot)$, into $V^{(l+1)}(\cdot)$. This recursion relation could be handled by computer and led to a *specific numerical estimate* for the exponents γ and ν for $d = 3$ dimensions, namely, $\gamma = 2\nu \simeq 1.217$, that were *quite different* from the classical values (and from the results of any previously soluble models like the *spherical model*[61]). On seeing those results I knew that a major barrier to progress had been overcome!

I returned from a year's sabbatic leave at Stanford University in the summer of 1971, by which time Ken Wilson's two basic papers were in print. Shortly afterwards, in September, again while walking to lunch as I recall, Ken Wilson discussed his latest results from the nonlinear recursion relation with me. Analytical expressions could be obtained by expanding $V^{(l)}(s)$ in a power series:

$$V^{(l)}(s) = r_l s^2 + u_l s^4 + v_l s^6 + \cdots . \tag{3.25}$$

If truncated at quadratic order one had a soluble model — the Gaussian model (or free-field theory) — and the recursion relation certainly worked *exactly* for that! But to have a nontrivial model, one had to start not only with r_0 (as, essentially, the temperature variable) but, as a minimum, one also had to include $u_0 > 0$: the model then corresponded to the well known $\lambda\varphi^4$ field theory. Although one might, thus, initially set $v_0 = w_0 = \cdots = 0$, all these higher order terms were immediately generated under renormalization; furthermore, there was no reason for u_0 to be small and, for this reason and others, the standard field-theoretic perturbation theories were ineffective.

[60]See R. Shankar in Cao (1999) and Shankar (1994).

[61]The spherical model, which is fully soluble, was devised by Berlin and Kac (1952); see also Montroll and Berlin (1951) and Helfand and Langer (1967). For accounts of the critical behavior of the spherical model, see Fisher (1966a), where long-range forces were also considered, and, *e.g.*, Stanley (1971), Baxter (1982), and Fisher (1983).

Now, I had had a long-standing interest in the effects of *the spatial dimensionality* d on singular behavior in various contexts:[62] so that issue was raised for Ken's recursion relation. Indeed, d appeared simply as an explicit parameter.[63] It then became clear that $d = 4$ was a special case in which the leading order corrections to the Gaussian model vanished. Furthermore, above $d = 4$ dimensions classical behavior reappeared in a natural way (since the parameters u_0, v_0, $\cdots$ all then became irrelevant). These facts fitted in nicely with the known special role of $d = 4$ in other situations.[64]

For $d = 3$, however, one seemed to need the infinite set of coefficients in Eq. (3.25) which all coupled together under renormalization. But I suggested that maybe one could treat the dimensional deviation, $\epsilon = 4 - d$, as a small, *nonintegral parameter* in analyzing the recursion relations for $d < 4$. Ken soon showed this was effective!

Indeed, using his analysis (Wilson, 1971b) and supposing the initial values, r_0 and u_0, are of order ϵ the recursion relations[63] become

$$r_{l+1} = 4\left[r_l + 3u_l(1+r_l)^{-1} - 9u_l^2\right] + O(\epsilon^3), \tag{3.26}$$

$$u_{l+1} = (1+\epsilon\ln 2)u_l - 9u_l^2 + O(\epsilon^3), \tag{3.27}$$

where, we may recall the $\ln 2$ would, in general, be $\ln b$. A fixed point is then simply a solution $r_l = r^*$ and $u_l = u^*$ independent of l. From the second relation we see there are just two fixed points: the Gaussian fixed point $r^* = u^* = 0$ and a new, nontrivial fixed point[65]

$$u = u^* = \tfrac{1}{9}\epsilon\ln 2 + O(\epsilon^2), \tag{3.28}$$

$$r = r^* = -\tfrac{4}{9}\epsilon\ln 2 + O(\epsilon^2). \tag{3.29}$$

In $d = 4$ spatial dimensions when $\epsilon = 0$, these two fixed points coincide. Linearizing for small deviations about the new fixed point leads to

$$\gamma = 2\nu = 1 + \tfrac{1}{6}\epsilon + O(\epsilon^2). \tag{3.30}$$

Furthermore, to order ϵ the results prove to be independent of b and, as hoped, universal. The critical value for r (which is effectively T_c) turns out to be $r_c = -4u_0 + O(\epsilon^2)$; as expected, r_c depends on the single-spin potential $V(s) \equiv V^{(0)}(s)$ through its variation with u_0.

A paper, entitled by Ken "Critical Exponents in 3.99 Dimensions," was shortly written, submitted, and published, it contained the first general expressions for

[62] Fisher and Sykes (1959), Fisher and Gaunt (1964), Fisher (1966a,b; 1967c; 1972). See also Helfand and Langer (1967) page 438, column 2.

[63] See the recursion relations (4.29) and (4.30) in Wilson (1971b). Note that the "2" in Eq. (4.30) should, in general, be replaced by b.

[64] See references in the Footnotes 61 and 62 and Larkin and Khmel'nitskii (1969), especially Appendix 2.

[65] Sometimes, these days, known as the Wilson–Fisher fixed point or "line" or "family," especially when continued up to $\epsilon = 2$ or down to $d = 2$, the evidence suggesting it remains well defined: see, *e.g.*, El-Showk *et al.* (2014b).

nonclassical exponents.[66] In addition, the analysis was extended to XY [or $O(n=2)$] models with planar two-component spins and to a version of Baxter's eight-vertex model.[67]

It transpired, however, that the perturbation parameter ϵ provided more — namely, a systematic way of ordering the infinite set of discrete recursion relations not only for the expansion coefficients of $V^{(l)}(s)$ but also for further multi-spin terms in the appropriate full space $\mathbb{H}$, involving spatial gradients, *etc.*[68] With that facility in hand, the previous approximations entailed in the phase-space cell analysis could be dispensed with; it became essential, however,[68] to then rescale the spins with the factor $b^{l(d-2+\eta)/2}$ where the exponent η had to be determined. Thus it could be established *exactly*, that $\eta = \frac{1}{54}\epsilon^2$, subject to corrections of order ϵ^3, while the result (3.30) for γ was correct to order ϵ.

In summary, Ken Wilson had seen that, by employing momenta or wave vectors $\mathbf{q}$ and labelling the spin variables, now re-expressed in Fourier space, as $\hat{s}_{\mathbf{q}}$, he could precisely implement his *momentum-shell renormalization group*[69] — subsequently one of the most-widely exploited tools in critical phenomena studies![70]

In essence the momentum-shell transformation is like decimation[71] except that the division of the variables in Eq. (3.21) is made in momentum space: for ferromagnetic or gas-liquid type critical points the set $\{\hat{s}^{<}_{\mathbf{q}}\}$ contains those 'long-wavelength' or 'low-momentum' variables satisfying $|\mathbf{q}| \le q_\Lambda/b$, where $q_\Lambda = \pi/a$ is the (ultraviolet) momentum cutoff implied by the lattice structure. Conversely, the 'short-wavelength', 'high-momentum' spin components $\{\hat{s}^{>}_{\mathbf{q}}\}$ having wave vectors lying in the momentum-space *shell*: $q_\Lambda/b < |\mathbf{q}| < q_\Lambda$, are integrated out. The spatial rescaling now takes the form

$$\mathbf{q} \Rightarrow \mathbf{q}' = b\mathbf{q}, \tag{3.31}$$

---

[66] See Wilson and Fisher (1972); Eqs. (9) and (10) are these relations. The first draft of this letter was written by Ken Wilson who graciously listed the authors in alphabetical order.

[67] See Baxter (1971, 1982). The four-vertex model was reformulated by Kadanoff and Wegner (1971) as an energy–energy-type coupling; see also Wu (1971).

[68] See Eq. (18) in Wilson and Fisher (1972) and the subsequent text.

[69] Wilson (1972), entitled: "Feynman-Graph Expansion for Critical Exponents."

[70] See, *e.g.*, Brézin, Wallace, and Wilson (1972), Wilson and Kogut (1974), the reviews Brézin, Le Guillou, and Zinn-Justin (1976), and Wallace (1976), and the texts Amit (1978) and Itzykson and Drouffe (1983), *etc.*

[71] A considerably more general form of RG transformation can be written as

$$\exp\big(\bar{\mathcal{H}}'[s']\big) = \mathrm{Tr}^{s}_{N}\big\{\mathcal{R}_{N',N}(s';s)\exp\big(\bar{\mathcal{H}}[s]\big)\big\},$$

where the trace is taken over the full set of original spins s. The $N' = N/b^d$ renormalized spins $\{s'\}$ are introduced via the RG kernel $\mathcal{R}_{N',N}(s';s)$ which incorporates spatial and spin rescalings, *etc.*, and which should satisfy a trace condition to ensure the partition-function-preserving property (see Footnote 57) which leads to the crucial free-energy flow equation: see Eq. (3.38) below. The decimation transformation, the momentum-shell RG, and other transformations can be written in this form.

as follows from Eq. (3.13); but in analogy to $\zeta(b)$ in Eq. (3.14), a *nontrivial spin rescaling factor* ("multiplicative-wave function renormalization" in QFT) is introduced via

$$\hat{s}_{\mathbf{q}} \Rightarrow \hat{s}'_{\mathbf{q}'} = \hat{s}_{\mathbf{q}} / \hat{c}[b, \bar{\mathcal{H}}] \,. \tag{3.32}$$

The crucially important rescaling factor $\hat{c}$ takes the form $b^{d-\omega}$ and must be *tuned* in the critical region of interest [which leads to $\omega = \frac{1}{2}(d - 2 + \eta)$: compare with Eqs. (3.4) and (3.6)]. It is also worth mentioning that by letting $b \to 1+$, one can derive a *differential* or continuous flow RG and rewrite the recursion relation Eq. (3.22) as[72]

$$\frac{d}{dl}\bar{\mathcal{H}} = \mathbb{B}\left[\bar{\mathcal{H}}\right] . \tag{3.33}$$

Such continuous flows are illustrated in Figs. 3.3 and 3.4. (If it happens that $\bar{\mathcal{H}}$ can be represented, in general only approximately, by a single coupling constant, say, g, then $\mathbb{B}$ reduces to the so-called beta-function $\beta(g)$ of QFT.)

For deriving ϵ expansions on the basis of the momentum shell RG, Feynman-graph perturbative techniques as developed for QFT prove very effective.[69,70] They enter basically because one can take $u_0 = O(\epsilon)$ small and they play a role both in efficiently organizing the calculation and in performing the essential integrals (particularly for systems with simple propagators and vertices).[73] Capitalizing on his field-theoretic expertise, Ken obtained, in only a few weeks after submitting the first article, *exact expansions* for the exponents ν, γ, and ϕ to order ϵ^2 (and, by scaling, for all other exponents).[74] Furthermore, the anomalous dimension — defined in Eq. (3.6) close to the beginning of our story — was calculated exactly to order ϵ^3. Here, with $\epsilon = 4 - d > 0$, is this striking result:

$$\eta = \frac{(n+2)}{2(n+8)^2}\epsilon^2 + \frac{(n+2)}{2(n+8)^2}\left[\frac{6(3n+14)}{(n+8)^2} - \frac{1}{4}\right]\epsilon^3 + O(\epsilon^4), \tag{3.34}$$

where the symmetry parameter n denotes the number of components of the microscopic spin vectors, $\vec{s}_{\mathbf{x}} \equiv (s^{\mu}_{\mathbf{x}})_{\mu=1,\ldots,n}$, so that one has just $n = 1$ for Ising spins.[75]

---

[72]See Wilson (1971a) and Footnote 58 above: in this form the RG semigroup can typically be extended to an Abelian group (MacLane and Birkhoff, 1967); but as already stressed, this fact plays a negligible role.

[73]Nickel (1978). Nevertheless, many more complex situations arise in condensed matter physics for which the formal application of graphical techniques without an adequate understanding of the appropriate RG structure can lead one seriously astray.

[74]Note that Wilson (1972) was received on 1 December 1971 while Wilson and Fisher (1972) carries a receipt date of 11 October 1971. Fisher and Pfeuty (1972) meanwhile submitted an order ϵ treatment of the anisotropic n-vector model on 17 November 1971 finding $\phi = 1 + n\epsilon/2(n+8)$. The first introductory textbook regarding the renormalization group, Pfeuty and Toulouse (1975), was sent to press in August 1974; the graduate text by Ma (1967a) soon followed.

[75]See, *e.g.*, Domb and Sykes (1962), Fisher (1967b, 1974b, 1983), Kadanoff *et al.* (1967), Stanley (1971), Fisher and Pfeuty (1972), Pfeuty and Toulouse (1975), Aharony (1976), and Patashinskii and Pokrovskii (1979). As discovered by Stanley (1968) the limit $n \to \infty$ corresponds precisely to the spherical model: see Footnote 61 above.

For completeness we quote also the results for γ and ϕ while as regards ν, the relation $(2-\eta)\nu = \gamma$ should be recalled.

$$\gamma = 1 + \frac{n+2}{2(n+8)}\epsilon + \frac{(n+2)(n^2+22n+52)}{4(n+8)^3}\epsilon^2 + O(\epsilon^3), \tag{3.35}$$

$$\phi = 1 + \frac{n\epsilon}{2(n+8)} + \frac{n(n^2+24n+68)}{4(n+8)^3}\epsilon^2 + O(\epsilon^3). \tag{3.36}$$

In fact these ϵ expansions have been extended to order ϵ^5 (and ϵ^6 for η).[76] Beyond that, Ken's analysis inspired the development of many further related expansions for critical exponents and other universal parameters.[77]

3.6. Flows, fixed points, universality and scaling

To complete my story — and to fill in a few logical gaps over which we have jumped — I should explain how Wilson's construction of RG transformations in the space $\mathbb{H}$ enables RG theory to accomplish the "tasks" set out above in Sec. 3.2. As illustrated in Fig. 3.4, the recursive application of an RG transformation $\mathbb{R}_b$ induces a *flow* in the space of Hamiltonians, $\mathbb{H}$. Then one observes that "sensible," "reasonable," or, better, "well-designed" RG transformations are *smooth*, so that points in the original physical manifold, $\bar{\mathcal{H}}^{(0)} = \bar{\mathcal{H}}(t,h)$, that are close, say in temperature, remain so in $\bar{\mathcal{H}}^{(1)} \equiv \bar{\mathcal{H}}'$, *i.e.*, under renormalization, and likewise as the flow parameter l increases, in $\bar{\mathcal{H}}^{(l)}$. Notice, incidentally, that since the spatial scale renormalizes via $\mathbf{x} \Rightarrow \mathbf{x}' = b^l\mathbf{x}$ one may regard

$$l = \log_b(|\mathbf{x}'|/|\mathbf{x}|), \tag{3.37}$$

as measuring, logarithmically, the scale on which the system is being described — one might speak of the *scale dependence of parameters*; but note that, in general, the *form* of the Hamiltonian is also changing as the "scale" is changed or l increases. Thus a partially renormalized Hamiltonian can be expected to take on a more-or-less generic, mesoscopic form: Hence it represents an appropriate candidate to give meaning to a Landau–Ginzburg or, now, adding Wilson's name, an LGW effective Hamiltonian.

---

[76] See Gorishny, Larin, and Tkachov (1984) for the initial result later corrected in Kleinert *et al.* (1991) but to little practical consequence for the summation procedures of Le Guillou and Zinn-Justin (1987), as shown by Guida and Zinn-Justin (1998).

[77] Especial mention should be made first of $1/n$ expansions, where n is the number of components of the vector order parameter (Abe, 1972, 1973; Fisher, Ma, and Nickel, 1972; Suzuki, 1972), and see Fisher (1974), and Ma (1976a); second, of coupling-constant expansions in fixed dimension: see Parisi (1973, 1974); Baker, Nickel, Green, and Meiron (1976); Le Guillou and Zinn-Justin (1977); Baker, Nickel, and Meiron (1978). Later Wilson himself undertook large scale Monte Carlo RG calculations for the simple-cubic Ising model (Pawley *et al.*. 1984). For other problems, modified *dimensionality expansions* have been made by writing $d = 8-\epsilon$, $6-\epsilon$, $4+\frac{1}{2}m-\epsilon$ $(m=1,2,\ldots)$, $3-\epsilon$, $2+\epsilon$, and $1+\epsilon$.

Thanks to the smoothness of the RG transformation, if one knows the free energy $f_l \equiv f\big[\bar{\mathcal{H}}^{(l)}\big]$ at the l-th stage of renormalization, then one knows the *original* free energy $f\big[\bar{\mathcal{H}}\big]$ and its critical behavior: explicitly one has[78]

$$f(t,h,\cdots) \equiv f\big[\bar{\mathcal{H}}\big] = b^{-dl} f\Big[\bar{\mathcal{H}}^{(l)}\Big] \equiv b^{-dl} f_l\Big(t^{(l)}, h^{(l)}, \cdots\Big). \tag{3.38}$$

Furthermore, the smoothness implies that all the universal critical properties are preserved under renormalization. Similarly one finds[79] that the critical point of $\bar{\mathcal{H}}^{(0)} \equiv \bar{\mathcal{H}}$ maps on to that of $\bar{\mathcal{H}}^{(1)} \equiv \bar{\mathcal{H}}'$, and so on, as illustrated by the bold flow lines in Fig. 3.4. Thus it is instructive to follow the *critical trajectories* in $\mathbb{H}$, *i.e.*, those RG flow lines that emanate from a physical critical point. In principle, the topology of these trajectories could be enormously complicated and even chaotic: in practice, however, for a well-designed or "apt" RG transformation, one most frequently finds that the critical flows terminate — or, more accurately, come to an asymptotic halt — at a *fixed point* $\bar{\mathcal{H}}^*$, of the RG: see Fig. 3.4. Such a fixed point is defined, via Eqs. (3.22) or (3.33), simply by

$$\mathbb{R}_b\big[\bar{\mathcal{H}}^*\big] = \bar{\mathcal{H}}^* \quad \text{or} \quad \mathbb{B}\big[\bar{\mathcal{H}}^*\big] = 0. \tag{3.39}$$

One then searches for fixed-point solutions: the role of the fixed-point equation is, indeed, roughly similar to that of Schrödinger's Equation $H\psi = E\psi$, for stationary states ψ_k of energy E_k in quantum mechanics (where ψ now denotes a wave function of appropriate variables).

Why are the fixed points so important? Some, in fact, are *not*, being merely *trivial*, corresponding to *no interactions* or to *all spins frozen*, *etc.* But the *non*trivial fixed points represent critical states; furthermore, the nature of their criticality, and of the free energy in their neighborhood, must, as explained, be *identical* to that of all those distinct Hamiltonians whose critical trajectories converge to the same fixed point! In other words, a particular fixed point *defines* a *universality class* of critical behavior which "governs," or "attracts" all those systems whose critical points eventually map onto it: see Fig. 3.4.

Here, then we at last have the natural explanation of *universality*: systems of quite different physical character may, nevertheless, belong to the domain of attraction of the *same* fixed point $\bar{\mathcal{H}}^*$ in $\mathbb{H}$. The distinct sets of inflowing trajectories reflect their varying physical content of associated irrelevant variables and the corresponding nonuniversal rates of approach to the asymptotic power laws dicated by $\bar{\mathcal{H}}^*$.

From each physical critical fixed point, there flow at least two "unstable" or outgoing trajectories. These correspond to one or more *relevant* variables, specifically, for the case illustrated in Fig. 3.4, to the temperature or thermal field, t, and the magnetic or ordering field, h. See also Fig. 3.3. If there are further relevant trajectories then, as mentioned above, one can expect *crossover* to different critical

[78]Recall the partition-function-preserving property set out in Footnote 57 above, which actually implies the basic relation Eq. (3.38).

[79]See Wilson (1971a), Wilson and Kogut (1974), and Fisher (1983).

behavior. In the space $\mathbb{H}$, such trajectories will then typically lead to distinct fixed points describing (in general) completely new universality classes.[80]

But what about *power laws* and *scaling*? The answer to this question was already sketched in Sec. 3.2; but we will recapitulate here, giving a few more technical details. However, trusting readers or those uninterested in the analysis may skip to the paragraph containing Eq. (3.46)!

That said, one must start by noting that the smoothness of a well-designed RG transformation means that it can always be expanded locally — to at least some degree — in a Taylor series.[81] It is worth stressing that it is this very property that fails for free energies in a critical region: to regain this ability, the large space of Hamiltonians is crucial. Near a fixed point satisfying Eq. (3.39) we can, therefore, rather generally expect to be able to *linearize* by writing

$$\mathbb{R}_b\big[\bar{\mathcal{H}}^* + gQ\big] = \bar{\mathcal{H}}^* + g\mathbb{L}_b Q + o(g) \tag{3.40}$$

as $g \to 0$, or in differential form,

$$\frac{d}{dl}\left(\bar{\mathcal{H}}^* + gQ\right) = g\mathbb{B}_1 Q + o(g). \tag{3.41}$$

Now $\mathbb{L}_b$ and $\mathbb{B}_1$ are *linear operators* (albeit acting in a large space $\mathbb{H}$). As such we can seek eigenvalues and corresponding "eigenoperators," say Q_k (which basic or "critical operators" will be partial Hamiltonians). Thus, in parallel to quantum mechanics, we may write

$$\mathbb{L}_b Q_k = \Lambda_k(b) Q_k \quad \text{or} \quad \mathbb{B}_1 Q_k = \lambda_k Q_k, \tag{3.42}$$

where, in fact, (by the semigroup property) the eigenvalues must be related by $\Lambda_k(b) = b^{\lambda_k}$. As in any such linear problem, knowing the spectrum of eigenvalues and eigenoperators or, at least, its dominant parts, tells one much of what one needs to know. Reasonably, the Q_k should form a basis for a general expansion

$$\bar{\mathcal{H}} \cong \bar{\mathcal{H}}^* + \sum_{k \geq 1} g_k Q_k. \tag{3.43}$$

Physically, the expansion coefficient g_k ($\equiv g_k^{(0)}$) then represents the thermodynamic field[82] conjugate to the critical operator Q_k which, in turn, will often be close to some combination of *local* operators. Indeed, in a characteristic critical-point problem one finds two *relevant operators*, say Q_1 and Q_2 with λ_1, $\lambda_2 > 0$. Invariably,

[80] A skeptical reader may ask: "But what if no fixed points are found? This can well mean, as it has frequently meant in the past, simply that the chosen RG transformation was poorly designed or "not apt." On the other hand, a fixed point represents only the simplest kind of asymptotic flow behavior: other types of asymptotic flow may well be identified and translated into physical terms. Indeed, near certain types of trivial fixed point, such procedures, long ago indicated by Wilson (1971a, Wilson and Kogut, 1974), *must* be implemented: see, *e.g.*, Fisher and Huse (1985). Limit cycles may also arise and be dealt with as shown by Glazek and Wilson (2002).

[81] See Wilson (1971a), Wilson and Kogut (1974), Fisher (1974b), Wegner (1972, 1976), Kadanoff (1976).

[82] Reduced, as expected, by the factor $1/k_B T$.

one of these operators can, say by its symmetry, be identified with the local energy density, $Q_1(\mathbf{r}) \cong E(\mathbf{r})$, so that $g_1 \cong t$ is the thermal field; the second then characterizes the order parameter, $Q_2(\mathbf{r}) \cong \Psi(\mathbf{r})$ with field $g_2 \cong h$. Under renormalization each g_k varies simply as $g_k^{(l)} \approx b^{\lambda_k l} g_k^{(0)}$.

Finally,[81] one examines the flow equation (3.38) for the free energy and for the correlations. The essential point is that the degree of renormalization, b^l, can be *chosen* as large as one wishes. When $t \to 0$, *i.e.*, in the critical region which it is our aim to understand, a good choice proves to be $b^l = 1/|t|^{1/\lambda_1}$, which clearly diverges to ∞. One then finds that Eq. (3.38) leads to the *basic scaling relation* Eq. (3.19) which we will rewrite here in greater generality as

$$f_s(t, h, \cdots, g_j, \cdots) \approx |t|^{2-\alpha} \mathcal{F}\left(\frac{h}{|t|^{\Delta}}, \cdots, \frac{g_j}{|t|^{\phi_j}}, \cdots \right). \tag{3.44}$$

This is the essential result: thus it is easy to see that it leads to the "collapse" of equation-of-state data,[83] *etc.*

Now, however, the critical exponents can be expressed directly in terms of the RG eigenexponents λ_k (for the fixed point in question). Specifically one finds

$$2 - \alpha = \frac{d}{\lambda_1}, \quad \Delta = \beta + \gamma = \beta\delta = \frac{\lambda_2}{\lambda_1}, \quad \phi_j = \frac{\lambda_j}{\lambda_1}, \quad \text{and} \quad \nu = \frac{1}{\lambda_1}. \tag{3.45}$$

Then, as already mentioned, the sign of a given ϕ_j and, hence, of the corresponding λ_j determines the *relevance* (for $\lambda_j > 0$), *marginality* (for $\lambda_j = 0$), or *irrelevance* (for $\lambda_j < 0$) of the corresponding critical operator Q_j (or "perturbation") and of its conjugate field g_j: this field might, but for most values of j will *not*, be under direct experimental control. As explained previously, all exponent relations (3.8), (3.20), *etc.*, follow from scaling, while the first and last of the Eqs. (3.45) yield the dimensionality-dependent *hyperscaling relation* Eq. (3.20).

When there are no marginal variables and the least negative ϕ_j is larger than unity in magnitude, a simple scaling description will usually work well and the Kadanoff picture almost applies. When there are *no* relevant variables and only one or a few *marginal variables*, field-theoretic perturbative techniques of the Gell-Mann–Low (1954), Callan–Symanzik[84] or so-called "parquet diagram" varieties[85] may well suffice (assuming the dominating fixed point is sufficiently simple to be well understood). There may then be little incentive for specifically invoking general RG theory. From the perspective of high energy physics this seems to be the current situation in QFT and it applies also in some condensed matter problems.[86]

[83] See Vicentini-Missoni (1972), Domb (1966) Figs. 1.10 and 6.6, Fisher (1967b) Fig. 18, pp. 710–712.

[84] See Wilson (1975), Brézin *et al.* (1976), Amit (1978), Itzykson and Drouffe (1989).

[85] As used effectively by Larkin and Khmel'nitskii (1969).

[86] See, *e.g.*, the case of dipolar Ising-type ferromagnets in $d = 3$ dimensions investigated experimentally by Ahlers, Kornblit, and Guggenheim (1975) following theoretical work by Larkin and Khmel'nitskii (1969), Fisher and Aharony (1973) and Aharony (see 1976, Sec. 4E).

Finally, for the set of critical operators Q_i, Q_j, $\cdots$ it is appropriate to follow Wilson (1969) and Kadanoff (1969) and define their *dimensions*[87] Δ_i, Δ_j, $\cdots$ via the correlation functions evaluated *at* criticality ($t = h = 0$), *i.e.* recalling Eq. (3.4),

$$G_{ij}^c(\mathbf{r}_1, \mathbf{r}_2) = \langle \Delta Q_i(\mathbf{r}_1) \Delta Q_j(\mathbf{r}_2) \rangle \sim |\mathbf{r}_1 - \mathbf{r}_2|^{-(\Delta_i + \Delta_j)} \tag{3.46}$$

for $|\mathbf{r}_1 - \mathbf{r}_2| \gg a$. It then becomes not unnatural to enquire about the nature and, more specifically, the dimensions, of products of operators like $Q_i(\mathbf{r}_1)Q_j(\mathbf{r}_2)$, where $\mathbf{r}_1$ and $\mathbf{r}_2$ are fairly *close* together. This then leads to the formulation of appropriate *operator algebras* — as independently proposed by Wilson (1969) and by Kadanoff (1969), while no doubt also in the mind of Polyakov (1969). In more concrete terms an algebra can be specified by its *operator product expansion* in the form

$$Q_i(\mathbf{r}_1)Q_j(\mathbf{r}_2) = \sum_k A_{ij,k}(\mathbf{r}_1 - \mathbf{r}_2) Q_k\left(\frac{\mathbf{r}_1 + \mathbf{r}_2}{2}\right), \tag{3.47}$$

in which the numerical coefficients $A_{ij,k}$ specify the structure of the algebra.

Perhaps rather optimistically, this raised the hope — see Kadanoff (1969, 1971) and Polyakov (1969) — that understanding the character of the algebra — subject, as it had to be, to various symmetries arising naturally in the problem at hand — might suffice to determine explicitly (or, at least, more directly) the actual values of the basic critical dimensions. If so, then all the critical exponents could be found! Thus, noting[87] $\Delta_1 = d - \lambda_1$, $\Delta_2 = d - \lambda_2$, *etc.*, the results (3.45) of our formulation may be rewritten, in terms *only* of the critical dimensions and the spatial dimensionality, as

$$\alpha = \frac{d - 2\Delta_1}{d - \Delta_1}, \quad \beta = \frac{\Delta_2}{d - \Delta_1}, \quad \gamma = \frac{d - 2\Delta_2}{d - \Delta_1}, \quad \delta = \frac{d}{\Delta_2} - 1, \quad \phi_j = \frac{d - \Delta_j}{d - \Delta_1}, \tag{3.48}$$

and, especially,

$$\nu = 1/(d - \Delta_1) \quad \text{and} \quad \eta = 2(\Delta_2 + 1) - d, \tag{3.49}$$

from which, following Kadanoff (1976), we have

$$\Delta_1 = d\frac{1 - \alpha}{2 - \alpha} \quad \text{and} \quad \Delta_2 = \frac{d}{\delta + 1} = \frac{1}{2}(d - 2 + \eta). \tag{3.50}$$

However, a further hypothesis regarding the algebraic character underlying the set of operators in fact proved necessary to fulfill the hope of fixing the exponent

[87] Clearly the dimensions, Δ_j, should not be confused with the *gap exponent*, $\Delta = \beta + \gamma$, appearing in Eqs. (3.44) and (3.45), or earlier in Eq. (3.19), which specifies how h scales with t. The notation Δ_j, or Δ_Ψ, *etc.*, originally employed by Polyakov (1970) seems currently favored: see El-Showk *et al.* (2014). Wilson (1969) originally used d_j while Kadanoff (1969) initially employed ν_Ψ, *etc.*, but later adopted $x_j = d - y_j$ or $x_h = d - y_h$, *etc.* (Kadanoff, 1976). Note, however, that the exponent ω introduced in Eq. (3.17) for Kadanoff's spin rescaling factor is, in fact, just Δ_2. In the present treatment the critical operator dimensions are simply related to the eigenvalues λ_j defined in Eq. (3.42).

values. This was the concept of *conformal covariance*, initially proposed by Polyakov (1970, 1974) and, in essence, fully verified for the two-dimensional Ising model. More recently this has led to a line of research for Ising-type models in $d = 3$ dimensions that promises most accurate values for the exponents: see El-Showk *et al.* (2014) who find $\Delta_1 = 1.41267(13)$ and $\Delta_2 = 0.518154(15)$.

3.7. Conclusions

My tale is now told: following Wilson's 1971 papers and the introduction of the ϵ-expansion in 1972 the significance of the renormalization group approach in statistical mechanics was soon widely recognized[88] and exploited by many authors interested in critical and multicritical phenomena and in other problems in the broad area of condensed matter physics, physical chemistry, and beyond. Some of these successes have already been mentioned in order to emphasize, in particular, those features of the full RG theory that are of general significance in the wide range of problems lying beyond the confines of quantum field theory.[89]

A further issue, originally motivated by Ken Wilson, is the relevance of renormalization group concepts to quantum field theory. I have addressed that only in various peripheral ways. Insofar as I am no expert in quantum field theory, that is not inappropriate; but perhaps one may step aside a moment and look at QFT from the general philosophical perspective of understanding complex, interacting

[88]Footnote 32 drew attention to the first international conference on critical phenomena organized by Melville S. Green and held in Washington in April 1965. Eight years later, in late May 1973, Mel Green, with an organizing committee of J.D. Gunton, L.P. Kadanoff, K. Kawasaki, K.G. Wilson, and the author, mounted another conference to review the progress in theory of the previous decade. The meeting was held in a Temple University Conference Center in rural Pennsylvania. The proceedings (Gunton and Green, 1974) entitled *Renormalization Group in Critical Phenomena and Quantum Field Theory*, are now mainly of historical interest. The discussions were recorded in full but most papers only in abstract or outline form. Whereas in the 1965 conference the overwhelming number of talks concerned experiments, now only J.M.H. (Anneke) Levelt Sengers and Guenter Ahlers spoke to review experimental findings in the light of theory. Theoretical talks were presented, in order, by P.C. Martin, Wilson, Fisher, Kadanoff, B.I. Halperin, E. Abrahams, Niemeijer (with van Leeuwen), Wegner, Green, Suzuki, Fisher and Wegner (again), E.K. Riedel, D.J. Bergman (with Y. Imry and D. Amit), M. Wortis, Symanzik, Di Castro, Wilson (again), G. Mack, G. Dell-Antonio, J. Zinn-Justin, G. Parisi, E. Brézin, P.C. Hohenberg (with Halperin and S.-K. Ma) and A. Aharony. Sadly, there were no participants from Russia, then the Soviet Union, but others included R. Abe, G.A. Baker, Jr., T. Burkhardt, R.B. Griffiths, T. Lubensky, D.R. Nelson, E. Siggia, H.E. Stanley, D. Stauffer, M.J. Stephen, B. Widom and A. Zee. As the lists of names and participants illustrates, many active young theorists had been attracted to the area, had made significant contributions, and were to make more in subsequent years.

[89]Some reviews that may be cited to illustrate applications are Fisher (1974b), Wilson (1975), Wallace (1976), Aharony (1976), Pokrovskii and Patashinskii (1979), Nelson (1983), and Creswick *et al.* (1992). Beyond these, attention should be drawn to the notable article by Hohenberg and Halperin (1977) that reviews dynamic critical phenomena, and to many articles on further topics in the Domb and Lebowitz series *Phase Transitions and Critical Phenomena*, Vols. 7 to 20 (Academic, London, 1983–2001).

systems. Then, I would claim, statistical mechanics is a central science of great intellectual significance — as just one reminder, the concepts of "spin-glasses" and the theoretical and computational methods developed to analyze them (such as "simulated annealing") have proved of interest in physiology for the study of neuronal networks and in operations research for solving hard combinatorial problems. In that view, even if one focuses only on the physical sciences, the land of statistical physics is broad, with many hills and dales, valleys and peaks to explore that are of relevance to the real world and to our ways of thinking about it. Within that land one may find an island, surrounded by water: but these days, more and broader bridges happily span the waters and communicate with the mainland! That island is devoted to what was "particle physics" and is now "high energy physics" or, more generally, to the deepest lying and, in that sense, the "most fundamental" aspects of physics. The reigning theory on the island is surely quantum field theory — the magnificent set of ideas and techniques that inspired the symposium that led to this article.[90] Those laboring on the island have built most impressive skyscrapers reaching to the heavens!

Nevertheless, from the global viewpoint of statistical physics — where many degrees of freedom, the ever-present fluctuations, and the diverse spatial and temporal scales pose the central problems — quantum field theory may be regarded as describing a rather special set of statistical mechanical models. As regards applications they have been largely restricted to $d = 4$ spatial dimensions [more physically, of course to $(3 + 1)$ dimensions] although in subsequent decades *string theory* dramatically changed that! The practitioners of QFT insist on the preeminence of some special symmetry groups, the Poincaré group, $SU(3)$, and so on, which are not all so "natural" at first sight — even though the role of guage theories as a unifying theme in modeling nature has been particularly impressive. But, of course, we know these special features of QFT are not matters of choice — rather, they are forced on us by our explorations of Nature itself. Indeed, as far as experiment tells us, there is only one high energy physics; whereas, by contrast, the ingenuity of chemists, materials scientists, and of Life itself, offers a much broader, multifaceted and varied panorama of systems to explore both conceptually and in the laboratory.

From this global standpoint, renormalization group theory represents a theoretical tool of depth and power. It first flowered luxuriantly in condensed matter physics, especially in the study of critical phenomena. But it is ubiquitous because of its potential for linking physical behavior across disparate scales; its ideas and techniques will continue to play a vital role in those situations where the fluctuations on many different physical scales truly interact. But it provides a valuable perspective — through concepts such as 'universality,' 'relevance,' 'marginality' and 'irrelevance,' even when scales are well separated! One can reasonably debate how vital renormalization group concepts are for quantum field theory itself. Certain

[90]See the Foreword above and Cao (1998).

aspects of the full theory do seem important because Nature teaches us, and particle physicists have learned, that quantum field theory is, indeed, one of those theories in which the different scales are connected together in nontrivial ways via the intrinsic quantum-mechanical fluctuations. However, in current quantum field theory, only certain facets of renormalization group theory play a pivotal role.[91] High energy physics did not have to be the way it is! But, even if it were quite different, we would still need renormalization group theory in its fullest generality in condensed matter physics and, one suspects, in future scientific endeavors.

Acknowledgments

Thanks are due to Professor Emeritus Alfred I. Tauber, Director (until 2010), and Professor Tian Yu Cao of the Center for Philosophy and History of Science at Boston University for their part in arranging the original symposium on quantum field theory and for inviting me to speak. Daniel M. Zuckerman kindly assisted with the Bibliography. Comments on the initial draft manuscript (based on my talk at the symposium) from colleagues and friends Stephen G. Brush, N. David Mermin, R. Shankar, David J. Wallace, B. Widom, and Kenneth G. Wilson were much appreciated. Stimulating interactions over many years have been enjoyed with Leo P. Kadanoff (including specifically for this article) and, but on fewer occasions, with Valery Pokrovskii.

Selected Bibliography

The reader is cautioned that this article is not intended as a systematic review of renormalization group theory and its origins. Likewise, this bibliography makes no claims of completeness; however, it includes those contributions of most significance in the personal views of the author (mainly prior to 1999). The reviews of critical phenomena and RG theory cited in Footnote 6 above contain many additional references. Further review articles appear in the series *Phase Transitions and Critical Phenomena*, edited by C. Domb and M.S. Green (later replaced by J.L. Lebowitz) and published by Academic Press, London (1972–2001): some are listed below. Introductory accounts in an informal lecture style are presented in Fisher (1965, 1983).

Abe, R., 1972, "Expansion of a Critical Exponent in Inverse Powers of Spin Dimensionality," *Prog. Theor. Phys.* **48**, 1414–15.

Abe, R., 1973, "Expansion of a Critical Exponent in Inverse Powers of Spin Dimensionality," *Prog. Theor. Phys.* **49**, 113–128.

[91] It is interesting to look back and read in Gunton and Green (1973) pp. 157–160, Wilson's thoughts in May 1973 regarding the "Field Theoretic Implications of the Renormalization Group" at a point just before the ideas of *asymptotic freedom* became clarified for non-Abelian gauge theory by Gross and Wilczek (1973) and Politzer (1973).

Aharony, A., 1973, "Critical Behavior of Magnets with Dipolar Interactions. V. Uniaxial Magnets in *d* Dimensions," *Phys. Rev. B* **8**, 3363–70; 1974, erratum *ibid.* **9**, 3946.

Aharony, A., 1976, "Dependence of Universal Critical Behavior on Symmetry and Range of Interaction," in *Phase Transitions and Critical Phenomena*, Vol. 6, edited by C. Domb and M.S. Green (Academic, London), pp. 357–424.

Ahlers, G., A. Kornblit, and H.J. Guggenheim, 1975, "Logarithmic Corrections to the Landau Specific Heat near the Curie Temperature of the Dipolar Ising Ferromagnet $LiTbF_4$," *Phys. Rev. Lett.* **34**, 1227–30.

Amit, D.J., 1978, *Field Theory, the Renormalization Group and Critical Phenomena* (McGraw-Hill Inc., London); see also 1993, the expanded Revised 2nd Edition (World Scientific, Singapore).

Bagnuls, C. and C. Bervillier, 1997, "Field-Theoretic Techniques in the Study of Critical Phenomena," *J. Phys. Studies* **1**, 366–382.

Baker, G.A., Jr., 1961, "Application of the Padé Approximant Method to the Investigation of some Magnetic Properties of the Ising Model," *Phys. Rev.* **124**, 768–774.

Baker, G.A., Jr., 1990, *Quantitative Theory of Critical Phenomena* (Academic, San Diego).

Baker, G.A., Jr. and N. Kawashima, 1995, "Renormalized Coupling Constant for the Three-Dimensional Ising Model," *Phys. Rev. Lett.* **75**, 994–997.

Baker, G.A., Jr. and N. Kawashima, 1996, "Reply to Comment by J.-K. Kim," *Phys. Rev. Lett.* **76**, 2403.

Baker, G.A., Jr., B.G. Nickel, M.S. Green, and D.I. Meiron, 1976, "Ising Model Critical Indices in Three Dimensions from the Callan–Symanzik Equation," *Phys. Rev. Lett.* **36**, 1351–54.

Baker, G.A., Jr., B.G. Nickel, and D.I. Meiron, 1978, "Critical Indices from Perturbation Analysis of the Callan–Symanzik Equation," *Phys. Rev. B* **17**, 1365–74.

Baxter, R.J., 1971, "Eight-vertex Model in Lattice Statistics," *Phys. Rev. Lett.* **26**, 832–833.

Baxter, R.J., 1982, *Exactly Solved Models in Statistical Mechanics* (Academic, London).

Benfatto, G. and G. Gallavotti, 1995, *Renormalization Group*, Physics Notes, edited by P.W. Anderson, A.S. Wightman, and S.B. Treiman (Princeton University, Princeton, NJ).

Berlin, T.H. and M. Kac, 1952, "The Spherical Model of a Ferromagnet," *Phys. Rev.* **86**, 821–831.

Bogoliubov, N.N. and D.V. Shirkov, 1959, *Introduction to the Theory of Quantized Fields* (Interscience, New York), Chap. VIII.

Brézin, E., J.C. Le Guillou, and J. Zinn-Justin, 1976, "Field Theoretical Approach to Critical Phenomena," in *Phase Transitions and Critical Phenomena*, Vol. 6, edited by C. Domb and M.S. Green (Academic, London), pp. 125–247.

Brézin, E., D.J. Wallace, and K.G. Wilson, 1972, "Feynman-Graph Expansion for the Equation of State near the Critical Point," *Phys. Rev. B* **7**, 232–239.

Buckingham, M.J. and J.D. Gunton, 1969, "Correlations at the Critical Point of the Ising Model," *Phys. Rev.* **178**, 848–853.

Burkhardt, T.W. and J.M.J. van Leeuwen (Eds.), 1982, *Real Space Renormalization* (Springer, Berlin).

Butera, P. and M. Comi, 1997, "N-Vector Spin Models on the Simple-Cubic and Body-Centered-Cubic Lattices: A study of the critical behavior of the susceptibility and of the correlation length by high-temperature series extended to order β^{21}," *Phys. Rev. B* **56**, 8212–40.

Campostrini, M., A. Pelissetto, P. Rossi, and E. Vicari, 2002, "25th Order High-Temperature Expansion Results for Three-Dimensional Ising-like Systems on the Simple Cubic Lattice," *Phys. Rev. E* **65**, 066127.

Cao, T.Y. (Ed.), 1999, *Conceptual Foundations of Quantum Field Theory* (Cambridge University Press, Cambridge).

Creswick, R.J., H.A. Farach, and C.P. Poole, Jr., 1992, *Introduction to Renormalization Group Methods in Physics* (John Wiley & Sons, Inc., New York).

Di Castro, C. and G. Jona-Lasinio, 1969, "On the Microscopic Foundation of Scaling Laws," *Phys. Lett. A* **29**, 322–323.

Di Castro, C. and G. Jona-Lasinio, 1976, "The Renormalization Group Approach to Critical Phenomena," in *Phase Transitions and Critical Phenomena*, Vol. 6, edited by C. Domb and M.S. Green (Academic, London), pp. 507–558.

Domb, C., 1960, "On the Theory of Cooperative Phenomena in Crystals, *Adv. Phys. (Philos. Mag. Suppl.)* **9**, 149–361.

Domb, C., 1996, *The Critical Point: A historical introduction to the modern theory of critical phenomena* (Taylor and Francis, London).

Domb, C. and D.L. Hunter, 1965, "On the Critical Behavior of Ferromagnets," *Proc. Phys. Soc.* **86**, 1147–51.

Domb, C. and M.F. Sykes, 1962, "Effect of Change of Spin on the Critical Properties of the Ising and Heisenberg Models, *Phys. Rev.* **128**, 168–173.

El-Showk, S., M.F. Paulos, D. Poland, S. Rychkova, D. Simmons-Duffin, and A. Vichi, 2014a, "Solving the 3d Ising Model with the Conformal Bootstrap II. c-minimization and precise critical exponents," *J. Stat. Phys.* **157**, 869–914.

El-Showk, S., M.F. Paulos, D. Poland, S. Rychkova, D. Simmons-Duffin, and A. Vichi, 2014b, "Conformal Field Theories in Fractional Dimensions," *Phys. Rev. Lett.* **112**, 141601.

Essam, J.W. and M.E. Fisher, 1963, "Padé Approximant Studies of the Lattice Gas and Ising Ferromagnet below the Critical Point," *J. Chem. Phys.* **38**, 802–812.

Fisher, D.S. and D.A. Huse, 1985, "Wetting Transitions: A functional renormalization group approach," *Phys. Rev. B* **32**, 247–256.

Fisher, M.E., 1959, "The Susceptibility of the Plane Ising Model," *Physica* **25**, 521–524.

Fisher, M.E., 1962, "On the Theory of Critical Point Density Fluctuations," *Physica* **28**, 172–180.

Fisher, M.E., 1964, "Correlation Functions and the Critical Region of Simple Fluids," *J. Math. Phys.* **5**, 944–962.

Fisher, M.E., 1965, "The Nature of Critical Points," in *Lectures in Theoretical Physics*, Vol. VIIC (University of Colorado Press, Boulder), pp. 1–159.

Fisher, M.E., 1966a, "Notes, Definitions and Formulas for Critical Point Singularities," in M.S. Green and J.V. Sengers (Eds.), 1966, cited below.

Fisher, M.E., 1966b, "Shape of a Self-Avoiding Walk or Polymer Chain," *J. Chem. Phys.* **44**, 616–622.

Fisher, M.E., 1967a, "The Theory of Condensation and the Critical Point," *Physics* **3**, 255–283.

Fisher, M.E., 1967b, "The Theory of Equilibrium Critical Phenomena," *Rep. Prog. Phys.* **30**, 615–731.

Fisher, M.E., 1967c, "Critical Temperatures of Anisotropic Ising Lattices II. General upper bounds," *Phys. Rev.* **162**, 480–485.

Fisher, M.E., 1969, "Phase Transitions and Critical Phenomena," in *Contemporary Physics*, Vol. 1, Proc. International Symposium, Trieste, 7–28 June 1968 (International Atomic Energy Agency, Vienna), pp. 19–46.

Fisher, M.E., 1971, "The Theory of Critical Point Singularities," in *Critical Phenomena, Proc. 1970 Enrico Fermi Internat. Sch. Phys., Course No. 51, Varenna, Italy*, edited by M.S. Green (Academic, New York), pp. 1–99.

Fisher, M.E., 1972, "Phase Transitions, Symmetry and Dimensionality," *Essays in Physics*, Vol. 4 (Academic, London), pp. 43–89.

Fisher, M.E., 1974a, "General Scaling Theory for Critical Points," *Proceedings of the Nobel Symp. XXIV, Aspenäsgården, Lerum, Sweden, June 1973*, in *Collective Properties of Physical Systems*, edited by B. Lundqvist and S. Lundqvist (Academic, New York), pp. 16–37.

Fisher, M.E., 1974b, "The Renormalization Group in the Theory of Critical Phenomena," *Rev. Mod. Phys.* **46**, 597–616.

Fisher, M.E., 1983, "Scaling, Universality and Renormalization Group Theory," in *Critical Phenomena*, edited by F.J.W. Hahne, Lecture Notes in Physics, Vol. 186 (Springer, Berlin), pp. 1–139.

Fisher, M.E., 1998, "Renormalization Group Theory: Its basis and formulation in statistical physics," *Rev. Mod. Phys.* **70**, 653–681; reprinted with revisions in Cao (1999), Part IV, Chap. 8, pp. 89–135.

Fisher, M.E. and A. Aharony, 1973, "Dipolar Interactions at Ferromagnetic Critical Points," *Phys. Rev. Lett.* **30**, 559–562.

Fisher, M.E. and R.J. Burford, 1967, "Theory of Critical Point Scattering and Correlations. I. The Ising model," *Phys. Rev.* **156**, 583–622.

Fisher, M.E. and J.-H. Chen, 1985, "The Validity of Hyperscaling in Three Dimensions for Scalar Spin Systems," *J. Phys. (Paris)* **46**, 1645–54.

Fisher, M.E. and D.S. Gaunt, 1964, "Ising Model and Self-Avoiding Walks on Hypercubical Lattices and 'High Density' Expansions," *Phys. Rev.* **133**, A224–A239.

Fisher, M.E., S.-K. Ma, and B.G. Nickel, 1972, "Critical Exponents for Long-Range Interactions," *Phys. Rev. Lett.* **29**, 917–920.

Fisher, M.E. and P. Pfeuty, 1972, "Critical Behavior of the Anisotropic n-Vector Model," *Phys. Rev. B* **6**, 1889–91.

Fisher, M.E. and M.F. Sykes, 1959, "Excluded-Volume Problem and the Ising Model of Ferromagnetism," *Phys. Rev.* **114**, 45–58.

Gell-Mann, M. and F.E. Low, 1954, "Quantum Electrodynamics at Small Distances," *Phys. Rev.* **95**, 1300–12.

Glazek, S.D. and K.G. Wilson, 2002, "Limit Cycles in Quantum Theories," *Phys. Rev. Lett.* **89**, 230401:1–4.

Gorishny, S.G., S.A. Larin, and F.V. Tkachov, 1984, "ϵ-Expansion for Critical Exponents: The $O(\epsilon^5)$ approximation," *Phys. Lett.* **101A**, 120–123.

Green, M.S. and J.V. Sengers (Eds.), 1966, *Critical Phenomena: Proceedings of a Conference held in Washington, D.C. April 1965*, N.B.S. Misc. Publ. **273** (U.S. Govt. Printing Off., Washington, 1 December 1966).

Griffiths, R.B., 1965, "Thermodynamic Inequality near the Critical Point for Ferromagnets and Fluids," *Phys. Rev. Lett.* **14**, 623–624.

Griffiths, R.B., 1972, "Rigorous Results and Theorems," in *Phase Transitions and Critical Phenomena*, Vol. 1, edited by C. Domb and M.S. Green (Academic, London), pp. 7–109.

Gross, D. and F. Wilczek, 1973, "Ultraviolet Behavior of Non-Abelian Gauge Theories," *Phys. Rev. Lett.* **30**, 1343–46.

Guida, R. and J. Zinn-Justin, 1998, "Critical Exponents of the N-Vector Model," *J. Phys.: Math. Gen.* **31**, 8103–21.

Gunton, J.D. and M.S. Green (Eds.), 1974, *Renormalization Group in Critical Phenomena and Quantum Field Theory: Proceedings of a Conference*, Chestnut Hill, Pennsylvania, 29–31 May 1973 (Temple University, Philadelphia).

Helfand, E. and J.S. Langer, 1967, "Critical Correlations in the Ising Model," *Phys. Rev.* **160**, 437–450.

Heller, P. and G.B. Benedek, 1962, "Nuclear Magnetic Resonance in MnF_2 near the Critical Point," *Phys. Rev. Lett.* **8**, 428–432.

Hille, E., 1948, *Functional Analysis and Semi-Groups*, 1948 (American Mathematical Society, New York).

Hohenberg, P.C. and B.I. Halperin, 1977, "Theory of Dynamic Critical Phenomena," *Rev. Mod. Phys.* **49**, 435–479.

Itzykson, D. and J.-M. Drouffe, 1989, *Statistical Field Theory* (Cambridge University Press, Cambridge).

Kadanoff, L.P., 1966, "Scaling Laws for Ising Models near T_c," *Physics* **2**, 263–272.

Kadanoff, L.P., 1969a, "Correlations along a Line in the Two-Dimensional Ising Model," *Phys. Rev.* **188**, 859–863.

Kadanoff, L.P., 1969b, "Operator Algebra and the Determination of Critical Indices," *Phys. Rev. Lett.* **23**, 1430–33.

Kadanoff, L.P., 1976, "Scaling, Universality and Operator Algebras," in *Phase Transitions and Critical Phenomena*, Vol. 5a, edited by C. Domb and M.S. Green (Academic, London), pp. 1–34.

Kadanoff, L.P. and H. Ceva, 1971, "Determination of an Operator Algebra for the Two-Dimensional Ising Model," *Phys. Rev. B* **3**, 3918–39.

Kadanoff, L.P., W. Götze, D. Hamblen, R. Hecht, E.A.S. Lewis, V.V. Palciauskas, M. Rayl, J. Swift, D. Aspnes, and J. Kane, 1967, "Static Phenomena near Critical Points: Theory and experiment," *Rev. Mod. Phys.* **39**, 395–431.

Kadanoff, L.P. and F.J. Wegner, 1971, "Some Critical Properties of the Eight-Vertex Model," *Phys. Rev. B* **4**, 3989–93.

Kaufman, B. and L. Onsager, 1949, "Crystal Statistics II. Short-range order in a binary lattice," *Phys. Rev.* **76**, 1244–52.

Kiang, C.S. and D. Stauffer, 1970, "Application of Fisher's Droplet Model for the Liquid-Gas Transition near T_c," *Zeits. Phys.* **235**, 130–139.

Kim, H.K. and M.H.W. Chan, 1984, "Experimental Determination of a Two-Dimensional Liquid-Vapor Critical-Point Exponent," *Phys. Rev. Lett.* **53**, 170–173.

Kleinart, H., J. Neu, V. Schulte-Frohlinde, K.G. Chetyrkin, and S.A. Larin, 1991, "Five-Loop Renormalization Group Functions of $O(n)$-Symmetric ϕ^4-Theory and ϵ-Expansions of Critical Exponents up to ϵ^5," *Phys. Lett. B* **272**, 39–44.

Koch, R.H., V. Foglietti, W.J. Gallagher, G. Koren, A. Gupta, and M.P.A. Fisher, 1989, "Experimental Evidence for Vortex-Glass Superconductivity in Y-Ba-Cu-O," *Phys. Rev. Lett.* **63**, 1511–14.

Landau, L.D. and E.M. Lifshitz, 1958, *Statistical Physics*, Course of Theoretical Physics, Vol. 5 (Pergamon, London), Chap. XIV.

Larkin, A.I. and D.E. Khmel'nitskii, 1969, "Phase Transition in Uniaxial Ferroelectrics," *Zh. Eksp. Teor. Fiz.* **56**, 2087–98; *Sov. Phys. JETP* **29**, 1123–28.

Le Guillou, J.C. and J. Zinn-Justin, 1977, "Critical Exponents for the n-Vector Model in Three Dimensions from Field Theory," *Phys. Rev. Lett.* **39**, 95–98.

Le Guillou, J.C. and J. Zinn-Justin, 1987, "Accurate Critical Exponents for Ising-like Systems in Non-Integer Dimensions," *J. Physique (Paris)* **48**, 19–24.

Ma, S.-K., 1976a, *Modern Theory of Critical Phenomena* (W.A. Benjamin, Inc., Reading, MA).

Ma, S.-K., 1976b, "The $1/n$ Expansion," in *Phase Transitions and Critical Phenomena*, Vol. 6, edited by C. Domb and M.S. Green (Academic, London), pp. 249–292.

MacLane, S. and G. Birkhoff, 1967, *Algebra* (MacMillan, New York), Chap. II, Sec. 9; Chap. III, Secs. 1, 2.

Montroll, E.W. and T.H. Berlin, 1951, "An Analytical Approach to the Ising Problem," *Commun. Pure Appl. Math.* **4**, 23–30.

Nelson, D.R. and M.E. Fisher, 1975, "Soluble Renormalization Groups and Scaling Fields for Low-Dimensional Spin Systems," *Ann. Phys. (N.Y.)* **91**, 226–274.

Niemeijer, Th. and J.M.J. van Leeuwen, 1976, "Renormalization Theory for Ising-like Systems," in *Phase Transitions and Critical Phenomena*, Vol. 6, edited by C. Domb and M.S. Green (Academic, London), pp. 425–505.

Nickel, B.G., 1978, "Evaluation of Simple Feynman Graphs," *J. Math. Phys.* **19**, 542–548.

Onsager, L., 1944, "Crystal Statistics I. A two-dimensional model with an order-disorder transition," *Phys. Rev.* **62**, 117–149.

Onsager, L., 1949, "Discussion Remark Following a Paper by G.S. Rushbrooke at the International Conference on Statistical Mechanics in Florence, *Nuovo Cimento Suppl. Series 9* **6**, 261.

Parisi, G., 1973, "Perturbation Expansion of the Callan–Symanzik Equation," in Gunton and Green (1974), cited above.

Parisi, G., 1974, "Large-Momentum Behaviour and Analyticity in the Coupling Constant," *Nuovo Cimento A* **21**, 179–186.

Patashinskii, A.Z. and V.L. Pokrovskii, 1966, "Behavior of Ordering Systems near the Transition Point," *Zh. Eksp. Teor. Fiz.* **50**, 439–447; *Sov. Phys. JETP* **23**, 292–297.

Patashinskii, A.Z. and V.L. Pokrovskii, 1979, *Fluctuation Theory of Phase Transitions* (Pergamon, Oxford).

Pawley, G.S., R.H. Swendsen, D.J. Wallace, and K.G. Wilson, 1984, "Monte Carlo Renormalization Group Calculations of Critical Behavior in the Simple-Cubic Ising Model," *Phys. Rev. B* **29**, 4030–40.

Pfeuty, P. and G. Toulouse, 1975, *Introduction au Group des Renormalisation et à ses Applications* (Univ. de Grenoble, Grenoble); translation, *Introduction to the Renormalization Group and to Critical Phenomena* (Wiley and Sons, London, 1977).

Politzer, H.D., 1973, "Reliable Perturbative Results for Strong Interactions?" *Phys. Rev. Lett.* **30**, 1346–49.

Polyakov, A.M., 1969, "Properties of Long and Short Range Correlations in the Critical Region," *Zh. Eksp. Teor. Fiz.* **57**, 271–283; *Sov. Phys. JETP* **30**, 151–157 (1970).

Polyakov, A.M., 1970, "Conformal Symmetry of Critical Fluctuations," *Zh. Eksp. Teor. Fiz. Pis. Red.* **12**, 538–541; *Sov. Phys. JETP Lett.* **12**, 381–383.

Polyakov, A.M., 1974, "Non-Hamiltonian Approach to Conformal Quantum Field Theory," *Zh. Eksp. Teor. Fiz.* **66**, 23–42; *Sov. Phys. JETP* **39**, 10–18.

Riesz, F. and B. Sz.-Nagy, 1955, *Functional Analysis* (F. Ungar, New York) Chap. X, Secs. 141, 142.

Rushbrooke, G.S., 1963, "On the Thermodynamics of the Critical Region for the Ising Problem," *J. Chem. Phys.* **39**, 842–843.

Schultz, T.D., D.C. Mattis, and E.H. Lieb, 1964, "Two-Dimensional Ising Model as a Soluble Problem of Many Fermions," *Rev. Mod. Phys.* **36**, 856–871.

Shankar, R., 1994, "Renormalization-Group Approach to Interacting Fermions," *Rev. Mod. Phys.* **66**, 129–192.

Stanley, H.E., 1968, "Spherical Model as the Limit of Infinite Spin Dimensionality," *Phys. Rev.* **176**, 718–722.

Stanley, H.E., 1971, *Introduction to Phase Transitions and Critical Phenomena* (Oxford University, New York).

Stell, G., 1968, "Extension of the Ornstein–Zernike Theory of the Critical Region," *Phys. Rev. Lett.* **20**, 533–536.

Suzuki, M., 1972, "Critical Exponents and Scaling Relations for the Classical Vector Model with Long-Range Interactions," *Phys. Lett.* **42A**, 5–6.

Symanzik, K., 1966, "Euclidean Quantum Field Theory. I. Equations for a scalar model," *J. Math. Phys.* **7**, 510–525.

Tarko, H.B. and M.E. Fisher, 1975, "Theory of Critical Point Scattering and Correlations. III. The Ising model below T_c and in a Field," *Phys. Rev. B* **11**, 1217–53.

Vicentini-Missoni, M., 1972, "Equilibrium Scaling in Fluids and Magnets," in *Phase Transitions and Critical Phenomena*, Vol. 2, edited by C. Domb and M.S. Green (Academic, London), pp. 39–78.

Wallace, D.J., 1976, "The ϵ-Expansion for Exponents and the Equation of State in Isotropic Systems," in *Phase Transitions and Critical Phenomena*, Vol. 6, edited by C. Domb and M.S. Green (Academic, London), pp. 293–356.

Wegner, F.J., 1972a, "Corrections to Scaling Laws," *Phys. Rev. B* **5**, 4529–36.

Wegner, F.J., 1972b, "Critical Exponents in Isotropic Spin Systems," *Phys. Rev. B* **6**, 1891–93.

Wegner, F.J., 1976, "The Critical State, General Aspects," in *Phase Transitions and Critical Phenomena*, Vol. 6, edited by C. Domb and M.S. Green (Academic, London), pp. 7–124.

Widom, B., 1965a, "Surface Tension and Molecular Correlations near the Critical Point," *J. Chem. Phys.* **43**, 3892–97.

Widom, B., 1965b, "Equation of State in the Neighborhood of the Critical Point," *J. Chem. Phys.* **43**, 3898–3905.

Wilson, K.G., 1969, "Non-Lagrangian Models of Current Algebra," *Phys. Rev.* **179**, 1499–1512.

Wilson, K.G., 1970, "Operator-Product Expansions and Anomalous Dimensions in the Thirring Model," *Phys. Rev. D* **2**, 1473–77.

Wilson, K.G., 1971a, "Renormalization Group and Critical Phenomena I. Renormalization group and Kadanoff scaling picture," *Phys. Rev. B* **4**, 3174–83.

Wilson, K.G., 1971b, "Renormalization Group and Critical Phenomena II. Phase-space cell analysis of critical behavior," *Phys. Rev. B* **4**, 3184–3205.

Wilson, K.G., 1972, "Feynman-Graph Expansion for Critical Exponents," *Phys. Rev. Lett.* **28**, 548–551.

Wilson, K.G., 1975, "The Renormalization Group: Critical phenomena and the Kondo problem," *Rev. Mod. Phys.* **47**, 773–840.

Wilson, K.G., 1983, "The Renormalization Group and Critical Phenomena" (1982, Nobel Prize Lecture) *Rev. Mod. Phys.* **55**, 583–600.

Wilson, K.G. and M.E. Fisher, 1972, "Critical Exponents in 3.99 Dimensions," *Phys. Rev. Lett.* **28**, 240–243.

Wilson, K.G. and J. Kogut, 1974, "The Renormalization Group and the ϵ Expansion," *Phys. Rep.* **12**, 75–200.

Wu, F.Y., 1971, "Ising Model with Four-Spin Interactions," *Phys. Rev. B* **4**, 2312–14.

Yang, C.N., 1952, "The Spontaneous Magnetization of a Two-Dimensional Ising Model," *Phys. Rev.* **85**, 808–816.

Zinn, S.-Y. and M.E. Fisher, 1996, "Universal Surface-Tension and Critical-Isotherm Amplitude Ratios in Three Dimensions," *Physica A* **226**, 168–180.

Zinn, S.-Y., S.-N. Lai, and M.E. Fisher, 1996, "Renormalized Coupling Constants and Related Amplitude Ratios for Ising Systems," *Phys. Rev. E* **54**, 1176–82.

Chapter 4

Early memories of Ken*,†

N. David Mermin
Department of Physics Cornell University Ithaca NY 14853 USA
ndm4@cornell.edu

I met Ken Wilson in 1952, when I was 17 and he was 16. We were both freshmen in the Harvard class of 1956, which produced an exceptionally large number of well-known physicists. As I remember Ken and I were in the same introductory German class, but it might have been the required freshman composition course, which met in the same building. What I remember for sure is that I got an A and Ken got a B. This was the only time in my life that I understood something better than he did.

What I remember most about Ken from 1952 is that although he was 16, he looked about 13. I don't think he looked 16 until he got tenure at Cornell. An important part of his youthful look was that when somebody said something that really pleased him, his face would lit up with such a sweet smile that it warmed your heart, like the first smiles of a baby.

Indeed, there was a performance at the 1964 Cornell Physics Department Christmas party, the year I got here, in which graduate students sang satirical songs about their Professors. What they sang about Ken was based on a song from South Pacific ("Bloody Mary is the one I love, now ain't that too damn bad!") It went "Kenny Wilson is the cutest Prof, but mesons ain't much fun." Mesons may indeed not have been much fun, but within ten years Ken was to provide theoretical physicists with the best opportunities for fun that they had since the invention of quantum mechanics.

Ken and I were math majors at Harvard. He had a formidable reputation as a mathematician and was at least two years ahead of me in the math curriculum. It

*This article was originally published in *J. Stat. Phys.* **157**, 625–627 (2014).
†Opening Talk at the Ken Wilson Memorial Symposium Cornell University, November 16, 2013

being Harvard, however, by general agreement, which I believe Ken subscribed to, he was only the second best mathematician in our class. I found this abundance of superior talent discouraging. To escape from mathematicians who were too smart to compete with, I stayed at Harvard for graduate school, but switched to physics. Out of the frying pan into the fire.

In 1959 Ken reappeared in Cambridge as a Junior Fellow. Much to my surprise, he too had become a physicist, irritatingly fast, at CalTech. But, also to my surprise, he seemed especially interested in electronic computers. This was still a time in which most of those who actually did computations used slide rules or, if we were really up to date, noisy electro-mechanical calculators. I thought it was a funny direction for such a talented mathematician to be moving in. Nobody had told me about von Neumann.

Ken was the first person I ever heard utter the phrase "electronic mail". "What's that?", I asked. He explained. What a silly idea, I thought. If such a thing had ever existed in the past, it would surely have been wiped out by the invention of the telephone. I think of this as my email moment. It was not the only email moment I had with Ken.

When I arrived at Cornell in 1964 as a new Assistant Professor. Ken had already been there for a year. I took his presence as a warning that I had gotten in over my head.

The two decades starting in the late 60s were a very special time to be in the Cornell Physics Department. A member of an external visiting committee evaluating the Department remarked to me after they had talked with a room full of graduate sudents: "They think they're in Valhalla!" Not one, not two, but three different future Nobel Prizes were being worked on.

There was the discovery of superfluidity in helium-3 by Dave Lee, Doug Osheroff, and Bob Richardson. There was an extended visit to Cornell by Tony Leggett, during which he put together, before our very eyes, his beautiful explanation of all the strange nuclear magnetic resonance signals observed in that superfluid.

And there was Ken.

Michael Fisher and Ben Widom ran a very lively weekly seminar on phase transitions. It started at lunchtime, and went on until everybody except Michael was exhausted. The only allowed medium of communication was chalk and blackboard. Interruptions of the speaker were strenuously encouraged.

In 1966 a preprint arrived from Leo Kadanoff. I volunteered to give a report on his manuscript at the Fisher–Widom seminar. I remember my title: "My Early Impressions of Kadanoff's Partial Justification of Widom's Conjecture on the Homogeneity of the Equation of State Near the Critical Point."

I didn't think much of Leo's partial justification. As I studied it, it struck me as a sloppy piece of work. He started with a well defined problem and then proceeded to butcher it, replacing detailed local configurations by gross averages, making unbelievably crude approximations to force those averages to look like the original problem, and then repeating the whole messy business over and over again.

In those days there was a wall between particle theorists and condensed matter theorists that was rarely crossed, but to my surprise Ken showed up for my talk. Even more to my surprise, he seemed to take Kadanoff seriously. About a week later he showed up at my office, ostensibly to talk about Kadanoff's paper. But what he actually had to say was that the crucial thing about the whole business was to think about a ball rolling up a hill with just enough energy to get exactly to the top. I humored him. "Yes Ken. Very nice Ken" I said as I ushered him out the door. It was another email moment for me.

The rest, of course, is history. I consider it one of the great experiences of my life to have had a front row seat at the revolution, and to have had Michael Fisher and Ben Widom at Cornell to explain to me over the next few years, with great skill and patience, what Ken had been trying unsuccessfully to tell me at the beginning.

A superb former Cornell professor of mathematics, Mark Kac, visited us often during the Valhalla years. He famously said "There are two kinds of geniuses: the ordinary and the magicians. An ordinary genius is a fellow whom you and I would be just as good as, if we were only many times better. There is no mystery as to how his mind works. Once we understand what they've done, we feel certain that we, too, could have done it. It is different with the magicians. They are...in the orthogonal complement of where we are and the working of their minds is for all intents and purposes incomprehensible. Even after we understand what they have done, the process by which they have done it is completely dark."

Mark was 20 years older than Ken and me, and although he clearly admired Ken, he was as suspicious of his theory of the critical point as I was of Kadanoff's "partial justification." Mark was as wrong as I was. Ken Wilson was a magician. It was a privilege to have known him.

Chapter 5

Ken Wilson — The early years*

R. Jackiw
Department of Physics, Massachusetts Institute of Technology, Cambridge, MA 02139, USA

Ken Wilson's lifetime achievements in fundamental theoretical physics are well known and are well documented in this memorial volume. Therefore, there is no purpose in my adding yet another appreciation of his seminal work. Rather, I shall describe Ken and some of his activities at the beginning of his career, when he was junior faculty at Cornell and I was his student — one of two in the first cohort of PhDs that he mentored. (The other was Gerald Estberg, long time faculty and now retired at the University of San Diego.)

I entered Cornell's physics PhD program, hoping to study with Hans Bethe. But he decided to leave elementary particle physics and remain with nuclear physics. Another eminent theorist, specializing in S-matrix/Regge theory, left Ithaca for the West Coast. Consequently a position was offered to Ken, and in 1963 he accepted, partially "because Cornell was a good university, was out in the country and [had] a good folk dancing group."[1] We graduate students were not familiar with his work because none was published. Evidently he got the job solely on his reputation among senior colleagues as a brilliantly unique quantum field theorist. Perhaps this disappointed some, who wished to follow the then-dominant S-matrix approach. But I was delighted, because my ambition was to master quantum field theory.

We were bemused by Ken's dedicated work habits: One could find him in office most of the time; otherwise he resided in a motel room. We were again bemused two years later when he won tenure after two publications. He attended our parties and other informal gatherings. His interactions were marked by very deliberate responses to conversational gambits. One frequently had to wait some moments before he responded; when an answer came it was complete — there was nothing more to say.

*This article was originally published in *Int. J. Mod. Phys. A* **29**, 1430008 (2014).

Ken's teaching style was methodical, addressing complicated matters in simple but opinionated fashion. He described his approach as

> "... not trying to state the final word on the physically (sic) meaning of quantum fields. Rather, I... present [an] intuitive understanding of the ... so-called 'asymptotic condition'. Rather than go through the formal mathematics involved in this work, I ... replace the rigorous but formal approach by an intuitive hypothesis used in an intuitive way, to obtain the same results. The value ... is that it is obtained without ... introducing ideas which are physically misleading and mathematically absurd. ('interaction representation' and the 'adiabatic hypothesis')"[2]

This attitude led to a clear but leisurely course presentation. By the end of the first semester of quantum field theory, we managed to quantize the free scalar field and discuss interactions, with no Feynman diagram in sight.

I presented myself to Ken and he agreed to direct my thesis research. He suggested that I study the renormalization group by reading the Gell-Mann–Low paper[3] and the Bogoliubov–Shirkov[4] text. Evidently already in 1963 Ken was thinking about the renormalization group. This choice came as a surprise to me because prevailing sentiment at that time maintained that nothing physically interesting can be gotten by renormalization group arguments, especially by techniques employed in the Soviet school.[5]

In fact Ken was an aficionado of the renormalization group from very early days. Already in his 1961 PhD thesis, he used that formalism to solve the Low equation. The thesis also exhibits Ken's reliance on numerical, computer assisted calculations — another feature of his mature work.[6]

When I was ready to begin research, Ken suggested that I use renormalization group methods to determine the large momentum behavior of the vertex (3-point) function in spinor electrodynamics. We hoped that rederiving known partial results would check the new approach, and that new results would demonstrate the power of the renormalization group in new settings. Let me explain.

The vertex function, depicted in Fig. 5.1, describes the propagation of an off mass-shell electron (solid lines) with the emission of an off mass-shell photon (dashed line). The 4-momenta are, respectively p, q and $k = p - q$. The on shell electron mass is m; the photon carries an infrared regulator mass μ.

$\Gamma^\alpha(p,q)$ is a 4×4 matrix, but the leading term may be isolated as $\Gamma^\alpha(p,q) = \gamma^\alpha \Gamma(p^2, q^2, k^2) + \cdots$. The task is to study Γ for large k^2. The answer, far off mass-shell, $|k^2| \gg |p^2|$, $|q^2| \gg m^2$, was found by Sudakov[7]:

$$\Gamma(p^2, q^2, k^2) \sim \exp\left[-\frac{\alpha}{2\pi} \ln\left|\frac{p^2}{k^2}\right| \ln\left|\frac{q^2}{k^2}\right|\right] \quad \text{(Sudakov)}\,.$$

Rederiving this with the help of the renormalization group would validate that technique, and then the on mass-shell formula for $\Gamma(m^2, m^2, k^2)$ could also be found.

Unfortunately I did not succeed. I could not find a defensible renormalization group argument for determining the large k^2 asymptote of $\Gamma(p^2, q^2, k^2)$ off or on

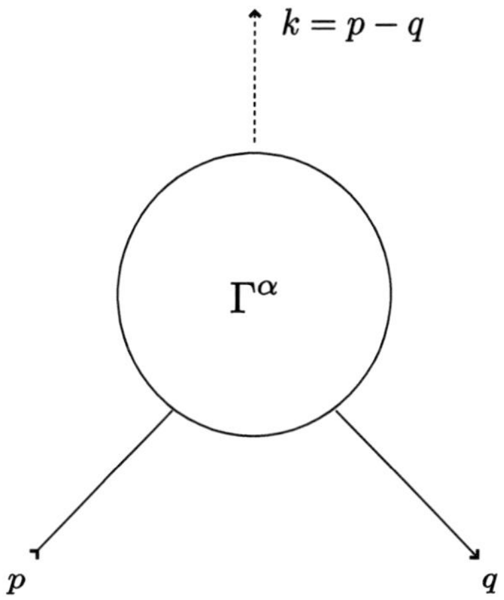

Fig. 5.1. Vertex (3-point) function.

mass-shell. After some futile struggle with the problem, I reported my failure to Ken. I was afraid that he would lose interest, once the renormalization group was abandoned.

Fortunately he was open to other methods. After a few days he told me to take a different, eikonal-type approach: In a Feynman diagram expansion of $\Gamma^\alpha(p,q)$, a generic propagator should be decomposed as

$$\frac{1}{r^2 - M^2 + i\varepsilon} = \frac{1}{2w_r}\left\{\frac{1}{r^0 - w_r + i\varepsilon} + \frac{1}{-r^0 - w_r + i\varepsilon}\right\},$$

$$w_r = \sqrt{\mathbf{r}^2 + M^2},$$

Decomposition of Feynman propagator.

Further analysis for large momenta shows that only one of the two terms in the decomposition dominates. With this observation it becomes possible to sum all relevant graphs. The Sudakov formula is reproduced and the on mass-shell asymptote is found[8]:

$$\Gamma(m^2, m^2, k^2) \sim \exp - \left\{\frac{\alpha}{4\pi} \ln^2 \frac{k^2}{\mu^2}\right\} \quad \text{(on mass-shell)},$$

$$m^2 = p^2\,, \quad q^2 \ll |k^2|.$$

Note that the on mass-shell formula is not merely Sudakov's expression evaluated at $p^2 = q^2 = m^2$; numerical factors differ. (With the advent of Effective Field Theory, a combination of eikonal and renormalization group methods can achieve both on shell and off shell results.[9])

While I was completing my research, there appeared a paper by S. Weinberg,[10] in which he proposed modified Feynman rules for calculating amplitudes in the infinite momentum frame. These are very similar to Ken's suggestion. When I went to inform Ken, he had already seen the paper, and with a smile called my attention to it. Never did he claim any priority in this matter — the subject simply was outside his interest, yet he could make a crucial contribution.

Ken was very supportive in my career. Upon my graduation he (and Bethe) secured for me a Harvard Junior Fellowship. It gave me great pleasure that he too held one, just before me. One time when I saw him, he was revising a paper on his short distance expansion — a technique with which he hoped to analyze the behavior of quantum fields, but had not yet come to fruition. I asked him how long would he remain with the subject without establishing useful results. He answered that he wouldn't give up for a decade. But he didn't have to wait that long. In 1969 he published the first of his renowned papers, "Non-Lagrangian Models of Current Algebra." In it he announced a new approach to quantum field theory:

> "What is proposed here is a new language for describing the short-distance behavior of fields in strong interactions. One talks about operator-product expansions for products of operators near the same point, instead of equal-time commutators. One discusses the dimension of an operator instead of how it is formed from products of canonical fields. Analyses of divergences in radiative corrections, etc., are carried out in position space rather than momentum space. Furthermore, one has qualitative rules for the strength of $SU(3) \times SU(3)$-symmetry-violating corrections at short distances... the hypotheses of this paper have the elegance of simplicity, once one is used to the language."[11]

He worked out several illustrative examples of physical processes by his methods. Most pleasing to me is the fact that he devoted a long discussion to the chiral anomaly, which appeared the same year.[12,13]

The Boston Joint Theoretical Physics Seminar met on Wednesdays. The day before Thanksgiving might have been sparsely attended; but this was not so, because the perennial speaker was Ken, who would be coming to visit his Boston area family for the holiday. As the years progressed, the audiences grew larger and larger, and the largest room had to be used.

References

1. K. G. Wilson, *Nobel Lectures in Physics, 1981–1990*, ed. G. Ekspong (World Scientific, Singapore, 1993).
2. K. G. Wilson, Quantum field theory class notes, Cornell (1963).
3. M. Gell-Mann and F. E. Low, *Phys. Rev.* **95**, 1300 (1954).
4. N. N. Bogoliubov and D. V. Shirkov, *Introduction to the Theory of Quantized Fields* (Interscience, New York, 1959).
5. K. Huang and F. E. Low, *Sov. Phys. JETP* **19**, 579 (1964).
6. K. G. Wilson, Caltech Ph.D. thesis (1961).
7. V. V. Sudakov, *Sov. Phys. JETP* **3**, 65 (1956).
8. R. Jackiw, *Ann. Phys. (N.Y.)* **48**, 292 (1968).

9. I. Stewart and J. Thaler, private communication.
10. S. Weinberg, *Phys. Rev.* **150**, 1313 (1966).
11. K. G. Wilson, *Phys. Rev.* **179**, 1499 (1969).
12. J. S. Bell and R. Jackiw, *Il Nuovo Cimento A* **60**, 47 (1969).
13. S. L. Adler, *Phys. Rev.* **177**, 2426 (1969).

Chapter 6

Kenneth Wilson in Moscow

A. M. Polyakov
Joseph Henry Laboratories
Princeton University
Princeton, New Jersey 08544, USA

I try to recreate the scientific atmosphere during Wilson's visits to Moscow in the 60s and 70s.

By the end of the 60s a few people, separated by the iron curtain, worked on the theory of critical phenomena. The brilliant mind of Kenneth Wilson and his visits to Moscow played a very important role in the subsequent development. In this article I shall try to reconstruct the atmosphere of these years. I have some "insider information" (restricted to the Russian side of the curtain) which may be interesting.

It seems to me that the modern development of the subject started with the work by Patashinskii and Pokrovsky.[1] The rumors go that by the end of the 1950s Landau already realized that his theory of phase transition is incomplete and that the fluctuation corrections were important, but I have never seen any written evidence of that and so my story begins with Ref. 1.

It was an ambitious and complicated paper. The revolutionary assumption there was the statement that in the Dyson equation $G^{-1} = G_0^{-1} - \Sigma[G]$ the contribution from the free Green function (the first term on the right) is negligible. Analogously, the bare contributions to the equations for the vertex parts are also irrelevant. This assumption immediately leads to "universality" — some small variations of the hamiltonian do not change the critical behavior, since they perturb only the bare Green functions. Some years later, Wilson introduced the concept of "irrelevant operators" which generalizes and reformulates the Patashinski–Pokrovsky idea. The next step in the paper was to guess that the equations are scale-invariant and to look for the power-like solutions for the Green functions. Unfortunately, they wanted too much and made some incorrect assumptions which fixed the critical exponents. However, in 1966 they quickly realized that and wrote another paper[2] which phenomenologically introduced scale invariance with anomalous dimensions, this time in a completely correct way. It was unclear, however, how this picture was related to the QFT approach. The issue was clarified in the different context (Regge calculus) by Gribov and Migdal.[3] They realized that with the correct treatment, the power-like propagator is consistent with QFT for arbitrary exponents, which are eventually determined by some bootstrap conditions. After that, in 1967–1968, the papers Refs. 8 and 7 applied these ideas to critical phenomena. An interesting by-product of these works was the realization that these phenomena are described by a relativistic field theory (after Wick's rotation).

There was another extremely important work by Larkin and Khmelnitsky: Ref. 9. They have solved in 1969 the four-dimensional critical theory, using the leading logs summation, equivalent to the Gell-Mann–Low renormalization group. This work was just a half step from the ε-expansion (all one had to do was to replace the logarithm in the Larkin–Khmelnitsky solution by a power, $\log p \Longrightarrow p^{\varepsilon}/\varepsilon$), but this had to wait for two years when it was discovered by Wilson and Fisher, together with the way to calculate the higher order corrections.

In 1969 I tried to apply these ideas to deep inelastic scattering, realizing that anomalous dimensions should break Bjorken's scaling in a multifractal way. At that time I ordered in the public library the latest *Physical Review* (which usually took a couple of weeks to get) and experienced a strong shock, finding very similar ideas in the article by some Kenneth Wilson, the name unknown to me. I still remember walking like a zombie through the center of Moscow where the library was located.

And then, I think in the 1969 or 1970, Ken visited Moscow. At that time Sasha Migdal and I were passionately interested in what he was doing. We had our own approach based on the bootstrap ideas and Ken's renormalization group didn't look promising to me. We spent hours with Ken discussing these matters. His approach at that time was based on the approximate recursion formula. Trying to understand it, I derived it by some crude truncation of Feynman's diagrams. Ken liked the derivation (and generously included it in his later review), but I thought it just showed that the recursion formula was too primitive. However, it later helped Ken to develop a general approach to the renormalization group and epsilon expansion.

In spite of our different "ideologies", I was very impressed by the power and depth of Ken's arguments, and learned a lot of subtle things from our discussions. One example was the operator product expansions (OPE). In 1969 they have been introduced in the various forms by Wilson,[4] L. Kadanoff[5] and myself.[6] While the general ideas were the same in these three papers, the deepest version definitely belongs to Ken. Namely, he traced the relation of the OPE to the canonical commutations relation. It impressed me very much and I started to think that field theory in general should be defined by means of the OPE, the associativity of which must restrict possible theories. This procedure is analogous to the classification of simple Lie algebras (one of the most beautiful parts of mathematics in my view), but infinitely more subtle. These dreams became more realistic with the 1970 discovery of the conformal 3-point function, which made the bootstrap equations concrete. I spent a lot of time trying the conformal field theory (CFT) approach for gravity, but so far unsuccessfully. Ken was not very enthusiastic about CFT. It was my impression (which may be wrong) that he valued his version of renormalization group much more than the OPE and was not very interested in their relations to each other. I had, and still have, the opposite view and expect some big surprises in the structures of higher-dimensional field theories.

Speaking of field theories, I should add that in the 60s the high energy theorists believed that field theory is a wrong way to approach Nature. Landau had expressed this point of view a decade before, but some people became more Catholic than the Pope. Not only the leading theorists in Russia were sceptical about conformal field theory, but at the 1970 Kiev Conference C. N. Yang expressed strong disagreement with my comment on the relation between critical phenomena and scale invariant QFT (this amusing exchange can be found in the proceedings). It was heart warming for Sasha and me to see that Ken Wilson did not share these prejudices, and that the idea that particle physics and critical phenomena are related was as natural for him as it was for us.

In the 70s Wilson's methods, renormalization group and ε-expansion, became tremendously popular and effective. They were easy to use in numerical simulations (this feature was very important for Ken), they also gave a nice qualitative picture of a system. The terms like "UV fixed point" or "irrelevant operator", introduced by Ken, became a part of the physics dictionary. Still if we talk about exact analytic results, Wilson's renormalization group is fully equivalent to the one by Gell-Mann and Low. But I don't think that it bothered Ken. In our discussions he said something like "Why should you care to get exact solutions? After all, from the computer point of view, the special functions are no different from any expression which you can calculate with good precision." (I am not sure that this was the exact wording, but I hope that the meaning of the phrase is accurate.) I disagreed, saying that if, say, a Bessel function appears in my calculation, it unites my problem with the innumerable other theories. Sasha Migdal at that time held the views very close to Ken's. Among many other things, he later improved Ken's formulation of the renormalization group. Of course, it is senseless to discuss who was right in this disagreement.

Next comes the theory of quark confinement and lattice gauge theory. This was also a major development. For the first time the precise formulation of confining gauge theory was given. Basically, Ken understood that what keeps quarks together are the quantized Faraday flux lines, forming a string. He then introduced the confinement criterion in terms of expectation values of the phase factor, now called the Wilson loop. This led to many efforts to derive confinement from the first principles. In the quasi-abelian case (the so-called compact QED) this was indeed possible and Ken appreciated these results. But the general problem of confinement is still unsolved. On the other hand, the lattice gauge theory became an immensely popular and useful tool for calculating physical properties of hadrons. The history of this outstanding development was very nicely described by Ken himself.[10] I can only add that in the 1972 dissertation by Vadim Berezinsky, the U(1) lattice gauge theory was explicitly written down. It served as an inspiration for my non-Abelian contribution to the subject, mentioned by Ken.

At about the same time, 't Hooft proposed $1/N$ expansion and conjectured that the lines of the planar Feynman diagram will become dense and form something like the string world-sheet. It is important to distinguish these two mechanisms (which are often confused in the literature). Electric flux lines are not directly related to the propagator lines in Feynman's diagrams and in QCD the diagrams do not become dense (in the matrix models they do, but this is another story). The modern gauge/strings duality is of the Wilson type. Still, the large N expansion gives us control over the topology of the world-sheet, as well as a good phenomenological approximation.

In 1979 in New York I was fortunate to have another scientific discussion with Ken. I was anticipating it with great excitement, especially because I had a number of new results, like a crude version of the non-critical string theory and gauge/strings duality, which, I hoped, should interest Ken. Discussion with him always led to new insights.

Unfortunately, this time Ken was not interested at all. Our conversation was fruitless. Perhaps I was unable to clearly communicate my ideas. And perhaps Ken was changing his views on science. Be it as it may, but the resonance was not there. It is sad to say that, but that was our last scientific interaction.

I would like to thank Boris Altshuler, Édouard Brézin, Igor Klebanov, Slava Rychkov and Victor Yakhot for comments and advice. This work was partially supported by the NSF grant PHY-1314198.

References

[1] A. Patashinski and V. Pokrovskii, *ZHETF* **46**, 994 (1964)
[2] A. Patashinski and V. Pokrovskii, *ZHETF* **50**, 439 (1966)
[3] V. Gribov and A. Migdal, *ZHETF* **55**, 1498 (1968)
[4] K. Wilson, *Phys. Rev.* **179**, 1499 (1969)
[5] L. Kadanoff, *Phys. Rev. Lett.* **23**, 1430 (1969)
[6] A. Polyakov, *ZHETF* **57**, 271 (1969)
[7] A. Migdal, *ZHETF* **55**, 1964 (1968)
[8] A. Polyakov, *ZHETF* **55**, 1026 (1968)
[9] A. Larkin and D. Khmelnitsky, *ZHETF* **56**, 2087 (1969)
[10] K. G. Wilson, *Nucl. Phys. B* (Proc. Suppl.) **140**, 3 (2005)

Chapter 7

Wilson's renormalization group: A paradigmatic shift*

E. Brézin

Laboratoire de Physique Théorique, Unité Mixte de Recherche 8549 du Centre National de la Recherche Scientifique et de l'École Normale Supérieure, Paris France

brezin@lpt.ens.fr

A personal and subjective recollection, concerning mainly Wilson's lectures delivered over the spring of 1972 at Princeton University (summary of a talk at Cornell University on November 16, 2013 at the occasion of the memorial Kenneth G. Wilson conference).

Keywords: Renormalization group; critical phenomena; quantum field theory.

7.1. Why was it so difficult to accept Wilson's viewpoint?

In 1971–72 I was on leave from Saclay, employed by the theory group at the physics department of Princeton University. My background at the time was mainly field theory. Just before this I had been working with C. Itzykson on the use of classical approximations in QED, dealing with problems such as pair creation by an oscillating electromagnetic field, but I was aware that the understanding of the intriguing features of scaling laws and universality near a critical point was a difficult non-perturbative problem. Wilson was invited in the fall of 1971 to give a talk on his recently published articles *Renormalization group and Kadanoff scaling picture* [1], *Phase space cell analysis of critical behavior* [2] but Ken said that he could not explain his ideas in one talk. The University[1] had the good idea of inviting him to talk as much as he needed and Ken ended up giving 15 lectures in the spring term of 1972, which resulted in the well-known 1974 Physics Reports *The renormalization group and the ϵ-expansion* by Wilson and Kogut (who had been

*This article was originally published in *J. Stat. Phys.* **157**, 644–650 (2014).

[1]Thanks to David Gross as confirmed at the Cornell Wilson memorial conference.

taking notes throughout the lectures), one of the most influential articles of the last decades [3].

Wilson's ideas were so different from the standard views at the time, that they were not easy to accept. His style was also completely different: for him field theory was not a set of abstract axioms but a tool to perform practical calculations even at the expense of oversimplified models whose relationship with reality was tenuous. Let us try to understand why Ken's ideas, nowadays so inbred in the training of any young physicist that they seem nearly obvious, were so difficult to accept. Let us first review the conventional wisdom at the time.

- Renormalization was a technique to remove perturbatively the ultra-violet singularities for those theories which were known to be renormalizable, i.e. theories for which this removal could be done at the expense of a finite number of parameters.
- Why should one consider such theories? Presumably because they were the only ones for which practical calculations are conceivable. For instance in QED, the minimal replacement $p \to p - eA$ provides such a theory but gauge invariance alone would allow for more couplings, such as a spin current coupled to the field strength, which would ruin the renormalizability.
- As a result one obtained with renormalized QED a theory which, a priori, was potentially valid at all distances from astronomical scales down to vanishingly small ones.
- The renormalization group in its conventional formulation was applicable only to renormalizable theories ; it allowed one to understand leading logarithms, at a given order in perturbation theory, from lower orders.

So what is it which was surprising in Wilson's approach?

- Critical points such as liquid-vapor, Curie points in magnets, Ising models, etc., are a priori classical problems in statistical mechanics. Why would a quantum field theory be relevant?
- Why renormalization in a theory in which there is a momentum cut-off $\Lambda = a^{-1}$, in which a is a physical short distance such as a lattice spacing, or a typical interatomic distance, and there is no reason to let a go to zero?
- The momentum space reduction of the number of degrees of freedom that Wilson used introduced singularities in the Hamiltonian generated by the flow. It was not quite clear that the procedure could be systematized.
- Irrelevance of most couplings was of course a fundamental piece of the theory but it looked a priori purely dimensional. For instance if one adds to a $g_4\varphi^4$ theory in four dimensions a coupling $(g_6/\Lambda^2)\varphi^6$ one obtains a flow equation for the dimensionless g_6 which concludes at its irrelevance as if the $1/\Lambda^2$ sufficed:

$$\Lambda\frac{\partial}{\partial\Lambda}g_6 - 2g_6 = \beta_6(g_4, g_6)$$

However the inclusion of the g_6/Λ^2 into a Feynman diagram with only g_4 vertices produces immediately an extra Λ^2 which cancels the previous one. So why is g_6 irrelevant?
- Although it is now physically very clear, it was not easy to understand that the critical surface had codimension two for an ordinary critical point, in others words that there are two and only two relevant operators in the large space of allowed coupling parameters.
- Even more difficult to take was the statement that four-dimensional renormalizable theories such as φ^4 or QED, i.e. non asymptotically free theories (an anachronistic name), if extended to all energy scales could only be free fields. How could one believe this, given the extraordinary agreement of QED with experiment?

I must admit that my initial difficulties at grasping the views exposed in Ken's lectures led me, in a friendly collaboration with David Wallace, to check their consistency. Amazed to see that the theory defeated our skepticism, we ended up working out with Ken the critical equation of state [4].

7.2. Wilson's renormalization group

It would take hundreds of pages to list all that we have learnt from Wilson's RG applied to critical exponents. The basic questions of scaling and universality were immediately made transparent, all the critical indices could be expressed in terms of two of them (plus a leading irrelevant operator for corrections to scaling, plus cross-over exponents). Wilson insisted that RG was a computational tool even in the absence of small parameters, but he provided also new parameters. Together with Michael Fisher, he introduced the famous epsilon-expansion [5]; he was also the first to realize that critical properties could also be computed in a $1/n$ expansion (he pointed out that in a $(\vec{\varphi}^2)^2$ theory non bubble diagrams were subleading in the large n-limit), he contributed to Monte-Carlo RG, etc. I would like here to insist simply on the conceptual revolutions brought by Wilson's RG:

- Field theory applied to the real world had to be considered as a long distance limit of yet unknown physics, even if the distances at which field theory is applied in particle physics go down to less than 10^{-6} femtometers.
- Renormalizability emerged as a consequence of considering this long distance limit, and not because the theory should be applicable down to vanishingly small distances.
- Non asymptotically free theories such as φ^4, QED or the Weinberg-Salam model are merely *effective* theories. If one tried to push them down to vanishingly small distances they would be forced to be only free field theories. These theories carry their own limitation, without any indication on the underlying physics.

It is interesting to try to track down the maturation of his ideas. He introduced first the operator product expansion in an unpublished Cornell report of 1964 *On products of quantum field operators at short distances.* In his Nobel lectures Ken wrote *I submitted the paper for publication; the referee suggested that the solution of the Thirring model might illustrate this expansion.*[2] *Unfortunately, when I checked out the Thirring model, I found that while indeed there was a short distance expansion for the Thirring model, my rules for how the coefficient functions behaved were all wrong, in the strong coupling domain. I put the preprint aside, awaiting resolution of the problem.* He computed indeed the anomalous dimensions of the operators in the Thirring model (Operator-product expansions and anomalous dimensions in the Thirring model, [6]: *It is shown that* ψ *has dimension* $1/2 + \frac{\lambda^2/4\pi^2}{1-\lambda^2/4\pi^2}$.) Clearly this had forced Ken to consider non-canonical dimensions of operators and the possibility that this could spoil the use of the OPE. Right after this he published his paper on . "Anomalous dimensions and breakdown of scale invariance in perturbation theory" [7]. Other scientists at the time were also realizing that the Gell-Mann and Low RG could be interpreted as asymptotic restoration of scale invariance broken by the regularization of quantum fluctuations, in particular C. Callan (*Broken scale invariance in scalar field theory* [8])and K. Symanzik (*Small distance behaviour analysis and Wilson expansions* [9]).

The application of these ideas to critical phenomena was a natural illustration of Ken's views and he acknowledges in his 1971, Phys. Rev. B papers quoted hereabove the role of Kadanoff's block spin picture [10] and his conversations with Ben Widom and Michael Fisher at Cornell. However Ken clearly was not very fond of the ϵ or $1/n$ expansions, although he had invented them. He regarded his RG as a tool that is applicable in the absence of any small parameter and thus the pseudo-small ϵ or $1/n$ were going back to directions that he didn't like. Likewise he regarded the field theoretic method such as the Callan–Symanzik equations as a very special case of a general method. To prove his point the participants of the 1973 Cargèse summer school (which I co-organized) remember that Ken arrived with a brilliant solution of the Kondo problem [11] (Renormalization group-critical phenomena and Kondo problem, Rev. Mod. Phys. 47, 773 1975) and his Nobel lectures he stated *[...] the solution of the Kondo problem is the first example where the full renormalization group (as the author conceives it) has been realized: the formal aspects of the fixed points, eigenoperators, and scaling laws will be blended with the practical aspect of numerical approximate calculations of effective interactions to give a quantitative solution [...] to a problem that previously had seemed hopeless.*

It is also at this Cargèse school that he presented for the first time lattice gauge theories, which turned QCD calculations of mass spectra into a problem of statistical mechanics which remains as the main tool for understanding quantitatively the predictions of QCD.

[2]It was reported at the Cornell conference quoted above that the referee was Arthur Wightman.

7.3. Quantitative aspects

If there is no doubt forty years later that the RG has led to a complete and extremely beautiful qualitative understanding of critical fluctuations, succeeding to explain universality, scaling laws and many new phenomena such as the geometry of polymers, percolation, etc, the quantitative aspects are often less clear. Indeed the connection with the three dimensional physical world remains a difficult non-perturbative problem and various numerical strategies have to be devised. How systematic are some of these methods, such as real space RG, living with the size limitations of numerical simulations, has left room for analytical methods and I think that renormalized field theory has played a significant role. Indeed the theory deals with correlations extending to distances much larger than the lattice spacing $x \gg a$, a correlation length $\xi \gg a$ and a finite ratio x/ξ. The effective renormalizable theory, the $a \to 0$ limit, provides exactly all the correlation functions in this limit. There are several strategies.

For instance in the "minimal subtraction scheme"

$$\beta_d(g) = -(4-d)g + \beta_4(g)$$
$$\gamma_d(g) = \gamma_4(g)$$

and all calculations in dimension d follow from computations in the critical (i.e. massless) theory in 4D. High order calculations in the ϵ or $1/n$ expansions follow from such techniques. However these expansions do not converge, no matter how small are ϵ or $1/n$. Using Lipatov's instanton method [12, 13] one can find the behavior of the perturbative series at order k for large k: for instance the successive terms of the ϵ expansion for the anomalous dimension of the field grow asymptotically like $\epsilon^k k!(-1/3)^k k^{7/2}$ (Ising like systems). This understanding led Zinn-Justin and co-workers [14, 15], using mathematical techniques such as Borel transform plus conformal maps, to accurate values of the critical exponents.

Parisi proposed to work directly in three dimensions [16], perturbation theory then makes sense only in the massive case, i.e. away from T_c. Again using the Lipatov type asymptotic estimate one can obtain at the end a good quantitative agreement with known values [14, 15, 17].

The point here is that it seems that one can transform the divergent perturbation series into seemingly convergent algorithms. Initially the impression was that one should better stop at some low order in the ϵ expansion if one did one want to run into absurdities and that was the best the theory could do. But now this says that the accuracy of the theory is not a priori limited: if comparison with experiment required better theoretical estimates, computing one more order in perturbation theory, although long and painful, would still improve the theoretical accuracy.

7.4. A crucial test of RG

In 1969 in a remarkable pre-Wilson article, Larkin and Khmel'nitskii [18] made two interesting points

1. in 4D mean field theory is violated by powers of logarithms, and they computed explicitly those logs with a parquet approximation (i.e. sum of leading logs at each order) which is equivalent to a one-loop RG
2. if one considers an Ising-like magnet with strong dipolar forces, instead of the usual short-range exchange forces, then it is in 3D and not in 4D, that one should observe those logarithmic violations of mean field scaling.

They predicted for instance that the specific heat, which in mean field theory remains finite at T_c with a simple step when one crosses the critical temperature, would diverge instead as

$$C = A_{\pm} \left| \log \left| \frac{T - T_c}{T_c} \right| \right|^{1/3}$$

with

$$\frac{A_+}{A_-} = \frac{1}{4}$$

Note that this is a prediction which does not require any approximation scheme; it is a pure test of RG very similar in spirit to the logarithmic deviations to scaling in deep inelastic electron scattering which provided the early tests of QCD [19]. The experiment was performed in 1975 by G.Ahlers *et al.* [20] and their finding was: *Quantitative experimental results for the logarithmic corrections to the Landau specific heat are reported for the dipolar Ising ferromagnet* $LiTbF_4$ *near its Curie temperature* $T_c = 2.885K$. *The power of the leading logarithmic term is found to be* 0.34 ± 0.03, *and the corresponding amplitude ratio is* 0.24 ± 0.01. *These results are in agreement with the predicted values of* $1/3$ *and* $1/4$, *respectively.*[3]

[3]We realized with Zinn-Justin in 1975 [21] that the correspondance between short range forces in 4D, a φ^4-theory with propagator $1/q^2$, and 3D with strong dipolar forces a φ^4-theory with propagator $1/q_\perp^2 + q_z^2/q_\perp^2$, would not extend beyond one-loop. A two-loop RG would give for the former

$$C_{shortrange,\ d=4} = A_{\pm} |\log|t||^{1/3} \left(1 - \frac{25}{81} \frac{|\log|\log|t||}{|\log|t||}\right)$$

and for the latter

$$C_{dipolar,\ d=3} = A_{\pm} |\log|t||^{1/3} \left(1 - \frac{1}{243}\left(41 + 108 \log\frac{4}{3}\right) \frac{|\log|\log|t||}{|\log|t||}\right)$$

A log-log is so slowly varying that this prediction has remained out of reach for experiments.

7.5. A few remarks concerning RG and conformal invariance

The idea of asymptotic conformal invariance of strong interactions was present at an early stage in the work of Mack, Kastrup, Salam, Symanzik and others [22]. Wilson himself in his 1970 paper *Broken scale invariance and anomalous dimensions* [23] refers to this asymptotic conformal symmetry. At the same period at the Landau Institute Polyakov [24] and Migdal [25] realized that the same conformal symmetry should hold near a critical point, and devised "bootstrap" equations to determine the theory. However this approach was temporarily abandoned for a number of reasons

- The self-consistent bootstrap equation for the coupling constant did not contain any small parameter ... until the ϵ-expansion was discovered.
- This self-consistent equation was shown to be equivalent to the fixed point equation of RG [26]
- The full conformal group in generic dimension D has $(D+1)(D+2)/2$ parameters, but if the generators of invariance under rotations, translations, and dilatation are conserved, the remaining D generators are conserved as well. Therefore there was not much interest to go beyond the standard RG.

The point of view changed drastically for $D = 2$ after the work of Belavin, Polyakov, Zamolodchikov [27] who pointed out that this was an infinite symmetry. The influence of this article on the last 30 years has been of course considerable. However it is only recently that the use of conformal symmetry plus crossing symmetry and unitarity was revived for three dimensional physics [28]. Using this symmetry and bounds coming from unitarity the authors determine an allowed domain in the plane $\Delta_\sigma, \Delta_\epsilon$, the dimensions of the spin and energy operators. For reasons which do not seem yet well understood the 3D Ising exponents lie on this boundary at a well-defined location. The authors also match their method with the ϵ-expansion and the agreement is impressive.

The theory is thus far from finished but without Wilson's RG which solved the problem of critical phenomena, and modified completely the understanding of particle physics, we would still be in the stone age.

References

1. Wilson, K.G.: Renormalization group and Kadanoff scaling picture. Phys. Rev. **B4**, 3174 (1971)
2. Wilson, K.G.: Phase space cell analysis of critical behavior. Phys. Rev. **B4**, 3184 (1971)
3. Wilson, K.G., Kogut, J.: The renormalization group and the ϵ -expansion. Phys. Rep. **12**, 75–200 (1974)
4. Brezin, E.: DJ Wallace and KG Wilson, Feynman-graph expansion for the equation of state near a critical pointPhys. Rev. Lett. **29**, 591 (1972). (Phys. Rev. B **7**, 232 1973)

5. Wilson, K.G., Fisher, M.E.: Critical exponents in 3.99 dimensions. Phys. Rev. Lett. **28**, 240 (1972)
6. Wilson, K.G.: Operator-product expansions and anomalous dimensions in the Thirring model. Phys. Rev. **D2**, 1473 (1970)
7. Wilson, K.G.: Anomalous dimensions and breakdown of scale invariance in perturbation theory. Phys. Rev. **D2**, 1478 (1970)
8. Callan, C.: Broken scale invariance in scalar field theory. Phys. Rev. **D2**, 1541 (1970)
9. Symanzik, K.: Small distance behaviour analysis and Wilson expansions. Comm. Math. Phys. **23**, 49 (1971)
10. Kadanoff, L.P.: Scaling laws for Ising models near Tc. Physics **2**, 263 (1966)
11. Wilson, K.G.: Renormalization group — critical phenomena and Kondo problem. Rev. Mod. Phys. **47**, 773 (1975)
12. Lipatov, L.N.: Divergence of the perturbation-theory series and pseudo particles. Pis'ma Zh.Eksp. Teor. Fiz. **25**, 116 (1977)
13. Brezin, E., Le Guillou, J.C., Zinn-Justin, J.: Perturbation theory at large order. Phys. Rev. D **15**, 1544 (1977)
14. Le Guillou, J.C., Zinn-Justin, J.: Critical exponents from field theory. Phys. Rev. B **21**, 3976 (1980)
15. Guida, R., Zinn-Justin, J.: Critical exponents of the N-vector model. J. Phys. A **31**, 8103 (1998). and references therein
16. Parisi, G.: Field-theoretic approach to second-order phase transitions in two- and three-dimensional systems. J. Stat. Phys. **23**, 49 (1980)
17. Brezin, E., Parisi, G.: Critical exponents and large-order behavior of perturbation theory. J. Stat. Phys. **19**, 269 (1978)
18. Larkin, A.I., Khmel'nitskii, D.E., Eksp, Zh.: Phase transitions in uniaxial ferroelectrics. Teor. Fiz. **56**, 647 (1969)
19. Gross, D.J.: Applications of the renormalization group to high-energy physics In: Proceedings of Les Houches, session XXVIII, 1975, 142–250 (1976).
20. Ahlers, G., Kornblit, A., Guggenheim, H.J.: Logarithmic Corrections to the Landau Specific Heat near the Curie Temperature of the Dipolar Ising Ferromagnet LiTbF4. Phys. Rev. Lett. **34**, 122 (1975)
21. Brezin, E., Zinn-Justin, J.: Critical behaviour of uniaxial systems with strong dipolar interactions. Phys. Rev. **B13**, 251 (1976)
22. Mack, G.: Introduction to conformal invariant quantum field theory in two and more dimensions in nonperturbative quantum field theory. NATO ASI Ser. **185**, 353–383 (1988)
23. Wilson, K.G.: In: Proceedings of the Midwest Conference on Theoretical Physics, University of Notre-Dame (1970).
24. Polyakov, A.M.: Conformal symmetry of critical fluctuations. ZhETF Pis. Red. **12**, 538 (1970)
25. Migdal, A.A.: Conformal invariance and bootstrap. Phys. Lett. **B37**, 386 (1971)
26. Mack, G.: Conformal invariance and short distance behavior in quantum field theory Springer lectures notes, Strong interaction physics (1973).
27. Belavin, A.A., Polyakov, A.M., Zamolodchikov, A.B.: Infinite conformal symmetry in two-dimensional quantum field theory. Nucl. Phys. B **241**, 333 (1984)
28. El-Showk, S., Paulos, M.F., Poland, D., Rychkov, S., Simmons-Duffin, D., Vichi, A.: Solving the 3D Ising model with the conformal bootstrap. Phys. Rev. D **86**, 025022 (2012)

Chapter 8

Kenneth Wilson — renormalization and QCD*

Franz J. Wegner

Institute for Theoretical Physics, Ruprecht-Karls-University, Philosopher 19, D-69120 Heidelberg, Germany

wegner@tphys.uni-heidelberg.de

Kenneth Wilson had an enormous impact on field theory, in particular on the renormalization group and critical phenomena, and on QCD. I had the great pleasure to work in three fields to which he contributed essentially: Critical phenomena, gauge-invariance in duality and QCD, and flow equations and similarity renormalization.

Keywords: Renormalization group; critical phenomena; gauge-invariant models; duality; similarity transformation; flow equations.

8.1. Introduction

I am grateful that several times Ken and I could work on similar problems. We owe him a lot in critical phenomena, where pursuing ideas of Leo Kadanoff he developed a general renormalization scheme for critical phenomena and found a calculational scheme to determine critical exponents and scaling functions (Sec. 8.2).

A second time our work had some connection. I had generalized the Kramers–Wannier duality of the two-dimensional Ising model to higher-dimensional models. This procedure yields gauge-invariant models. Wilson generalized these ideas to non-Abelian gauge theories and developed a theory for the confinement of quarks (Sec. 8.3).

In 1993/1994 Głazek and Wilson and independently the present author developed the idea of a canonical transformation which diagonalizes a many-particle Hamiltonian (Sec. 8.4).

*This article was originally published in *Int. J. Mod. Phys. A* **29**, 1430043 (2014).

I will shortly report on these developments and imbed them in the state the field had reached before and add some results grown out of these developments. I apologize to all I do not mention in this article. There would be simply too many to be cited. See also my article in *Journal of Statistical Physics.*[59]

8.2. Renormalization group and critical phenomena

The oldest theories predicting critical phenomena were molecular field theories: Probably the first one was given by Van der Waals[53] in 1873 for the gas–liquid transition. Curie[10] (1895) and Weiss[62] (1905) formulated theories for the ferromagnetic transition, and Landau[35] in 1937 gave a general field theoretic concept for phase transitions of second order, which in its simplest approximation gave also molecular field behavior. They all predicted that the order parameter behaves below T_c like $\propto (T_c - T)^\beta$ with $\beta = 1/2$.

However, already in 1900 Verschaffelt[54] observed $\beta = 0.34\ldots$ for the difference of the density of the liquid and the vapor of isopentane on the basis of data by Young.[73] Along the isotherm he found $\rho - \rho_c \propto (P - P_c)^{1/\delta}$ with $\delta = 4.259$ instead of $\delta = 3$ by Van der Waals. More and more experiments showed that molecular field theory did not give the correct description of the behavior close to the critical point.

The exact solution of two models showed however, that critical behavior might differ from molecular field type: Onsager[43] showed that the two-dimensional Ising model yields a logarithmic singularity of the specific heat and Yang[72] obtained a critical exponent $\beta = 1/8$ for the spontaneous magnetization. Berlin and Kac[3] developed the spherical model, which below dimension 4 showed partly a different behavior. The three-dimensional model yields still $\beta = 1/2$, but the specific heat shows a kink instead of a jump. The susceptibility $\chi \propto (T - T_c)^{-\gamma}$ yields $\gamma = 2$ in contrast to the molecular field value $\gamma = 1$. If the dimension d of the system is considered a continuous variable, then the exponents depend on d between $d = 2$ and $d = 4$. Above four dimensions this model shows molecular behavior.

Many critical exponents were introduced. Inequalities between them were derived.[17,23,24,48] Widom[64] suggested that the difference of various thermodynamic quantities from their critical values, e.g. that of the chemical potential, density, and temperature in a fluid, are related by homogeneous functions. This explains equalities between critical exponents, which had been guessed before, and also equal exponents above and below the critical temperature.

This homogeneity picture was confirmed by Kadanoff's cell model.[33] In this model he replaces the spins within each cell by its total spin and considers the new effective interaction between these cell spins. The critical point is related to the fixed point of this mapping. Small deviations from the fixed-point interaction grow (or decay) with certain factors from one length scale to the other. This allows the singular part of the free energy to be written as a homogeneous function in the sense of Widom. This basic idea connects interactions at different length scales and carries already the idea of the renormalization group.

A quite different approach used high-temperature expansions for n-vector models (these are n-component unit vectors). Thus quantities like the specific heat and the susceptibility were expanded in powers of the inverse temperature and from these series one estimated the critical temperature and the behavior close by. Reviews on such calculations and estimates are given in volume 3 of the Domb-Green series[14] on critical phenomena. In 1967 the knowledge in critical phenomena, but also the fascinating question *How can we understand critical phenomena?* had reached a level, at which Michael Fisher,[19] P. Heller,[27] and Leo Kadanoff *et al.*[34] summed up their knowledge in reviews.

The idea of universality emerged as for example expressed by Michael Fisher, David Jasnow, Michael Wortis, and Bob Griffiths.[18,25,32] Accordingly the exponents depend only on the dimension d of the system, the dimension n of the (easy components of the) order parameter, and on the range of interaction, if it decreases slowly with distance. Stanley showed[50] that the $n \to \infty$-limit yields the results of the spherical model.

The calculations of critical exponents in isotropic and anisotropic classical Heisenberg models by Jasnow and Wortis[32] inspired Eberhard Riedel and myself to derive homogeneity relations for this model,[46] where we introduced the crossover exponent and argued how the temperature region showing anisotropic behavior shrinks with the anisotropy.

That renormalization ideas are useful to explain critical phenomena, was pointed out also by Larkin and Khmelnitskii.[37] They obtained power laws multiplied by fractional powers of logarithms for the critical behavior due to marginal operators at the upper critical dimension $d_c = 4$ for the n-vector model and at $d_c = 3$ for uniaxial ferroelectrics. Di Castro and Jona-Lasinio showed that the multiplicative renormalization group provides scaling laws.[11]

There were still many open questions and doubts in 1970 as one can see from the report[36] on the midwinter conference 1970.

Wilson analyzed Kadanoff's cell model in his first paper[65] and made clear in particular that more than two couplings may or should be considered, and he discussed the occurrence of irrelevant operators.

The important breakthrough contains his second paper,[66] where he gives a very intuitive approximation for the elimination of the short wavelength degrees of freedom and thus obtains an explicite recursion formula for the effective Hamiltonian, from which he obtains critical exponents for the three-dimensional Ising model, which where close to measured ones.

I became really aware of these papers, when Ken Wilson's and Michael Fisher's paper[70] on the $4 - \epsilon$ expansion of critical exponents of the Ising- and the x–y-model appeared. I was fascinated that here was a theory which easily fulfilled the Ginzburg criterion.[22] From then on I worked on $4 - \epsilon$ expansions,[26,56,60] partially with colleagues at Brown University, and I investigated general consequences of the Kadanoff–Wilson renormalization group picture.[55,57]

Wilson's early papers on this subject and the review by Wilson and Kogut[71] integrated over the short wavelength components of the order parameter in the Lagrangian. Wilson's work gave a clear understanding of critical phenomena and simultaneously the possibility for explicit calculations starting from the upper critical dimension four. However, the basic idea is not restricted to expansions around upper and lower critical dimensions. During a visit at Cornell I had a lively discussion with Ken Wilson and Michael Fisher on the ϵ-expansion for the cross-over exponent.[21,56] Shortly later Eberhard Riedel and myself investigated the tricritical point[47,61] with upper critical dimension $d = 3$.

Very fast it became clear, that the Landau[35] theory of phase transitions, which at first glance gave only molecular field exponents is very useful as field theory and became a basic theory for critical phenomena. Wilson[67] and Brézin, Wallace, and Wilson[6] applied Feynman-graph techniques to this theory and one switched to dimensional regularization. See the review by Brézin, Le Guillou, and Zinn-Justin.[9] However, in many cases functional renormalization is very useful.[30,42,44,52,63,71]

Much of this work as well as that of other scientists is reported in volume 6 of the Domb–Green series,[14] which after Melville Green's death was continued by Cyril Domb and Joel Lebowitz.[15] $1/n$-expansions for the n-vector model were performed by Abe,[1] Ma,[38] Suzuki,[51] Ferrell and Scalapino,[16] and also by Wilson[68] starting from the exactly known $n = \infty$-limit of the n-vector model[50] as the spherical model.[3] Shang-keng Ma gave a review.[39]

Wilson's description of critical phenomena also spread to critical dynamics. A summary of the various dynamical universality classes is given in the review[29] by Hohenberg and Halperin. Other aspects of critical phenomena are finite-size scaling (review by Barber),[2] critical surfaces and interfaces (reviews by Binder, Diehl and Jasnow),[5,12,31] and wetting (review by Dietrich).[13]

Mermin and Wagner[40] had shown that continuous symmetries cannot be broken in $d \leq 2$ dimensions at finite temperatures provided the interaction is not too long-range. Migdal[41] and Polyakov[45] started an expansion for critical exponents from this lower dimensionality two, work which has been continued by e.g. Brézin, Hikami, and Zinn-Justin.[7,8,28] I became much interested in this expansion, since Lothar Schäfer and myself[49,58] could map the mobility edge problem of the Anderson model of particles on disordered lattices on a matrix-model, which can be investigated in $2 + \epsilon$ dimensions.[4]

Wilson gave a beautiful review of the use of the renormalization group in critical phenomena and he solved the long standing problem of the s-wave Kondo Hamiltonian by a nondiagrammatic computer method.[69]

Wilson's work in critical phenomena had an enormous impact on this field, since simultaneously it allowed explicite calculations and gave an intuitive picture of the renormalization group procedure.

Michael Fisher[20] has given a review article on the renormalization group with emphasis on Wilson's accomplishment.

8.3. Duality and quantum chromodynamics

In 1970–1971 I thought about duality of models with Ising-spins in dimensions larger than two, similar to the Kramers–Wannier duality[78] of the two-dimensional Ising model. It occurred to me that such dual models could be expressed as gauge-invariant Ising-models, which may also show a transition, but without a local order parameter.[80]

I introduced models $M_{d,n}$ on a d-dimensional cubic lattice. For $n = 1$ the spins are located at the lattice sites of a primitive hypercubic lattice. The interaction is the sum of the product of two spins of nearest neighbors. For $n = 2$ spins are located on the links between nearest neighbors of the lattice. The interaction is a sum of the product of the four spins at the perimeter of elementary squares (plaquettes). This model shows local gauge-invariance, since flipping all the spins nearest to a lattice point leaves the energy of the system invariant. Generally $M_{d,n}$ consists of Ising-spins located in the centres of $(n-1)$-dimensional hypercubes. The interaction sums the products of all Ising-spins around elementary n-dimensional hypercubes. A duality-relation exists for the free-energy between models $M_{d,n}$ and $M_{d,d-n}$. Kramers and Wannier derived this for $d = 2$, $n = 1$. In three dimensions duality holds between the conventional $M_{3,1}$-Ising model and the plaquette model $M_{3,2}$.

The gauge-invariant Model $M_{4,2}$ is self-dual in four dimensions. It has the same transition temperature as the ordinary two-dimensional Ising-model. Instead of a local order-parameter the product of spins along a closed loop show a different behavior in both phases. At high temperature it obeys an area law, at low temperatures a perimeter law,

$$\left\langle \prod_{i\in\text{loop}} S_i \right\rangle = \begin{cases} \exp\left(-\dfrac{a}{\alpha(T)}\right) & T > T_c, \\ \exp\left(-\dfrac{l}{\xi(T)}\right) & T < T_c, \end{cases} \tag{8.1}$$

where a is the enclosed area and l the length of the perimeter.

Ken Wilson made the enormous step to generalize such models to continuous non-Abelian gauge-theories[82] for gluons and quarks, which allowed him to describe the confinement of quarks, a description, which became very important in high-energy physics. Other important contributions were by Balian, Drouffe, and Itzykson.[74–76]

Since the degrees of freedom are located at the links for $n = 2$ they are well suited for vector fields such as the electromagnetic vector potential and for gluons. It may be mentioned that the fields of discretized Maxwell equations (e.g. in accelerator physics) are located at appropriate points of the lattice and dual lattice.[81]

The correlation (8.1) may be considered as time-integrals of an effective interaction. Temperature acts as coupling strength. Then depending on the coupling strength the potential (between quark and antiquark) increases linearly in distance or approaches a finite value at large distances, indicating a transition between a confining and a nonconfining phase.

The transition temperature for the Ising gauge model was confirmed by Creutz, Jacobs, and Rebbi[77] numerically to the precision allowed by hysteresis effects at the first-order transition. Many papers of this subject are compiled in a review volume by Rebbi.[79]

At the occasion of 30 years of lattice QCD Wilson gave a retrospect on the discovery and the development of this field[83] at the XXII International Symposium on lattice field theory, which also contains quite a number of remarks related to other work by him.

8.4. Flow equations and similarity renormalization

Our work came very close for a third time. Głazek and Wilson[109,110] (GW) and myself[144] (W) developed independently equations for a unitary Hamiltonian flow, which brings the Hamiltonian in diagonal or block-diagonal form. It is also called similarity transformation or similarity renormalization. We expressed our surprise that we had similar ideas basically at the same time, when we once met (probably at a workshop in Les Houches).

Critical phenomena deal usually with classical fields. Here however, commutator relations are taken into account, which restricts the transformations to canonical ones. Common to both procedures is that one approaches the (block-)diagonal form continuously. The flow equation may be written

$$\frac{\mathrm{d}H(\ell)}{\mathrm{d}\ell} = [\eta(\ell), H(\ell)] \tag{8.2}$$

with the generator $\eta(\ell)$ of the unitary transformation and ℓ the flow-parameter. One starts with $H(0) = H$ and reaches diagonalization or block-diagonalization in the limit $\ell \to \infty$.

The main difference is that (GW) in their original formulation eliminate the off-diagonal matrix elements between states of an energy difference larger than some ΔE completely, which shrinks continuously during the flow, whereas (W) suggests a smoother elimination procedure, given by

$$\eta(\ell) = [H_{\mathrm{d}}(\ell), H(\ell)], \tag{8.3}$$

where H_{d} is the diagonal part of the Hamiltonian or some other appropriate H_{d} derived from H. (GW) working in high-energy physics take care of the renormalization of ultraviolet divergencies, whereas (W) starts from solid-state models on a lattice, where such divergencies no longer show up. Both procedures aim at the infrared problem. These methods are apt for fermionic and bosonic and even spin models.

Later it was realized that the mathematicians Brockett, Chu, Driessel[97–99] performed similar transformations in information theory. They call the method double bracket flow and isospectral flow, resp. The notion double bracket flow becomes obvious, when one inserts (8.3) into (8.2). The two-beam coupling in photorefractive media itself obeys the flow equation scheme.[84]

It is special for fermionic solid-state systems that the important physics takes place at the Fermi edge which in dimension $d > 1$ is no longer restricted to one or two points in momentum space, but extends over a $(d-1)$-dimensional region. Thus Shankar[139,140] eliminates states away from the Fermi edge and keeps only those very close to it. Similar elimination ideas not for the Hamiltonian, but for irreducible vertices in the form of Polchinski equations[44] were introduced by Zanchi and Schulz,[148] Halboth and Metzner,[116] and Salmhofer and Honerkamp.[137]

The flow equation scheme was applied in many cases in high-energy physics, nuclear physics and atomic physics. I mention some work in light front QED[114,119] and in light front QCD[96,106,147] and in effective nuclear interactions [90,91] and for nuclear few body problems.[88,89]

Wilson was fascinated by the idea of limit cycles, which he investigated together with Głazek.[112,113] The connection between asymptotic freedom and limit cycles has been studied in .[107,108] Already in 1970 Ken Wilson was interested in systems invariant under a change of energy scale.[146] Consequently Głazek and Wilson investigated systems whose states had energies and interactions increasing by powers of some factors $b > 1$.[107,111] An infrared limit cycle in QCD was suggested by Braaten and Hammer.[92] A nice example for limit cycles[85–87,135] are the Efimov states,[100,101,142] first investigated for helium-3 and tritium. Such states are of interest in the physics of ultracold atoms,[93] when they are tuned to large scattering lengths[129] close to the Feshbach resonance.[104]

There are numerous applications in solid-state physics. The smoothness of the transformation yields smoother results for the elimination of the electron–phonon interaction in superconductors [133,134] than the transformation by Fröhlich.[105] The flow equation result comes close to those by Eliashberg,[102,103] but in a simpler way. Similar smooth results are obtained for the Anderson impurity model[123] in contrast to the Schrieffer–Wolf transformation.[138] A model closely related to the mechanism of dissipation is the spin-boson model, which in this framework was first treated by Kehrein, Mielke and Neu.[124–127] Various aspects of the Kondo model can be investigated by this method.[118,121,141,143] Also the dynamics of spins on certain lattices can be investigated.[94,95,128,136]

Electronic systems in $d > 1$ dimensions can be brought to the form described by Landau's Fermi liquid theory[130–132] by means of flow equations.[120,122] The Hubbard model was treated in the weak-coupling limit,[115,117] where the Hamitonian was not diagonalized, but brought to a block-diagonal form, for which molecular field theory is exact. Thus various instabilities (antiferromagnetism, d-wave-superconductivity, Pomeranchuk instability) can be read from this block-diagonal form. Many applications in solid state theory can be found in the short review[145] and in the book by Kehrein.[122]

8.5. Concluding remark

Ken Wilson's field theoretic work in critical phenomena and renormalization and in lattice QCD had an enormous impact. He was honored by the Nobel Prize. Wilson was also engaged in physics education. I am glad to have met Ken Wilson.

Acknowledgment

I am indebted to Stan Głazek for providing me with several useful comments.

References

Critical Phenomena and Renormalization Group

1. R. Abe, *Prog. Theor. Phys.* **48**, 1414 (1972).
2. M. N. Barber, Finite size scaling, in *Phase Transitions and Critical Phenomena*, Vol. 8 (1983), p. 145.
3. T. H. Berlin and M. Kac, *Phys. Rev.* **86**, 821 (1952).
4. W. Bernreuther and F. J. Wegner, *Phys. Rev. Lett.* **57**, 1383 (1986).
5. K. Binder, Critical behavior at surfaces, in *Phase Transitions and Critical Phenomena*, Vol. 8 (1983), p. 1.
6. E. Brézin, D. J. Wallace and K. G. Wilson, *Phys. Rev. Lett.* **29**, 591 (1972).
7. E. Brézin and J. Zinn-Justin, *Phys. Rev. Lett.* **36**, 691 (1976).
8. E. Brézin and J. Zinn-Justin, *Phys. Rev. B* **14**, 3110 (1976).
9. E. Brézin, J. C. LeGuillou and J. Zinn-Justin, Field theoretical approach to critical phenomena, in *Phase Transitions and Critical Phenomena*, Vol. 6 (1976), p. 126.
10. P. Curie, Propriétés magnétiques des corps à diverses températures (Magnetic properties of bodies at different temperatures), Dissertation, Paris (1895).
11. C. di Castro and G. Jona-Lasinio, *Phys. Lett. A* **29**, 322 (1969).
12. H. W. Diehl, Field-theoretic approach to critical behavior at surfaces, in *Phase Transitions and Critical Phenomena*, Vol. 10 (1986), p. 75.
13. S. Dietrich, Wetting phenomena, in *Phase Transitions and Critical Phenomena*, Vol. 12 (1988), p. 1.
14. C. Domb and M. S. Green (eds.), *Phase Transitions and Critical Phenomena* (Academic Press), Vols. 1–6.
15. C. Domb and J. L. Lebowitz (eds.), *Phase Transitions and Critical Phenomena* (Academic Press), Vols. 7–20.
16. R. A. Ferrell and D. J. Scalapino, *Phys. Rev. Lett.* **29**, 413 (1972).
17. M. E. Fisher, *J. Math. Phys.* **5**, 944 (1964).
18. M. E. Fisher, *Phys. Rev. Lett.* **16**, 11 (1966).

19. M. E. Fisher, *Rep. Prog. Phys.* **30**, 615 (1967).
20. M. E. Fisher, *Rev. Mod. Phys.* **70**, 653 (1998).
21. M. E. Fisher and P. Pfeuty, *Phys. Rev. B* **6**, 1889 (1972).
22. V. L. Ginzburg, *Sov. Phys. Solid State* **2**, 1824 (1960).
23. R. B. Griffiths, *Phys. Rev. Lett.* **14**, 623 (1965).
24. R.B. Griffiths, *J. Chem. Phys.* **43**, 1958 (1965).
25. R. B. Griffiths, *Phys. Rev. Lett.* **24**, 1479 (1970).
26. M. K. Grover, L. P. Kadanoff and F. J. Wegner, *Phys. Rev. B* **6**, 311 (1972).
27. P. Heller, *Rep. Prog. Phys.* **30**, 731 (1967).
28. S. Hikami and E. Brézin, *J. Phys. A: Math. Gen.* **11**, 1141 (1978).
29. P. C. Hohenberg and B. I. Halperin, *Rev. Mod. Phys.* **49**, 435 (1977).
30. A. Houghton and F. J. Wegner, *Phys. Rev. A* **8**, 401 (1973).
31. D. Jasnow, Renormalization group of interfaces, in *Phase Transitions and Critical Phenomena*, Vol. 10 (1986), p. 269.
32. D. Jasnow and M. Wortis, *Phys. Rev.* **176**, 739 (1968).
33. L. P. Kadanoff, *Physics* **2**, 263 (1966).
34. L. P. Kadanoff, W. Götze, D. Hamblen, R. Hecht, E. A. S. Lewis, V. V. Palciauskas, M. Rayl, J. Swift, D. Aspnes and J. Kane, *Rev. Mod. Phys.* **39**, 395 (1967).
35. L. D. Landau, *Phys. Z. Sowjet.* **11**, 26 (1937) [*JETP* **7**, 19 (1937)].
36. G. E. Laramore, *J. Stat. Phys.* **2**, 107 (1970).
37. A. I. Larkin and D. E. Khmelnitskii, *Zh. Eksp. Teor. Fiz.* **56**, 2087 (1969) [*Sov. Phys. JETP* **29**, 1123 (1969)].
38. S.-K. Ma, *Phys. Rev. Lett.* **29**, 1311 (1972).
39. S.-K. Ma, The $1/n$ expansion, in *Phase Transitions and Critical Phenomena*, Vol. 6 (1976), p. 249.
40. N. D. Mermin and H. Wagner, *Phys. Rev. Lett.* **17**, 1133 (1966).
41. A. A. Migdal, *Zh. Eksp. Teor. Fiz.* **69**, 1457 (1975) [*JETP* **42**, 743 (1975)].
42. T. Morris, *Int. J. Mod. Phys. A* **9**, 2411 (1994).
43. L. Onsager, *Phys. Rev.* **65**, 117 (1944).
44. J. Polchinski, *Nucl. Phys. B* **231**, 269 (1984).
45. A. M. Polyakov, *Phys. Lett. B* **59**, 79 (1975).
46. E. Riedel and F. Wegner, *Z. Phys.* **225**, 195 (1969).
47. E. K. Riedel and F. J. Wegner, *Phys. Rev. Lett.* **29**, 349 (1972).
48. G. S. Rushbrooke, *J. Chem. Phys.* **39**, 842 (1963).
49. L. Schäfer and F. Wegner, *Z. Phys. B* **38**, 113 (1980).
50. H. E. Stanley, *Phys. Rev.* **176**, 718 (1968).
51. M. Suzuki, *Phys. Lett. A* **42**, 5 (1972).
52. N. Tetradis and C. Wetterich, *Nucl. Phys. B* **422**, 541 (1994).
53. J. D. van der Waals, Over de continuiteit van den gas — en vloeistoftoestand (On the continuity of the gaseous and the liquid state), Dissertation, Leiden (1873).
54. J. E. Verschaffelt, *Proc. Kon. Akad. Amst.* **588** (1900) [*Comm. Phys. Lab. Leiden* **55** (1900)].
55. F. J. Wegner, *Phys. Rev. B* **5**, 4529 (1972).
56. F. J. Wegner, *Phys. Rev. B* **6**, 1891 (1972).
57. F. J. Wegner, The critical state, general aspects, inin *Phase Transitions and Critical Phenomena*, Vol. 6 (1976), p. 7.
58. F. Wegner, *Z. Phys. B* **35**, 207 (1979).
59. F. J. Wegner, *J. Stat. Phys.* **157**, 628 (2014).
60. F. J. Wegner and A. Houghton, *Phys. Rev. A* **10**, 435 (1974).
61. F. J. Wegner and E. K. Riedel, *Phys. Rev. B* **7**, 248 (1973).

62. P. Weiss, *J. Phys. Theor. Appl.* **6**, 661 (1905).
63. C. Wetterich, *Phys. Lett. B* **301**, 90 (1993).
64. B. Widom, *J. Chem. Phys.* **43**, 3898 (1965).
65. K. G. Wilson, *Phys. Rev. B* **4**, 3174 (1971).
66. K. G. Wilson, *Phys. Rev. B* **4**, 3184 (1971).
67. K. G. Wilson, *Phys. Rev. Lett.* **28**, 548 (1972).
68. K. G. Wilson, *Phys. Rev. D* **7**, 2911 (1973).
69. K. G. Wilson, *Rev. Mod. Phys.* **47**, 773 (1975).
70. K. G. Wilson and M. E. Fisher, *Phys. Rev. Lett.* **28**, 240 (1972).
71. K. G. Wilson and J. Kogut, *Phys. Rep.* **12**, 75 (1974).
72. C. N. Yang, *Phys. Rev.* **85**, 808 (1952).
73. S. Young, *Phil. Mag.* **33**, 153 (1892).

Lattice Gauge Theories and Quantum Chromodynamics

74. R. Balian, J. M. Drouffe and C. Itzykson, *Phys. Rev. D* **10**, 3376 (1974).
75. R. Balian, J. M. Drouffe and C. Itzykson, *Phys. Rev. D* **11**, 2098 (1975).
76. R. Balian, J. M. Drouffe and C. Itzykson, *Phys. Rev. D* **11**, 2104 (1975).
77. M. Creutz, L. Jacobs and C. Rebbi, *Phys. Rev. Lett.* **42**, 1390 (1979).
78. H. A. Kramers and G. H. Wannier, *Phys. Rev.* **60**, 252 (1941).
79. C. Rebbi, *Lattice Gauge Theories and Monte Carlo Simulations* (World Scientific, 1983).
80. F. J. Wegner, *J. Math. Phys.* **12**, 2259 (1971).
81. T. Weiland, *Archiv für Elektronik und Übertragungstechnik* **31**, 116 (1977).
82. K. G. Wilson, *Phys. Rev. D* **10**, 2445 (1974).
83. K. G. Wilson, *Nucl. Phys. B* (*Proc. Suppl.*) **140**, 3 (2005).

Flow Equations and Similarity Renormalization

84. D. Z. Anderson, R. W. Brockett and N. Nutall, *Phys. Rev. Lett.* **82**, 1418 (1999).
85. P. F. Bedaque, H.-W. Hammer and U. van Kolck, *Phys. Rev. Lett.* **82**, 463 (1999).
86. P. F. Bedaque, H.-W. Hammer and U. van Kolck, *Nucl. Phys. A* **646**, 444 (1999).
87. P. F. Bedaque, H.-W. Hammer and U. van Kolck, *Nucl. Phys. A* **676**, 357 (2000).
88. S. K. Bogner, R. J. Furnstahl and R. J. Perry, *Phys. Rev. C* **75**, 061001 (2007).
89. S. K. Bogner, R. J. Furnstahl, R. J. Perry and A. Schwenk, *Phys. Lett. B* **649**, 488 (2007).
90. S. K. Bogner, A. Schwenk, T. T. S. Kuo and G. E. Brown, Renormalization group equation for low momentum effective nuclear interactions, arXiv:nucl-th/0111042.
91. S. K. Bogner, T. T. S. Kuo and A. Schwenk, *Phys. Rep.* **386**, 1 (2003).
92. E. Braaten and H.-W. Hammer, *Phys. Rev. Lett.* **91**, 102002 (2003).
93. E. Braaten and H.-W. Hammer, *Ann. Phys.* (*N.Y.*) **322**, 120 (2007).
94. W. Brenig, *Phys. Rev. B* **67**, 064402 (2003).
95. W. Brenig and A. Honecker, *Phys. Rev. B* **65**, 140407 (2002).
96. M. Brisudova and R. Perry, *Phys. Rev. D* **54**, 1831 (1996).
97. R. W. Brockett, *Linear Algebra Appl.* **146**, 79 (1991).
98. M. T. Chu, *Fields Inst. Commun.* **3**, 87 (1994).
99. M. T. Chu and K. R. Driessel, *SIAM J. Numer. Anal.* **27**, 1050 (1990).
100. V. N. Efimov, *Phys. Lett. B* **33**, 563 (1970).
101. V. N. Efimov, *Yad. Fiz.* **12**, 1080 (1970) [*Sov. J. Nucl. Phys.* **12**, 589 (1971)].

102. G. M. Eliashberg, *Zh. Eksp. Teor. Fiz.* **28**, 966 (1960) [*Sov. Phys. JETP* **11**, 696 (1960)].
103. G. M. Eliashberg, *Zh. Eksp. Teor. Fiz.* **29**, 1437 (1960) [*Sov. Phys. JETP* **12**, 1000 (1960)].
104. H. Feshbach, *Ann. Phys.* (*N.Y.*) **5**, 357 (1958).
105. H. Fröhlich, *Proc. R. Soc. A* **215**, 291 (1952).
106. S. D. Głazek, *Phys. Rev. D* **63**, 116006 (2001).
107. S. D. Głazek, *Phys. Rev. D* **75**, 025005 (2007).
108. S. D. Głazek and R. J. Perry, *Phys. Rev. D* **78**, 045011 (2008).
109. S. D. Głazek and K. G. Wilson, *Phys. Rev. D* **48**, 5863 (1993).
110. S. D. Głazek and K. G. Wilson, *Phys. Rev. D* **49**, 4214 (1994).
111. S. D. Głazek and K. G. Wilson, *Phys. Rev. D* **57**, 3558 (1998).
112. S. D. Głazek and K. G. Wilson, *Phys. Rev. Lett.* **89**, 230401 (2002).
113. S. D. Głazek and K. G. Wilson, *Phys. Rev. B* **69**, 094304 (2004).
114. E. L. Gubankova and F. J. Wegner, *Phys. Rev. D* **58**, 025012 (1998).
115. I. Grote, E. Körding and F. Wegner, *J. Low Temp. Phys.* **126**, 1385 (2002).
116. C. J. Halboth and W. Metzner, *Phys. Rev. B* **61**, 7364 (2000).
117. V. Hankevych, I. Grote and F. Wegner, *Phys. Rev. B* **66**, 094516 (2002).
118. W. Hofstetter and S. Kehrein, *Phys. Rev. B* **63**, 140402 (2001).
119. B. D. Jones, R. Perry and S. D. Głazek, *Phys. Rev. D* **55**, 6561 (1997).
120. A. Kabel and F. Wegner, *Z. Phys. B* **103**, 555 (1997).
121. S. Kehrein, *Phys. Rev. Lett.* **95**, 056602 (2005).
122. S. Kehrein, *The Flow Equation Approach to Many-Particle Systems*, Springer Tracts in Modern Physics, Vol. 217 (Springer, 2006).
123. S. Kehrein and A. Mielke, *J. Phys. A*: *Math. Gen.* **27**, 4259 (1994) [Erratum: *ibid.* **27**, 5705 (1994)].
124. S. Kehrein and A. Mielke, *Phys. Lett. A* **219**, 313 (1996).
125. S. Kehrein and A. Mielke, *Ann. Phys.* (*Berlin*) **6**, 90 (1997).
126. S. Kehrein and A. Mielke, *J. Stat. Phys.* **90**, 889 (1998).
127. S. Kehrein, A. Mielke and P. Neu, *Z. Phys. B* **99**, 269 (1996).
128. C. Knetter and G. S. Uhrig, *Eur. Phys. J. B* **13**, 209 (2000).
129. T. Kraemer, M. Mark, P. Waldburger, J. G. Danzl, C. Chin, B. Engeser, A. D. Lange, K. Pilch, A. Jaakkola, H.-C. Nägerl and R. Grimm, *Nature* **440**, 315 (2006).
130. L. D. Landau, *Sov. Phys. JETP* **3**, 920 (1956).
131. L. D. Landau, *Sov. Phys. JETP* **5**, 101 (1957).
132. L. D. Landau, *Sov. Phys. JETP* **8**, 70 (1959).
133. P. Lenz and F. Wegner, *Nucl. Phys. B* **482**, 693 (1996).
134. A. Mielke, *Ann. Phys.* (*Berlin*) **6**, 215 (1997).
135. R. F. Mohr, R. J. Furnstahl, R. J. Perry, K. G. Wilson and H.-W. Hammer, *Ann. Phys.* **321**, 225 (2006).
136. C. Raas, A. Bühler and G. S. Uhrig, *Eur. Phys. J. B* **21**, 369 (2001).
137. M. Salmhofer and C. Honerkamp, *Prog. Theor. Phys.* **105**, 1 (2001).
138. J. R. Schrieffer and P. A. Wolff, *Phys. Rev.* **149**, 491 (1966).
139. R. Shankar, *Physica A* **177**, 530 (1991).
140. R. Shankar, *Rev. Mod. Phys.* **66**, 129 (1994).
141. B. Thimmel, Flussgleichungen für das Kondo-Modell, Diploma thesis, Heidelberg (1996).
142. L. H. Thomas, *Phys. Rev.* **47**, 903 (1935).

143. E. Vogel, Flussgleichungen für das Kondo-Modell, Diploma thesis, Heidelberg (1997).
144. F. Wegner, *Ann. Phys. (Berlin)* **3**, 77 (1994).
145. F. Wegner, *J. Phys. A: Math. Gen.* **39**, 8221 (2006).
146. K. G. Wilson, *Phys. Rev. D* **2**, 1438 (1970).
147. K. G. Wilson, T. S. Walhout, A. Hadrindranath, W. M. Zhang, R. J. Perry and S. D. Głazek, *Phys. Rev. D* **49**, 6720 (1994).
148. D. Zanchi and H. J. Schulz, *Phys. Rev. B* **61**, 13609 (2000).

Chapter 9

Lattice gauge theory and the large N reduction*

Tohru Eguchi

Department of Physics and Research Center for Mathematical Physics, Rikkyo University, Tokyo 171-8501, Japan

I recall my encounter with Ken Wilson's lattice gauge theory and review the present status of the idea of large N reduction.

Keywords: Lattice gauge theory; large N reduction; quenched model; twisted model; adjoint fermions.

PACS numbers: 11.15.−q, 11.15.Ha, 11.90.+t

9.1. Lattice gauge theory

It is an honor for me to contribute this essay to the volume in memory of the late professor Kenneth Wilson. Wilson has been a great hero of theoretical physics of our time who turned around the theory of renormalization in quantum field theories from a technical device to a powerful machinery to study the scaling behavior of quantum fields. In late 1960s it was a challenge for us students to assimilate his writings on the operator product expansion, the renormalization group, etc.

Unfortunately, I did not have a close personal acquaintance with Ken: our closest point of encounter was via his formulation of lattice gauge theory.

His fundamental paper on lattice QCD has appeared in 1974.[1] The paper contained several key ideas which were quite novel to us: exponential of an integral of a gauge potential along infinitesimal line elements (link variables, $U_\mu(x) = \exp\{aA_\mu(x)\}$) plays the fundamental role in his formulation rather than the gauge potentials themselves. Thus the Lie group elements instead of Lie algebra elements appear in the action of lattice gauge theory. Furthermore there is no gauge fixing procedure in the lattice formulation. Gauge symmetry is left unfixed and the theory

*This article was originally published in *Int. J. Mod. Phys. A* **29**, 1430036 (2014).

keeps exact gauge invariance. Actually the gauge fixing procedure is necessary in continuum theory only when we develop perturbation theory. If we attempt to study the gauge theory as a whole nonperturbatively, there is no need to introduce gauge fixing and break gauge invariance. This was the stand point of Wilson's formulation.

He has introduced a famous observable which is now widely called as the Wilson loop amplitude

$$\langle W(L)\rangle = \langle \mathrm{Tr}[U_\mu(x)U_\nu(x+\mu)\cdots U_\lambda(x-\lambda)]\rangle \tag{9.1}$$

for a contour on the lattice $L = (x, x+\mu, x+\mu+\nu, \ldots, x-\lambda)$.

Wilson loop amplitude is defined to be an average over holonomies of gauge fields and serves to distinguish different phases of gauge theory depending on its behavior: $\langle W(L)\rangle \approx \exp(-R)$ (perimeter law) or $\langle W(L)\rangle \approx \exp(-R^2)$ (area law). Here R denotes the spatial size of the loop.

I was very much impressed by the elegant idea of characterizing the confining phase of the theory by the area law where the loop in space–time is interpreted as the trajectory of a massive quark–antiquark pair.

Despite this, it took some time for me get convinced of the validity of the lattice approach. I had a misunderstanding that the cancellation of longitudinal and temporal degrees of freedom of gauge field takes place only in a Lorentz invariant theory. However, after a while when I listened to the lectures by Susskind at SLAC on his Hamiltonian formulation of lattice QCD, I became convinced of the lattice approach and became a great fan of lattice formulation.

9.2. Large N reduction

Another major idea at that time in the study of gauge theory is the large N limit of 't Hooft.[2] It is well known that in gauge theories there is no simple way to approximate the theory without violating the gauge invariance. We may, for instance, sum over an infinite number of ladder diagrams, however, we still have to sum all the remaining Feynman diagrams in order to keep gauge invariance. Not even a single diagram can be dropped and we can maintain the gauge symmetry. This is the origin of the fundamental difficulty in solving the gauge theory.

The only exception to this rule is the method of large N expansion by 't Hooft. In $SU(N)$ gauge theory if we take the combination

$$\lambda = g^2 N \tag{9.2}$$

fixed (g^2 denotes the couplng constant), we can classify the Feynman diagrams by their orders in N. When one considers the Wilson loop amplitude, for instance, one finds an expansion

$$\langle W(L)\rangle = \sum_{h=0} N^{1-2h} f_h(\lambda). \tag{9.3}$$

Here h denotes the number of handles in the Feynman diagrams. In the limit of large N with λ kept fixed, we may just keep the planar amplitude $f_0(\lambda)$ and drop all the other nonplanar contributions. We maintain gauge invariance by the planar approximation.

A very special feature of the large N limit is its factorization property. For instance, in the case of a product of two Wilson loops L_1 and L_2 its expectation value is given by the disconnected component

$$\langle W(L_1) \cdot W(L_2) \rangle = \langle W(L_1) \rangle \langle W(L_2) \rangle. \tag{9.4}$$

In fact it is easy to check that when gluons are exchanged between the loops L_1, L_2 we obtain an amplitude of order $O(Ng^2) \approx 1$ and dominated by the disconnected piece, $\Pi_{i=1,2} \langle W(L_i) \rangle \approx O(N^2)$.

The above behavior [Eq. (9.4)] indicates that the large N limit is a kind of mean field approximation of gauge theory. In fact, we have

$$\langle (W(L) - \langle W(L) \rangle)^2 \rangle = 0 \tag{9.5}$$

and some special field configuration seems to dominate the path integration. This fictitious field configuration of large N limit is sometimes called the master field.

Note that in the large N limit strong force is exhausted to form bound states of mesons and baryons and there exist no remaining forces acting between them. Thus the large N limit is the best place to study the issue of confinement in gauge theory.

I became impressed by the factorization equation (9.4), (9.5) and wanted to figure out its implications. Equation (9.4) says that the amplitude does not depend on the relative distance between the loops L_1, L_2 but depend only on their shapes. It occurred to me that in the large N limit, link variable $U_\mu(x)$ somehow forgets about its location x while it remembers the direction μ. This is like assuming translation invariance for the master field. Thus Kawai and myself [3] tried to replace the action of lattice theory by that of a very simple model, the reduced model, consisting of only four matrices U_μ, $\mu = 1, 2, 3, 4$,

$$S_{\text{red}} = \frac{1}{g^2} \sum_{\mu,\nu} \text{Tr}(U_\mu U_\nu U_\mu^\dagger U_\nu^\dagger). \tag{9.6}$$

This is a rather drastic simplification and we have completely eliminated the space–time degrees of freedom. Space–time is supposed to reappear or emerge from the color degrees of freedom of link matrices U_μ.

In order to check how much our reduced model (9.6) is able to reproduce the original gauge theory, we have studied Schwinger–Dyson (SD) equations for Wilson loop amplitude. When we compare equations of the reduced model (9.6) with those of the original theory, we find extra terms appear in the reduced model. Consider, for instance, the case of a contour

$$C = (x, x + \mu, \ldots, y - \nu, y, y + \alpha, \ldots, z - \beta, z, z + \alpha, \ldots, x - \sigma, x), \tag{9.7}$$

where links (y, α) and (z, α) appear. In the reduced model the same variable U_α is assigned to these links even for $y \neq z$. Then we create an additional term in SD equation of reduced model

$$\begin{aligned} W(C_1, C_2) &= \langle \mathrm{Tr}(U_\alpha \cdots U_\beta)\, \mathrm{Tr}(U_\alpha \cdots U_\sigma U_\mu \cdots U_\nu) \rangle \\ C_1 &= (y, y+\alpha, \ldots, z-\beta, z), \\ C_2 &= (z, z+\alpha, \ldots, x-\sigma, x, x+\mu, \ldots, y-\nu, y). \end{aligned}$$

Using the factorization property

$$W(C_1, C_2) = \langle \mathrm{Tr}(U_\alpha \cdots U_\beta) \rangle \langle \mathrm{Tr}(U_\alpha \cdots , U_\sigma U_\mu \cdots U_\nu) \rangle. \tag{9.8}$$

Note that since C_1 and C_2 are open paths for $y \neq z$, there is at least one direction ρ for which U_ρ and $U_\rho^\dagger$ appear different number of times in both the sequences $(\alpha, \ldots, \beta)$ and $(\alpha, \ldots, \sigma, \mu, \ldots, \nu)$. The SD equations of reduced model coincides with those of the parent theory if amplitudes for open loops all vanish.

Action of the reduced model has the following symmetry:

$$U_\mu \to S U_\mu S^{-1}, \tag{9.9}$$
$$U_\mu \to e^{i\theta_\mu} U_\mu. \tag{9.10}$$

Equation (9.9) is the remnant of the local gauge symmetry of the original gauge theory while Eq. (9.10) is the symmetry under the center of the gauge group $U(N)$ (Z_N in the case of $SU(N)$).

As we have noted above, open paths possess different number of U_ρ and $U_\rho^\dagger$ for some ρ. Hence the center symmetry $U(1)^d$ prohibits all amplitudes of open paths acquiring nonzero values. Thus if the center symmetry is preserved, the set of SD equations of reduced model will coincide with those of the original gauge theory and the reduced model correctly reproduces the amplitudes of large N limit.

In the strong coupling region this is certainly what happens. Link variables U_μ fluctuates randomly and its eigenvalues are uniformly distributed over the unit circle. The center symmetry is maintained. However, it was soon noticed by various authors using numerical simulations (I heard Ken was one of the first to notice) that in the weak coupling region the eigenvalues of U_μ tend to attract each other which creates a clump in the eigenvalue distribution. Then the center symmetry becomes spontaneous broken and we lose the equivalence of the reduced model to the parent theory. In order to restore the equivalence of the reduced model to the standard gauge theory, it is necessary to introduce a mechanism which protects the center symmetry in the weak coupling region. (For a nice survey of these attempts, see Ref. 4).

9.3. Quenching

Bhanot, Heller and Neuberger[5,6] and Parisi[7] proposed the following modification of the large N reduction:

$$M(x)_{ij} \Rightarrow e^{i(k_\mu^i - k_\mu^j)x_\mu} M_{ij}, \quad 0 \le k_\mu^i, \quad k_\mu^j < 2\pi, \quad i,j = 1,\dots,N \tag{9.11}$$

for a matrix field $M(x)$. Here k_μ^i, k_μ^j are quenched random variables. Then the action of a matrix ϕ^4 field theory, for instance,

$$S = \sum_x \left\{ \sum_\mu \frac{1}{2} \operatorname{Tr} |M(x+\mu) - M(x)|^2 + \frac{m^2}{2} \operatorname{Tr} M(x)^2 + \frac{g}{N} \operatorname{Tr} |M(x)|^4 \right\} \tag{9.12}$$

is reduced to

$$S_{\text{QEK}} = \frac{1}{2} \sum_{i,j} |M_{ij}|^2 \left[2d + m^2 - 2 \sum_\mu \cos(k_i^\mu - k_j^\mu) \right] + \frac{g}{N} \operatorname{Tr} |M|^4. \tag{9.13}$$

If we identify the combination

$$p_\mu = k_\mu^i - k_\mu^j \tag{9.14}$$

as the momentum of the propagator of M_{ij} and integrate over p_μ, we recover the standard Feynman amplitudes of ϕ^4 perturbation theory. k_μ^i, k_μ^j are associated with the lines carrying color index i and j in the double line representation of the M_{ij} propagator. In the case of planar diagrams, it is easy to see that the number of integration momenta agrees with the number of color loops and we can express the momentum p_μ as a difference of k_μ^i, k_μ^j consistently. Momentum conservation is guaranteed by the flow of color indices.

Value of momentum runs over the Brillouin zone when the color index runs from 1 to N. If we denote by L the size of the four-dimensional periodic lattice, we find the relation

$$N = L^4 \tag{9.15}$$

in the quenching prescription.

From the point of numerical simulation quenching method is time-consuming since there is an extra work of integration over quenched variables as compared with ordinary simulations. Equation (9.14) is, however, a beautiful formula which captures the essence of planar perturbation theory.

9.4. Twisting

Twisting[8] is a method of large N reduction applicable to fields in adjoint representations of the gauge group: gauge field and also fermions in adjoint representations.

One considers a reduced action multiplied by a phase factor $Z_{\mu\nu}$,

$$S_{\mathrm{TEK}} = \beta \sum_{\mu \neq \nu} \mathrm{Tr}\big(Z_{\mu\nu} U_\mu U_\nu U_\mu^\dagger U_\nu^\dagger\big), \tag{9.16}$$

where

$$Z_{\mu\nu} = \exp \frac{2\pi i n_{\mu\nu}}{N}, \quad n_{\mu\nu} = L, \quad \mu > \nu, \tag{9.17}$$

$$N = L^2. \tag{9.18}$$

In this model the large N reduction takes place via the twisting matrices Γ_μ,

$$U_\mu(x) = D(x) U_\mu D(x)^\dagger, \quad D(x) = \prod_\nu (\Gamma_\nu)^{x_\nu}. \tag{9.19}$$

These matrices provide the so-called twist-eaters

$$\Gamma_\mu \Gamma_\nu = Z_{\nu\mu} \Gamma_\nu \Gamma_\mu \tag{9.20}$$

and give the ground state configuration of the twisted action

$$S = \beta \sum_{\mu \neq \nu} \mathrm{Tr}\big(Z_{\mu\nu} \Gamma_\mu \Gamma_\nu \Gamma_\mu^\dagger \Gamma_\nu^\dagger\big) = \beta d(d-1) \tag{9.21}$$

in d dimensions.

Twisting matrices are given explicitly as

$$\Gamma_1 = P = \begin{bmatrix} 0 & 1 & & & \\ & 0 & 1 & & \\ & & 0 & 1 & \\ & & & \cdots & \\ & & & & 1 \\ 1 & & & & 0 \end{bmatrix},$$

$$\Gamma_2 = Q = \begin{bmatrix} 1 & & & & \\ & e^{2\pi i/N} & & 0 & \\ & & \cdot & & \\ & & & \cdot & \\ & 0 & & & e^{2\pi i(N-1)/N} \end{bmatrix}$$

in two dimensions and by their tensor product in four dimensions

$$\Gamma_0 = Q \otimes Q, \quad \Gamma_1 = QP \otimes Q,$$
$$\Gamma_2 = P \otimes Q, \quad \Gamma_3 = 1 \otimes P.$$

Twisted reduced action of matrix ϕ^4 theory in two dimensions, for instance, is given by

$$S = \frac{1}{2} \sum_x \sum_\mu |M(x+\mu) - M(x)|^2 + \sum_x \frac{1}{2} m^2 \,\mathrm{Tr}\, M(x)^2 + \sum_x \frac{\lambda}{N} \,\mathrm{Tr}\, M(x)^4$$

$$\Rightarrow S_{\text{TEK}} = 2\,\mathrm{Tr}\,M^2 - \mathrm{Tr}\,MPMP^\dagger - \mathrm{Tr}\,MQMQ^\dagger + \frac{1}{2}m^2\,\mathrm{Tr}\,M^2 + \frac{\lambda}{N}\,\mathrm{Tr}\,M^4.$$

If we introduce color-singlet field $M(p,q)$,

$$M_{ij} = \frac{1}{\sqrt{N}}\sum_{p=1}^{N} M(p,q)\exp\left(\frac{\pi i}{N}(i+j)p\right), \quad q = i-j,$$

one can express the kinetic term as

$$(S_{\text{TEK}})_{\text{kin}} = \frac{1}{2}\sum_{p,q=1}^{N}\left(4\sin^2\frac{\pi p}{N} + 4\sin^2\frac{\pi q}{N} + m^2\right)|M(p,q)|^2. \tag{9.22}$$

We identify the size of the periodic lattice L as

$$N = L. \tag{9.23}$$

In the case of four dimensions, we instead have

$$N = L^2. \tag{9.24}$$

The interaction vertices have a complex phase dependence

$$\mathrm{Tr}\,M^4 = \frac{1}{N}\prod_{i=1}^{4}\sum_{p_i=1}^{N}\sum_{q_i=1}^{N}\delta_{\sum p_i,0}\delta_{\sum q_i,0}M(p_1,q_1)M(p_2,q_2)M(p_3,q_3)M(p_4,q_4) \times\exp\left[\frac{-\pi i}{N}\left(2\sum_{i=1}^{3}\sum_{j=i+1}^{4}q_ip_j + \sum_{i=1}^{4}q_ip_i\right)\right]. \tag{9.25}$$

It is possible to show[8,9] that the above phase factor cancels exactly each other in the case of arbitrary planar Feynman diagrams: thus the twisted reduced model correctly reproduces the planar amplitudes of the original theory. On the other hand, the phase factors do not cancel in nonplanar diagrams and suppress amplitudes by order $O(N^2)$ due to the rapid oscillation of phase under integration.

When one expresses the momentum in the phase factor (9.25) in terms of derivatives, it becomes exactly the same form as the Moyal product which appears in field theories of noncommutative space–time[10,11] (in this case, the fuzzy torus). Thus the twisted reduced model is in fact one of the first examples of field theories of noncommutative geometry.

9.5. Adjoint Fermions

Fermions in the adjoint representations of the gauge group is coupled to a pair of link variables U_μ and $U_\mu^\dagger$ and preserves the center symmetry. We may couple the gauge and adjoint fermion system and consider its large N reduction.[12]

Various authors have computed the one-loop effective potential on the eigenvalues of a link variable generated by the integration over adjoint fermion field. We use periodic boundary conditions for fermions. In the case of N_f flavors of fermions, one has

$$V_{\text{eff}}(\Omega) = \frac{N_f - 1}{24\pi^2 L^4}\left\{\frac{8\pi^4}{15}N_c^2 - \sum_{i,j=1}^{N_c} |v_i - v_j|^2(|v_i - v_j| - 2\pi)^2\right\}, \tag{9.26}$$

where v_i are eigenvalues of the link variable.

This formula seems to fit nicely with our expectations: when $N_f = 1$, the theory has an $\mathcal{N} = 1$ supersymmetry and no potential term is generated. While $N_f \geq 2$, the effective potential has a repulsive interaction between eigenvalues. There is no clumping of eigenvalues and the center symmetry will be unbroken and this is consistent with the results of numerical simulations (in the case of small fermion masses).

On the hand, when $N_f = 0$, there is an attractive force among eigenvalues which will trigger the spontaneous breaking of the center symmetry in the original reduced model.

At the moment the system of adjoint fermions seems to give a most clear case for the large N reduction. Numerically, there is no sign of $U(1)$ breaking in the whole range of coupling constants and the original EK reduction works without quenching or twisting for this system.

9.6. Discussions

Actually there has been a bad news several years ago for twisted (also quenched) model of large N reduction: several authors have discovered the breaking of $U(1)$ (center) symmetry[13–15] in twisted (and quenched) reduced model of intermediate coupling constants in numerical simulation with higher values of $N(\geq 100)$. This was somewhat surprising since the results of all the preceding simulations showed the unbroken center symmetry.

Possible explanation of this phenomenon was as follows: there may exist $U(1)$-breaking semistable extrema in twisted theory and the system gets trapped in these configurations. It will be quite possible for nonperturbative fluctuation of N^2 degrees of freedom may overcome $O(N)$ gap of ground state action of twisted theory. We need a fine tuning of twist parameters in order to avoid such unwanted, nonperturbative configurations.

See Ref. 16, for instance, for a recent result of twisted theory with refined twist parameters

$$n_{\mu\nu} = kL \tag{9.27}$$

with $k = 9$, $L = 17$, $N = 289$. With these parameters the theory preserves $U(1)$ symmetry for all values of couplings and the value of the string tension agrees precisely with that obtained from the standard theory with gauge groups $SU(N)$, $N = 3, 5, 6, 8$ and extrapolated in N. Thus the theory appears to work with these parameters. In general, there a restricted range of values for k and $\bar{k}$ defined as $k\bar{k} = 1 \pmod{L}$ for the system to work.

Here the issue is far from being settled and much better understanding of the structure of twisted theory is needed. Twisted theory seems to possess complex hidden structures and also exhibit odd behaviors of field theories of noncommutative space–time.

Finally, we would like to draw attention to an interesting remark [17] on the system of gauge theory coupled to adjoint fermions. Here we consider fermions with the periodic boundary condition which corresponds to computing the twisted partition function

$$Z(L) = \mathrm{tr}(-1)^F e^{-LH} = \int dM[\rho_B(M) - \rho_F(M)]e^{-LM}. \tag{9.28}$$

Here $\rho_B(M)$ and $\rho_F(M)$ are the densities of bosonic and fermionic states with mass M.

If we assume the density of states having the standard Hagedorn form of exponential growth

$$\rho(M) \approx \frac{1}{M}\left(\frac{T_H}{M}\right)^a e^{\frac{M}{T_H}}, \tag{9.29}$$

we expect $Z(L)$ to suffer a phase transition at

$$\frac{1}{L} = T_H. \tag{9.30}$$

In the case of thermodynamic partition function, on the other hand, we use anti-periodic boundary condition for fermions and one has

$$Z(\beta) = \mathrm{tr}\, e^{-\beta H} = \int dM[\rho_B(M) + \rho_F(M)]e^{-\beta M} \tag{9.31}$$

which possesses the finite temperature deconfining transition at

$$T_c = T_H. \tag{9.32}$$

Thus we expect the analogous transition (9.30) in the theory of adjoint fermions. We have, however, strong evidence of numerical simulations for the absence of such a transition in the adjoint fermion system.

In order to solve this puzzle, the authors of Refs. 17 suggest a strong cancellation between the density of states ρ_B and ρ_F which eliminate the phase transition. In fact, in the case of one flavor of adjoint fermion, the system possesses $\mathcal{N} = 1$ supersymmetry which gives a complete cancellation of ρ_B, ρ_F except for the ground states. In the case of more flavors $N_f \geq 2$, there should also be a strong cancellation between ρ_B, ρ_F or the match between bosonic and fermionic states. This would imply the existence of some fermionic symmetry which is different from supersymmetry.

We want to propose a different explanation for the absence of phase transition (9.30). We recall that at the large N limit gauge theory becomes simply a noninteracting gas of mesons and baryons and there is no force acting between them. In particular, there are no bound states of hadrons like nuclei. Then the state density will drop sharply from exponential to a power

$$\rho(M) \approx M^n. \tag{9.33}$$

Then there will be no phase transition in the twisted partition function $Z(L)$.

Furthermore, also in the purely bosonic theory like the large N gauge theory,

$$Z(\beta) = \int dM\, \rho_B(M) e^{-\beta M}, \tag{9.34}$$

we do not have a phase transition.

Thus we propose the absence of deconfining transition and the low density of states in the large N limit of gauge theory. It is curious to see if refined versions of reduced models may be able to provide some insight into this proposal.

Acknowledgments

I am very grateful to Dr. M. Okawa for discussions on the present status of the large N reduction. This research is supported in part by JSPS KAKENHI grant Nos. 25400273, 22224001 and 23340115.

References

1. K. Wilson, *Phys. Rev. D* **10**, 2445 (1974).
2. G. 't Hooft, *Nucl. Phys. B* **72**, 461 (1974).
3. T. Eguchi and H. Kawai, *Phys. Rev. Lett.* **48**, 1063 (1982).
4. S. Das, *Rev. Mod. Phys.* **59**, 235 (1987).
5. G. Bhanot, U. Heller and H. Neuberger, *Phys. Lett. B* **113**, 47 (1982).
6. G. Bhanot, U. Heller and H. Neuberger, *Phys. Lett. B* **115**, 237 (1982).
7. G. Parisi, *Phys. Lett. B* **112**, 463 (1982).
8. A. Gonzalez-Arroyo and M. Okawa, *Phys. Rev. D* **27**, 2397 (1983).
9. T. Eguchi and R. Nakayama, *Phys. Lett. B* **122**, 59 (1983).
10. N. Seiberg and E. Witten, *J. High Energy Phys.* **9909**, 032 (1999), arXiv:hep-th/99008142.
11. C.-S. Chu and P.-M. Ho, *Nucl. Phys. B* **550**, 151 (1999), arXiv:hep-th/9812219.
12. P. Kovtun, M. Ünsal and L. Yaffe, *J. High Energy Phys.* **0706**, 019 (2007), arXiv:hep-th/0702021.
13. T. Ishikawa and M. Okawa, talk at *JPS Meeting* (2003).
14. M. Teper and H. Vairinhos, *Phys. Lett. B* **652**, 359 (2007), arXiv:hep-th/0612097.
15. B. Bringoltz and S. R. Sharpe, *Phys. Rev. D* **78**, 034507 (2008), arXiv:0805.2146.
16. A. Gonzalez-Arroyo and M. Okawa, *Phys. Lett. B* **718**, 1524 (2013), arXiv:1206.0049.
17. G. Basar, A. Cherman, D. Dorigoni and M. Ünsal, *Phys. Rev. Lett.* **111**, 121601 (2013), arXiv:1306.2960.

Chapter 10

The lattice gauge field Lagrangian and Hamiltonian

Belal E. Baaquie
Department of Physics
National University of Singapore, Singapore 117542

The lattice gauge theory proposed by Wilson is discussed. The lattice gauge field Lagrangian and Hamiltonian are analyzed and shown to have many features quite different from the continuum formulation. Gauge-fixing for both the Lagrangian and Hamiltonian is discussed and a one-loop calculation is explicitly carried out to provide an exemplar of lattice perturbation theory.

10.1. Introduction

A memorial volume for Kenneth G. Wilson, or Ken as he preferred to be addressed, is testament to the seminal contributions made by Ken to theoretical physics. Ken's genius lies in having changed the very conception of what constitutes a quantum field theory — a bedrock of contemporary theoretical physics.

Up to the 1960's the foundation of quantum field theory was far from clear, especially since the procedure of renormalization, by which infinities were removed from physical quantities, seemed *ad hoc* if not outright wrong. As observed by Ken "the intuitive ideas (of renormalization) were encased in a thick shell of formalism" [1]. But since experiments supported the results obtained from quantum field theory, it could not be set aside that easily.

Ken demonstrated that renormalization, far from being an *ad hoc* procedure, is at the very essence of quantum field theory and is a reflection of the infinitely many coupled length scales that constitute a quantum field theory. He further demonstrated that a quantum field theory is renormalizable because of a fixed point at infinite momentum; the existence of this fixed point has been termed as Wilson's vision of physics at infinite momentum.

I arrived at Cornell University in 1972 as a fresh graduate student and Ken's fame had already spread far and wide. Ken was completing his work on applying the renormalization group to critical phenomena and working on the Kondo problem.

So it came as a bit of a surprise to many when, in 1973, he published his ground breaking article titled "Confinement of Quarks" [2], and which ushered in the lattice formulation of gauge fields and fermions. What intrigued me the most was the path integral formulation of the fermion degrees of freedom, although I did not have a chance to study this till much later.

In 1973, I went to meet Ken to ask him if I could do my PhD under his supervision. He asked me only one question: "Were you an undergraduate at Caltech?" to which my answer was affirmative. I finished my PhD in the summer of 1976, and for the duration I was his student, I worked on Ken's formulation of gauge fields. I would study at my own pace without any pressure to produce results. I used to see Ken about twice a week and always found him encouraging and pleasant — always willing to help and offer guidance. Ken's attention to detail was remarkable; he took pains to get all of them correct. There was no place for carelessness in his work.

Ken was a person not given to small talk and our discussions were always straight to the point. On one occasion, there was a discussion on the role of competition in physics and Ken remarked that one needs to compete if one is studying problems that others are studying and one needs not compete if one can do something all by oneself. This truism was exemplified by Ken himself, whose research was always pioneering and ground breaking. During the period I was Ken's student, there was great ferment and progress in physics, and Ken was at the center of the great breakthroughs of that era, from the understanding of quantum field theory to its applications to critical phenomenon and to high energy physics.

Being a student of Ken was one of my most exciting, intellectually enriching and life changing experience. Ken was at the peak of his creative genius and his presence had a tremendous impact on all those close to him. We all struggled to emulate Ken's work style that, to put it simply, meant that one had to study for all of one's waking hours. We often marveled at Ken's iron clad discipline and relentless effort that he could muster to attack a problem. I learnt more from observing his approach to physics than any specific calculation. The memories of Ken are still strong, and the period at Cornell under his supervision continue to remain fresh.

Ken was a person of strong convictions and with a deep and intuitive understanding of physics. The insights he had seemed almost like magic since often, even after knowing of his derivations, many of us could not figure out how he had arrived at the solution in the first place. I recall when the charm quark was observed in 1974, Ken had a discussion with a colleague who was still not convinced that a fourth quark had been found — and Ken was absolutely unyielding.

After being awarded the Nobel Prize in Physics in 1981, Ken extended his interest to the field of pedagogy and later, to the study of technological innovations. The last time I met Ken was in 2010, in Singapore, on the occasion of Murray Gell-Mann's 80th birthday and he was studying the factors that lead to the emergence of new technologies. He had many ideas on how nascent and fledgling technologies should be organized and managed. Ken was of the view that a multi-billion dollar international prize should be created and be awarded to the inventors of new technologies. He was of the view that this would motivate large teams of innovators to take interest in winning the prize — and thus being a positive force spurring the invention of new technologies.

Gauge fields defined on a finite lattice are free from all infinities. The finite lattice provides both an ultraviolet and infrared cutoff and makes the path integral into a finite dimensional multiple integral. This is the reason that the lattice formulation is ideally suited for numerical studies and which indeed is an extensive field in of itself. It is the view of many theorists that a rigorous and axiomatic formulation of quantum field theory is to define it on a finite lattice. The limits of taking the size of the lattice to infinity and taking the lattice spacing to zero are necessary for defining a relativistic quantum field theory.

The lattice formulation of gauge fields in Euclidean time, and of quantum fields in general, reveals a clear connection of quantum field theory with classical statistical mechanics. In particular, Ken's work on the renormalization group showed that *classical* systems that are critical and undergoing a second order phase transition are mathematically equivalent to renormalizable quantum field theories.

Ken's work in applying the mathematics of quantum field theory and of renormalization theory to explain classical phase transitions has greatly expanded the domain of quantum field theory beyond the study of quantum systems. Calculus originated in Newton's study of dynamics and has gone on to have universal applications in almost all quantitative branches of knowledge. *Quantum mathematics* is the application of the mathematics of quantum mechanics and quantum field theory to branches of knowledge beyond quantum systems and, similar to calculus, seems to be the mathematics of the future. Ken's work has greatly enriched the field of quantum mathematics by showing that the mathematics of quantum field theory can describe any system, be it classical or quantum, that is constituted by infinitely many coupled degrees of freedom.

10.2. Lattice gauge field

There are two defining features of the lattice theory: firstly, the gauge degrees of freedom are compact group elements of the gauge group and secondly, that the finite lattice spacing provides an ultraviolet regularization of the quantum field theory.

The lattice gauge field is quantized on a discrete lattice embedded in a four-dimensional Euclidean spacetime. The reason for going to a lattice is twofold. Firstly, the lattice provides an ultraviolet cutoff, and hence there are no ultraviolet divergences in the theory. Secondly, using the lattice as a cutoff allows one to formulate the cutoff theory so that we have exact local gauge invariance for the lattice gauge field.

Consider a finite and periodic four-dimensional Euclidean lattice with spacing of a and with N^4 lattice sites. Let n specify the lattice site and $\hat{\mu}$ the directions on the

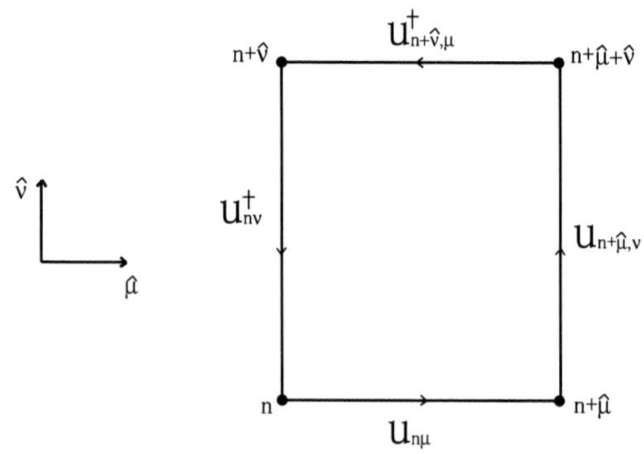

Fig. 10.1. The plaquette $W_{n\mu\nu}$.

lattice. The local gauge degrees of freedom are the finite group elements belonging to the gauge group G, which for definiteness is taken to be SU($\mathcal{N}$). The gauge field action functional is defined by (Tr signifies trace)

$$\mathcal{A} = \frac{1}{2g_0^2}\sum_n \sum_{\mu\neq\nu} \mathrm{Tr}(W_{n\mu\nu})$$

where g_0 is the *dimensionless* bare coupling constant. The plaquette $W_{n\mu\nu}$, shown in Figure 10.1, is defined by

$$W_{n\mu\nu} = U_{n\mu}U_{n+\hat{\mu},\nu}U^{\dagger}_{n+\hat{\nu},\mu}U^{\dagger}_{n\nu}.$$

Gauge transformation for the lattice is defined by

$$U_{n\mu} \to \Phi_n U_{n\mu}\Phi^{\dagger}_{n+\hat{\mu}}, \tag{10.2.1}$$

where Φ_n is also an element of the gauge group, and yields

$$W_{n\mu\nu} \to \Phi_n W_{n\mu\nu}\Phi^{\dagger}_n.$$

Note that $\mathcal{A}$ is invariant under local gauge transformations. The gauge field theory is quantized by integrating $e^{\mathcal{A}}$ over all possible values for $U_{n\mu}$, i.e.,

$$Z(g_0^2) = \prod_n \prod_\mu \int dU_{n\mu}e^{\mathcal{A}} = \int DUe^{\mathcal{A}} < \infty \tag{10.2.2}$$

where $dU_{n\mu}$ is the invariant Haar measure on the SU($\mathcal{N}$) group space.

The partition function is perfectly finite since each group integration $dU_{n\mu}$ is over a compact space. There is no need to fix a gauge to make the theory finite. In the strong coupling regime for which $g_0 \gg 1$, one can expand $\exp\{\mathcal{A}\}$ as a power series in the action $\mathcal{A}$; the compact integrations $\int DU$ then exhibit the area law for the Wilson loops and lead to the confinement of quarks.

10.3. The weak coupling approximation

The Lagrangian of the lattice gauge theory is studied for its weak coupling behavior, namely in the limit of $g_0 \to 0$ [3]. This limit is of central importance since it is well known that non-Abelian gauge fields are asymptotically free and hence have $g_0 \to 0$ at short distances. As one separates the quarks, the coupling constant grows continuously and the gauge field becomes strongly coupled at large distances.

Let $B^a_{n\mu}$ be the local lattice spacetime gauge field, ϕ^a_n be a local scalar field, and let $f^a_{n\mu\nu}$ be the local gauge field tensor. Then

$$W_{n\mu\nu} = e^{if^a_{n\mu\nu}X^a} \quad ; \quad U_{n\mu} = e^{iB^a_{n\mu}X^a} \quad ; \quad \Phi_n = e^{i\phi^a_n X^a} \tag{10.3.1}$$

where $\{X^a\}$ are the generators of the Lie group $SU(\mathcal{N})$ and

$$[X^a, X^b] = iC^{abc}X^c$$
$$\mathrm{Tr}(X^a X^b) = \delta^{ab}/s^2.$$

For the fundamental representation, the generators are given by $\mathcal{N} \times \mathcal{N}$ Hermitian matrices with $s^2 = 2$. The structure constants C^{abc} are completely anti-symmetric in the group indices a, b, c for compact Lie groups, which includes $SU(\mathcal{N})$. The group index a has the range of $1, 2, \ldots, \mathcal{N}^2 - 1$ for $SU(\mathcal{N})$.

$B^a_{n\mu}$ and $f^a_{n\mu\nu}$ are bounded variables which take values in the compact parameter space of $SU(\mathcal{N})$. We consider the case when $B^a_{n\mu} \simeq 0$. Using the equation $e^A e^B = e^{A+B+[A,B]/2+\cdots}$ yields

$$\begin{aligned} f^a_{n\mu\nu} = \Delta_\mu B^a_{n\mu} - \Delta_\nu B^a_{n\mu} - \frac{1}{2}C^{abc}\Big(& B^b_{n+\hat{\nu},\mu}B^c_{n+\hat{\mu},\nu} + B^b_{n+\hat{\nu},\mu}B^c_{n,\nu} + B^b_{n\mu}B^c_{n+\hat{\mu},\nu} \\ & - B^b_{n\mu}B^c_{n\nu} - B^b_{n\mu}B^c_{n+\hat{\nu},\mu} - B^b_{n+\hat{\mu},\nu}B^c_{n,\nu}\Big) + O(B^3) = -f^a_{n\nu\mu} \end{aligned}$$

where repeated *group indices* are summed over and $\Delta_\mu h_n = h_{n+\hat{\mu}} - h_n$ is the finite lattice derivative. In general, the lattice field tensor $f^a_{n\mu\nu}$ is an infinite power series of the $\{B^a_{n\mu}, B^a_{n+\hat{\mu},\nu}, B^a_{n+\hat{\nu},\mu}, B^a_{n\nu}\}$ variables and is an analytic function of these variables as a consequence of the group multiplication law.

For the four-dimensional lattice with spacing a, the continuum limit is obtained by defining the continuum gauge field A_μ as follows

$$A_\mu(x) = sg_0 B^a_{n\mu} \quad ; \quad x = na$$

that yields the Yang–Mills action and field tensor

$$\mathcal{A} = -\frac{1}{4}\sum_{\mu\nu;a}\int d^4x (F^a_{\mu\nu})^2 \quad ; \quad F^a_{\mu\nu} = \partial_\mu A^a_\nu - \partial_\nu A^a_\mu - g_0 C^{abc} A^b_\mu A^c_\nu.$$

From Eq. (10.2.1), the gauge transformation on the $B^a_{n\mu}$ variables is given by

$$e^{iB^a_{n\mu}(\phi)X^a} = e^{i\phi^a_n X^a} e^{iB^a_{n\mu}X^a} e^{-i\phi^a_{n+\hat{\mu}}X^a}.$$

Using the equation

$$\exp(A)\exp(B) = \exp\left\{A + B + \frac{1}{2}[A,B] + \frac{1}{12}[[A,B],B] - \frac{1}{12}[[A,B],A] + \cdots\right\}$$

we obtain

$$B^a_{n\mu}(\phi) = B^a_{n\mu} - \Delta_\mu \phi^a_n - \frac{1}{2}C^{abc}(\phi^b_n + \phi^b_{n+\hat{\mu}})B^c_{n\mu} + \frac{1}{2}C^{abc}\phi^b_n\phi^b_{n+\hat{\mu}}$$
$$+ \frac{1}{12}C^{abe}C^{cde}B^b_{n\mu}B^c_{n\mu}(\phi^d_n - \phi^d_{n+\hat{\mu}}) + O(\phi^3, B^3\phi). \qquad (10.3.2)$$

10.4. Gauge-fixing the Lagrangian

For concreteness, we examine the effect of the gauge transformation on the path integral of the action functional. The action is gauge invariant under this transformation, that is, $\mathcal{A}[W(\Phi)] = \mathcal{A}[W]$, independent of the $\{\Phi_n\}$ variables.

Although the path integral is well defined, it is not suited for Feynman perturbation expansion since all the gauge field variables $B^a_{n\mu}$ are not constrained to be small as $g_0 \to 0$ due to gauge invariance. The gauge-fixing term is introduced to control the divergence due to gauge invariance. This means, in terms of the original variables $B^a_{n\mu}$, that the action has added to it a term that necessarily breaks gauge invariance. To leave the gauge-invariant sector unchanged by the gauge-fixing procedure we further add the counter-term. The counter-term is a gauge-invariant functional of the gauge field and is evaluated from the gauge-fixing term via a path integral.

Let A_α be the gauge-fixing term and A_c be the counter-term. The modified action is defined as

$$\mathcal{A}' = \mathcal{A} + A_\alpha + A_c.$$

The actions $\mathcal{A}$ and $\mathcal{A}'$ give the same gauge-invariant physics.

10.4.1. *Zero mode*

In spite of gauge-fixing the lattice gauge field, note that for any finite lattice, there is a *single* degree of freedom that is still unconstrained by the gauge fixing. This is because on the lattice, there are only $N^4 - 1$ independent gauge transformations and hence even after gauge-fixing there is *one* degree of freedom that is left unconstrained. To keep track of the single variable that is not constrained by the gauge fixing, define the *single* lattice site $N \equiv (N, N.N, N)$. The *single* degree of freedom variable $B^a_{N\mu}$ with $U_{N\mu} = e^{iB^a_{N\mu}X^a}$ takes values in the compact group space, that is $B^a_{N\mu} \in G$; in particular note that $B^a_{N\mu} \notin [-\infty, +\infty]$.

In the case of the Abelian Maxwell gauge field, both for the lattice and continuum, the zero mode of the gauge field decouples from the theory (in the absence of fermions) since the Abelian theory is linear.

Although this single variable may seem unimportant and an artifact of the lattice, this is far from true. The single variable $B^a_{N\mu}$ will later be mapped, via a discrete Fourier transform, to the zero momentum degree of freedom, namely the zero mode.

It is a remarkable feature of the lattice gauge field on a finite lattice that the zero mode can be separated from the other degrees of freedom. This identification of the zero mode is not possible even for an infinite lattice since the momentum index becomes continuous and hence the zero mode is part of the continuum degrees of freedom. The fact that the zero mode can be isolated is the fundamental basis for a program of studying the mass renormalization of the lattice gauge field for a finite lattice and is further discussed in Section 10.6.

10.4.2. *Gauge-fixed path integral*

One has a wide choice as to what functional of the field variables A_α should be. The only necessary condition is that

$$Z' = \prod_{n\neq N}\prod_{\mu,a}\int_{-\infty}^{\infty} dB^a_{n\mu}\mu(B^a_{n\mu})\int_G \prod_{\mu} dU_{N\mu}e^{A'} < \infty \ ; \ N \equiv (N, N.N, N)$$

where $\mu(B^a_{n\mu})$ is the Haar measure.

On the finite lattice, since there are only $N^4 - 1$ independent gauge transformations, we choose $\Phi_N = \mathbb{I}$. To define A_c we introduce the following notation

$$U^{(\phi)}_{n\mu} = \Phi_n U_{n\mu}\Phi^\dagger_{n+\hat{\mu}} \ ; \ \Phi_N = \mathbb{I}$$

$$D\Phi = \prod_{n\neq N} d\Phi_n \ ; \ DU = \prod_n\prod_\mu dU_{n\mu}.$$

Define A_c by

$$e^{A_c[U]} = 1/\int D\Phi e^{A_\alpha[U^{(\phi)}]} \quad : \quad \text{gauge invariant.}$$

Note the identity

$$1 = \int D\Phi e^{A_\alpha[U^{(\phi)}]} \Big/ \int D\Phi' e^{A_\alpha[U^{(\phi')}]}.$$

Let $K[U]$ be an arbitrary *gauge-invariant* function. Then

$$\int DUK[U] = \int DUK[U]\frac{\int d\Phi e^{A_\alpha[U^{(\phi)}]}}{\int d\Phi' e^{A_\alpha[U^{(\phi')}]}}$$

$$= \left(\int D\Phi\right)\int DUK[U]e^{A_\alpha[U^{(\phi)}]}e^{A_c[U]}.$$

Since the gauge group is compact, we have

$$\int D\Phi = 1.$$

The lattice gauge field defines a completely finite quantum field theory solely due to the compactness of the gauge group.

Perform the *inverse* gauge transformation on $\{U_{n\mu}\}$ variables, namely

$$U'_{n\mu} = \Phi_n^\dagger U_{n\mu}\Phi_{n+\hat{\mu}} \quad \Rightarrow \quad DU' = DU$$
$$\Rightarrow K[U]e^{A_c[U^{(\phi)}]} = K[U']e^{A_c[U']} \;\; ; \;\; e^{A_c[U^{(\phi)}]} = e^{A_c[U']}.$$

We hence have the following

$$\int DUK[U] = \left(\int D\Phi\right)\left\{\int DU'K[U']e^{A_\alpha[U']+A_c[U']}\right\}$$
$$= \int DUK[U]e^{A_\alpha[U]+A_c[U]}.$$

We thus see that $e^{A_\alpha+A_c}$ leaves the gauge-invariant sector *unchanged*. Hence, in particular,

$$Z(g_0) = \int DUe^{\mathcal{A}[U]} = \int DUe^{\mathcal{A}+A_\alpha+A_c}. \tag{10.4.1}$$

Note that the result Eq. (10.4.1) is valid exactly for the lattice theory. This formulation reduces to the continuum formulation in the weak coupling approximation.

10.5. The Faddeev–Popov counter-term

We now choose a specific A_α and calculate A_c for it. Consider the following gauge-fixing term

$$e^{A_\alpha[B]} = \prod_{n,a}{}' \delta(s_n^a - t_n^a) \;\; ; \;\; \prod_n{}' = \prod_{n\neq N}$$

where $\{t_n^a\}$ are fixed numbers and

$$s_n^a = \sum_\mu \Delta_\mu B_{n+\hat{\mu},\mu}^a \;\; ; \;\; \sum_n s_n^a = 0. \tag{10.5.1}$$

Recall that $B_{n\mu}^a(\phi)$ is defined by

$$\exp[iB_{n\mu}^a(\phi)X^a] = \Phi_n U_{n\mu}\Phi_{n+\hat{\mu}}^\dagger$$

and hence

$$s_n^a(\phi) = \sum_\mu \Delta_\mu B_{n+\hat{\mu},\mu}^a(\phi). \tag{10.5.2}$$

Note that there are only $N^4 - 1$ independent variables for the s_n^a.

The combined effect of $e^{A_\alpha + A_c}$ is to leave the gauge-invariant sector unchanged and yields

$$Z(g_0) = \int DU e^{\mathcal{A}+A_\alpha+A_c} = \int DU e^{\mathcal{A}}.$$

Since $Z(g_0)$ is independent of $\{t_n^a\}$ and it can be shown that

$$Z(g_0) = \int DU e^{\mathcal{A}+A_c+A_\alpha} \equiv \int DU e^{\mathcal{A}'}$$

where

$$A_\alpha = -\frac{\alpha}{2}\sum_{n,a}{}'(s_n^a)^2 \quad ; \quad \sum_n{}' = \sum_{n\neq N}$$

and

$$e^{-A_c} = \int D\Phi \prod_{n,a}{}' \delta(s_n^a(\phi) - s_n^a). \tag{10.5.3}$$

$A_\alpha + A_c$ is not gauge invariant, but it has a lower symmetry, which is the BRST symmetry.

In summary

$$\mathcal{A}' = \mathcal{A} + A_\alpha + A_c.$$

Let $\alpha = O(1/g_0^2)$; then the modified action $\mathcal{A}' = \mathcal{A} + A_\alpha + A_c$ restricts all the variables (except $B_{N\mu}^a$) to be $O(g_0)$ and is suited for studying the weak coupling sector using Feynman diagrams.

10.5.1. *Lattice Faddeev–Popov ghost action*

For a finite lattice we have that

$$e^{-A_c} = \int D\Phi \prod_{n,a}{}' \delta(s_n^a(\phi) - s_n^a)$$

where, from Eq. (10.5.1)

$$s_n^a = \sum_\mu \Delta_\mu B_{n+\hat{\mu},\mu}^a.$$

For $B_{n\mu}^a(\phi)$ defined by

$$\exp[iB_{n\mu}^a(\phi)X^a] = \Phi_n U_{n\mu} \Phi_{n+\hat{\mu}}^\dagger$$

and hence, from Eq. (10.5.2)

$$s_n^a(\phi) = \sum_\mu \Delta_\mu B_{n+\hat{\mu},\mu}^a(\phi).$$

Define $u_n^a = u_n^a(\phi)$ by

$$u_n^a \equiv s_n^a(\phi) - s_n^a$$

where $s_n^a = s_n^a(\phi)|_{\phi=0}$.

We make a change of variable from $\{\phi_n^a\}$ to $\{u_n^a\}$ to evaluate Eq. (10.5.8). The δ functions make $u_n^a(\phi) = 0$; this in turn implies $\phi_n^a = 0$ as the unique solution for which $u_n^a = 0$ (as long as $B_{n\mu}^a \simeq 0$). Then from Eq. (10.5.8)

$$e^{-A_c[B]} = \int D\Phi \prod_{n,a}{}' \delta(s_n^a(\phi) - s_n^a) = \int D\Phi \prod_{n,a}{}' \delta(u_n^a). \tag{10.5.4}$$

Let[1]

$$u_n^a = \sum_k e^{ikn} u_k^a \quad ; \quad B_{n\mu}^a = \sum_k e^{ikn} B_{k\mu}^a \quad ; \quad \phi_n^a = \sum_k e^{ikn} \phi_k^a.$$

We analyze the variable $u_n^a = u_n^a(\phi)$. From Eqs. (10.5.2) and (10.3.2)

$$\begin{aligned} u_k^a &= \sum_n e^{-ikn} u_n^a \\ &= \sum_\mu |1 - e^{ik_\mu}|^2 \phi_k^a + \frac{1}{2} C^{abc} \sum_{k,\mu} (1 - e^{-ik_\mu})(1 + e^{-iq_\mu}) B_{k-q,\mu}^b \phi_q^c \\ &\quad + \frac{1}{12} C^{abc} C^{cde} \sum_{p,q} \sum_\mu (1 - e^{-ik_\mu})(1 - e^{-iq_\mu}) B_{k-p-q,\mu}^b B_{p\mu}^c \phi_q^d + O(B^3\phi, B^2\phi^2). \end{aligned}$$

Note from above equation that $u_{k=0}^a = 0$; i.e., it is not coupled to the ϕ_n^a. We can hence redefine $u_{k=0}^a$ to be

$$u_{k=0}^a = \phi_{k=0}^a.$$

Then, from above[2]

$$\begin{aligned} u_k^a &= d_k \phi_k^a + \sum_q (M^{ab}(k,q) + L^{ab}(k,q)) \phi_q^b \\ &\equiv \sum_q T^{ab}(k,q) \phi_q^b \end{aligned} \tag{10.5.5}$$

where

$$M^{ad}(k,q) = \frac{1}{2} C^{abc} \sum_{k,\mu} (1 - e^{-ik_\mu})(1 + e^{-iq_\mu}) B_{k-q,\mu}^b \tag{10.5.6}$$

---

[1] To define the Fourier transform of the variables let h_n be any arbitrary function of n. Owing to the torus structure of the lattice, we have $h_{n+N} = h_n$: periodic in all the coordinates with period N. Hence h_n can be expanded in terms of the basis functions $\{e^{ik_\mu n_\mu}\}$, $k_\mu = 0, 2\pi/N, \ldots, 2\pi(N-1)/N$. That is

$$h_n = \sum_k e^{ikn} h_k \equiv \frac{1}{N^4} \left(\prod_\mu \sum_{k_\mu = 0}^{2\pi(N-1)/N} e^{ik_\mu n_\mu} \right) h_k$$

$$h_k = \sum_n e^{-ikn} h_n \quad ; \quad \delta_{k,q} = N^4 \prod_{i=0}^{2} \delta_{k_i + q_i}.$$

[2]

$$d_k = 1 \text{ if } k = 0 \quad ; \quad d_k = \sum_\mu |1 - e^{ik_\mu}|^2 \text{ if } k \neq 0.$$

and

$$L^{ad}(k,q) = \frac{1}{12}C^{abe}C^{cde}\sum_{p,q}\sum_{\mu}(1-e^{-ik_\mu})(1-e^{-iq_\mu})B^b_{k-p-q,\mu}B^c_{p\mu}.$$

Equation (10.5.5) yields

$$Du = \det(T)D\phi.$$

Due to the delta functions in Eq. (10.5.4), $D\Phi = D\phi$. Hence the counter-term, from Eq. (10.5.4), is given by

$$e^{A_c[B]} = \det(T).$$

Using the property det(AB) = det(A) det(B) and that d_k is independent of the gauge field $\{B^a_{n\mu}\}$ upto an overall constant, we have the following:

$$\begin{aligned}
e^{A_c[B]} &= \det\left(1+\frac{1}{d}M+\frac{1}{d}L\right) \\
&\simeq \det\left(1+\frac{1}{d}M\right)\det\left(\frac{1}{d}L\right) \\
&= \det\left(1+\frac{1}{d}M\right)\exp\left[\mathrm{Tr}\left(\frac{1}{d}L\right)\right]. \qquad (10.5.7)
\end{aligned}$$

We evaluate

$$\begin{aligned}
\mathrm{Tr}\left(\frac{1}{d}L\right) &= \sum_a\sum_k\frac{1}{d_k}L^{aa}(k,k) \\
&= -\frac{\mathcal{N}}{12}\left\{\sum_k\frac{1}{d_k}\sum_\mu|1-e^{ik_\mu}|^2\right\}\times\sum_{q,a,\nu}B^a_{-q\nu}B^a_{q\nu}
\end{aligned}$$

using the identities

$$\sum_k\frac{1}{d_k}|1-e^{ik_\mu}|^2 = \frac{1}{4} \quad\text{and}\quad C^{abc}C^{\alpha bc} = \mathcal{N}\delta^{a\alpha}$$

and obtain

$$\mathrm{Tr}\left(\frac{1}{d}L\right) = -\frac{\mathcal{N}}{48}\sum_k\sum_{a,\mu}B^a_{-k\mu}B^a_{k\mu} = -\frac{\mathcal{N}}{48}\sum_n\sum_{a,\mu}(B^a_{n\mu})^2.$$

Note that $\mathrm{Tr}[(1/d)L]$ is completely local. Hence, we conclude from Eq. (10.5.7) that

$$e^{A_c[B]} = \det(d_k\delta^{ab}\delta_{k,q}+M^{ad}(k,q))\times\exp\left[-\frac{\mathcal{N}}{48}\sum_{\mathcal{N}\mu a}(B^a_{n\mu})^2\right]+O(g_0^3). \qquad (10.5.8)$$

Equation (10.5.8) is the answer for the Faddeev–Popov counter-term to $O(B^2)$. The determinant in the expression for $A_c[B]$ can be represented by fermion integration variables, namely the ghost fermion field $c^a_n, c^{a\dagger}_n$ and yields the following

Faddeev–Popov counter-term

$$e^{A_c[B]} = \prod_{n,a} \int dc_n^a dc_n^{a\dagger} \exp\{c^\dagger(d+M)c\} \times \exp\left[-\frac{\mathcal{N}}{48} \sum_{\mathcal{N}\mu a} (B_{n\mu}^a)^2\right] \tag{10.5.9}$$

$$(d+M)^{ab}(k,q) = d_k \delta^{ab} \delta_{k,q} + M^{ad}(k,q).$$

Unlike the continuum case, for the ghost action the lattice gauge field yields an infinite order polynomial in the lattice gauge field $B_{n\mu}^a$. In particular, the extra local term $\sum_{n\mu a}(B_{n\mu}^a)^2$ is absent in the continuum formulation. This term is *quadratically divergent* and plays an important role in ensuring that there is no mass renormalization necessary for the lattice gauge field.

10.6. One-loop mass renormalization

To illustrate a specific feature of the lattice regularization of the gauge field, the computation of mass renormalization to one-loop is summarized. If the net result of the one-loop computation is that there is no mass term for the gauge field, then this is due to the fact that there is no mass renormalization for the lattice gauge field, and it remains massless in spite of the nonlinear interactions. The absence of mass renormalization is due to exact gauge invariance on the lattice that forbids the appearance of a mass term. The gauge field can generate a spectrum of states for which there is a mass gap implying the existence of a massive gauge boson. This massive state is a low energy collective bound state of the gauge field and its mass does not arise from a mass term in the action.

Recall that an important feature of the lattice regularization is that there are N^4 degrees of freedom (integration variables) and hence one can isolate every single momentum mode of the lattice gauge field. In particular, the zero momentum mode of the gauge field directly contains the mass term of the field: if the gauge field does not have a bare mass and is massless, the zero mode is unconstrained and if not then the field has a quadratic mass divergence, leading to an inconsistent theory.

An efficient method for computing the quadratic mass divergence is to isolate the zero mode and then integrate out *all* the non-zero momentum degrees of freedom. The net effect of this procedure, which in effect yields the effective action, is to provide the requisite mass renormalization. To carry out this computation, consider the following Fourier expansion

$$B_{n\mu}^a = \sum_k e^{ikn} B_{k\mu}^a = \theta^a + \sum_{k\neq 0} e^{ikn} B_{k\mu}^a \equiv \theta^a + F_{n\mu}^a.$$

Note that the field $F^a_{n\mu}$ is a short-hand for the non-zero Fourier modes $B^a_{k\mu}, k \neq 0$, and

$$\sum_n F^a_{n\mu} = 0.$$

θ^a_μ is the zero mode given by

$$\theta^a_\mu = \frac{1}{N^4} B^a_{k=0,\mu} = \sum_n B^a_{n\mu}.$$

The path integral yields the following expression for $W(\theta)$, the zero mode effective action

$$Z = \prod_{\mu,a} \int_G d\theta^a_\mu e^{W(\theta)} \quad ; \quad \int DF = \prod_{\mu,a} \prod_n \int_{-\infty}^{+\infty} dF^a_{n\mu}$$

$$e^{W(\theta)} = \int DF \mu[\theta + F] e^{\mathcal{A}[\theta+F]+A_c[\theta+F]+A_\alpha[\theta+F]}.$$

Note that the effective action $W(\theta)$ is being evaluated using the background field method and this method can be executed directly in the continuum. The result, if one uses dimensional regularization, can be read off directly from the action since every term that yields a quadratic divergence is rendered zero by dimensional regularization. The case for the lattice is quite different, since all the quadratic divergences have been regulated by the lattice cutoff; hence one needs to compute all the quadratically divergent terms to verify that their sum indeed cancels exactly.

The effective action $W(\theta)$ is given by the following expansion in powers of θ (the pre-factor of $\mathcal{N}$ from SU($\mathcal{N}$) and the volume factor N^4 have been factored out for future convenience)

$$W(\theta) = -\delta m \cdot \frac{1}{4} \mathcal{N} N^4 \left(\sum_{\mu,a} \theta^a_{n\mu} \right)^2 + O(\theta^3).$$

δm is given by a perturbation expansion in g_0 and we will compute only the leading term, namely

$$\delta m = O(1) + O(g_0^2).$$

Note that δm is quadratically divergent and has contributions from the following sources:

- the measure term $\mu[\theta + F]$;
- the gauge field action $\mathcal{A}[\theta + F]$;
- the gauge-fixing term $A_\alpha[\theta + F]$;
- the Faddeev–Popov counter-term $A_c[\theta + F]$.

Hence

$$\delta m = m_\mu + m_a + m_c + m_\alpha + O(g_0^2).$$

All the variables $F^a_{n\mu}$ are constrained by the gauge-fixing term so that

$$F^a_{n\mu} = O(g_0).$$

The gauge-fixing term A_α chosen in Section 5 makes it independent of θ and hence we have

$$A_\alpha[B] = A_\alpha[\theta + F] = A_\alpha[F] \quad : \quad \text{independent of } \theta.$$

Hence

$$m_\alpha = 0.$$

The measure term for SU($\mathcal{N}$) yields, as shown in Figure 10.2(a),

$$\mu[B] = \exp\left\{-\frac{\mathcal{N}}{24}\sum_{n,\mu,a}(B^a_{n\mu})^2\right\} = \exp\left\{-\frac{\mathcal{N}}{12}N^4\left(\sum_{\mu,a}\theta^a_{n\mu}\right)^2\right\} + O(g_0^2)$$

and hence

$$m_\mu = \frac{1}{6}.$$

10.6.1. *Determination of m_c for $A_c[\theta + F]$*

From Eq. (10.5.8) we have

$$e^{A_c[B]} = \det(d_k\delta^{ab}\delta_{k,q} + M^{ad}(k,q)) \times \exp\left[-\frac{\mathcal{N}}{48}\sum_{n,\mu a}(B^a_{n\mu})^2\right] + O(g_0^3)$$

where, from Eq. (10.5.6)

$$M^{ad}(k,q) = \frac{1}{2}C^{abc}\sum_\mu(1 - e^{-ik_\mu})(1 + e^{-iq_\mu})B^b_{k-q,\mu}.$$

To extract the mass renormalization term m_c, note that since the matrix $M[B]$ depends linearly on B, we have

$$A_c[B] = A_c[\theta + F] = A_c[\theta] + O(F): \quad \text{independent of } F.$$

Hence, the leading $O(1)$ contribution of the counter-term A_c to mass renormalization *decouples* from the functional integral $\int DF$ and depends only on the ghost fields $c, c^\dagger$ and is shown in the Feynman diagrams given in Figure 10.2(b).

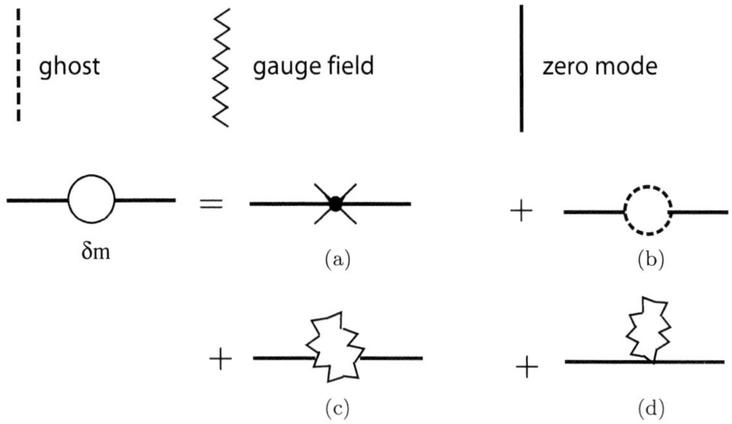

Fig. 10.2. Feynman diagrams for the computation of the lowest-order mass renormalization of the gauge field quantum. The external lines correspond to the zero mode θ^a. (a) is the contribution m_μ from the measure term μ. (b) shows the contribution m_c from the Faddeev–Popov ghost action A_c. (c) and (d) are the contributions that yield m_a from the nonlinear gauge field action $\mathcal{A}$.

The contribution of the Faddeev–Popov counter-term to mass renormalization can be evaluated from the determinant $\det(d_k\delta^{ab}\delta_{k,q}+M^{ad}(k,q))$ by setting $B_{n\mu}=\theta_\mu$. This yields

$$\begin{aligned} e^{A_c[\theta]} &= \exp\left[-\frac{\mathcal{N}N^4}{8}I\sum_{n,\mu a}(\theta^a_\mu)^2\right]\times\exp\left[-\frac{\mathcal{N}N^4}{48}\sum_{n,\mu a}(\theta^a_\mu)^2\right]+O(g_0^3)\\ &= \exp\left[-m_c\frac{\mathcal{N}N^4}{4}\sum_{n,\mu a}(\theta^a_\mu)^2\right] \end{aligned}$$

where

$$m_c=\frac{1}{12}+\frac{1}{2}I. \tag{10.6.1}$$

Define the lattice constants

$$I=\sum_{k\neq 0}\frac{|e^{ik_1}-e^{-ik_1}|^2}{d_k^2}, \tag{10.6.2}$$

$$K=\sum_{k\neq 0}\frac{1-\cos k_1\cos k_2}{d_k^2}\quad;\quad J=\sum_{k\neq 0}\frac{1}{d_k}. \tag{10.6.3}$$

Note that lattice constant I arises from Figure 10.2(b) and the two propagators yield the denominator d_k^2 whereas the ghost vertices yield the factor $|e^{ik_1}-e^{-ik_1}|^2$.

10.6.2. *Determination of m_a for $\mathcal{A}[\theta + F] + A_\alpha[F]$*

To determine the last piece of mass renormalization, namely m_a, we collect the results we have obtained so far. The effective action is given by

$$e^{W(\theta)} = \int DF\mu[\theta + F]e^{\mathcal{A}[\theta+F]+A_c[\theta+F]+A_\alpha[\theta+F]}$$
$$= \mu[\theta]e^{A_c[\theta]}\int DFe^{\mathcal{A}[\theta+F]+A_\alpha[F]}.$$

The gauge field action, to leading order in g_0, can be represented schematically as follows

$$e^{\mathcal{A}[\theta+F]} = e^{\mathcal{A}_0[F]}\{1 + (\theta F)^2 + (\theta)^2F^2 + O(g_0^2)\}$$

where $\mathcal{A}_0[F] = O(F^2)$ is the quadratic part of the gauge field action $\mathcal{A}[\theta + F]$.

Feynman perturbation expansion is based on the quadratic action $\mathcal{A}_0[F]+A_\alpha[F]$. Performing the functional integral $\int DF$ yields two terms, shown by the Feynman diagrams in Figures 10.2(c) and (d). The computation yields

$$\int DFe^{\mathcal{A}[\theta+F]+A_\alpha[F]} = \exp\left\{-\frac{1}{2}\left(12J - \frac{7}{4} - 6I - 12K\right)\frac{\mathcal{N}N^4}{4}\sum_{\mu a}(\theta_\mu^a)^2\right\}$$

and hence

$$m_a = \frac{1}{2}(12J - \frac{7}{4} - 6I - 12K).$$

Note that lattice constant K arises from Figure 10.2(c) and the two propagators yield the denominator d_k^2 whereas the non-Abelian vertices yield the factor $1 - \cos k_1 \cos k_2$. The lattice constant J arises from the tadpole diagram given in Figure 10.2(d).

10.6.3. *$\delta m = 0$ to leading order*

For mass renormalization

$$\delta m = m_\mu + m_c + m_a$$

we obtain, collecting all the results, the leading result order

$$\delta m = -\left(\frac{5}{8} + \frac{5}{2}I + 6K - 6J\right) + O(g_0^2). \tag{10.6.4}$$

The identity

$$1 = \sum_k{}' \frac{d_k^2}{d_k^2}$$

yields using the four dimensional hyper cubic symmetry of the summation, the following

$$1 = \sum_k{}' \frac{1}{d_k^2}[16(\cos^2(k_1) - 1) - 48(1 - \cos k_1 \cos k_2) + 16d_k]$$
$$= -4I - 48K + 16J$$

Hence, from above, we have

$$I = 4J - 12K - \frac{1}{4}$$

and Eq. (10.6.4) simplifies to

$$\delta m = 4(6K - J).$$

It can be shown that the four dimensional hyper-cubic lattice constants obey the identity

$$J = 6K + O(e^{-N^4}).$$

Hence we obtain that

$$\delta m = 0 \quad : \quad \text{zero mass renormalization.}$$

The absence of a quadratic mass divergence in the lattice theory is expected to follow from exact gauge invariance for the lattice: the gauge invariant action does not generate any terms that are not gauge invariant, and that include a mass term that breaks gauge invariance. It is worthwhile to note that the absence of the quadratic mass divergence is realized in a nontrivial manner by the remarkable cancellation of lattice constants.

Starting from the expression

$$E[c_I^a \Delta_\sigma B^a_{n-\hat{\sigma},\sigma} e^{\mathcal{A}+A_\alpha+A_c}]$$

where the ghost fields $c, c^\dagger$ have been defined in Eq. (10.5.9), one can derive the Slavnov–Taylor identities for the lattice gauge field using the BRST symmetry. For the propagator, the Slavnov–Taylor identity yields

$$(1 - e^{ik_\mu})(1 - e^{ik_\nu})D^{ab}_{k\mu\nu} = \frac{1}{\alpha}\delta^{ab} \tag{10.6.5}$$

$$\text{where} \quad D^{ab}_{k\mu\nu} = E[B^a_{-k\mu} B^b_{k\nu} e^{\mathcal{A}+A_\alpha+A_c}].$$

From Eq. (10.6.5) we have that $D_k \sim 1/k^2$ for $k \simeq 0$. Hence we conclude that to all orders in perturbation theory there is no mass renormalization for the gauge field. The one-loop computation shows that if there was a quadratic divergence, then perturbation theory would have consistently evaluated it.

The absence of a quadratic divergence does not mean that the gauge field cannot have a finite mass. It is expected that the pure gauge field does in fact have a non-zero renormalized mass that can be physically observed. The pure gauge field's mass is due to nonlinear and strong coupling effects and does not arise from the ultra-violet degrees of freedom; but instead is expected to arise from the infra-red degrees of freedom.

10.7. The lattice gauge field Hamiltonian

The Hamiltonian for QCD (quantum chromodynamics) has been widely studied using the lattice and continuum formulations [4].

The mathematical treatment of gauge fixing the Yang–Mills Hamiltonian goes back to Schwinger [5]. The paper by Christ and Lee [6] gives a clear and complete treatment of gauge fixing the continuum gauge-field Hamiltonian.

The lattice Hamiltonian [7] is regulated to all orders and could be used for calculations involving two loops or higher. If we want to analyze the lattice Hamiltonian using the weak coupling approximation, it is necessary to fix a gauge, for example, the Coulomb gauge. Gauge fixing the lattice Hamiltonian essentially involves only the lattice gauge field and the quarks enter only through the quark-color-charge operator. So we will essentially study the gauge field and introduce the quark fields only when necessary.

Gauge fixing the lattice Hamiltonian is very similar in spirit to gauge fixing the continuum Hamiltonian; this similarity can be clearly seen in the action formulation. For the Hamiltonian we will basically follow the treatment given by Christ and Lee [6]. There are, however, significant differences between the lattice and continuum Hamiltonians both for the kinetic operator and the potential term.

The lattice gauge field is defined using finite-group elements of SU($\mathcal{N}$) as the fundamental degrees of freedom whereas the continuum uses only the Lie algebra of SU($\mathcal{N}$). This difference will introduce a lot of extra complications. Given an appropriate generalized interpretation of the basic symbols, it will turn out, however, that the form of the gauge-fixed continuum and lattice Hamiltonians are very similar.

10.8. Lattice chromo-electric operator

Consider a d-dimensional Euclidean *spatial* lattice with spacing a; let $U_{ni}, i = 1, 2, \ldots, d$, be the SU($\mathcal{N}$) link degree of freedom from lattice site n to $n+\widehat{i}$ ($\widehat{i}$ is the unit lattice vector in the ith direction) and let ψ_n, $\bar{\psi}_n$ be the lattice quark field. The Hamiltonian for SU($\mathcal{N}$) lattice gauge field in the temporal axial gauge is given by

$$H = H_{YM}[U] + H_F[\bar{\psi}, \psi, U],$$

where

$$H_{YM} = -\frac{g^2}{2a}\sum_{n,i} \nabla^2(U_{ni}) - \frac{1}{ag^2}\sum_{n,ij} \mathrm{Tr}(U_{ni}U_{n+\widehat{i},j}U^\dagger_{n+\widehat{j},i}U^\dagger_{nj})$$

and H_F is the quark-gauge-field part. Note that ∇^2 is the SU($\mathcal{N}$) Laplace–Beltrami operator. The Hamiltonian acts only on gauge-invariant wave functionals Ψ. Gauge transformation is given by

$$U_{ni} \to U_{ni}(\Phi) \equiv \Phi_n U_{ni} \Phi^\dagger_{n+\widehat{i}} \tag{10.8.1}$$

and the wave functional Ψ is invariant under Eq. (10.8.1), that is,

$$\Psi[U] = \Psi[U(\Phi)]. \tag{10.8.2}$$

Choose canonical coordinates B^a_{ni} given by

$$U_{ni} = \exp(iB^a_{ni}X_a).$$

Then we have, suppressing the lattice and vector indices and summing on repeated non-Abelian indices,

$$E^L_a(U) = e^L_{ab}(U)\frac{\partial}{i\partial B^b} \equiv \frac{\delta}{i\delta_L B^a} \tag{10.8.3}$$

$$E^R_a(U) = e^R_{ab}(U)\frac{\partial}{i\partial B^b} \equiv \frac{\delta}{i\delta_R B^a}.$$

Note[3] that

$$e^L_{ab}(U) = e^R_{ba}(U).$$

$E^L_a(U)$ is *left-invariant* in the sense that

$$E^L_a(VU) = E^L_a(U) \quad ; \quad V \in \mathrm{SU}(\mathcal{N})$$

and $E^L_a(U)$ is *right-invariant* in the sense that

$$E^L_a(UV) = E^L_a(U) \quad ; \quad V \in \mathrm{SU}(\mathcal{N}).$$

The operators E^R_a and E^L_a are the lattice *chromo-electric* field operators, which are first-order Hermitian differential operators with the commutation equation

$$\begin{aligned}
&[E^L_a, E^L_b] = iC_{abc}E^L_c \\
&[E^R_a, E^R_b] = -iC_{abc}E^R_c \\
&E^R_a(U) = R_{ab}(U)E^L_b(U), \quad R_{ab}(U) = \mathrm{Tr}(X_aUX_bU^\dagger), \\
&[E^R_a, E^L_b] = 0,
\end{aligned}$$

where R_{ab} is the adjoint representation. Recall X_a are the generators and C_{abc} the structure constants of $\mathrm{SU}(\mathcal{N})$.

In canonical coordinates the gauge transformation is given by

$$\Phi_n = \exp(i\phi^a_n X_a)$$

and

$$e^{i\phi^a_n E^R_a(U)}U = e^{i\phi^a_n X_a}U = \Phi_n U \quad ; \quad e^{i\phi^a_n E^L_a(U)}U = Ue^{i\phi^a_n X_a} = U\Phi_n. \tag{10.8.4}$$

The gauge transformation, from Eq. (10.8.4), is hence given by

$$\exp\left(i\sum_{ni}\left\{\phi^a_n\left(E^R(U_{ni}) - E^L_a(U_{n-\hat{\imath},i})\right)\right\}\right)\Psi[U] = \Psi\Big[U(\Phi)\Big]. \tag{10.8.5}$$

[3]Explicit expressions for $e^{L(R)}_{ab}$ can be given but will not be required for our derivations.

Since, from Eq. (10.8.2), $\Psi[U] = \Psi[U(\Phi)]$ we have, from Eq. (10.8.5), Gauss's law given by

$$\left[\sum_i \left\{E_a^R(U_{ni}) - E_a^L(U_{n-\hat{i},i})\right\}\right] |\Psi\rangle = 0. \tag{10.8.6}$$

Gauss's law confirms our identification $E_a^L(U_{ni})$ as the chromo-electric operator of the gauge field corresponding to the link variable U_{ni}.

The quark field is introduced using the color charge operator. The operator $\rho_{na}(\bar{\psi}, \psi, U)$ is the lattice quark-color-charge operator and satisfies

$$[\rho_{na}, \rho_{mb}] = iC_{abc}\rho_{nc}\delta_{nm}\,.$$

Hence, in the presence of the *chromo-electric charge operator* ρ_{na}, Gauss's law yields

$$\begin{aligned} 0 &= \left[\sum_i [R_{ab}(U_{ni})E_b^L(U_{ni}) - E_a^L(U_{n-\hat{i},i})] - \rho_{na}\right] |\Psi\rangle \\ &\equiv \left[\sum_{m,i} D_{nmi}^{ab} E_b^L(U_{mi}) - \rho_{na}]\right] |\Psi\rangle, \end{aligned} \tag{10.8.7}$$

where D_{nmi}^{ab} is the lattice covariant *backward* derivative.

Let $|n, a\rangle$ be a ket vector of lattice site n and non-Abelian index a; then, from Eq. (10.8.7) we have the real matrix D_i given by

$$\begin{aligned} D_{nmi}^{ab} &= \langle n, a|D_i|m, b\rangle \\ &= R_{ab}(U_{ni})\delta_{nm} - \delta_{ab}\delta_{n-\hat{i}m}. \end{aligned} \tag{10.8.8}$$

We see from above that D_i performs a finite rotation R_{ab} on the ket vector and then displaces it in the backward direction.

We write the full Hamiltonian of the gauge field coupled to fermions as sum of the kinetic and potential energy, that is

$$H = K(U) + P[\bar{\psi}, \psi, U], \tag{10.8.9}$$

where

$$K = -\frac{g^2}{2a}\sum_{n,i} \nabla^2(U_{ni}). \tag{10.8.10}$$

It is known that[4]

$$-\nabla^2(U) = \sum_a E_a^L(U)E_a^L(U). \tag{10.8.11}$$

[4]Note that we also have

$$-\nabla^2(U) = \sum_a E_a^R(U)E_a^R(U).$$

10.9. Gauge-fixed chromo-electric operator

We can see from Gauss's law that all the U_{ni}'s are not required to describe the gauge-invariant wave functional Ψ. We gauge transform U_{ni} to a new set of variables V_{ni} which are constrained; the constrainted variables V_{ni} will decouple from Gauss's law.

Consider the *change of variables* from $\{U_{ni}\}$ to $\{\Phi, V_{ni}\}$, with $\{V_{ni}\}$ having one constraint for each n. That is,

$$\psi_n = \Phi_n \zeta_n, \quad \bar{\psi}_n = \bar{\zeta}_n \Phi_n^\dagger, \tag{10.9.1}$$

$$U_{ni} = \Phi_n V_{ni} \Phi_{n+\hat{i}}^\dagger. \tag{10.9.2}$$

The new gauge degrees of freedom $\{V_{ni}\}$ *constrained* by choosing the Coulomb gauge for the lattice gives

$$\mathcal{X}_n^a(V) \equiv \operatorname{Im} \sum_i \operatorname{Tr} X_a(V_{ni} - V_{n-\hat{i},i}) = 0. \tag{10.9.3}$$

In canonical coordinates we have

$$V_{ni} = \exp(iA_{ni}^a X_a), \quad \Phi_n = \exp(i\phi_n^a X_a).$$

For small variation $A^a + dA^a$, we have

$$\begin{aligned} V(A + dA) &= V(A)\left[1 + V^\dagger(A)\frac{\partial V(A)}{\partial A^a} dA^a\right] \\ &= V(A)[1 + iX_a f_{ab}^R(A) dA^b], \end{aligned} \tag{10.9.4}$$

where

$$f_{ab}^R = -i\operatorname{Tr}\left[V^\dagger \frac{\partial V}{\partial A^a} X_b\right] = f_{ba}^L.$$

Define

$$\delta_L A^a = f_{ab}^L(A) dA^b \quad ; \quad \delta_R A^a = f_{ab}^R(A) dA^b. \tag{10.9.5}$$

Then

$$\begin{aligned} V(A + dA) &= V(A)(1 + iX_a \delta_R A^a) \\ &= (1 + iX_a \delta_L A^a) V(A). \end{aligned} \tag{10.9.6}$$

It can be shown that

$$e_{a\alpha}^L f_{\alpha b}^L = \delta_{ab} \quad ; \quad e_{a\alpha}^R f_{\alpha b}^R = \delta_{ab} \tag{10.9.7}$$

and hence matrix $e_{a\alpha}$ is the inverse of $f_{a\alpha}$.

Under the change of variables from U_{ni} to V_{ni}, given in Eqs. (10.9.1) and (10.9.2), the potential energy $P(\bar{\psi}, \psi, U)$ in Eq. (10.8.9) can be expressed as a function of only V_{ni}. For the kinetic energy K we need the expression for E_a^L. Note, using the chain rule and formula (10.9.7),

$$\begin{aligned}\frac{\partial}{\partial B^b_{mj}} &= \sum_{n,i} \frac{\partial A^a_{ni}}{\partial B^b_{mj}} \frac{\partial}{\partial A^a_{ni}} + \sum_n \frac{\partial \phi^a_n}{\partial B^b_{mj}} \frac{\partial}{\partial \phi^a_n} \\ &= \sum_{n,i} f^R_{a\alpha}(A_{ni}) \frac{\partial A^\alpha_{ni}}{\partial B^b_{mj}} e^L_{\alpha\beta}(A_{ni}) \frac{\partial}{\partial A^\beta_{ni}} + \cdots . \end{aligned} \tag{10.9.8}$$

Therefore, from Eqs. (10.8.3), (10.9.5) and (10.9.8),

$$E_b^L(U_{mj}) = \frac{\delta}{i\delta_L B^b_{mj}} = \sum_{n,i} \frac{\delta_R A^a_{ni}}{\delta_L B^b_{mj}} \frac{\delta}{i\delta_L A^a_{ni}} + \sum_n \frac{\delta_R \phi^a_n}{\delta_L B^b_{mj}} \frac{\delta}{i\delta_L \phi^a_n}.$$

We now evaluate the coefficient functions of the above equation. The constraint given in Eq. (10.9.3) is valid under variations of A^a_{ni} to $A^a_{ni} + dA^a_{ni}$; hence

$$0 = \mathcal{X}^a_n(A) \tag{10.9.9}$$

$$= \mathcal{X}^a_n(A + dA). \tag{10.9.10}$$

Hence, from Eqs. (10.9.9) and (10.9.10),

$$\sum_{m,i} \Gamma^{ab}_{nmi}(A)\delta_R A^b_{mi} = 0, \tag{10.9.11}$$

where for constraint (10.9.2) we have

$$\begin{aligned}\Gamma^{ab}_{nmi} &= \langle n, a|\Gamma_i|m, b\rangle \\ &= \delta\mathcal{X}^a_n/\delta_L A^b_{mi} \\ &= \omega^{ab}_{ni}\delta_{nm} - \omega^{ab}_{n-\hat{\imath},i}\delta_{n-\hat{\imath},m},\end{aligned}$$

and from Eq. (10.9.3)

$$\omega^{ab}_{ni} = \mathrm{Tr}(X_a V_{ni} X_b + X_b V^\dagger_{ni} X_a).$$

The constraint (10.9.11) on δA^a_{ni} determines $\delta\phi/\delta B$. Consider from Eq. (10.9.2), the following variation:

$$V_{ni}(A + dA) = \Phi^\dagger_n(\phi + d\phi)U_{ni}(B + dB)\Phi_{n+\hat{\imath}}(\phi + d\phi),$$

and which yields, from Eq. (10.9.6)

$$\delta_R A^a_{ni} = \delta_R \phi^a_{n+\hat{\imath}} - R_{ab}(V^\dagger_{ni})\delta_R\phi^b_n + R_{ab}(\Phi^\dagger_{n+\hat{\imath}})\delta_R B^b_{ni} \tag{10.9.12}$$

$$\equiv \sum_m \mathscr{D}^{ab}_{nmi}\delta_R\phi^b_m + R_{ab}(\Phi^\dagger_{n+\hat{\imath}})\delta_R B^b_{ni}\,. \tag{10.9.13}$$

From Eqs. (10.9.12) and (10.9.13), we have the lattice covariant *forward* derivative operator $\mathscr{D}_i$ given by

$$\begin{aligned}\mathscr{D}^{ab}_{nmi} &= \langle n, a|\mathscr{D}_i|m, b\rangle \\ &= \delta_{ab}\delta_{n+\hat{\imath},m} - R_{ab}(V^\dagger_{ni})\delta_{nm}.\end{aligned}$$

From Eqs. (10.9.11) and (10.9.13), we have

$$\sum_{m,i,b} \langle n,a|\Gamma_i \mathscr{D}_i|m,b\rangle \delta_R \phi^b_m + \sum_{m,i,b} \langle n,a|\Gamma_i R^T_i|m,b\rangle \delta_R B^b_{mi} = 0\,, \tag{10.9.14}$$

where T stands for transpose and

$$\langle n,a|R_i|m,b\rangle = \delta_{nm} R_{ab}(\Phi_{n+\hat{i}})\,.$$

Hence, from Eq. (10.9.14) we have

$$\frac{\delta_R \phi^a_n}{\delta_L B^b_{mj}} = -\left\langle n,a \left| \frac{1}{\Gamma\cdot\mathscr{D}} \Gamma_j R^T_j \right| m,b \right\rangle\,, \tag{10.9.15}$$

where $(\Gamma\cdot\mathscr{D})^{-1}$ is the inverse of operator $\sum_i \Gamma_i \mathscr{D}$. We also have from Eqs. (10.9.13) and (10.9.15)

$$\frac{\delta_R A^a_{ni}}{\delta_L B^b_{mj}} = -\left\langle n,a \left| \left[\mathscr{D}_i \frac{1}{\Gamma\cdot\mathscr{D}} \Gamma_j R^T_j - R^T_j \delta_{ij} \right] \right| m,b \right\rangle\,. \tag{10.9.16}$$

Hence, from Eqs. (10.9.9), (10.9.15), and (10.9.16),

$$\begin{aligned} \frac{\delta}{\delta_L B^b_{mj}} &= \sum_{n,i} \left\langle n,a \left| \left[\delta_{ij} - \mathscr{D}_i \frac{1}{\Gamma\cdot\mathscr{D}} \Gamma_j \right] R^T_j \right| m,b \right\rangle \frac{\delta}{\delta_L A^a_{ni}} \\ &\quad - \sum_n \left\langle n,a \left| \frac{1}{\Gamma\cdot\mathscr{D}} \Gamma_j R^T_j \right| m,b \right\rangle \frac{\delta}{\delta_L \phi^a_n}\,. \end{aligned} \tag{10.9.17}$$

Equation (10.9.17) provides the solution for expressing the unconstrained chromo-electric operator $\delta/\delta_L B$ in terms of the new constrained operator $\delta/\delta_L A$ and the gauge transformation $\delta/\delta_L \phi$. In essence, this solves the problem of gauge fixing the lattice Hamiltonian. Note that from Eq. (10.9.16) we have the identity

$$\sum_{n,i} \langle l,c\,|\Gamma_i|\,n,a\rangle \frac{\delta_R A^a_{ni}}{\delta_L B^b_{mj}} = 0$$

as expected from Eq. (10.9.11). We also have from Eq. (10.9.11)

$$\sum_{m,i} \langle n,a\,|\Gamma_i|\,m,b\rangle \frac{\delta}{\delta_L A^b_{mi}} = 0\,. \tag{10.9.18}$$

Hence, from Eqs. (10.8.3) and (10.9.18)

$$\left[\frac{\delta}{\delta_L A^b_{ni}}, A^b_{mj} \right] = \left[\delta_{mm}\delta_{ij}\delta_{ac} - \left\langle n,a \left| \Gamma^T_i \frac{1}{\Gamma\cdot\Gamma^T} \Gamma_j \right| m,c \right\rangle \right] e^L_{cb}(A_{mj})\,.$$

10.10. Gauss's law

We check that constrained variables V_{ni} decouple from Gauss's law. Recall from Eqs. (10.8.8) and (10.9.17), we have

$$\sum_{m,j} \langle l,c\,|D_j|\,m,b\rangle \frac{\delta}{\delta_L B^b_{mj}} = \sum_{n,ij} \left\langle n,a \,\middle|\, \left[\delta_{ij} - \mathscr{D}_i \frac{1}{\Gamma\cdot\mathscr{D}}\Gamma_j\right] R^T_j D^T_j \,\middle|\, l,c \right\rangle \frac{\delta}{\delta_L A^a_{ni}} - \sum_{n,ij} \left\langle n,a \,\middle|\, \frac{1}{\Gamma\cdot\mathscr{D}}\Gamma_j R^T_j D^T_j \,\middle|\, l,c \right\rangle \frac{\delta}{\delta_L \phi^a_n}\,. \quad (10.10.1)$$

From the definitions of D_i and $\mathscr{D}_i$ given in Eqs. (10.8.8) and (10.9.14), respectively, we have the crucial operator identity

$$R^T_j D^T_j = \mathscr{D}_j R^T\,, \quad (10.10.2)$$

where

$$\langle n,a\,|R|\,m,b\rangle = \delta_{nm} R_{ab}(\Phi_n)\,.$$

Hence, from Eqs. (10.10.1) and (10.10.2) we see that the first term in (10.10.1) is zero and we have

$$\sum_{m,j} D^{cb}_{lmj} \frac{\delta}{\delta_L B^b_{mj}} = -R_{cb}(\Phi_l)\frac{\delta}{\delta_L \phi^b_l} = \frac{\delta}{\delta_R \phi^c_l}\,. \quad (10.10.3)$$

We see that V_{ni} has decoupled from Gauss's constraint, and we have from Eqs. (10.8.7) and (10.10.3)

$$\left[\frac{\delta}{i\delta_R \phi^a_n} + \rho_{na}\right]|\Psi\rangle = 0\,. \quad (10.10.4)$$

Solving Eq. (10.10.4), we have from Eq. (10.9.1)

$$\Psi(\bar\psi,\psi,U) = \exp\left[-i\sum_n \rho_{na}\phi^a_n\right]\Psi(\bar\zeta,\zeta,V)\,, \quad (10.10.5)$$

since, using Eq. (10.8.7)

$$\frac{\delta}{i\delta_R\phi^a}\exp(i\phi^\alpha\rho_\alpha) = \rho_a \exp(i\phi^\alpha\rho_\alpha)\,.$$

The change of variables from $\{U_{ni}\}$ to $\{V_{ni},\Phi_n\}$ has a Jacobian given by the Faddeev–Popov determinant, and can be shown to be equal to

$$\mathscr{J}^{-1}[V] = \prod_n \int d\Phi_n \prod_{n,a} \delta[\mathcal{X}^a_n(\Phi_n V_{ni} \Phi^\dagger_{n+\hat{i}})]\,. \quad (10.10.6)$$

For weak coupling, Faddeev–Popov Jacobian $\mathscr{J}[V]$ has been evaluated to be $O(A^2)$ in Section 5. Hence we have (suppressing the fermion variables) for some gauge-invariant operator G and gauge-invariant state $|\Psi\rangle$, from Eqs. (10.9.1) and (10.10.5),

$$\langle\Psi|G|\Psi\rangle = \prod_{n,i}\int dU_{ni}\Psi^*[U]G[U,\delta/\delta U]\Psi[U]$$

$$= \prod_{n,i}\int dV_{ni}\prod_{n,a}\delta[\mathcal{X}_n^a(V)]\left[\Psi^*[V]\mathscr{J}^{1/2}[V]\exp\left[i\sum_n\phi_n^a\rho_{na}\right]\right]$$

$$\times\left(\mathscr{J}^{1/2}[V]\widehat{G}[V,\delta/\delta V]\mathscr{J}^{-1/2}[V]\right)\left[\exp\left[-i\sum_n\phi_n^a\rho_{na}\right]\mathscr{J}^{1/2}[V]\Psi[V]\right].$$

Hence, the effective wave functional defined by absorbing the Jacobian is given by

$$\widetilde{\Psi}[V] = \mathscr{J}^{1/2}[V]\Psi[V]$$

and the effective operator is

$$\widetilde{G} = \mathscr{J}^{1/2}[V]\exp\left[i\sum_n\phi_n^a\rho_{na}\right]\times G\exp\left[-i\sum_n\phi_n^a\rho_{na}\right]\mathscr{J}^{-1/2}[V], \tag{10.10.7}$$

such that

$$\langle\Psi|G|\Psi\rangle = \langle\widetilde{\Psi}|\widetilde{G}|\widetilde{\Psi}\rangle. \tag{10.10.8}$$

10.11. Gauge-fixed lattice Hamiltonian

We need to evaluate the kinetic operator given from Eqs. (10.8.3), (10.8.10) and (10.8.11), and as (summing on all repeated indices)

$$K = \frac{\delta}{i\delta_L B_{ni}^a}\frac{\delta}{i\delta_L B_{ni}^a}. \tag{10.11.1}$$

Let us symbolically write the transformation (10.9.17) as

$$\frac{\delta}{\delta_L B_p} = L_{pq}\frac{\delta}{\delta_L C_q}. \tag{10.11.2}$$

Then from Eqs. (10.11.1) and (10.11.2)

$$K = L_{pq}\frac{\delta}{i\delta_L C_q}\left[L_{pq'}\frac{\delta}{i\delta_L C_{q'}}\right] \tag{10.11.3}$$

$$= \frac{1}{L}\frac{\delta}{i\delta_L C_q}\left[LL_{qp}^T L_{pq'}\frac{\delta}{i\delta_L C_{q'}}\right], \tag{10.11.4}$$

where

$$L = \det\|L_{ab}\|. \tag{10.11.5}$$

For the transformation given by Eq. (10.9.17) we have

$$L = \mathscr{J}[V]. \tag{10.11.6}$$

and the Jacobian $\mathscr{J}$ is given by Eq. (10.10.6). The choice of operator ordering given by Eq. (10.11.4) allows for further simplifications. Recall that from Eq. (10.9.18) that $\delta/\delta_L A^a_{ni}$ is 'transverse'; using this equation and Eq. (10.11.4), we have

$$K = \frac{1}{\mathscr{J}} \frac{\delta}{i\delta_L A^a_{ni}} \left[\mathscr{J} \frac{\delta}{i\delta_L A^a_{ni}} \right] + \frac{1}{\mathscr{J}} \left[\frac{\delta}{i\delta_L \phi^a_n} + \frac{\delta}{i\delta_L A^{a'}_{n'i}} \mathscr{D}^{a'a}_{n'ni} \right]$$
$$\times \mathscr{J} \left\langle n, a \left| \frac{1}{\Gamma \cdot \mathscr{D}} \Gamma_j \Gamma^T_j \frac{1}{\mathscr{D}^T \cdot \Gamma^T} \right| m, b \right\rangle \left[\frac{\delta}{i\delta_L \phi^b_m} + \mathscr{D}^{Tbb'}_{mm'k} \frac{\delta}{i\delta_L A^{b'}_{m'k}} \right]. \tag{10.11.7}$$

The effective Hamiltonian, using Eq. (10.10.7), is given by

$$\widetilde{H} = \mathscr{J}^{1/2} e^{i\phi^a_n \rho_{na}} H e^{-i\phi^a_n \rho_{na}} \mathscr{J}^{-1/2}.$$

Note that

$$e^{i\phi^a_n \rho_{na}} \frac{\delta}{i\delta_L \phi^b_m} e^{-i\phi^a_n \rho_{na}} = -\rho_{mb}.$$

We hence have the final expression for the gauge-fixed lattice Hamiltonian given by

$$\widetilde{H} = \frac{g^2}{2a} \left\{ \mathscr{J}^{-1/2} \frac{\delta}{i\delta_L A^a_{ni}} \left[\mathscr{J} \frac{\delta}{i\delta_L A^a_{ni}} \mathscr{J}^{-1/2} \right] \right.$$
$$+ \mathscr{J}^{-1/2} \left[\frac{\delta}{i\delta_L A^{a'}_{n'i}} \mathscr{D}^{a'a}_{n'ni} - \rho_{na} \right] \mathscr{J} \left\langle n, a \left| \frac{1}{\Gamma \cdot \mathscr{D}} \Gamma \cdot \Gamma^T \frac{1}{\mathscr{D}^T \cdot \Gamma^T} \right| m, b \right\rangle$$
$$\left. \left[\mathscr{D}^{Tbb'}_{mm'k} \frac{\delta}{i\delta_L A^{b'}_{m'k}} - \rho_{mb} \right] \mathscr{J}^{-1/2} \right\} + P(\bar{\zeta}, \zeta, V). \tag{10.11.8}$$

The quark-color charge ρ_{na} has the instantaneous non-local and non-Abelian lattice Coulomb potential given by $(\Gamma \cdot \mathscr{D})^{-1} \Gamma \cdot \Gamma^T (\mathscr{D}^T \cdot \Gamma^T)^{-1}$.

The wave functionals depend on only the constrained variables V_{ni}, i.e.,

$$\widetilde{\Psi} = \widetilde{\Psi}(\bar{\zeta}, \zeta, V).$$

Recall that we have from Eq. (10.9.19) the commutation equation

$$\left[\frac{\delta}{\delta_L A^a_{ni}}, A^b_{mj} \right] = \left[\delta_{nm} \delta_{ij} \delta_{ac} - \left\langle n, a \left| \Gamma^T_i \frac{1}{\Gamma \cdot \Gamma^T} \Gamma_j \right| m, c \right\rangle \right] e^L_{cb}(A_{mj}).$$

Equations (10.11.8)–(10.11.9) completely define the gauge-fixed Hamiltonian for the SU($\mathcal{N}$) lattice gauge field. The redundant gauge degrees of freedom $\{\Phi_n\}$ have completely decoupled from the system, as expected. The expression for $\widetilde{H}$ in Eq. (10.11.8) is exact, and is equally valid for strong and weak couplings. Comparing Eqs. (10.11.1) and (10.11.7), we see that the coordinates $\{U_{ni}\}$ are analogous to Cartesian coordinates for the gauge field whereas coordinates $\{V_{ni}\}$ are analogous to curvilinear coordinates.

10.12. Summary

The two underpinnings of the lattice gauge field — namely that it is defined directly using the finite group element of the gauge group and secondly that it has a lattice cutoff — introduce many new features that are absent in the continuum formulation. In the case of the Lagrangian, the one-loop mass renormalization produced a number of quadratically divergent terms that exactly canceled due to identities involving the lattice constants.

For the case of the Hamiltonian, the finite group element lead to the chromo-electric field operators being differential operators on the group manifold. Gauge fixing necessitates the removal of the extra degrees of freedom from the kinetic operator and requires properties of the lattice chromo-electric field operator that do not arise in the continuum formulation.

The lattice gauge field introduced by Wilson has stood the test of time — continuing to provide new results — and is one of his lasting legacies to the edifice of theoretical physics.

Acknowledgment

As a graduate student, it was a pleasure to have had interacted with H. R. Krishnamurthy, Michael Peskin, Bambi Hu, Miguel Levy, John Kogut, Steve Shenker, Don Lewis, Danny Rosenhouse and Gyan Bhanot.

References

[1] Kenneth G. Wilson and John Kogut. The renormalization group and the ϵ expansion. *Physics Reports*, 12 (2):75–199, 1974.
[2] Kenneth G. Wilson. Confinement of quarks. *Physical Review D*, 10(8):2445, 1974.
[3] Belal E. Baaquie. Gauge fixing and mass renormalization in the lattice gauge theory. *Physical Review D*, 16(8):2612, 1977.
[4] John Kogut and Leonard Susskind. Hamiltonian formulation of Wilson's lattice gauge theories. *Physical Review D*, 11(2):395, 1975.
[5] Julian Schwinger. Non-abelian gauge fields. Relativistic invariance. *Physical Review*, 127(1):324, 1962.
[6] N. H. Christ and T. D. Lee. Operator ordering and Feynman rules in gauge theories. *Physical Review D*, 22(4):939, 1980.
[7] Belal E. Baaquie. Gauge fixing the $SU(n)$ lattice-gauge-field hamiltonian. *Physical Review D*, 30(10):2774–2779, 1985.

Chapter 11

Ken Wilson: Solving the strong interactions*

Michael E. Peskin
SLAC Stanford University, Menlo Park CA 94025, USA
mpeskin@slac.stanford.edu

Ken Wilson's ideas on the renormalization group were shaped by his attempts to build a theory of the strong interactions based on the concepts of quantum field theory. I describe the development of his ideas by reviewing four of Wilson's most important papers.

Keywords: Strong interactions; renormalization group; scale invariance.

11.1. Introduction

Ken Wilson is best known for his contributions to statistical mechanics. His breakthroughs in this field, including the computation of critical exponents and the solution of the Kondo problem, have had wide influence. Wilson began his career, however, as an elementary particle physicist. His ambition was to "solve the strong interactions", that is, to find a predictive theory of the subnuclear strong interactions. The ideas that he developed profoundly influenced our understanding of that problem, just as they provided tools and insights for problems in statistical physics.

In this article, I will review the development of Wilson's ideas on the strong interactions through a review of four of his most important papers [1–4]. I recommend these papers to all students of theoretical physics. All four read like explorations of realms previously unknown. They give insight into the problems Wilson sought to address with his initial concepts of the renormalization group. And, both for particle physicists and for condensed matter physicists, they illustrate how issues in each domain gave insight into the other.

*This article was originally published in *J. Stat. Phys.* **157**, 651–665 (2014).

11.2. The fixed source problem

Quantum field theory (QFT) had some of its greatest successes in the late 1940's, with the development of quantum electrodynamics and the successful explanation of the electron magnetic moment and the Lamb shift. The resulting euphoria led to the idea that QFT could be used to build a theory of the strong interactions based on a Lagrangian for pion-nucleon interactions. The new technology of Feynman diagrams assisted calculation (as Feynman recalled memorably in his Nobel Prize lecture [5]). However, it did not produce a better understanding of the nuclear forces. Pion exchange did not lead in any clear way to the observed phenomenology of nucleon-nucleon scattering. It could not account for the nucleon and meson resonances that began to be discovered.

Most importantly, the theory had few concrete predictions. It was stymied by the fact that the strong interactions are strong, while the methods developed for quantum electrodynamics relied on weak-coupling perturbation theory. Strong coupling in QFT implies that states with an arbitrarily large number of interacting quanta play an essential role. Feynman diagrams, which introduce additional quanta one by one, cannot easily give insight into this strong-coupling limit.

A relatively simple problem that encapsulated the difficulties of QFT is the fixed source problem. This is the problem of a static or infinitely heavy nucleon with two states

$$|p\rangle \qquad |n\rangle \tag{11.1}$$

interacting at its location $\vec{x} = 0$ with a pion field

$$\pi^a(x) = \left(\pi^+(x), \pi^0(x), \pi^-(x)\right) . \tag{11.2}$$

Lee [8] showed that a truncated version of this model, with only the π^+ field, could be solved exactly. However, the full problem allows complex intermediate states with many virtual pions, shown in Fig. 11.1a as loops coupling to the nucleon. Further, each loop has an ultraviolet divergence. The diagram shown in Fig. 11.1b has the value

$$-ig^3 \int \frac{d^4k}{(2\pi)^4} \frac{1}{(k^0 + p^0)(k^0 + p^0 + q^0)(k^2 - m_\pi^2)} , \tag{11.3}$$

and is logarithmically divergent. So there is no clear way even to compute the first loop sensibly, much less to limit the number of loops relevant to the final answer.

By the mid-1950's, the search for the theory of the strong interactions had turned away from QFT to other methods, essentially phenomenological techniques such as dispersion relations and more fundamental proposals based on the analytic properties of scattering amplitudes. Geoffrey Chew proposed that there was a unique

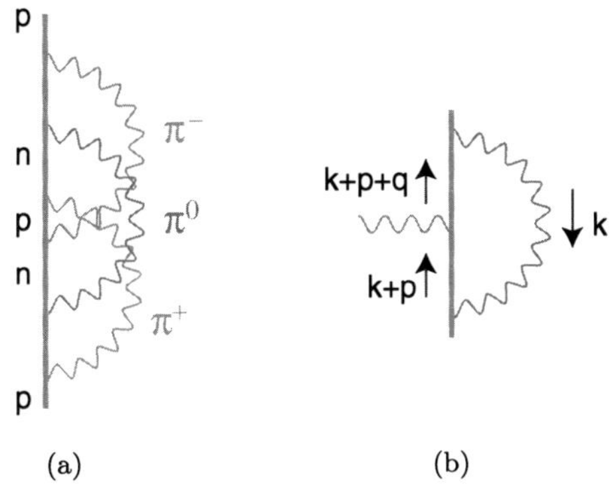

Fig. 11.1. Feynman diagrams for the fixed source problem: (a) a diagram illustrating the complexity of the problem; (b) a divergent one-loop diagram.

analytic S-matrix that could be discovered by deep analysis. As late as 1968, he stated:

> There exists at present no mechanical framework consistent with both quantum and relativistic principles. The chief candidate is local Lagrangian field theory, but countless theoretical studies have suggested insuperable pathologies in the concept of interaction between fields at a point of space-time. [6]

Even for those who tried to build up the theory of strong interactions from symmetry principles, the infinities of quantum field theory posed a barrier to taking this theory completely literally. For example, Murray Gell-Mann's paper that introduced the method of current algebra — one of the most important theoretical methods used in particle physics in the 1960's — includes the following statement:

> ... we use the method of abstraction from a Lagrangian field theory model. In other words, we construct a mathematical theory of strongly interacting particles, which may or may not have anything to do with reality, find suitable algebraic relations that hold in the model, postulate their validity, and then throw away the model. We may compare this process to a method sometimes employed in French cuisine: a piece of pheasant is cooked between two slices of veal, which are then discarded. [7]

Ken Wilson was Gell-Mann's student at Caltech from 1957 to 1961. He chose the fixed source problem described above as the topic of his thesis research. In his thesis, he threw at this problem the full arsenal of mathematical methods developed in the 1950's, with minimal success. This investigation proved Wilson's skills and promise, but it not make much headway toward the solution.

Many first-rank theorists find their thesis problem overreaching and frustratingly difficult. Usually, the solution is to pick another problem that yields more easily to their talents. This was not Wilson's style. He would continue to struggle with the fixed-source problem for many more years.

11.3. Momentum slicing

Wilson's first published work on the fixed-source problem did not appear until 1965. It is the first of the four papers that are the subject of this review:

> "Model Hamiltonians for Local Quantum Field Theory", Phys. Rev. **140**, B445 (1965) [1]

This paper had no immediate impact, because it enunciated a point of view that ran against the main current of theoretical particle physics in the 1960's. Wilson's boldly stated attitude is that there is no mysticism about QFT. The way to understand its issues to reduce problems in QFT to ordinary quantum-mechanical problems that can be solved by the standard methods of atomic physics. As Wilson writes in this paper:

> The Hamiltonian formulation of quantum mechanics has been essentially abandoned in the investigations of the interactions of π mesons, nucleons, and strange particles. This is a pity. The Hamiltonian approach has several advantages over the kind of approach (using dispersion relations) presently in use. One advantage is that all properties of a system are uniquely determined ... A second advantage is the existence of many approximation schemes ... A third advantage is that one can often analyze a Hamiltonian intuitively. [1]

For the neglect of QFT, Wilson blamed the problem of infinities. To rectify this, what was needed was a direct assault on that problem.

The infinities of QFT arise from the fact that the quantum excitations represented by the legs of Feynman diagrams may have any momenta and, in particular, momenta taking arbitrarily high values. Wilson's approach to the infinities was to lay out these momenta in an orderly set of regions that could be analyzed one by one. He called this concept "momentum slicing".

In [1], the full momentum space available to pions in the fixed source problem is replaced by the set of intervals

$$0 < |k| < m_\pi \ , \quad \frac{1}{2}\Lambda < |k| < \Lambda \ , \quad \frac{1}{2}\Lambda^2 < |k| < \Lambda^2 \ , \quad \ldots, \quad \frac{1}{2}\Lambda^n < |k| < \Lambda^n \ , \tag{11.4}$$

The slicing of momentum space is illustrated in Fig. 11.2. This severe reduction of the allowed phase space not only sharpens the problem of infinities but also reverses the standard viewpoint. At first sight, it is the low-momentum degrees of freedom that are the most important for the physics of pion-nucleon interactions encoded by the fixed-source problem. The appearance of high momenta is an intrusion that needs to be controlled. However, if the system defined by (11.4) is studied using the standard approximation schemes of quantum mechanics, the opposite is true. The most important terms in the Hamiltonian are those in the highest momentum interval. This part of the Hamiltonian must be diagonalized before lower-momentum intervals can be studied.

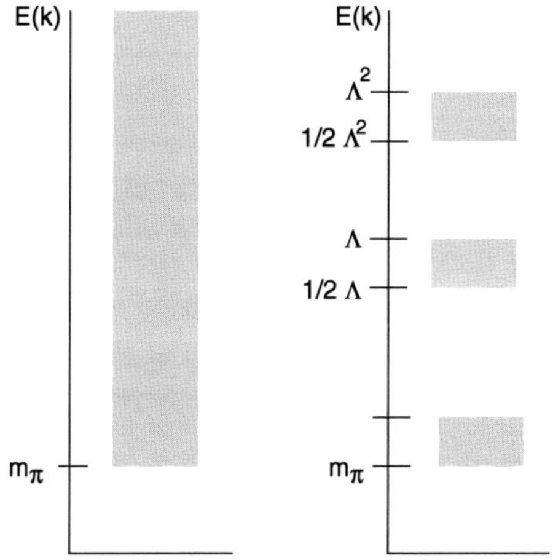

Fig. 11.2. The slicing of momentum space used by Wilson in [1]. The vertical axis gives the energies associated with the selected momentum states, on a logarithmic scale.

Standard ideas of quantum-mechanical perturbation theory dictate how this diagonalization should be done. The problem of pions in the highest momentum interval should be solved first, and the ground state of this system found. This ground state configuration of the pion modes with momenta of order Λ^n can then be used as the starting point for an analysis of the pion modes with momenta of order Λ^{n-1}. It is only at the end of this process that momenta of the order of m_π come into play.

The diagonalization of the Hamiltonian is then naturally structured as an iteration. The modes at $|k| \sim \Lambda^n$ primarily affect the modes at $|k| \sim \Lambda^{n-1}$ by modifying their coupling to the nucleon. This gives a recursion equation

$$g_{n-1} = f(g_n) \tag{11.5}$$

The effect of modes at large momentum scales on the pion modes at $|k| \sim m_\pi$ is then encapsulated in the evolution of the coupling constant from scale to scale that results from this evolution. At each stage of the evolution, the higher momenta are said to be "integrated out" and are removed from Hamiltonian. This is the essence of Wilson's concept of the renormalization group.

The restriction of momenta to the domains (11.4) is of course an extreme truncation of the original problem. To solve the fixed source problem quantitatively by momentum slicing, it is necessary to consider regions that are continguous, without gaps, and to integrate out each high-momentum region down to its boundary in an accurate way. In [9], Wilson addressed this question as a matter of principle, proving that the method led to a solution to the fixed source problem with all infinities eliminated.

However, Wilson did not stop there. It happens that the fixed source problem is related to a puzzle that appeared in the theory of magnetism, the Kondo problem [10]. This is the problem of a fixed magnetic impurity coupling to a free gas of electrons. For ferromagnetic coupling, the impurity behaves as a weakly coupled free spin interacting with the electrons. But for antiferromagnetic coupling, however weak in the underlying theory, the ground state contains a strong binding of the impurity to an electron that essentially quenches its magnetism. By introducing additional operators and corresponding couplings that transform under a multiparameter recursion equation, Wilson was able to integrate out shells in the electron momentum sufficiently accurately to identify the transition energy from weak to strong coupling. The full story of this calculation is outside the scope of this review, but it is described lucidly in [11]. Wilson's calculation of the strong-coupling scale in this model

$$\frac{T_K}{4\pi T_0} = 0.1032 \pm 0.0005 \tag{11.6}$$

was later verified by an exact solution of the Kondo problem, using Bethe's ansatz, by Andrei and Lowenstein [12, 13].

11.4. The operator product expansion

The idea of integration out has a more general consequence for the description of operators in QFT. Integrating out momentum modes with $|k| \sim \Lambda$ corresponds to the solution of the quantum theory for point separatons of order $|x - y| \sim \pi/\Lambda$. Local operators placed more closely together that this distance cannot be considered separately after integration out of this mometum shell. They must merge to become single operators located at some intermediate point, as shown in Fig. 11.3.

Formalizing this intuition gives Wilson's Operator Product Expansion. The idea of the operator product expansion appeared fully formed in the literature in the second paper of this review

"Non-Lagrangian Models of Current Algebra", Phys. Rev. **179**, 1499 (1969). [2]

The concept is expressed by the statement that all expectation values of a pair of QFT operators located at points x and y together with operators located at points z far from x, y can be computed by the replacement

$$\mathcal{O}_A(x)\mathcal{O}_B(y) = \sum_C \mathcal{C}_{ABC}(x-y)\mathcal{O}_C(y) \ , \tag{11.7}$$

where the sum over C runs over all operators in the QFT with appropriate quantum numbers, and $\mathcal{C}_{ABC}(x - y)$ is a c-number coefficient function.

The relation is especially simple in theories in which the QFT dynamics at distances smaller than $|x - y|$ is independent of any intrinsic length scale. Then, for

Fig. 11.3. Generation of a composite operator by integrating out high-momentum degrees of freedom.

scalar operators

$$\mathcal{C}_{ABC}(x-y) = \frac{c_{ABC}}{|x-y|^{d_A+d_B-d_C}}, \tag{11.8}$$

where c_{ABC} and d_A, d_B, d_C are numbers. For operators with spin, c_{ABC} is replaced by a number times an appropriate Lorentz structure.

The quantities d_i are called the dimensions of the operators and reflect the scaling of operator matrix elements with changes of distance scale. In Wilson's original conception, these dimensions were integers, as in free field theory. An anonymous referee (now known to be Arthur Wightman [14]) called Wilson's attention to the Thirring model, an exactly solved model in (1+1) dimensions, and prompted Wilson to make a serious study of this model. In the Thirring model, operator dimensions can be arbitrary real numbers. In [2], the idea that the values of operator dimensions could have a nontrivial influence on physical phenomena was presented for the first time.

The paper [2] applied these ideas to one of the most important problems being considered at that time, the nature of products of currents. Such products appear in the structure of the weak interactions, in analyses of the properties of pions and kaons using the methods introduced in [7], and in the analysis of deep inelastic electron scattering. In 1967, the results of the SLAC-MIT electron scattering experiment and their interpretation by Bjorken [15] and Feynman [16] pointed to models of the structure of the proton with free-field behavior at short distances for the proton constituents. Theoretical analysis of these experiments required the high-momentum asymptotic behavior of a pair of electromagnetic currents.

One of the properties of the Thirring model that was striking to Wilson in this context is that the current algebra of the model remains unchanged as the operator spectrum of the model is distorted by the effects of strong interactions. In our (3+1)-dimensional world, conserved currents would remain operators of dimension $d = 3$ while the dimensions of other operators would shift. Typically, results derived from Gell-Mann's current algebra for strong interaction matrix elements depend not only on the algebra but also on the behavior of short-distance limits. The operator product expansion gave a systematic way to analyze this issue.

The result of the paper that seems most striking from our modern point of view is the explanation Wilson gives for the $\Delta I = \frac{1}{2}$ rule, the fact that $\Delta I = \frac{1}{2}$ weak decays of K mesons and strange baryons, for example $K^0 \to \pi^+\pi^-$, go more than 100 times faster than $\Delta I = \frac{3}{2}$ decays such as $K^+ \to \pi^+\pi^0$. Wilson suggested that different operators, $\mathcal{O}_C$ in (11.7), in the product of W boson currents contributed to these two amplitudes, and that the difference in the amplitudes arises from the different factors

$$(m_K/m_W)^{6-d_C}. \tag{11.9}$$

in their operator product coefficients. This was the first suggestion of a qualitative effect on physics caused by dynamically-generated differences in operator dimensions. In 1974, Gaillard and Lee analyzed the operator product of W boson currents in the gauge theory of strong interactions Quantum Chromodynamics (QCD), to be described below, and showed that this effect does account for a large part, if not all, of the $\Delta I = \frac{1}{2}$ enhancement [17].

11.5. Scale invariance at short distances

Wilson discussed the results of the paper [2] in the context of a vision for the structure of a QFT description of strong interactions. The infinities of the theory would be tamed by the principle that the recursion described in Sect. 11.3 would have converged to a set of couplings that did not change with scale. Wilson describes a "skeleton theory" for strong interactions that is exactly scale invariant. To build a realistic model with nonzero hadron masses, this theory would be perturbated by mass terms or other operators with dimensionful coefficients. Wilson notes the idea of Kastrup [18] and Mack [19] that these terms might arise from the spontaneous breaking of a scale symmetry.

I have already pointed out that the idea that the strong interactions are described at short distances by a scale-invariant free field theory was very much in the air at this time. Both current algebra and the parton description of deep inelastic scattering rested on this foundation. However, it was recognized that this foundation could not be realized in any interacting QFT model.

Wilson's ideas cut through the haze surrounding this question. They suggested a framework for a model that could actually arise from QFT. However, the exact form of that model was still obscure. Wilson ends the paper [2] with the statement:

> It is hard to imagine that one could have a complete formula ... without having a complete solution of the hadron skeleton theory. The prospects for obtaining such a solution seem dim at present. [2]

11.6. The paper with three errors

If all one knows about the underlying scale-invariant theory of strong interactions is that it arises from renormalization group recursion, one can at least analyze the possibilities for the structure of such a theory by examining the renormalization group equations more closely. Wilson presented such an analysis in the third paper reviewed here,

> "Renormalization Group and Strong Interactions", Phys. Rev. **D3**, 1818 (1970) [3].

This paper concentrates on the case of one coupling constant evolving according to a continuous renormalization group equation. In modern notation, this equation is

$$\frac{dg(\mu)}{d\log\mu} = \beta(g(\mu)), \tag{11.10}$$

where μ is a momentum scale, $g(\mu)$ is the dimensionless coupling constant of the strong interaction theory, and $\beta(g)$ describes its evolution with scale. Wilson did not consider this equation familiar to his auidience. Rather, much of the paper is devoted to deriving the equation in perturbative QFT, beginning from the original treatment of Gell-Mann and Low [20]. Wilson gives as his primary reference for the renormalization group the textbook of Bogoliubov and Shirkov [21], though his explanation is a highly processed version of the one found there.[1]

Wilson's approach to (11.10) was to analyze it as the equation for a general dynamical system. Possible asymptotic behaviors for a dynamic system include a fixed point and a limit cycle. Wilson described fixed-point solutions to (11.10) with the example of a β function of the form shown in Fig. 11.4. At momenta M high enough that mass parameters could be ignored, the coupling constant g would take some value g_M on the horizontal axis. For values $g_M < g_1$, the value would then increase at larger scales, coming close asymptotically to the value g_1, which would be a fixed point of the renormalization group. If the value of g_M were in the range $g_1 < g_M < g_2$. the value of $g(\mu)$ would decrease to the fixed point g_1.

An alternative picture discussed by Wilson is one in which, at very high momentum scales, the weak and electromagnetic couplings become as large at the strong

[1]Wilson's students of the 1970's could be recognized by the fact that they carried the massive grey book [21] everywhere. I dropped mine in the snow on a bus trip to New York City, and it was never the same after that.

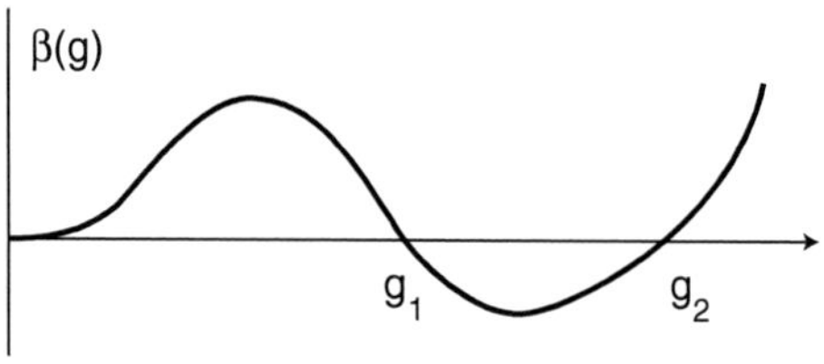

Fig. 11.4. An example of a renormalization group function $\beta(g)$.

interaction coupling. The high-momentum value of g is then determined by properties of this unified theory. At lower scales, where the strong interactions can be treated in isolation from the weak and electromagnetic interactions, g takes on the renormalization group evolution described by $\beta(g)$. In that case, the low-momentum behavior might be evolution to an infrared-asymptotic fixed point of the renormalization group, such as g_2. The observed strong interaction coupling would be modified slightly from g_2 by effects of the mass terms.

Wilson analyzes one more possibility. In principle, if there is more than one coupling constant, the high-energy asymptotic behavior of the renormalization group equation might be a limit cycle. In that case, the values of the coupling constants would perpetually oscillate, with a regular period in $\log \mu$. This would be directly observable as an oscillating behavior of the cross section for e^+e^- annihilation to hadrons.

This paper was eye-opening for many theorists at the time [22]. It introduced the idea of qualitative analysis of a QFT through visualization of its renormalization group flows, an idea that is now a standard method both in particle physics and in statistical mechanics.

Still, Wilson is said to have refered to this work as "the paper with three errors". In hindsight, the omissions are easy to find. Limit cycles of the renormalization group have never played a role in particle physics (though examples do exist for discrete renormalization group transformations [23]). The idea that the low energy value of the strong interaction coupling constant would be an infrared-stable fixed point of the renormalization group was also not reallized. This idea might still be relevant in the theory of the top quark mass [24,25]. Most importantly, though, Wilson assumed that the β function must be positive in the low-g region of Fig. 11.4, where it is computed in perturbation theory. He writes that, negative $\beta(g)$ "violates the Källén–Lehmann representation for the photon propagator" [3]. The last secret of the strong interactions was hidden here, as I will explain in a moment.

11.7. Statistical mechanics and quantum field theory

It was in this same period that Wilson became involved with problems of the theory of phase transitions. Wilson's interaction with David Mermin, Michael Fisher, Ben

Widom, and other statistical mechanics experts at Cornell will be covered by other contributions to this volume. The street ran both ways. The converse of the idea that statistical mechanics problems can be modelled by QFT is the idea that QFT can be given a foundation by constructions taken from statistical mechanics.

From our previous discussion, the ingredients are all in place. A lattice with spacing a in d dimensions can represent the space that results when quantum states at large momenta are integrated out down to $|k| \sim \pi/a$. Integration out potentially leads to a complicated Hamiltonian with many nonzero operator coefficients. However, most of these operators have high dimension and so do not affect physics at energy scales of the order of particle masses.

The cleanest connection of this type is between lattice statistical mechanics problems on a d-dimensional lattice and Euclidean QFT in d dimensions. It follows from the axioms of QFT [26] that operator expectation values on Lorentzian spacetime can be analytically continued to a Euclidean spacetime with

$$x^2 = (x^0)^2 + (x^1)^2 + (x^2)^2 + (x^3)^2. \tag{11.11}$$

Continuation to Euclidean space carries the time translation operator

$$U(t) = e^{-iHt} \to T(t_E) = e^{-L_E t_E}\ , \tag{11.12}$$

where L_E can be identified with the Lagrangian of the analytically continued problem. The object on the right is the transfer matrix of a statistical mechanics problem. Setting $t_E = a$ gives the evolution from one lattice spacing to the next. The complete partition function is

$$Z = \mathrm{tr}[T(a)^N] \tag{11.13}$$

for a lattice of length Na.

Wilson's student Ashok Suri worked out these connections in detail and explained them in his 1969 Ph.D. thesis [27]. That thesis became a basic reference document for the developments to follow.

11.8. Quantum chromodynamics

The missing piece in the story of the scale invariance of strong interactions popped out in the spring of 1973 with the announcement by Politzer, Gross, and Wilczek that non-Abelian Yang-Mills theory is asymptotically free [28, 29]. By this, I mean that the renormalization group equation in this theory has a negative β function at small values of g, causing the coupling constant to run to zero for large momenta. In (3+1)-dimensions, non-Abelian gauge theories are absolutely exceptional in allowing negative values of the β function [30, 31]. The indefinite metric spaces used in the quantization of gauge fields allow them to slip through a crack in the argument that the β function is always positive.

A theory with asymptotic freedom would be asymptotically scale-invariant, and, in fact, free, at short distances. However, in certain observables, one could still find large effects of operator dimensions, now scaling with powers of logarithms of momenta

$$\left(\log \frac{m_K^2}{\Lambda^2} \Big/ \log \frac{m_W^2}{\Lambda^2}\right)^{\gamma_A+\gamma_B-\gamma_C} \tag{11.14}$$

rather than (11.9). The combination of these two features in the same package made this an ideal solution to all of the problems of constructing a theory of strong interactions. Almost immediately, the idea of building interactions from gauge fields converged with other aspects of strong interaction phenomenology to pick out the Yang-Mills gauge group $SU(3)$, with quarks belonging to the fundamental **3** representation [32]. This theory was named Quantum Chromodynamics (QCD). Today, a wealth of evidence supports the claim that QCD is the fundamental description of the strong interactions.

Quantum chromodynamics included a picture of strongly interacting particles as bound states of more fundamental spin $\frac{1}{2}$ particles, quarks. The quark model explained the mass spectrum and quantum numbers of the baryons and mesons. It gave a basis for current algebra and the structure of hadronic weak interactions. But, together with these successes came a puzzle: The quark model required the electric charge assignments $+\frac{2}{3}$ for the u quark and $-\frac{1}{3}$ for the d and s quarks. Yet, no particles with fractional electric charge had been seen in nature. By 1973, extensive searches had been done, all with negative results. Searches for free fractionally-charged particles are reviewed in [33].

11.9. Lattice gauge theory

Wilson had not been studying gauge theories, or any weak-coupling proposal for the nature of the skeleton theory of strong interactions. However, now a question arose that he was uniquely positioned to answer: How do non-Abelian gauge theories behave when their coupling constants are taken to be strong? The answer to this question was given in the final paper for this review

"Confinement of Quarks", Phys. Rev. **D10**, 2445 (1974) [4]

That answer turned out to be profound. The following important results were summarized in the introduction to this paper:

> A new mechanism which keeps quarks bound will be proposed in this paper. The mechanism applies to gauge theories only.
> By analogy to the solid-state situation one can think of the transition from zero to nonzero photon mass as a change of phase.
> ... the strong-coupling expansion ... has the same general structure as the relativistic string model of hadrons ... [4]

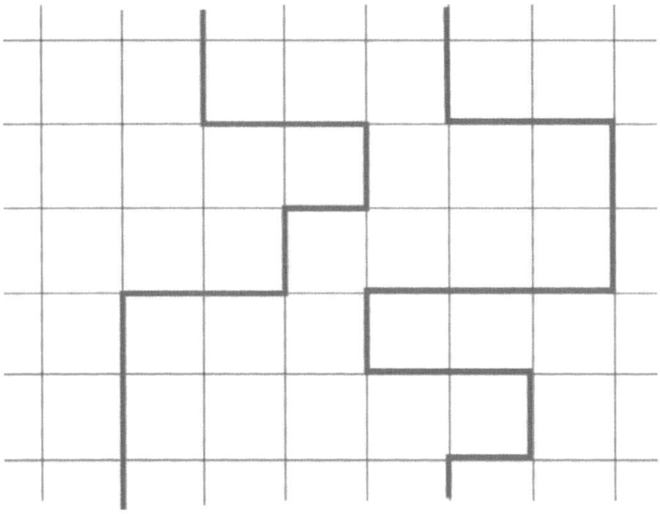

Fig. 11.5. Path of a heavy quark-antiquark pair in a lattice space-time. Euclidean time runs upward.

The correspondence between continuum QFT and lattice statistical mechanics was the crucial tool for this investigation. By now Wilson had thoroughly assimilated the idea of lattice-regulated QFT. He writes:

> The model discussed in this paper is a gauge theory set up on a four-dimensional Euclidean lattice. The inverse of the lattice spacing serves as an ultraviolet cut-off. The use of a Euclidean space ... instead of a Lorentz space is not a serious restriction.

Other members of our particle physics community took decades to get used to this idea. Today, gauge theory on a Euclidean lattice is a proven numerical tool for calculations in the low-momentum region of QCD [34].

To begin the description of a local gauge symmetry on a lattice, we might start from the description of a matter particle in a lattice QFT. A path of a heavy particle in Euclidean spacetime has the form shown in Fig. 11.5. A quantum particle travelling on paths of this type can be realized by a scalar field transforming under a global symmetry, which we associate with the particle number,

$$\phi_n \to e^{i\alpha}\phi_n \ , \tag{11.15}$$

where α is a global parameter. A discretized derivative of the field can be defined as

$$\Delta_\mu \phi_n = \frac{1}{a}(\phi_{n+a\hat{\mu}} - \phi_n) \tag{11.16}$$

A lattice QFT with the Lagrangian

$$L_E = \sum_n a^3 \left[\frac{1}{2T}|\Delta_\mu \phi_n|^2 + \frac{1}{2}m^2|\phi_n|^2\right] \tag{11.17}$$

is invariant under the symmetry. Expanding the partition function in the parameter $1/T$ is a controlled expansion in the lattice-regulated theory. This expansion is analogous to the high-temperature expansion of lattice statistical models. It corresponds to a sum of graphs with paths of the form shown in the figure. A similar treatment can be given for fermions on the lattice. To implement this, Wilson described the integral over fermionic variables invented by Berezin [35], still, at that time, quite unfamiliar as a QFT tool.

The generalization to a local gauge symmetry raises new issues. The symmetry transformation is now

$$\phi_n \to e^{i\alpha_n}\phi_n, \tag{11.18}$$

where α_n is independent at each lattice site. The derivative (11.16) now no longer has a linear transformation law under the symmetry group.

The remedy for this problem is to generalize the lattice derivative with an additional element

$$\Delta_\mu \phi_n = \frac{1}{a}(\phi_{n+a\hat{\mu}} - U_{n+a\hat{\mu},n}\phi_n), \tag{11.19}$$

where U_{n_1,n_2}, with (n_1, n_2) neighboring lattice sites, has the transformation

$$U_{n_1,n_2} \to e^{i\alpha_{n_1}} U_{n_1,n_2} e^{-i\alpha_{n_2}} \tag{11.20}$$

Minimally, U_{n_1,n_2} can be taken to be a unitary matrix representing an element of the gauge group, with the identification $U_{n_2,n_1} = U^\dagger_{n_1,n_2}$. The statistical sum over U_{n_1,n_2} is an integral over the gauge group for each link of the lattice.

In terms of continuum variables, a quantity with the same transformation law as U_{n_1,n_2} is the exponential of the line integral of the vector potential

$$U_{n_1,n_2} \equiv \exp\left[ig\int_{n_2}^{n_1} dx^\mu A_\mu\right] \tag{11.21}$$

Making this identification and expanding for the case in which $A_\mu(x)$ varies slowly over a lattice spacing, we find

$$\Delta_\mu \phi = (\partial_\mu - igA_\mu)\phi \ , \tag{11.22}$$

the standard gauge-covariant derivative. A gauge-invariant quantity built purely from the U_{n_1,n_2} is

$$\mathrm{tr}[V_{n,\mu,\nu}] = \mathrm{tr}\left[U_{n,n+\hat{\nu}}U_{n+\hat{\nu},n+\hat{\mu}+\hat{\nu}}U_{n+\hat{\mu}+\hat{\nu},n+\hat{\mu}}U_{n+\hat{\mu},n}\right] \tag{11.23}$$

In the continuum, $V_{n,\mu,\nu}$ can be identified as the exponential of a line integral around the elementary square,

$$V_{n,\mu\nu} \equiv \exp\left[ig\oint dx^\mu A_\mu\right] = \exp\left[ig\int d^2 s^{\mu\nu} F_{\mu\nu}\right] \tag{11.24}$$

I have written these formulae for the case of an Abelian gauge group, but they go through with only minor modifications for a non-Abelian gauge group. The lattice

QFT action

$$L_E = \sum_{n,\mu,\nu} \frac{1}{g^2} \mathrm{tr}\left[V_{n,\mu,\nu} + V^{\dagger}_{n,\mu,\nu}\right] \tag{11.25}$$

then gives the usual gauge field action $\mathcal{L} = -\frac{1}{4}(F_{\mu\nu})^2$. Similar lattice QFTs were constructed by Wegner [36], for the case of discrete gauge symmetry, and by Polyakov [37].

The lattice Lagrangian (11.25) has the amazing property of possessing a straightforward expansion in powers of $1/g^2$. One simply needs to expand the exponential of (11.25) in a power series. Each factor of $\mathrm{tr}[V_n]$ brings down four factors of $U_{n,n+\hat{\mu}}$, which are then integrated over the gauge group. This $1/g^2$ expansion realized one of Wilson's longstanding goals, to directly compute the structure of a QFT in the limit of strong coupling.

To understand the implications of this expansion, go back to Fig. 11.5. In the gauge theory, each link of the path of the heavy quark and antiquark acquires a factor $U_{n,n+\hat{\mu}}$. Each term $\mathrm{tr}[V_n]$ is a product of four factors of $U_{n,n+\hat{\mu}}$ arranged around an elementary square of the lattice. We might imagine this as a tile placed on the square. These factors must come together to prevent the integrals over the gauge group from giving a zero result. Indeed,

$$\int dg U(g)_{ij} = 0 \quad \text{while} \quad \int dg U(g)_{ij} U^{\dagger}(g)_{k\ell} = c\delta_{i\ell}\delta_{kj} \; . \tag{11.26}$$

A term in the strong-coupling perturbation theory is nonzero only if each link has a matching number of factors of U and $U^{\dagger}$. The nonzero terms correspond to tilings of the region between the quark and antiquark paths, as shown in Fig. 11.6.

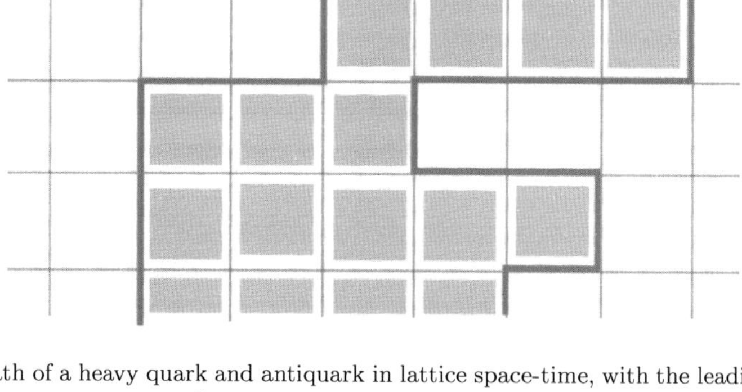

Fig. 11.6. Path of a heavy quark and antiquark in lattice space-time, with the leading terms from the gauge theory at strong coupling.

We find that the amplitude for propagation of a heavy quark-antiquark pair is nonzero only if the entire region between the paths of these particles is spanned by tiles. If the quark and antiquark are far apart, there must be gauge excitation covering every interval between them. This led Wilson to the conclusion that the strong-coupling gauge theory gives a potential between quarks and antiquarks of the form

$$V(|\vec{x}_q - \vec{x}_{\bar{q}}|) \sim k|\vec{x}_q - \vec{x}_{\bar{q}}| \tag{11.27}$$

rising linearly with distance.

For general values of the coupling, the qualitative behavior of this potential would depend on the long-range order in the gauge degrees of freedom. At strong coupling

$$\langle U_{n,\hat{\mu}} \rangle = 0 \ , \tag{11.28}$$

and we find the quark-confining potential (11.27). At weak coupling, at least in electrodynamics, there is an expansion about

$$U_{n,\hat{\mu}} \approx 1 \tag{11.29}$$

that leads to the usual Coulomb potential. In non-Abelian gauge theories, it is plausible that the renormalization-group flow makes the coupling strong enough, at sufficiently large distances, that the strong coupling region is reached and the theory is confining. This statement is not yet proven rigorously, but it is supported by a wealth of numerical data [34].

11.10. Afterword

I am ending this review almost exactly at the point where my career intersects the story. I came to Cornell as a graduate student in the fall of 1973. The first particle theory seminar that I attended was Wilson's seminar at Cornell on the lattice gauge theory results that I have just described. A year later, the discovery of the J/ψ resonance led to striking evidence for all aspects of the picture I have described here: the quark model, asymptotic freedom at short distances, a linear confining potential at large distances, even the direct experimental verification of the number 3 in the $SU(3)$ gauge group [38]. The rout of QCD was on.

By the time I finished graduate school, QCD was already an established theory. I became interested in what I felt would be the next problem ripe for solution, the physics of the spontaneous breaking of symmetry responsible for the properties of the subnuclear weak interactions. Thirty-five years later, that problem is still an open one, although the recent discovery of the Higgs boson at the Large Hadron Collider [39, 40] surely provides an important piece of the puzzle.

Every student of physics seeks to emulate his or her thesis advisor. Having Ken Wilson as an advisor sets a very high standard.

The experience of working with Ken instilled some values that continue to guide my approach to physics. First, even when approaching the fundamental equations of nature, a physicist should dismiss mysticism. The universe is essentially mechanical. There is a Hamiltonian; solve it. For better or worse, I find the current Standard Model of particle physics too lacking in explanatory power, and too lacking in specific mechanisms that might explain the fact and the consequences of its spontaneous symmetry breaking.

Second, a physicist should have a vision, and pursue it to the end. We are not all as blessed with genius as Ken Wilson, but the mountains are there nevertheless. Ken always climbed straight up.

Acknowledgments

I am grateful to Belal Baaquie, Edouard Brézin and John Cardy for encouragement to write this review of Wilson's work, and to Jeevak Parpia and Peter Lepage for the organization of a symposium devoted to Ken Wilson at which this material was first presented. As to the content of the paper, I owe much to all of my teachers and colleagues at Cornell in the 1970's. This work was supported by the U.S. Department of Energy under contract DE-AC02-76SF00515.

References

1. Wilson, K.G.: Model Hamiltonians for local quantum field theory. Phys. Rev. **140**, B445 (1965)
2. Wilson, K.G.: Non-Lagrangian models of current algebra. Phys. Rev. **179**, 1499 (1969)
3. Wilson, K.G.: The renormalization group and strong interactions. Phys. Rev. D **3**, 1818 (1971)
4. Wilson, K.G.: Confinement of quarks. Phys. Rev. D **10**, 2445 (1974)
5. Feynman, R.P.: The development of the space-time view of quantum electrodynamics. Science **153**, 699 (1966)
6. Chew, G.F.: "Bootstrap": a scientific idea? Science **161**, 762 (1968)
7. Gell-Mann, M.: The symmetry group of vector and axial vector currents. Physics **1**, 63 (1964)
8. Lee, T.D.: Some special examples in renormalizable field theory. Phys. Rev. **95**, 1329 (1954)
9. Wilson, K.G.: A model of coupling constant renormalization. Phys. Rev. D **2**, 1438 (1970)
10. Kondo, J.: Resistance minimum in dilute magnetic alloys. Prog.Theor. Phys. **32**, 37 (1964)
11. Wilson, K.G.: The renormalization group: critical phenomena and the Kondo problem. Rev. Mod. Phys. **67**, 773 (1975)
12. Andrei, N.: Diagonalization of the Kondo Hamiltonian. Phys. Rev. Lett. **45**, 379 (1980)
13. Andrei, N., Lowenstein, J.H.: Scales and scaling in the Kondo model. Phys. Rev. Lett. **46**, 356 (1981)
14. Schweber, S.: personal communication

15. Bjorken, J.D.: Asymptotic sum rules at infinite momentum. Phys. Rev. **179**, 1547 (1969)
16. Feynman, R.P.: Photon–Hadron Interactions. W. A. Benjamin, Reading (1972)
17. Gaillard, M.K., Lee, B.W.: Delta I = 1/2 rule for nonleptonic decays in asymptotically free field theories. Phys. Rev. Lett. **33**, 108 (1974)
18. Kastrup, H.A.: Infrared approach to large angle scattering at very high energies. Phys. Rev. **147**, 1130 (1966)
19. Mack, G.: Partially conserved dilatation current. Nucl. Phys. B **5**, 499 (1968)
20. Gell-Mann, M., Low, F.E.: Quantum electrodynamics at small distances. Phys. Rev. **95**, 1300 (1954)
21. Bogoliubov, N.N., Shirkov, D.V.: Introduction to the Theory of Quantized Fields. Interscience, New York (1959)
22. Gross, D.J.: Asymptotic freedom and the emergence of QCD. In: Hoddeson, L., Brown, L., Riordan, M., Dresden, M. (eds.) The Rise of the Standard Model. Cambridge University Press, Cambridge (1997) [hep-ph/9210207]
23. Glazek, S.D., Wilson, K.G.: Limit cycles in quantum theories. Phys. Rev. Lett. **89**, 230401 (2002) [Erratum-ibid. 92, 139901 (2004)] [hep-th/0203088]
24. Pendleton, B., Ross, G.G.: Mass and mixing angle predictions from infrared fixed points. Phys. Lett. B **98**, 291 (1981)
25. Hill, C.T.: Quark and Lepton masses from renormalization group fixed points. Phys. Rev. D **24**, 691 (1981)
26. Streater, R.F., Wightman, A.S.: PCT, Spin and Statistics, and All That. W. A. Benjamin, Reading (1964)
27. Suri, A.: Cornell University, Ph.D. Thesis (1969)
28. Politzer, H.D.: Reliable perturbative results for strong interactions? Phys. Rev. Lett. **30**, 1346 (1973)
29. Gross, D.J., Wilczek, F.: Ultraviolet behavior of nonabelian gauge theories. Phys. Rev. Lett. **30**, 1343 (1973)
30. Zee, A.: Study of the renormalization group for small coupling constants. Phys. Rev. D **7**, 3630 (1973)
31. Coleman, S.R., Gross, D.J.: Price of asymptotic freedom. Phys. Rev. Lett. **31**, 851 (1973)
32. Fritzsch, H., Gell-Mann, M., Leutwyler, H.: Advantages of the color octet gluon picture. Phys. Lett. B **47**, 365 (1973)
33. Jones, L.W.: A review of Quark search experiments. Rev. Mod. Phys. **49**, 717 (1977)
34. Kronfeld, A.S.: Twenty-first century lattice gauge theory: results from the QCD Lagrangian. Annu. Rev. Nucl. Part. Sci. **62**, 265 (2012) http://arxiv.org/abs/1203.1204 arXiv:1203.1204 [hep-lat]
35. Berezin, F.A.: The Method of Second Quantization. Academic Press, New York (1966)
36. Wegner, F.J.: Duality in generalized Ising models and phase transitions without local order parameters. J. Math. Phys. **12**, 2259 (1971)
37. Polyakov, A.M.: Unpublished
38. Appelquist, T., Barnett, R.M., Lane, K.D.: Charm and beyond. Annu. Rev. Nucl. Part. Sci. **28**, 387 (1978)
39. Aad, G., et al. [ATLAS Collaboration]: Observation of a new particle in the search for the standard model Higgs boson with the ATLAS detector at the LHC. Phys. Lett. B **716**, 1 (2012) http://arxiv.org/abs/1207.7214 arXiv:1207.7214 [hep-ex]
40. Chatrchyan, S., et al. [CMS Collaboration]: Observation of a new boson at a mass of 125 GeV with the CMS experiment at the LHC. Phys. Lett. B **716**, 30 (2012) http://arxiv.org/abs/1207.7235 arXiv:1207.7235 [hep-ex]

Chapter 12

Renormalization group approach to quantum Hamiltonian dynamics*

Stanisław D. Głazek

Institute of Theoretical Physics, Faculty of Physics, University of Warsaw, ul. Pasteura 5, 02-093 Warsaw, Poland

Ken Wilson developed powerful renormalization group procedures for constructing effective theories and solving a broad class of difficult physical problems. His insights allowed him to later advance the Hamiltonian approach to quantum dynamics of particles and fields in the Minkowski space–time, motivated by QCD. The latter advances are described in this article, concluding with a remark on Ken's related interest in difficult systemic issues of society.

Keywords: Quantum field theory; Hamiltonian; renormalization group; QCD; front form; asymptotic freedom; limit cycle.

PACS numbers: 11.10.−z, 11.10.Gh, 12.38.−t, 01.75.+m

12.1. Introduction

Ken Wilson's early research in elementary particle and condensed matter physics produced a number of brilliant results that are famous for their depth of concept, breadth of scope, and usefulness in application. His renormalization group approach to dynamical problems provided physicists with a new way of understanding their theories and equipped them with powerful computational tools. A number of articles in this volume describe and explain these results. Ken's later involvement in studies of relativistic Hamiltonian dynamics in quantum field theory is less known. This article provides information about Ken's work that became available to me through a collaboration with him in physics that began in 1990.

I met Ken in the fall of 1990 at The Ohio State University (OSU) Department of Physics in Columbus, Ohio. I was invited there to give a seminar on the inclusion of

*This article was originally published in *Int. J. Mod. Phys. A* **30**, 1530023 (2015).

quark and gluon vacuum condensates in Hamiltonian description of heavy quarkonia, in the form of dynamics that Dirac called the front form (FF).[1] Subsequently, I visited OSU for the 1991 academic year as a postdoctoral fellow. Ken's towering reputation as a physicist motivated many young researchers to learn from him and I was among them. My goal was to understand "renormalization," which I did not understand sufficiently to renormalize the FF Hamiltonian of QCD.

At that time, the OSU nuclear and particle theory groups were interested in advancing the theory of quarks and gluons by starting from first principles and producing a suitable Hamiltonian. The Hamiltonian was needed for defining quantum states of hadrons as its eigenstates and for solving the corresponding eigenvalue problems. Solutions for the hadronic eigenvalues and associated quark and gluon wave functions are needed for calculation of the scattering amplitudes that involve hadrons in the initial or final states. The Hamiltonian could also be used to calculate the evolution of physical systems including strong interactions.

There were two issues to deal with. One was understanding hadrons, such as nucleons or π-mesons, in a relativistic way that would allow one to describe them in all kinds of physical situations. The other was to derive nuclear theory from particle theory; that is, a derivation of nuclear forces, mainly how protons and neutrons interact, from QCD. However, the initial focus was on the issue of developing a Hamiltonian approach to QCD without distinguishing any one of these two purposes.

The OSU group answered the question of which form of Hamiltonian dynamics one should use by choosing the FF.[1] This form requires some introduction, needed for appreciation of the depth of Ken's contribution.

12.2. The infinite momentum frame and front form of dynamics

The theory of strong interactions is intimately related to the picture that an inertial observer might imagine when thinking about a hadron that, in her frame of reference, moves very fast. When the hadron momentum increases to infinity, the observer's frame of reference is called the infinite momentum frame (IMF). One imagines that, according to the special theory of relativity, the internal clock of a fast moving hadron goes very slowly and the hadron constituents, called partons because they are parts of a hadron, including the assumed carriers of interactions, appear practically frozen until the moment when the hadron collides with some target. The associated picture of the hadron is called the parton model.[2,3] In the parton model, an observer imagines a moving hadron as a collection of constituents each of which carries some fraction of the hadron total momentum P. The hadron structure is hence described in the IMF by the probability distributions $f_i(x)$ of finding a constituent of type i carrying the fraction x of the momentum P that tends to infinity. A potential concern about this way of thinking is that it assumes a classical special relativity description of hadronic structure that could be false due to its actually quantum nature. However, measured cross sections could be interpreted using the parton model quite well. Ken's early interest in theory of the parton picture is documented in the form of a qualitative model that he invented using elements of the scaling picture he previously developed for other purposes.[4]

Once QCD was discovered in 1973 to be asymptotically free,[5,6] physicists could use its Feynman rules and calculate scattering amplitudes for partons identified as quarks or gluons.[7] It was also understood that[8] "The rate of approach to asymptopia must be determined by non-perturbative methods. This problem, as well as the understanding of the low-energy and on-mass-shell behavior of the theory, requires the development of new theoretical techniques." Ken contributed to the development of required techniques in a spectacular way in 1974 by formulating gauge theories on the lattice.[9] However, in 1990 the strong interaction theory still posed serious problems, which were the subject of discussion at OSU in the context of the FF of Hamiltonian dynamics.

The FF of dynamics can be seen as relevant to the parton model by looking at the expression for energy of a freely moving parton of mass m and momentum components $p^z = xP$ and $p^\perp = k^\perp$, where $\perp$ denotes the two directions transverse to the z-axis. Field-theoretic consequences of this behavior of energy as a function of momentum attracted the attention of many researchers.[10–25] In the limit of $P \to \infty$, one obtains

$$E = \sqrt{m^2 + \vec{p}^{\,2}} \to |x|P + \frac{m^2 + k^{\perp\,2}}{2|x|P} + \cdots , \tag{12.1}$$

where the three dots indicate terms vanishing faster than $1/P$ when $P \to \infty$. By adding and subtracting the energy and z-component of momentum, for $x > 0$, and neglecting terms that vanish faster than $1/P$, one obtains

$$p^+ = E + p^z = 2xP\,, \quad p^- = E - p^z = \frac{m^2 + k^{\perp\,2}}{2xP} . \tag{12.2}$$

The leading term in energy is conserved by interactions if the z-component of momentum is conserved. The off-energy-shell development of quantum dynamics appears in the variable p^-.

To see a connection of these results with the FF of Hamiltonian dynamics, one can write the mass-shell condition for the parton momentum, $p^2 = m^2$, in the form $p^+p^- - k^{\perp\,2} = m^2$ and evaluate

$$p^- = \frac{m^2 + k^{\perp\,2}}{p^+} . \tag{12.3}$$

Comparison with Eqs. (12.2) shows that the leading term of p^- in the IMF corresponds to the exact p^- in the on-mass-shell four-momentum of a free particle. The wave function of the free particle reads

$$\psi_p(t, \vec{x}) = Ne^{-ipx} , \tag{12.4}$$

where N is a normalization constant,

$$px = Et - \vec{p}\vec{x} = \frac{1}{2}p^- x^+ + \frac{1}{2}p^+ x^- - k^\perp x^\perp \,, \tag{12.5}$$

and $x^\pm = t \pm z$, with $x^\perp = (x, y)$. Thus, when one traces the evolution of a physical system using the variable x^+ instead of time t, then p^- plays the role of energy while p^+ and $p^\perp$ play the roles of kinematical momentum variables (e.g., see Ref. 12).

In this way of describing physical systems, the space–time hyperplanes of fixed values of x^+ play the same role as that played by hyperplanes of fixed values of t in describing evolution of physical systems using time. A hyperplane of fixed x^+ is swept by the front of a plane wave of light moving along the z-axis toward decreasing values of z. This is why Dirac[1] called this form of dynamics the FF, distinguishing it from the standard form, which he called the instant form (IF). In the FF, the role of Hamiltonian that generates the evolution is played by $P^- = H - P^z$, which is the operator that naturally describes the dynamics of partons, or quarks and gluons in the IMF. Namely, the quantum dynamics off energy shell with changes of energy measured according to Eq. (12.1) in the IMF, appears similar to the FF quantum dynamics with changes of "energy" measured using p^- in Eq. (12.3). The similar appearance does not imply identity because the IMF limit of $P \to \infty$ applies only for finite transverse momenta and positive fractions x.

Since the FF of quantum Hamiltonian dynamics uses hyperplanes associated with fronts of plane waves of light for collecting quantum data, the FF Hamiltonian formulation of QCD was often called the light-front QCD, or LF QCD. Many authors had already contributed to its development[26–38] by the time the OSU group identified its core interest in defining and solving LF QCD.

12.3. The FF connection between partons and the constituent quark model

The constituent quark model (CQM) picture of hadrons, which says that mesons are quark–antiquark pairs and baryons are three quarks, serves as a basis for the classification of hadrons in particle data tables.[39] The CQM predates QCD, and the IMF picture of hadrons made of QCD quarks and gluons is different from the CQM. In QCD, hadrons are made of partons that are understood as quanta of the quark and gluon fields. Hadrons are made not only of two or three quarks but also of innumerably many gluons and quark–antiquark pairs.

In view of the complexity of QCD, the simplicity of the CQM is stunning and thus its explanation and improvement using the LF QCD became of interest in the discussions at OSU. Ken's own words about it can be heard in a video recording of the discussion he led in 1994 at the workshop "Theory of Hadrons and Light-Front QCD" held near Zakopane, Poland.[40] The FF of dynamics was a preferred tool for seeking a connection between the IMF picture and the CQM picture of hadrons because of the boost symmetry of the FF.

Consider two free particles of four-momenta p_1 and p_2, each of mass m, whose total momentum is P. In distinction from previous IF consideration in the IMF, let P be ordinary, finite momentum and let the particles be described using the FF. One can parametrize the particle momenta by assigning them fractions x and $1-x$ of P and introducing a relative transverse momentum $k^\perp$, by writing

$$p_1^+ = xP^+ \,, \qquad p_1^\perp = xP^\perp + k^\perp \,, \tag{12.6}$$

$$p_2^+ = (1-x)P^+ \,, \quad p_2^\perp = (1-x)P^\perp - k^\perp \,. \tag{12.7}$$

In this notation,

$$P^- = p_1^- + p_2^- = (\mathcal{M}^2 + P^{\perp 2})/P^+ \,, \tag{12.8}$$

$$\mathcal{M}^2 = \frac{m^2 + k^{\perp 2}}{x(1-x)} \,. \tag{12.9}$$

The invariant mass of the pair, $\mathcal{M}^2 = P^2$, does not depend on the total momentum components P^+ and $P^\perp$; it is only a function of the fractions x and the relative transverse momentum of partons, $k^\perp$. This simplification does not occur in the IF of dynamics because the IF energy involves a square root, absent in the FF energy. In the FF Hamiltonian of QCD, interactions are added to the invariant mass squared, i.e., to $\mathcal{M}^2$. The total kinematical momentum of a physical system is thus separated from the internal dynamics of its parts that determine the mass of the system.

The two-particle example illustrates the general feature of the FF of Hamiltonian dynamics: it allows one to describe formally physical systems using relative motion variables, such as x and $k^\perp$ above, independently of how fast the system as a whole moves. One can say that it happens to be so because the front hyperplane described by the equation $x^+ = 0$ is preserved by a seven-parameter subgroup of the Poincaré group, in contrast to the instant hyperplane $t = 0$ that is preserved only under action of a six-parameter subgroup (consisting of three translations and three rotations). The extra symmetry of the front that makes the difference is the boost along the z-axis.

So, if one imagines that a meson in its rest frame is built of a quark and an antiquark, the two constituents could serve equally well as two partons of the meson in the IMF, with the same wave function of variables x and $k^\perp$. These variables are invariant under the boost. Consequently, the kinematical (i.e., holding irrespective of interactions) boost symmetry of the FF of Hamiltonian dynamics became the reason for hoping that the LF QCD Hamiltonian eigenvalue problem may provide simultaneous understanding of hadrons both in the rest frame and in the IMF. Could LF QCD provide such a connection as a feature of a rigorous theory? It was clear that the LF QCD, as a dynamical theory, should provide additional insights in comparison to the early attempts at simultaneous understanding of current and constituent quarks,[41] but how could it actually be done?

Ken expressed the shortcomings of theory as he saw it at the beginning of the 1990s by saying[40] that there existed formulations of QCD that one could call exact, such as the complete set of Dyson–Schwinger equations or a continuum limit of lattice QCD, and in every such formulation simplifying assumptions are made to produce specific models, such as truncating the set of equations or using a strong coupling limit. While exact formulations should match each other, there was no necessity for matching of different models. If the LF QCD Hamiltonian were to provide an exact formulation of QCD, it should be mathematically clear how it could describe confinement and chiral symmetry breaking in a complete quantum theory, including a solution to the problem of constructing its ground state, called vacuum.

12.4. Severity of renormalization problems

In defining LF QCD, one starts from its canonical Hamiltonian, H, in gauge $A^+ = 0$. The free energy terms count p^- of quark and gluon field quanta. The quark–gluon interaction terms cause quarks to emit and absorb gluons, or gluons turn into or emerge from quark pairs. The gluon interaction terms are of two kinds. In one of them, involving three gluons, a gluon can turn into two or vice versa. In the other, involving four gluons, one gluon changes to three, two to two, or three to one. These features are also known in the IF dynamics. Additional terms that result from solving constraints cause quarks and gluons to interact at a distance through their color charge densities.

One distinct feature of the FF Hamiltonian is that it does not contain terms that only create or only annihilate quarks or gluons. Such terms are absent if one demands that all field quanta carry only positive amounts of momentum p^+. Demanding $p^+ > \epsilon^+$ is natural when one thinks about free quanta of mass m, for which

$$p^+ = \sqrt{m^2 + \vec{p}^{\,2}} + p^z > 0\,. \tag{12.10}$$

Arbitrarily small m is sufficient to secure this result for as long as the three-momentum is limited by some arbitrarily large ultraviolet cutoff parameter, say Δ, so that $\epsilon^+ = m^2/(2\Delta)$. Note that the condition $p^+ > \epsilon^+$ for all quanta implies also that the number of quanta in any state of total P^+ is limited from above by the ratio P^+/ϵ^+. This means that more quanta can be found in states with greater P^+. Similarly, the allowed number of quanta increases to infinity in the Hamiltonian eigenstates of finite P^+ when $\epsilon^+ \to 0$.

If the Hamiltonian H is invariant with respect to translations on the front, and canonical quantization secures this by defining the Hamiltonian as an integral of its density over the entire front $x^+ = 0$, then all terms in H conserve p^+. This means that the sum of + components of momenta of quanta created by an interaction term must be the same as the sum for quanta annihilated by the same term.

If a term were only to create quanta, the sum of momenta p^+ of the created quanta would have to be zero, and this cannot be the case if each of the created quanta has a positive p^+. The same reasoning explains the absence of terms that could only annihilate quanta. Thus, the regulated canonical LF Hamiltonian of QCD contains only normal-ordered terms in which at least one annihilation operator and at least one creation operator appear. Hence, the Hamiltonian annihilates the bare vacuum state $|0\rangle$, defined as the state annihilated by all annihilation operators. The state $|0\rangle$ is an eigenstate of H with eigenvalue zero. It could be considered a candidate for a ground state, unless interactions were strong enough to produce eigenstates with negative eigenvalues, corresponding to tachyons.

A simple vacuum eigenstate does not exist in the IF of dynamics, where canonical Hamiltonians of quantum field theory create bare quanta out of the bare vacuum $|0\rangle$. The only limitation on the IF three-momenta of created quanta is that they sum to zero. So, the created quanta could spontaneously emerge from the empty space with unlimited IF energy as long as they flew out in different directions. In the IMF, where xP could be finite or vanish despite that $P \to \infty$, one could have quanta with very small positive and negative values of x. In the context of considerations using the IMF, the nature of the vacuum problem was partly captured in the phrase "you can't outrun the vacuum,"[25] construed in analogy to "you can't outrun the long arm of the law" in a song made famous by Kenny Rogers.

While the regulated FF Hamiltonian annihilated $|0\rangle$, as an operator it was sensitive to the regularization that enforced the condition $p^+ > \epsilon^+$. Since regularization required "renormalization" in order to eliminate dependence of predictions for observables on the cutoff parameters, questions concerning the vacuum in LF QCD were related to questions concerning "renormalization."

Severity of the vacuum problem in the IF of quantum field dynamics and its significance for perturbative rules of renormalization have been stressed by Dirac in 1965.[42] Dirac was concerned that in perturbative calculus "...all the terms except the first few involve divergent integrals. Among the Feynman diagrams there are some that involve v–v (vacuum to vacuum) transitions.... They are the worst terms, as they prevent one from getting a solution to represent even the vacuum state. The usual procedure of physicists in quantum field theory is to neglect such terms altogether, and to excuse themselves by saying that such terms could not correspond to anything observable. This neglect involves a drastic departure from logic. It changes the whole character of the theory, from logical deduction to a mere setting up of working rules." The severity of the problem led Dirac to say that "The Schrödinger wave functions involve infinities, associated with v–v Feynman diagrams, which destroy all hope of logic." When the interaction Hamiltonian of a local theory acts on a normalizable state, it creates a state that is not normalizable. Dirac postulated that relativistic quantum mechanics of fields, such as QED, required an understanding of the associated mathematics that was still missing. According to Dirac, one needed to revise the quantum mechanics concept of states as vectors in a Hilbert space because the vacuum and all other states of field quanta built from the vacuum in a relativistic theory were not elements of a Hilbert space.

The central issue for Dirac was that when one imposed cutoffs on three-momenta of field quanta in a Hamiltonian in order to make the states in its domain normalizable, one eliminated an infinite range of momentum variables that were required for respecting principles of special relativity in the theory. Namely, a Lorentz transformation can change momentum by an amount greater than any cutoff. To avoid violation of the Lorentz symmetry one had to send the cutoffs to infinity. But this meant that no matter how small a coupling constant was, the interaction in no time caused infinite effects and one could not apply principles of perturbation theory in a systematic analysis.

Dirac's article concerning conceptual difficulties with the diverging QED vacuum problem[42] appeared in August 1965 and Wilson's first article on renormalization appeared in October 1965.[43] We cannot ask the authors anymore about the extent to which they were in contact with each other, but one can infer from the available texts that they had similar attitudes toward "renormalization." I did not know about the existence of Ken's article before Ken mentioned it himself as the place where I should start my learning about his renormalization group procedure. Ken said he understood renormalization by studying models such as the one described in his 1965 article.[43] Before the OSU group started working on LF QCD, which means for about twenty five years between 1965 and 1990, Ken's 1965 article was cited only about ten times and five of these citations were by Ken himself. Ken's sequel to the 1965 article appeared in 1970,[44] for the purpose of checking if his understanding of renormalization in the 1965 paper could be confirmed quantitatively with definite numerical precision.

In 1990, the OSU group had discussed divergences in the Tamm–Dancoff type[45–48] of approach to FF Hamiltonian dynamics.[49] The Tamm–Dancoff approach was based on studying dynamics of a limited number of quanta included in the space of states. Robert Perry, Avaroth Harindranath, and Ken Wilson thought about a method in which "In each order of Tamm–Dancoff theory old counterterms migrate to new sectors of Fock space, and the mass-renormalization condition determines a new counterterm in the lowest sector." Increasing the order of calculation meant that the number of quanta included in the space of states increased. The remaining conceptual difficulty was that a sequence of calculations with an increasing upper limit on the number of quanta, such as 2, 3, etc. had to be dealt with, while the momentum cutoff parameters were being made extreme. If kinematical momenta were extreme, states of extreme free energies were involved, and among such states there also were many with large numbers of quanta that *a priori* should be included, and they were not. For strong couplings, the role of states with large numbers of quanta in the Tamm–Dancoff type of approach would have to be understood numerically. The increasing number of quanta led to increasing demands on computer memory and speed and one had to hope that some regularities might appear when the upper limit on the number of quanta increased. One was thus back to the problem of determining the role of interactions among the states with large numbers of quanta, each of which carried small p^+, especially the question of how such states could produce effects associated with vacuum in the IF of dynamics.[50]

Another distinct feature of the FF of Hamiltonian dynamics that motivated interest was the interaction at a distance. In the IF, the most important example is provided by the Coulomb potential term in the Hamiltonian of QED. It originates from solving the constraint equation of Gauss's law.[51] The problem with the Coulomb potential is that it acts instantaneously at a distance and appears to violate the rule that interactions cannot propagate faster than light. To recover this rule in QED, one has to combine effects of the instantaneous Coulomb interaction with effects of exchange of transverse photons. Electric current conservation and time ordering of events matter in obtaining covariant scattering amplitudes. In the IF bound-state dynamics, which cannot be reduced to on-energy-shell scattering amplitudes, the situation is more complicated than in perturbatively evaluated scattering amplitudes.[52]

In the FF of dynamics in $A^+ = 0$ gauge, the constraint analogous to the Gauss's law is considerably different; instead of inversion of the three-dimensional Laplace operator, it only involves inversion of derivatives with respect to one space coordinate, x^-. As a result, one obtains a potential between two charges located at points x and y on the front that is linear in $|x^- - y^-|$ and acts instantaneously in x^+, if and only if $x^\perp = y^\perp$. This action-at-a-distance is singular but does not by itself violate the rule that interactions cannot propagate faster than light. Namely, two points on the front can be connected by a light signal when their transverse positions are the same no matter how much their positions differ in their coordinate x^-. In addition to the action-at-a-distance between quark color charge densities, one also obtains the action-at-a-distance between gluon color charge densities, and between quark and gluon densities. Thus, the canonical FF Hamiltonian of QCD appears to contain a seed for confining interactions.

On the one hand, the linear potential that acts at a distance in agreement with rules of special relativity and appears to be a seed of confinement is a very attractive feature of the FF Hamiltonians of gauge theories. On the other hand, it is a singular interaction, local in transverse direction and non-local and increasing in the direction of x^-. In Abelian theory, such as LF QED, the linear term is canceled for massless photons. In non-Abelian theory, such as LF QCD, the cancellation may be incomplete because gluons interact with themselves (photons do not) and the coupling constant in the bound-state dynamics is large. Therefore, it makes sense to seek a precise construction of the FF Hamiltonian of QCD and find out what comes out. However, in order to understand and overcome canonical singularities, one needs a powerful renormalization strategy for Hamiltonians.

An approximate rule states that the total magnitude of difficulty in constructing a theory cannot be changed. If some element of a theory is simplified then some other must get more complicated. The shift of difficulties with vacuum and confinement to the renormalization difficulty meant that the renormalization of FF Hamiltonians was not to be easy. The FF Hamiltonian formulation treated x^+ or p^- differently than the remaining three front coordinates. Among the latter, one had to deal with kinematical momentum variables of two kinds, p^+ and $p^\perp$, or position variables of two kinds, x^- and $x^\perp$. The IF Hamiltonian formulation of a theory also involved a different treatment of time coordinate, x^0, or energy, p^0, in comparison to how it treated spatial variables, kinematical momenta $\vec{p}$ or space coordinates $\vec{x}$, but it took advantage of simple rotational symmetry in space. Symmetries to exploit among the FF variables p^+ and $p^\perp$, or x^- and $x^\perp$ are much less familiar through the present day.

When Ken constructed his ground-breaking models of renormalization in the IF of dynamics,[43,44] he introduced shells of momentum. Quanta in a shell number n could have momentum $\vec{p}$ of any direction and $|\vec{p}| \sim \Lambda^n$, with a large number Λ, in some units. In the FF dynamics, such shells were not available. One had to understand the difference between p^+ and p^z. For example, $p^+ > 0$ while p^z could have any sign. Correspondingly, the FF phase-space analysis would have to be different from the IF phase-space analysis. Ken developed FF power counting using wavelets, which later became a key part of a paper on LF QCD.[53] His engagement in an approximate wavelet analysis had deep roots in his long-term interest in understanding strong interactions. Already in his IF model of renormalization,[43] Ken observed that "There is scant hope that we will ever understand the Hamiltonians of relativistic theories if we await an exact solution of them." Further, in conclusion of his 1970 paper,[44] he expressed his hope that "...the renormalization theory of the model can be generalized to relativistic field theory...."

The FF bound-state eigenvalue problems were necessarily relativistic due to the use of front foliation of space–time. In momentum variables, the eigenvalue equations involved three-dimensional integrals of a different type than in the IF. Use of the non-relativistic approximation, known to work in QED for atoms, was of limited utility when the divergences originated from extreme values of $p^\perp$, corresponding to the ultraviolet region in the IF, or extremely small values of the ratio p^+/P^+, where p^+ was a constituent momentum and P^+ was the hadron momentum. This ratio corresponded to the Bjorken variable x in deep inelastic lepton–proton scattering.[54] One was naturally led to thinking that the issues of parton density saturation at small x and small p^+ singularities were closely related and at the same time one was enticed by the possibility of understanding the relationship using renormalization concepts to eliminate the FF cutoff dependence.

Since the simplified models that Ken used to understand renormalization in the IF of Hamiltonian dynamics[43,44] were translated to the FF,[55] it became clear that the momentum variables and shells he used did not apply to the absolute momenta but to the relative-motion variables. In the context of Ken's models, these were momenta of mesons with respect to the center of mass of a set of mesons and a very heavy source particle, in the limit where the mass of the source particle was much greater than the cutoff mass scale. One could seek a systematic method for studying LF QCD taking advantage of the FF translation of Ken's models. The idea of using relative energy changes instead of absolute values of energy resurfaced later in the similarity renormalization group procedure.[56] However, the most impressive aspect of the methodology that Ken used in his models was — from my point of view — the idea of simplification for the purpose of understanding. This idea turned out central to discussions that Ken and I had over the years, concerning different issues in various contexts. Ken set a great example to follow by putting stress on *simplify* in research on difficult problems.

12.5. Simplify to understand

Questions of vacuum, chiral symmetry breaking and confinement of color were identified by the OSU group as inseparably connected with the issue of how to renormalize Hamiltonians of relativistic quantum field theory. But there were also pressing phenomenology issues that one could see as intimately related to the FF of Hamiltonian dynamics.

It had been discovered[57] that the structure functions of polarized protons corresponded to the parton distributions which appeared to indicate that the proton spin is not carried just by three quarks, two u and one d, as suggested by the CQM classification of hadrons that is used in the particle data tables.[39] It was also found that the proton structure functions extracted from lepton–nucleus scattering cross-sections involved significant corrections dependent on the nucleus mass number (for a review, see e.g. Ref. 58), as if the mechanism of binding of quarks in individual nucleons was visibly related to the mechanism of binding nucleons in nuclei.

I had some experience with extracting neutron properties from proton and deuteron data and I hoped to contribute to the ongoing discussion of new findings. At some point it occurred to me that I should talk with Ken about the proton spin. I went to his office as usual and asked him a question concerning how one should deal with the proton spin data using theory. Ken looked at me and said, with his typical smile, "I do not want to talk about it." I was surprised, since so many particle and nuclear physicists did talk about it. I insisted that he explain his reluctance. Ken responded by saying that he did not want to talk about it because he did not want to think about it. I stood there not knowing what to make of his response and Ken noticed that I did not understand why he was reluctant to pursue the matter. His real explanation was illuminating. He said "I would have to think about many other things first."

The issue of proton spin emerges from interpretation of data using various theoretical concepts that are not yet fully understood as elements of a mathematically well-defined theory, certainly not to the degree that Dirac or Wilson would demand. One would have to resolve difficult conceptual problems, at least qualitatively, in order to imagine the kind of computational scheme that might be used to systematically interpret proton and nuclear data. Without such a systematic approach, the data could make conflicting impressions on researchers, who would miss both a physical picture and a precise mathematical method needed to quickly narrow the questions and extract definite answers from data. What Ken was saying sounded to me like a warning that in order to understand the complex spin observables one would have to have thought through a number of problems that Ken knew had no solution yet and were still far away from being understood from his point of view. He would prefer to focus on the impeding problems he saw rather than sail on the high seas of paradoxes one would have to deal with if these problems remained unsolved, or not placed clearly somewhere on the conceptual ladder of assumptions that are made while interpreting data. There was nothing wrong with data puzzling physicists, but Ken wanted first to solve the problems that were preventing us from using the theory.

Ken thought that we should stop thinking in terms of many details of quantum field theory and instead consider the scales of energy that are involved. I think this starting point reflects the rule Ken was proud of using: if one tackles a difficult problem, one ought to first simplify it as much as one can, which means as far as possible without losing the crux. Once the problem is brought to its bare-bone structure, there is a chance for understanding what it really is about and how to go about solving it. Only then one starts adding the bells and whistles that increase the complexity as required but no longer prevent one from doing approximate computations that one knows how to do, for comparing the theory with experiment.

The problem of binding of quarks and gluons in nucleons and nuclei is an example of a difficult problem. Hence, a lot of pruning is required in the case of lepton–nucleon and lepton–nucleus high-energy scattering. Ken knew this problem as not accepting any piecemeal approaches. Moreover, research on core problems typically turns out to produce unexpected results of unpredictable value. Sometimes the resulting value is high. This is why the risks, associated with embarking on a long and difficult research path, are worth taking.

12.6. Simple model for renormalization of Hamiltonians

In order to find a method for renormalizing Hamiltonians as singular as the canonical FF Hamiltonian of QCD, we needed to make a suitable model. Here I use the model Ken and I have found most instructive. I first use the model to illustrate the renormalization group strategy that Ken invented in 1965[43] and later famously developed[59] before coming to OSU. The model also shows why the 1965 strategy fails in the case of strongly bound states and how one can change the renormalization group procedure in order to deal with the problem (see below).

Consider a Hamiltonian H_F whose normalized eigenstates $|n\rangle$, with integers $n = M, M+1, \ldots, N-1, N$, correspond to eigenvalues $E_n = b^n$ in some arbitrary units, with $b > 1$. The subscript F originates in the word *free*. Let the eigenstates of H_F form the basis in the model space of states. Each and every basis state corresponds to a different scale of energy as measured by H_F. Let M be large and negative and N large and positive. Thus, the model infrared energy cutoff is $\epsilon = E_M = b^M$ and its ultraviolet energy cutoff is $\Lambda = E_N = b^N$. In total, the space of states has dimension $N - M + 1$.

Let the full Hamiltonian of the model, H, include the interaction H_I,

$$H = H_F + H_I \,, \tag{12.11}$$

and let the matrix elements of H_I in the free basis be

$$H_{Imn} = \langle m|H_I|n\rangle = -g\sqrt{E_m E_n} \,, \tag{12.12}$$

where $g > 0$ is a dimensionless coupling constant. The resulting quantum mechanical system is asymptotically free, includes a bound state and can be generalized to include a limit cycle, as will be discussed later.

The eigenvalue problem for H,

$$H|\psi\rangle = E|\psi\rangle \,, \tag{12.13}$$

expressed in terms of wave function coefficients ψ_n in the free basis, where

$$|\psi\rangle = \sum_{n=M}^{N} \psi_n |n\rangle \,, \tag{12.14}$$

is equivalent to a system of $N - M + 1$ linear equations

$$E_m \psi_m - g\sqrt{E_m} \sum_{n=M}^{N} \sqrt{E_n}\,\psi_n = E\psi_m \,, \tag{12.15}$$

where $m = M, M+1, \ldots, N-1, N$ and E denotes the eigenvalues one wants to compute.

The spirit of Ken's renormalization group (RG) procedure is to first solve the dynamical problem for states corresponding to the highest available energy scale, $E_N = \Lambda$. In the model, this means first solving the equation with $m = N$ for ψ_N. The result is

$$\psi_N = \frac{g\sqrt{E_N}}{E_N - E - gE_N} \sum_{n=M}^{N-1} \sqrt{E_n}\,\psi_n \,, \tag{12.16}$$

whereby the highest energy component is expressed by $N - M$ lower energy components. By inserting this solution into the set of remaining $N - M$ equations, one obtains a new set of equations that is smaller by one equation than the original set,

$$E_m\psi_m - g\sqrt{E_m}\sum_{n=M}^{N-1}\sqrt{E_n}\psi_n - \frac{g^2 E_N}{E_N - E - gE_N}\sqrt{E_m}\sum_{n=M}^{N-1}\sqrt{E_n}\psi_n = E\psi_m\,. \tag{12.17}$$

This set resembles the original one except for two features. It is cutoff at energy scale Λ/b instead of Λ and it contains a new coupling constant,

$$g_{N-1} = g + \frac{g^2 E_N}{E_N - E - gE_N}\,. \tag{12.18}$$

The phrase "coupling constant" is used here with some liberty since g_{N-1} is a function of the eigenvalue E and it only appears in the role of a coupling constant in the reduced set of equations. The eigenvalue dependence will be addressed below and in the next section.

In the spirit of Wilsonian RG procedure, the cutoffs Λ and Λ/b can be denoted by Λ_N and Λ_{N-1}, respectively. The initial coupling constant g can be denoted by g_N. The new notation foresees that one is to consider a sequence of smaller cutoffs $\Lambda_n = b^n$ and corresponding coupling constants g_n with $n < N$, by repeating the step of reducing by one the number of energy scales explicitly included in the eigenvalue problem.

In realistic theories, the initial Hamiltonian is not specially designed for simplicity as the model Hamiltonian is. As a result of analogous steps of cutoff reduction, one obtains sets of equations that differ in more ways from the original set than just by the values of the cutoff and one coupling constant. However, the model simplicity allows one to identify the inherent difficulty of the bound state problem in asymptotically free theories.

The coupling constant formula in Eq. (12.18) can be simplified to

$$g_{N-1} = g_N + \frac{g_N^2}{1 - g_N - E/E_N}\,, \tag{12.19}$$

where one sees that for eigenvalues that are negligible in comparison with the cutoff Λ_N, one has

$$g_{N-1} - g_N = \frac{g_N^2}{1 - g_N}\,. \tag{12.20}$$

This result illustrates in a simple fashion the Gell-Mann and Low finding[60] that the change of a dimensionless coupling constant due to change of a large energy scale may only be a function of the coupling constant itself. Again, if the model were not specially designed, one would have to consider more than one coupling constant. In quantum field theory, one expects many couplings to be involved.[44,59,61–64]

For as long as one considers cutoffs $\Lambda_n = b^n$ that are much greater than the eigenvalues E one wants to compute, the recursion formula of Eq. (12.20), or

$$g_{n-1} = \frac{g_n}{1 - g_n}, \tag{12.21}$$

is valid, with a solution

$$g_n = \frac{g_N}{1 - g_N(N - n)}. \tag{12.22}$$

Recalling that $n = \ln \Lambda_n / \ln b$ for all n, this result reads

$$g_\lambda = \frac{g_\Lambda}{1 - g_\Lambda \ln(\Lambda/\lambda)}, \tag{12.23}$$

where the scale λ denotes the cutoff Λ_n, $g_\lambda = g_n / \ln b$ and $g_\Lambda / \ln b$. Thus,

$$g_\Lambda = \frac{g_\lambda}{1 + g_\lambda \ln(\Lambda/\lambda)}, \tag{12.24}$$

which demonstrates that the coupling constant tends to zero for the cutoff Λ tending to infinity and thus exhibits that the model is *asymptotically free*. This result also illustrates the possibility of using some energy scale λ to fix the value of a dimensionless coupling constant, called *dimensional transmutation.*[65] One can use Eq. (12.24) to express g_Λ in terms of $g_0 = g_{\lambda_0}$ and insert it into Eq. (12.23) to obtain

$$g_\lambda = \frac{g_0}{1 + g_0 \ln(\lambda/\lambda_0)}. \tag{12.25}$$

The number g_0 can be adjusted to data using the Hamiltonian that corresponds to λ_0. Thus required number can be written in the form

$$g_0 = \frac{1}{\ln(\lambda_0/\Lambda_{\text{model}})}, \tag{12.26}$$

which leads to

$$g_\lambda = \frac{1}{\ln(\lambda/\Lambda_{\text{model}})}, \tag{12.27}$$

where Λ_{model} appears in the role resembling the role of Λ_{QCD} in perturbative calculations of Green's functions in QCD.

In the limit of $b \to \infty$, the eigenvalue of order λ that would be observable in the model world, is of the form

$$E_\lambda = E_n(1 - g_\lambda). \tag{12.28}$$

This is a simple illustration of the possibility that one may measure the flow of a coupling constant with energy scale, which is analogous to the flow of a strong coupling constant as a function of scale in observables explained using QCD.

The continuum limit of the model, $b \to 1$, requires consideration of a continuous spectrum. This limit of the model corresponds to a two-dimensional Schrödinger equation for two particles interacting through a δ-potential.

The bound state in the model corresponds to the eigenvalue E smaller than 0. From Eq. (12.15) it follows that in this case

$$\psi_m = C\,\frac{g_N\sqrt{E_m}}{E_m + |E|}\,, \tag{12.29}$$

where the constant C is given by

$$C = \sum_{n=M}^{N} \sqrt{E_n}\,\psi_n\,. \tag{12.30}$$

The constant C appears on both sides of this relation and can be removed to obtain

$$1 = g_N \sum_{n=M}^{N} \frac{1}{1 + |E|/E_n}\,. \tag{12.31}$$

It is visible that the sum grows linearly with N for $E_N \gg |E|$, which means linearly with $\ln\Lambda$. Therefore, demanding that E matches the finite bound-state data value $E_{\rm exp}$ in the model world forces one to consider g_N on the order of the inverse of $\ln\Lambda$. The sum in Eq. (12.31) has a well-defined limit when the infrared cutoff $\epsilon = b^M$ tends to 0, by letting M go to negative infinity.

Equation (12.31) shows that the bound state energy eigenvalue is sensitive to dynamical contributions from all high-energy scales in the model. The ultraviolet renormalization of g_N must be carried out precisely in order to obtain precise results for the binding energy.

In principle, one could use the bound-state eigenvalue solution of Eq. (12.31) to trade g_N for the function of Λ and $|E_{\rm exp}|$ that secures agreement of the theoretical bound-state eigenvalue E with the experimental value $E_{\rm exp}$ in the model world. Any one of the solutions given in Eq. (12.28) could also be used for this purpose when b is very large. However, fixing the coupling constant as a function of Λ using one eigenvalue does not *a priori* imply that any other eigenvalue should also become independent of Λ. If such a result is obtained without knowledge of the reason, it may be perceived as a special feature of a special theory. Such theories are called renormalizable and are distinguished by this feature from all other theories in which fixing a small number of coupling constants using a small number of observables does not ensure that all other observables are independent of the cutoff.

However, when one uses Eq. (12.24) that is valid for all eigenvalues that are negligible in comparison to λ, so that

$$g_N = \frac{g_n}{1 + g_n(N-n)}\,, \tag{12.32}$$

then the bound-state eigenvalue Eq. (12.31) becomes

$$1 = \frac{g_n}{1 + g_n(N-n)} \sum_{n=M}^{N} \frac{1}{1 + |E|/E_n}\,, \tag{12.33}$$

where the dependence on N, or $\ln\Lambda$, cancels out irrespective of the finite value of g_n that is required to obtain $E_{\rm exp}$ as a solution for E. This means also that all other eigenvalues that are negligible in comparison with Λ_n do not depend on Λ. Moreover, it becomes apparent in the model that Wilsonian RG procedure guarantees by construction that the eigenvalues also do not depend on the finite cutoff scale $\lambda = \Lambda_n$, as long as the latter is large enough.

It can be seen in Eq. (12.33) that the coupling constant g_n and the eigenvalue E are related to each other in a way that includes contributions from all high-energy states in the theory. Moreover, every energy scale of order b^n with $n > n_E \sim \ln |E|/\ln b$ contributes practically the same amount to this relationship. This means that, working with a finite cutoff, one has to know precisely how to adjust the coupling constant to obtain all small eigenvalues of interest independent of the finite cutoff.

At this point one might ask how to choose λ in theories with asymptotic freedom in order to carry out precise computations. The smaller λ the smaller the dimension of the space of states in which the renormalized Hamiltonian with cutoff Λ_n is to be diagonalized. In the model, the dimension is $n - M + 1$. On the other hand, the smaller the cutoff λ the larger the ratio E/λ and the less accurate the formula for g_λ that neglects this ratio. Inclusion of this ratio in realistic quantum field theories requires consideration of Hamiltonian operators that depend on their eigenvalues in a complex way and hence must satisfy difficult to satisfy self-consistency conditions between the eigenvalue to be found and the one that stands in the operator to be diagonalized. Theories with asymptotic freedom additionally confuse the situation in a way analogous to what happens in our model (see below).

Equation (12.27) shows that the coupling constant g_λ increases when the ratio of λ to Λ_{model} decreases from some large value toward one. The increase of the coupling constant also occurs in the RG transformation of Eq. (12.21) with the properly reinstated ratio E/E_n. For a bound state,

$$g_{n-1} = \frac{g_n}{1 - g_n + |E|/E_n} \left(1 + |E|/E_n\right). \tag{12.34}$$

In this formula, the ratio $|E|/E_n$ in the denominator is added to the difference $1 - g_n$ that decreases in size when g_n increases toward 1. In other words, the negative interaction energy increases in size and reduces the value of energy from which the eigenvalue is subtracted in the RG recursion denominator. Eventually, the ratio $|E|/E_n$ is not compared with 1 but with a considerably smaller number $1 - g_n$. Therefore, asymptotic freedom forces one to invent a method for properly handling the ratio E/E_n in the RG recursion. In theories such as QED, the situation is different. The coupling constant analogous to g_n decreases in magnitude for decreasing n.

Ken designed a way to handle the dependence on E in his RG transformation in 1970,[44] a few years before the discovery of asymptotic freedom. In his calculation, a parameter corresponding to b^{-1} was used as a small parameter to make a perturbative expansion. Ken's parameter had to be smaller than $10^{-6}/4$ for proving that the RG solution for finite-cutoff Hamiltonians converged when the cutoff number analogous to N was increasing. One can apply a similar method in the model discussed here and eliminate the eigenvalue E from the RG recursion. However, one is interested primarily in handling theories in the continuum limit which corresponds to the limit of $b \to 1$ instead of $b \to \infty$. Assuming that $1 < b \sim 1$ prevents one from using the concept of widely separated scales of energy.

One can study the model using expansion in a small coupling constant g_n for some finite value of n and try to extrapolate to a physical value of g_n. A perturbative formula allows one to eliminate E from the RG transformation.[44] However, the perturbative formula involves energy denominators that contain a difference between an energy below and an energy above the running cutoff, causing large terms when $b \to 1$. In the case of a Hamiltonian as complex as QCD, it is unlikely that one can easily control such terms because even in our simple model, which is free from overwhelming problems caused by the degeneracy of states that is characteristic for QCD, it is not easy to trace their role in obtaining accurate solutions for the bound-state eigenvalue when the model is extrapolated as a function of the effective coupling constant toward 1. More generally, one may say that the strongly bound states cannot be easily separated into their high-energy and low-energy parts that could be matched using perturbation theory.

12.7. Similarity renormalization group procedure

Ken and I had concluded that the RG idea of integrating out high-energy components in order to obtain effective Hamiltonians for low-energy components of the eigenstates had to be replaced by some better idea that would avoid the dependence on unknown eigenvalues and did not produce large terms due to small energy-denominators in perturbation theory. We also considered that perturbation theory would be the initial tool in application of any hopeful method to asymptotically free theories. We invented a candidate and showed that it was sufficiently good for proving renormalizability of the simple model described in the previous section to all orders of perturbation theory in g for any value of $b > 1$.[66,67] We called the new procedure similarity renormalization group (SRG).

The SRG abandons the procedure of integrating out high-energy components and instead uses a procedure of integrating out transitions with large virtual changes of energy. Using the changes of energy rather than energies themselves, which referred to internal dynamics of a physical system without including its motion as whole, was also suggested by Ken's fixed source models[43,44] in their FF version.[55] In the model Hamiltonian eigenvalue problem discussed in the previous section, instead of expressing the high-energy components by the low-energy components and an eigenvalue, one performs a change of basis in a way designed so that in the new basis the interaction Hamiltonian matrix does not have any significant matrix elements between states that differ in their free energy by more than a finite parameter, say λ. The eigenvalues are not changed but the wave functions of eigenstates are changed because the new basis states are adjusted to the interaction. The change of basis is designed to be independent of the eigenvalues. Its perturbative expansion never generates energy denominators that are smaller than λ.

Imagine the Hamiltonian matrix $H_{mn} = \langle m|H|n\rangle$ in the model discussed in the previous section. After the change of basis in the SRG procedure, one obtains a matrix $H_{mn\lambda}$ of a band-diagonal form, in which, roughly speaking and ignoring some details, the matrix elements with $|E_m - E_n| > \lambda$ are reduced to zero. Matrix elements between states of finite free energies in the band may depend on λ but by definition should not depend on the upper energy cutoff Λ. This condition is so strong that, in the limit of $\Lambda \to \infty$, no finite eigenvalue of the resulting theory can depend on the cutoff Λ in any finite order of perturbation theory used to calculate observables.

Besides perturbative expansion in powers of g for observables, one can also use the expansion in powers of g to calculate the SRG transformation. Before renormalization, this transformation generates matrix elements in $H_{mn\lambda}$ that depend on Λ. To remove this dependence, one adds Λ-dependent counterterms to the Hamiltonian H that corresponds to $\lambda = \infty$. These counterterms can be found order by order and they are responsible for the change of g to g_N described in the previous section. The diverging parts of the counterterms are thus found without knowing any solution for observables. The finite parts of the counterterms require fixing by comparison of theoretical predictions with data, which in the model means that one evaluates some eigenvalue of the Hamiltonian with an unknown parameter Λ_{model} and adjusts Λ_{model} by demanding that some eigenvalue matches the required data value.

Diagonalization of the Hamiltonian including counterterms can be performed using the matrix $H_{mn\lambda}$ instead of H_{mn}. This is useful because typical band-diagonal Hamiltonian matrices have eigenstates whose wave functions have significant values only for the basis states whose free energies differ from the eigenvalue by amounts not exceeding the energy width λ of the Hamiltonian. This means that one can obtain approximate solutions of the theory by numerically diagonalizing relatively small submatrices of the entire Hamiltonian matrix, which we called *window* Hamiltonians, while the submatrices, or window Hamiltonians are calculated in perturbation theory. Perturbative calculations of the window Hamiltonians of width λ are carried out using expansion in powers of g_λ instead of $g = g_N$.

A beautifully simple equation for smoothly bringing finite Hamiltonian matrices to band-diagonal form has been proposed by Wegner[68,69] and one could use his equation in the SRG procedure. Wegner later found that similar equations had been used by mathematicians for different computational purposes.[70–72] Only a slightly altered Wegner equation was sufficient for convergence of perturbative calculations of small window Hamiltonians in the asymptotically free model discussed in the previous section.[73]

The SRG procedure for Hamiltonians can be seen as qualitatively different from the RG procedure based on integrating out high-energy states when one considers relativity. Integrating out high-energy components down to a finite cutoff implies that the effective Hamiltonian describes states of a limited range in energy, which also means that the eigenvalues are of limited size. However, the Lorentz symmetry requires that energies of states can change from observer to observer and the change may be greater than any finite cutoff. With the SRG procedure the situation is different. Instead of a limit on the energy, one introduces a limit on changes of energy. This limitation can take the form of a limitation on changes of invariant masses, which are invariant with respect to the free Lorentz transformation. Such a situation is found in the FF of Hamiltonian dynamics. The boost symmetry corresponds to an energy shift up or down along the band of the Hamiltonian matrix. The energies change but the range of virtual changes of invariant masses remains limited by the SRG parameter λ.

12.8. Light-front QCD, nuclear physics and the limit cycle

In the first half of 1990s, Ken's involvement in developing FF of Hamiltonian dynamics in quantum field theory ranged from general considerations[49] to discussions of specific renormalization issues[74–76] and eventually to formulation of a computational scheme for defining and solving LF QCD in the Minkowski space using the weak coupling expansion.[77] In the latter paper, the QCD bound-state problem was formulated so that one could utilize a weak coupling expansion in analogy to QED, where quantum mechanics of bound states is combined with weak coupling techniques for calculating radiative corrections.

In order to work around non-perturbative field-theoretic effects, the bare Hamiltonian is supplied with constituent mass terms. The masses cause the coupling constant to never become large enough to invalidate perturbative calculations of the effective Hamiltonians. Stabilization of the theory requires addition of an artificial potential with a coefficient that vanishes for a physical value of the coupling constant. Such setup is guided by the simplicity of the CQM picture of hadrons. As described in previous sections, the price of this setup is the need for solving complex renormalization problems that arise in the canonical FF Hamiltonian of QCD supplied with the mass terms and stabilizing potentials. This difficulty is approached using the SRG procedure. The renormalization is first outlined using a phase space analysis, using Ken's power counting and wavelets. Then the SRG examples of precise calculations are shown for second-order terms. Further computations are sketched, including the initial non-relativistic non-perturbative calculations of hadron masses which specify the stabilizing potential (whereby the binding energies are secured to be of fourth order in powers of the coupling constant), the SRG calculations of radiative corrections, and an indication of the extension of perturbation theory required for studying the strong-coupling dynamics of bound states.

In addition to working on these issues with faculty members at OSU, Ken also devoted his time to working with postdocs and graduate students.[78–80] A simple

Hamiltonian model of a bound state computation in asymptotically free theory was described in 1998.[81] The simplicity of the model allowed for a numerical study of the accuracy of perturbative expansion for window Hamiltonians in the SRG procedure. It was intended to help future students in identifying directions for research. Ken continued to speak with enthusiasm about LF QCD in 2004 in public,[82] as well as privately.

An article reviewing research on QCD and other field theories on the light front in 1990s,[83] refers to eleven articles on these subjects co-authored by Ken. Twenty years later, Ken's work on these subjects still forms an important part of the research program of International Light-Cone Advisory Committee (ILCAC),[84,85] of which Ken was a member in the 1990s.

The program outlined in Ref. 77 has not been completed yet. It requires research in new directions implied by weaknesses in available formulations of QCD. The constituent quark model, parton model, and confinement mechanism still await understanding in a mathematically well-defined theory with proper degrees of freedom in the Minkowski space.[86] In Ken's hierarchy of precisely formulated physical theories, the *status quo* of strong interaction theory was difficult to accept. He led the program of Ref. 77 thinking about the new generations of physicists who will carry the torch.

The impact of Ken's research from the 1990s is also evident in the progress made by nuclear physicists, who have used the SRG to develop a new approach to nucleon-nucleon interactions, generating the three-body forces that the SRG procedure systematically yields along with other interactions.[87] Instead of using potentials with large cutoffs on energy, such as in Refs. 88 and 89, they use calculated effective potentials that have small cutoffs on energy changes. These effective potentials exhibit universal behavior that unifies different models of nuclear interactions.[90–93]

In his 2004 account of the origins of lattice gauge theory,[82] Ken recalled that, in his 1971[94] classification of possible RG behaviors of quantum field theories, he missed asymptotic freedom because the only examples of beta functions he knew at that time were positive at small g and, he said, "it never occurred to me that gauge theories could have negative beta functions for small g." In the same 1971 paper, Ken found that, in the presence of more than one coupling constants, they may keep varying indefinitely as functions of the cutoff, moving along a fixed circular trajectory in the space of coupling constants. Ken thus identified the RG limit cycle as an *a priori* possibility. I think that in the future the RG limit cycle may turn out to be quite relevant to physics (see below), even if it is not being thoroughly researched today in particle physics[95] because perturbative analyses of the Standard Model and its contemplated extensions do not lead to a limit cycle.

Ken gave a talk about a limit cycle in nuclear physics in the Institute for Nuclear Theory at the University of Washington in 2000, stimulated by the work of Bedaque, Hammer, and van Kolck,[96,97] who showed that a three-body Hamiltonian with δ-function potentials regulated by an ultraviolet cutoff Λ in momentum space is renormalizable and in the limit $\Lambda \to \infty$ a three-body coupling constant approaches a limit cycle. The associated geometric spectrum of bound states was identified before.[98,99] There also existed limit cycles in statistical systems[100,101] but they appeared rare. Listening to Ken's talk, I realized that in an appendix to our 1993 paper,[103] a similar behavior of the coupling constant with cutoff was obtained in a simple continuum model that I could solve analytically in an elementary form. We had discussed the model in 1993 between the two of us; its solution appeared strange. When the cutoff was lowered, the coupling constant would increase to infinity, jumping at some point to negative infinity and then rising again, until it would make the next jump. We did not recognize in 1993 that we had an example of a limit cycle and we practically forgot about it, preferring to focus on asymptotically free theories. However, motivated in 2000–2001 by the difficulty of RG analysis of the three-body problem, which was much more difficult than for our 1993 model, we decided to publish a discrete version of our model as a simple example of a limit cycle.[104] Ken's excitement about the model limit cycle increased even more when Braaten and Hammer[105] hypothesized that an infrared RG limit cycle exists in QCD for a special choice of the u and d quark masses, slightly greater than the phenomenological value of several MeV. We used our discrete model version to describe the RG universality in the case of limit cycles.[106] Ken was engaged in a collaboration on the three-body problem and Mohr, Furnstahl, Perry, Wilson, and Hammer published precise numerical results for limit cycles in the quantum three-body problem.[107]

The fact that Ken identified the limit cycle in his 1971 paper[94] may in the future make his missing of asymptotic freedom in that article not as great a blunder as he still thought it was in 2004.[82] An explanation follows.

The asymptotically free Hamiltonian of Eq. (12.11), whose matrix elements are

$$H_{mn} = (\delta_{mn} - g)\sqrt{E_m E_n}\,, \tag{12.35}$$

can be altered to

$$H_{mn} = [\delta_{mn} - g - ih\,\mathrm{sgn}(m-n)]\sqrt{E_m E_n}\,, \tag{12.36}$$

where sgn is a sign function equal zero at zero. The coupling constant h measures the strength of the skew symmetric imaginary part of the Hermitian Hamiltonian matrix. Applying the same procedure of eliminating components of highest energy, scale by scale, neglecting E/E_n, one obtains the RG recursion[104]

$$h_n = h\,, \tag{12.37}$$

$$g_{n-1} = \frac{g_n + h^2}{1 - g_n}\,. \tag{12.38}$$

This recursion for g_n becomes the same as the one in Eq. (12.21) when $h \to 0$. The solution for $h \neq 0$ is

$$\frac{g_n}{h} = \tan\left[\arctan\frac{g_N}{h} + (N-n)\arctan h\right], \tag{12.39}$$

or, in analogy to Eqs. (12.24) and (12.25) and setting $b = e$ for simplification,

$$g_\lambda = h \tan\left(\arctan \frac{g_0}{h} - \ln \frac{\lambda}{\lambda_0} \arctan h\right). \tag{12.40}$$

Corrections due to $E/E_n \neq 0$ are described in Ref. 106. It is visible that, when

$$h = \tan \frac{\pi}{p}, \tag{12.41}$$

with integer $p > 2$, the coupling constant is a periodic function of the logarithm of the cutoff $\lambda = e^n$ with period p. A precisely periodic behavior of a coupling constant, for negligible E/E_n, is called the limit cycle. If p is a rational number, one obtains a more complicated pattern of periodic behavior. When p is an irrational number, the coupling constant behaves chaotically.

A closer analysis of the same model, using the SRG procedure,[108] is free from dependence on the eigenvalue E. It shows that the coupling constant g_λ is a continuously varying function of the Hamiltonian energy width λ. The coupling constant h is constant as a function of λ, which means it stays as small as one wishes to make it in the initial Hamiltonian with cutoff Λ.

The coupling constant g_λ makes continuous jumps periodically in $\ln \lambda$. Every jump in g_λ, which occurs at every energy scale at which the argument of the tangent passes by $\pi/2$ plus an integer times π, corresponds to the bound state with binding energy on the order of λ at the jump. The corresponding spectrum is necessarily made of a collection of p geometrical series.

The point of explaining the limit cycle in the model is that, in the limit $h \to 0$, one obtains a model with asymptotic freedom. The limit $h \to 0$ corresponds to $p \to \infty$. The coupling constant g_λ returns to the same value when the cutoff λ changes to $e^p\lambda$. The maximal value of the coupling constant in a long cycle appears at λ that matches the magnitude of the binding energy of a bound state. There occurs only one bound state per cycle. For larger values of the cutoff than the bound state energy, say E, the coupling appears to tend to zero as if the model were asymptotically free, until the cutoff λ reaches giant values on the order of $e^{p/2}E$. Subsequently, i.e., for still larger λ, the coupling becomes negative instead of decreasing to zero, and eventually becomes sizably negative. At $\lambda \sim e^p E$, the coupling constant quickly rises to the same positive value it had for $\lambda \sim E$. At this point a new bound state becomes dynamically involved in the theory and the next cycle begins. In the limit of $h \to 0$, over the entire part of the single cycle in which the coupling constant is positive and decreases as an inverse of the logarithm of the width λ, the limit cycle model is indistinguishable from the model with asymptotic freedom.[108]

Asymptotic freedom appears embedded in a limit cycle in the model due to the inherent nature of the Hamiltonian whose matrix is Hermitian but not real. In quantum mechanics, such Hamiltonian matrices are allowed, but they do not occur in classical theories, which do not involve $i = \sqrt{-1}$. Since the space of imaginable real Hamiltonian matrices is of measure zero in the space of imaginable complex Hamiltonian matrices, the RG limit cycle and more chaotic behavior may be commonplace for quantum theories requiring renormalization.

Perhaps Ken's pointing out the possibility of the existence of limit cycles in 1971 will make the fact that he missed asymptotic freedom much less of a blunder than Ken thought in 2004.[82] Namely, it is by no means excluded that the currently known asymptotic freedom in gauge theories will turn out in some future theory to be a part of a limit cycle.

12.9. Conclusion

The image of Ken as a physicist would be largely incomplete without mention of his work on systemic issues of education. Ken's interest in education was stimulated by his own learning path in mathematics and physics, starting from his boyhood when he was encouraged by his grandfather[109] and father,[110] and later by Arnold Arons, a major contributor to physics education,[111] who advised Ken on his first summer research projects at the Woods Hole Oceanographic Institution.[112]

Some references to Ken's work in education are available on the memorial website for him at OSU.[113] Ken's main book on the subject, co-authored with Bennett Daviss, is *Redesigning Education.*[114] Ken believed that every student should have an opportunity to learn to the best of his or her ability. Ken and Jane Butler Kahle led *Project Discovery*, the Ohio statewide Systemic Initiative for improving science and mathematics teaching in the public schools.

Constance Barsky was Ken's close collaborator in education. Ken and Constance led the effort to create the issue of Daedalus *Education Yesterday, Education Tomorrow*,[115] in which they contributed the concluding article *Applied Research and Development: Support for Continuing Improvement of Education.* Quoting from the preface by Stephen Graubard, p. XXIII, "If there is no 'science of education' comparable to what exists in American medicine or agriculture, and the scholarship that would produce such learning is both controversial and suspect, what hope can be held out for fundamental change?" Ken believed in the need to develop the science of change. Ken and Constance concluded that "Most people have long thought that education is something anyone can do with minimum of experience. Today we have to think of education as demanding in multiple dimensions: as a science, as a design challenge, and as a performing art while still being an imperative for life in a democracy. Handed-down traditions are no longer enough."

Ken analyzed complex issues that matter in reform of education with stunning accuracy. The library of books and articles that he organized in his mind into a coherent view contained several thousand items. A review of Ken's approach to education is beyond the scope of this article, but a list of examples of deeply related issues can help students of his work orient themselves to the magnitude of the problems. So, in addition to Ken's experience in learning, research, and teaching in physics, the studies he found necessary for understanding what needs to be done for improving educational systems included: history of educational systems, with the prominent example of the work of Comenius from 1657[116]; history and philosophy of science, as shown in his review[117] of the work of Thomas Kuhn[118,119] and through collaboration with George Smith[120]; physics education, especially the results of Lillian McDermott and Physics Education Group at the University of Washington[121] that were used in *Project Discovery*; basic needs of understanding how children learn to read, shown by Marie Clay,[122] if one wants to create systemic conditions for them not to fail; needs of learning how to do it from *Reading Recovery*[123]; insights into how one lets and helps people unfold their creativity, provided by Henry Schaefer-Simmern[124]; information on new results concerning the central nervous system structure and function, such as the role of myelin[125]; the US risk of losing its leadership position if the educational system is not improved[126]; recent history and status of reform efforts[127]; relationships that lead to formation of settings relevant to reform of the educational systems and the difficulties of creating such settings, through working with Seymour Sarason[128,129]; examples of reasonable experimental reform efforts, such as *MicroSociety*[130]; misconceptions and errors concerning systemic issues[131]; regularities in development of professional expertise[132]; bold examples of innovation with large systemic implications, such as the invention of flight[133]; technological progress, e.g., Refs. 134 and 135; theories of technological revolutions and financial capital, such as by Carlota Perez[136]; Peter Drucker's insights concerning management of non-profit organizations[137] and rules of innovation and entrepreneurship,[138] including requirement of values in the education of top managers in the modern world of organizations, mainly through working with Joseph Maciariello[139]; and accumulation of systemic data, as illustrated in Ref. 140.

Ken was convinced that accomplished physicists, who know how to manage collaboration of free minds, can help in solving great societal problems thanks to their physics training. However, they need to learn a lot to become aware of the scope and difficulty of these problems before they will be able to tackle them successfully and thus make themselves useful as high-level managers of the required organizations of researchers that need the human drive and methodology known in physics. He thought that promising physicists should get considerable funding for research at their discretion, provided that they commit to learning about societal problems and training in top-level management concurrent with conducting research of their choice in physics.[141] We know that professionals learn as they want to. Currently, society is not accustomed to investing in physicists and their research for this purpose.

When a student once asked Ken at what point one should start doing research, still in college or afterwards, Ken responded "You start research at birth and you never stop." In Ken's departure, we have lost one of a few minds that are capable and willing to help in solving difficult problems.

References

1. P. A. M. Dirac, *Rev. Mod. Phys.* **21**, 392 (1949).
2. R. P. Feynman, *Phys. Rev. Lett.* **23**, 1415 (1969).
3. R. P. Feynman, *Photon-Hadron Interactions* (Addison-Wesley Longman, Reading, 1972).
4. K. G. Wilson, *Phys. Rev. Lett.* **27**, 690 (1971).
5. D. J. Gross and F. Wilczek, *Phys. Rev. Lett.* **30**, 1343 (1973).
6. H. D. Politzer, *Phys. Rev. Lett.* **30**, 1346 (1973).
7. H. D. Politzer, *Phys. Rept.* **14**, 129 (1974).
8. D. J. Gross and F. Wilczek, *Phys. Rev. D* **9**, 980 (1974).
9. K. G. Wilson, *Phys. Rev. D* **10**, 2445 (1974).
10. S. Weinberg, *Phys. Rev.* **150**, 1313 (1966) [Erratum: *ibid.* **158**, 1638 (1967)].
11. K. Bardakci and M. B. Halpern, *Phys. Rev.* **176**, 1686 (1968).
12. L. Susskind, *Phys. Rev.* **165**, 1535 (1968).
13. S.-J. Chang and S.-K. Ma, *Phys. Rev.* **180**, 1506 (1969).
14. J. B. Kogut and D. E. Soper, *Phys. Rev. D* **1**, 2901 (1970).
15. T.-M. Yan and S. D. Drell, *Phys. Rev. D* **1**, 2402 (1970).
16. H. Leutwyler, J. R. Klauder and L. Streit, *Nuovo Cim. A* **66**, 536 (1970).
17. J. D. Bjorken, J. B. Kogut and D. E. Soper, *Phys. Rev. D* **3**, 1382 (1971).
18. F. Rohrlich, *Acta. Phys. Austr.* **32**, 87 (1970).
19. R. A. Neville and F. Rohrlich, *Nuovo Cimento A* **1**, 625 (1971).
20. S.-J. Chang, R. G. Root and T.-M. Yan, *Phys. Rev. D* **7**, 1133 (1973).
21. S.-J. Chang and T.-M. Yan, *Phys. Rev. D* **7**, 1147 (1973).
22. T.-M. Yan, *Phys. Rev. D* **7**, 1760 (1973).
23. T.-M. Yan, *Phys. Rev. D* **7**, 1780 (1973).
24. S. J. Brodsky, R. Roskies and R. Suaya, *Phys. Rev. D* **8**, 4574 (1973).
25. J. B. Kogut and L. Susskind, *Phys. Rept.* **8**, 75 (1973).
26. E. Tomboulis, *Phys. Rev. D* **8**, 2736 (1973).
27. W. A. Bardeen and R. B. Pearson, *Phys. Rev. D* **13**, 547 (1976).
28. A. Casher, *Phys. Rev. D* **14**, 452 (1976).
29. C. B. Thorn, *Phys. Rev. D* **19**, 639 (1979).
30. C. B. Thorn, *Phys. Rev. D* **20**, 1934 (1979).
31. W. A. Bardeen, R. B. Pearson and E. Rabinovici, *Phys. Rev. D* **21**, 1037 (1980).
32. G. P. Lepage and S. J. Brodsky, *Phys. Rev. D* **22**, 2157 (1980).
33. V. A. Franke, Y. V. Novozhilov and E. V. Prokhvatilov, *Lett. Math. Phys.* **5**, 239 (1981).
34. V. A. Franke, Y. V. Novozhilov and E. V. Prokhvatilov, *Lett. Math. Phys.* **5**, 437 (1981).
35. G. P. Lepage, S. J. Brodsky, T. Huang and P. B. Mackenzie, *Particles and Fields 2*, eds. A. Z. Capri and A. N. Kamal (Plenum, New York, 1983).
36. S. D. Głazek, *Phys. Rev. D* **38**, 3277 (1988).
37. S. J. Brodsky and G. P. Lepage, *Perturbative Quantum Chromodynamics*, ed. A. H. Mueller (World Scientific, Singapore, 1989).
38. E. V. Prokhvatilov and V. A. Franke, *Sov. J. Nucl. Phys.* **49**, 688 (1989).
39. Particle Data Group (K. A. Olive *et al.*), *Chin. Phys. C* **38**, 090001 (2014).
40. K. G. Wilson, videorecorded discussion at the workshop "Theory of Hadrons and Light-Front QCD", Polana Zgorzelisko near Zakopane, Poland, August 15–25, 1994;

link "In Memoriam" on the workshop website at the University of Warsaw, http://www.fuw.edu.pl/~lfqcd.
41. H. J. Melosh, *Phys. Rev. D* **9**, 1095 (1974).
42. P. A. M. Dirac, *Phys. Rev.* **139**, B684 (1965).
43. K. G. Wilson, *Phys. Rev.* **140**, B445 (1965).
44. K. G. Wilson, *Phys. Rev. D* **2**, 1438 (1970).
45. I. Tamm, *J. Phys.* (*U.S.S.R.*) **9**, 449 (1945).
46. S. M. Dancoff, *Phys. Rev.* **78**, 382 (1950).
47. F. J. Dyson, *Phys. Rev.* **90**, 994 (1953).
48. F. J. Dyson, *Phys. Rev.* **91**, 421 (1953).
49. R. J. Perry, A. Harindranath and K. G. Wilson, *Phys. Rev. Lett.* **65**, 2959 (1990).
50. A. Casher and L. Susskind, *Phys. Rev. D* **9**, 436 (1974).
51. J. D. Bjorken and S. D. Drell, *Relativistic Quantum Fields* (McGraw-Hill, New York, 1965), Sec. 92.
52. R. P. Feynman, *Phys. Rev.* **76**, 769 (1949).
53. See Sec. 12.8 in this article.
54. J. D. Bjorken, *Phys. Rev.* **179**, 1547 (1969).
55. S. D. Głazek and R. J. Perry, *Phys. Rev. D* **45**, 3734 (1992).
56. See Sec. 12.7, in this article.
57. European Muon Collab. (J. Ashman *et al.*), *Nucl. Phys. B* **328**, 1 (1989).
58. D. F. Geesaman, K. Saito and A. W. Thomas, *Ann. Rev. Nucl. Part. Sci.* **45**, 337 (1995).
59. K. G. Wilson, *Rev. Mod. Phys.* **55**, 583 (1983).
60. M. Gell-Mann and F. E. Low, *Phys. Rev.* **95**, 1300 (1954).
61. R. J. Perry and K. G. Wilson, *Nucl. Phys. B* **403**, 587 (1993).
62. R. J. Perry, *Ann. Phys.* **232**, 116 (1994).
63. R. Oehme and W. Zimmermann, *Commun. Math. Phys.* **97**, 569 (1985).
64. W. Zimmermann, *Commun. Math. Phys.* **97**, 211 (1985).
65. S. Coleman and E. Weinberg, *Phys. Rev. D* **7**, 1888 (1973).
66. S. D. Głazek and K. G. Wilson, *Phys. Rev. D* **48**, 5863 (1993).
67. S. D. Głazek and K. G. Wilson, *Phys. Rev. D* **49**, 4214 (1994).
68. F. Wegner, *Ann. Phys.* **506**, 77 (1994).
69. F. Wegner, *Int. J. Mod. Phys. A* **29**, 1430043 (2014).
70. R. W. Brocket, *Linear Algebra Appl.* **146**, 79 (1991).
71. M. T. Chu, *Fields Inst. Commun.* **3**, 87 (1994).
72. M. T. Chu and K. R. Driessel, *SIAM J. Numer. Anal.* **27**, 1050 (1990).
73. S. D. Głazek and J. Młynik, *Phys. Rev. D* **67**, 045001 (2003).
74. D. Mustaki, S. Pinsky, J. Shigemitsu and K. Wilson, *Phys. Rev. D* **43**, 3411 (1991).
75. S. Głazek, A. Harindranath, S. Pinsky, J. Shigemitsu and K. Wilson *Phys. Rev. D* **47**, 1959 (1993).
76. O. Abe, K. Tanaka and K. G. Wilson, *Phys. Rev. D* **48**, 4856 (1993).
77. K. G. Wilson *et al.*, *Phys. Rev. D* **49**, 6720 (1994).
78. K. G. Wilson and D. G. Robertson, *Proceedings of the Fourth International Workshop on Light-Front Quantization and Non-Perturbative Dynamics: Theory of Hadrons and Light-Front QCD*, ed. S. D. Głazek (World Scientific, Singapore, 1995), pp. 15–28.
79. K. G. Wilson and M. M. Brisudova, *Proceedings of the Fourth International Workshop on Light-Front Quantization and Non-Perturbative Dynamics: Theory of*

Hadrons and Light-Front QCD, ed. S. D. Głazek (World Scientific, Singapore, 1995), pp. 166–181.
80. M. M. Brisudova, R. J. Perry and K. G. Wilson, *Phys. Rev. Lett.* **78**, 1227 (1997).
81. S. D. Głazek and K. G. Wilson, *Phys. Rev. D* **57**, 3558 (1998).
82. K. G. Wilson, *Nucl. Phys. B* (*Proc. Suppl.*) **140**, 3 (2005).
83. S. J. Brodsky, H.-C. Pauli and S. S. Pinsky, *Phys. Rept.* **301**, 299 (1998).
84. B. L. G. Bakker *et al.*, *Nucl. Phys. B* (*Proc. Suppl.*) **251–252**, 165 (2014).
85. International Light-Cone Advisory Committee, Inc., http://www.ilcacinc.org/about.html.
86. M. Gell-Mann, Quarks, color, and QCD in *The Rise of the Standard Model*, eds. L. Hodderson *et al.* (Cambridge University Press, Cambridge, 1999), p. 625.
87. S. K. Bogner, R. J. Furnstahl and R. J. Perry, *Phys. Rev. C* **75**, 061001 (2007).
88. D. R. Entem and R. Machleidt, *Phys. Rev. C* **68**, 041001 (2003).
89. E. Epelbaum, W. Glöckle and U. G. Meissner, *Nucl. Phys. A* **747**, 362 (2005).
90. R. J. Furnstahl, *Nucl. Phys. B* (*Proc. Suppl.*) **228**, 139 (2012).
91. R. J. Furnstahl and K. Hebeler, *Rept. Prog. Phys.* **76**, 126301 (2013).
92. B. Dainton, R. J. Furnstahl and R. J. Perry, *Phys. Rev. C* **89**, 014001 (2014).
93. M. D. Schuster, S. Quaglioni, C. W. Johnson, E. D. Jurgenson and P. Navratil, *Phys. Rev. C* **90**, 011301 (2014).
94. K. G. Wilson, *Phys. Rev. D* **3**, 1832 (1971).
95. M. E. Peskin, SLAC-PUB-15966, arXiv:1405.7086v3 [physics.hist_ph].
96. P. F. Bedaque, H. W. Hammer and U. van Kolck, *Phys. Rev. Lett.* **82**, 463 (1999).
97. P. F. Bedaque, H. W. Hammer and U. van Kolck, *Nucl. Phys. A* **646**, 444 (1999).
98. L. H. Thomas, *Phys. Rev.* **47**, 903 (1935).
99. V. Efimov, *Phys. Lett. B* **33**, 563 (1970).
100. D. A. Huse, *Phys. Rev. B* **24**, 5180 (1981).
101. D. Bernard and A. LeClair, *Phys. Lett. B* **512**, 78 (2001).
102. E. Braaten and H. W. Hammer, *Phys. Rev. A* **67**, 042706 (2003).
103. S. D. Głazek and K. G. Wilson, *Phys. Rev. D* **47**, 4657 (1993).
104. S. D. Głazek and K. G. Wilson, *Phys. Rev. Lett.* **89**, 230401 (2002) [Erratum: *ibid.* **92**, 139901 (2004)].
105. E. Braaten and H. W. Hammer, *Phys. Rev. Lett.* **91**, 102002 (2003).
106. S. D. Głazek and K. G. Wilson, *Phys. Rev. B* **69**, 094304 (2004).
107. R. F. Mohr, R. J. Furnstahl, R. J. Perry, K. G. Wilson and H. W. Hammer, *Ann. Phys.* **321**, 225 (2006).
108. S. D. Głazek, *Phys. Rev. D* **75**, 025005 (2007).
109. K. G. Wilson, videorecorded meeting with students of Faculty of Physics, Universty of Warsaw, at the Students Physics Club SKFiz, 15 November 2010, http://www.fuw.edu.pl/~lfqcd/KGWilson-SKFiz.
110. E. B. Wilson, Jr., *An Introduction to Scientific Research* (Dover Publications, Inc., New York, 1990).
111. A. B. Arons, *A Guide to Introductory Physics Teaching* (John Wiley & Sons, New York, 1990).
112. http://www.nobelprize.org/nobel_prizes/physics/laureates/1982/wilson-bio.html.
113. http://artsandsciences.osu.edu/news/remembering-theoretical-physicist-and-nobel-laureate-kenneth-g-wilson.
114. K. G. Wilson and B. Daviss, *Redesigning Education* (Teachers College Press, New York, 1996).
115. *Proceedings of the American Academy of Arts and Sciences*, Daedalus, Fall 1998, Vol. **127**, No. 4 (MIT Press, Cambridge, MA, 1998).

116. J. A. Comenius, *The Great Didactic*, originally published in Amsterdam in Latin, 1657 [trans. M. W. Keatinge, 1910, Kessinger Publishing Company (1992)], www.kessinger.net.
117. K. G. Wilson, *Phys. Today* **54**, 53 (2001).
118. T. S. Kuhn, *The Structure of Scientific Revolutions*, 2nd edn. (University of Chicago Press, Chicago, 1972).
119. T. S. Kuhn, *The Road since Structure: Philosophical Essays, 1970–1993, with an Autobiographical Interview*, eds. J. Conant and J. Haugeland (University of Chicago Press, Chicago, 2000).
120. I. B. Cohen and G. E. Smith (eds.), *The Cambridge Companion to Newton* (Cambridge Companions to Philosophy) (Cambridge University Press, Cambridge, 2002).
121. L. C. McDermott and the Physics Education Group at the University of Washington, *Physics by Inquiry: An Introduction to the Physical Sciences* (John Wiley & Sons, New York, 1996).
122. M. M. Clay, *Observing Young Readers: Selected Papers* (Heinemann, Portsmouth, NH, 1990).
123. C. A. Lyons and G. Su Pinnell, *Systems for Change in Literacy Education: A Guide to Professional Development* (Heinemann, Portsmouth, NH, 2001).
124. H. Schaefer-Simmern, *The Unfolding of Artistic Activity: Its Basis, Processes, and Implications* (University of California Press, Berkeley, 1962).
125. F. R. Douglas, *Science* **330**, 768 (2010).
126. The National Commission on Excellence in Education, *A Nation at Risk: The Imperative for Educational Reform*, A Report to the Nation and the Secretary of Education, United States Department of Education, April 1983; http://datacenter.spps.org/uploads/SOTW_A_Nation_at_Risk_1983.pdf.
127. M. Fullan, *The New Meaning of Educational Change*, 4th edn. (Teachers College Press, New York, 2007).
128. S. B. Sarason, *Revisiting "The Culture of the School and the Problem of Change"* (Teachers College Press, New York, 1996).
129. S. B. Sarason, *And What Do You Mean by Learning?* (Heinemann, Portsmouth, NH, 1996).
130. C. Cherniss, *School Change and the MicroSociety Program* (Corwin Press, Thousand Oakes, CA, 2006).
131. D. Ravitch, *The Death and Life of the Great American School System: How Testing and Choice Are Undermining Education* (Basic Books, New York, 2010).
132. K. Anders Ericsson (ed.), *Development of Professional Expertise: Toward Measurement of Expert Performance and Design of Optimal Learning Environments* (Cambridge University Press, Cambridge, 2009).
133. H. Combs and M. Caidin, *Kill Devil Hill: Discovering the Secret of the Wright Brothers*, Foreword by N. Armstrong (Ternstyle Press, Ltd., Denver, CO, 1979).
134. T. P. Hughes, *Networks of Power: Electrification in Western Society, 1880–1930* (Johns Hopkins University Press, Baltimore, MD, 1983).
135. T. P. Hughes, *American Genesis: A Century of Invention and Technological Enthusiasm* (Punguine Books, New York, 1990).
136. C. Perez, *Technological Revolutions and Financial Capital: The Dynamics of Bubbles and Golden Ages* (Edward Elgar Publishing Ltd., Glos, UK, 2003).
137. P. F. Drucker, *Managing the Non-Profit Organization: Principles and Practices* (Harper Business, New York, 1990).
138. P. F. Drucker, *Innovation and Entrepreneurship* (Harper Business, New York, 1993).

139. J. A. Maciariello and K. E. Linkletter, *Drucker's Lost Art of Management: Peter Drucker's Timeless Vision for Building Effective Organizations* (McGraw-Hill, New York, 2011).
140. T. Caplow, L. Hicks and B. J. Wattenberg, *The First Measured Century: An Illustrated Guide to Trends in America, 1900–2000* (The AEI Press, Washington, DC, 2001).
141. http://www.fuw.edu.pl/~lfqcd/KGWilson-UWColloquium/.

Chapter 13

RG for non-relativistic fermions

R. Shankar

Sloane Physics Lab, Yale University, New Haven CT 06520

I welcome this opportunity to pay tribute to one of my all time heroes. Wilson's obvious intellect was easy to admire, but he became as much my hero for that as for his uniform treatment of all those who interacted with him. From his Olympian heights he did not seem to be able to distinguish between ordinary people of varying abilities and answered the questions posed to him at face value with no particular regard to who posed it. This quality, by no means universal, made a tremendous impact on me during those impressionable years.

I met Ken for the first time when I was a Junior Fellow at Harvard and he dropped into my office to find out the location of someone else's office. I had no idea who he was and did not appreciate the fact that this was going to be the only occasion when I knew something he did not. Later in the day when I went to the colloquium and found that he was the speaker. As someone who had come from Berkeley with zero knowledge of quantum field theory I used to attend these events mainly as a form of blood sport, in the same spirit as a Roman at the coliseum, to watch our local gladiators decimate the speaker. Well, this one was different. When one of our experts asked him to go over an argument (which in retrospect I think had to do with why QED was not confining if QCD was) Ken did so, and when the inquisitor did not seem persuaded, assured him that it would all become clear if he thought about it some more at leisure. I was stunned by two things: his response and even more importantly, its placid acceptance. As soon as the talk ended I asked my friend to explain what I had just witnessed, and he said "You really don't know that Ken Wilson is our field-theory guru?"

Ken entered my life in earnest a few years later after I had switched to statistical mechanics and was learning about the Renormalization Group. The RG was a topic that absolutely baffled me in the context of field theory, where it boiled down to getting rid of infinities. I was coming around to the view that the only good integral was a divergent one. The modern approach of viewing the RG ushered by Kadanoff and Wilson, as a way of trading the original problem for another one with different degrees of freedom, that yielded the same answer in a certain asymptotic region, was a revelation to me. It was love at first bite. I first learned Kadanoff's way [1] of explaining Widom's scaling laws via block spins, then Wilson's sweeping extension of the RG to the space of all possible Hamiltonians [2] and finally the ε-expansion of Wilson and Fisher [3] that gave actual numerical answers to the age old question of critical exponents. This completed my indoctrination. I seriously considered adding a middle initial G to my name to openly flaunt my newfound faith.

As an aside, I should add that I initially made the common mistake of not going to the source. Then I took a shot at Ken's Physical Review papers and found them remarkably comprehensible. A central lesson that I still preach to errant practitioners is that singularities in the correlation functions should come from integration of the flow and not from singular terms in the action. The β-function should nowhere be singular if it is to be amenable to approximations, even at the critical point. Ken provided a marvelous analogy in his paper. Consider a particle that starts rolling down from the top of a hill, $x = 0$, obeying the equation $-dV/dx = mv$. How does t_h, the time it takes to get half way down the hill, behave as a function of its starting position x? This time diverges as the starting point approaches the top of the hill where the slope vanishes. The divergence in t_h is determined entirely by dV/dx near its zero at $x = 0$. In the same manner are diverging lengths and correlation functions determined by the flow near a zero of the β-function.

In 1985 I went to a workshop on the Monte Carlo Renormalization Group at Cornell. My student Ganpathy Murthy and I discussed our work on redundant operators. At some point the question of how to include more interactions in the flow to get better answers came up. Ken was against just adding a second neighbor coupling. He thoroughly disapproved of people who added terms till the answer agreed with the high-temperature expansion and then published. He thought one should add all possible couplings of a given range. When I asked him how many interactions would satisfy him he said with a straight face "Between 1500 and 2000".

13.1. RG for non-relativistic fermions

To a man with a hammer everything looks like a nail. Thus in the late 1980's I was walking around looking for RG applications everywhere. At that time the high-T_c stampede had just begun. Anderson [4] pointed out right away that the normal state of the superconductor was a non-Fermi liquid. This was a real possibility given that in one spatial dimension the slightest interaction kills the sharply defined quasi-particle, leaving behind just a cut, and only in three dimensions were Landau–Fermi liquids known to exist. So which way will two dimensions go? While experimentally the high-T_c materials certainly looked non-Fermi, I wanted to know *if a Fermi liquid was at all possible in two dimensions for non-zero interactions.* I decided to use the RG because it would provide an unbiased way to settle the matter (unlike, say, Hartree–Fock). This suited me well since I did not know enough to have a bias.

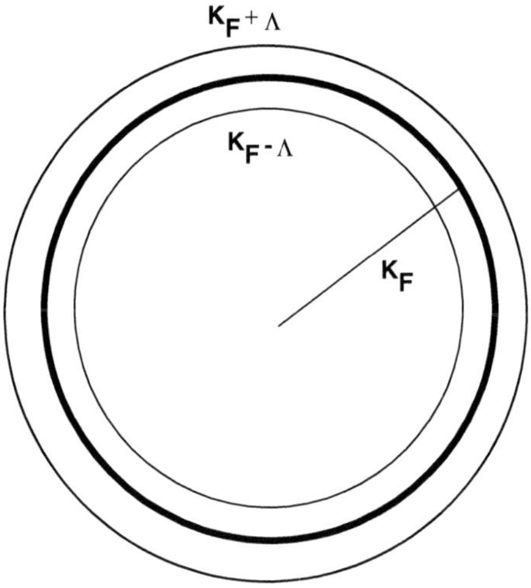

Fig. 13.1. The low energy region for fermions at finite density is an annulus of thickness 2Λ about the Fermi surface.

The general strategy of the RG is to zero-in on the variables that control the asymptotic behavior and integrate away the rest. Usually this corresponds to keeping just the long-wavelength, low-energy modes. In ϕ^4 theory for example, the action is near the origin in Euclidean momentum space. One starts with a ball of radius Λ and lets $\Lambda \to 0$ and keeps track of the couplings. When the couplings are expanded as functions of the external momenta, only the constant quartic term u_0 survives (near four dimensions) and becomes the *coupling constant* of the theory.

In adapting the RG to fermions at finite density I had to make some changes to suit the problem. The biggest difference was that the low energy states were not near the origin in momentum space, but near the Fermi surface. Let us focus on $d = 2$ which is easiest to explain and experimentally the most relevant. The region of interest is an annulus of thickness 2Λ, located symmetrically about the Fermi circle, as shown in Figure 13.1. The RG consists of reducing Λ to zero. Since the low energy region is a surface and not point, we will end up with *coupling functions* that vary over the surface. One can expect the flows of these functions to be complicated. What happened when I actually did it was a pleasant surprise. I will describe it in minimum detail since I have described it in pitiless detail elsewhere.

The Euclidean action for the free theory is

$$S_0 = \int_{-\infty}^{\infty} \int_{-\Lambda}^{\Lambda} \int_{0}^{2\pi} d\omega dk d\theta \; \bar{\psi}(\omega, k, \theta) \left[i\omega - k \right] \psi(\omega, k, \theta) \tag{13.1}$$

where k is the deviation from the Fermi momentum K_F. Factors actors like K_F in the measure have been absorbed in the measure, and the Fermi velocity has been set equal to unity.

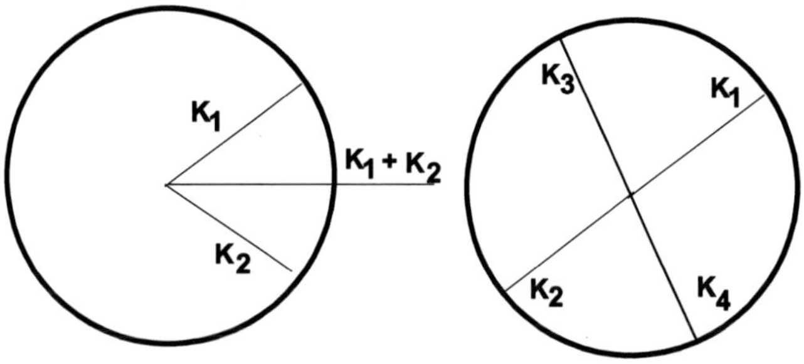

Fig. 13.2. Since the momenta come from the Fermi circle, the only way to conserve momentum is either (left) forward scattering or (right) Cooper scattering, where 1 and 2 are opposite as are 3 and 4, but in a different direction.

We see that the action is that of a Euclidean 1 + 1 dimensional field theory: ω is the frequency and k is the momentum. The angle θ is an internal isospin-like variable of which the energy is independent. (In the usual 1 + 1 dimensional theory [5], the Fermi surface is a pair of disjoint points labeled left and right.) Consequently the only interaction that will survive is the quartic one, which will be marginal. Furthermore, its dependence on ω and k is irrelevant. However it can depend on the external θ's:

$$u_0 = u_0(\theta_1, ..., \theta_4) \tag{13.2}$$

which is the new feature here. Obviously only three of the θ's are independent due momentum conservation. However, because the momenta come not from the plane but the Fermi circle, only two of them, say the incoming ones θ_1 and θ_2, can be specified freely, and the other two must be equal to them up to an irrelevant permutation. Thus only forward scattering amplitudes survive, as shown in the left half of Figure 13.2.

A subtle point: the fermions' momenta are not confined to the FS, but free to wiggle by an amount of order Λ. However the coupling functions may be evaluated on the FS since their dependence on k is irrelevant. So we have now

$$u_0 = u_0(\theta_1, \theta_2). \tag{13.3}$$

Finally rotational invariance tells us that

$$u_0 = u_0(\theta_1, \theta_2) = u_0(\theta_1 - \theta_2) \equiv F(\theta) \tag{13.4}$$

and so the asymptotic behavior of the fermions is controlled by a single function $u_0 \equiv F$. It is called the Landau parameter, because he arrived at it in his own way decades ago, without explicitly appealing to the RG.

There is an exception to the forward scattering restriction shown in the right half of Figure 13.2. If the incoming angles are exactly opposite, the final ones also have to be equal and opposite, but the final pair can be any angle θ relative to the initial, leading to another function $V(\theta)$ that survives the RG and describes the scattering of Cooper pairs.

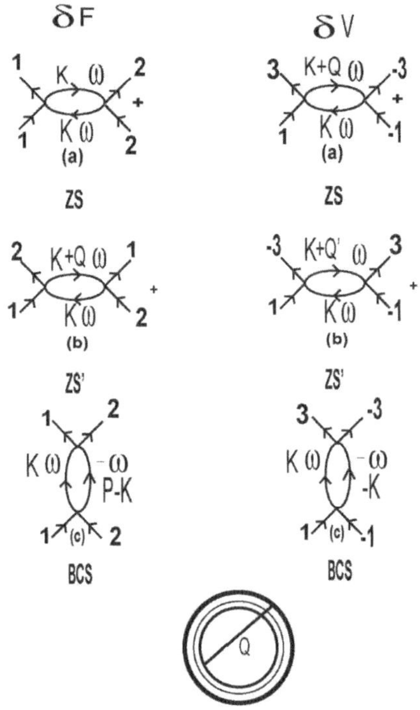

Fig. 13.3. The diagrams that contribute to the flow of F and V. Only ZS has a contribution of order $d\Lambda$ to δu but the diagram is killed by ω-integration because the poles lie on the same half-plane. The BCS diagram contributes to δV to order $d\Lambda$ which survives the ω-integration because the poles lie on opposite half-planes. The bottom figure shows that a momentum Q of order K_F determines the initial momentum on either shell being integrated up to a small multiplicity.

The fate of the marginal couplings has to be decided by a one-loop calculation. The diagrams for F and V appear in Figure 13.3.

These diagrams have the same topology as ϕ^4 field theory, but behave differently. In the diagram labeled ZS (for zero-sound) there is no transfer of momentum since $\theta_1 = \theta_3$ and no transfer of ω because the external legs can be chosen to have $\omega = 0$. This means the two propagators in the loop have poles in the same half-plane in ω and give zero under ω-integration, for we may close the contour the other way. (It is well known that in the ZS channel, only particle–hole diagrams can survive the ω-integration.) In the ZS diagram with crossed outgoing momenta, the two external legs at the left vertex restrict the loop variables to a region of order $d\Lambda^2/K_F$, where a factor $d\Lambda$ is from the shell width and a factor $d\Lambda/K_F$ is from the angular restriction. Finally, in the BCS diagram the two initial momenta again restrict the loop momenta to a region of size $d\Lambda^2/K_F$. So $du/dt = 0$, where $dt = d\Lambda/\Lambda$. Later I will argue that u does not flow to all orders.

The flow of the Cooper amplitude is similar with notable exception: the particle–particle bubble, with opposite ω and same energy (due to time-reversal invariance), makes a contribution of order $d\Lambda$ when the loop ω is integrated. Furthermore, the angle of the loop momenta can range freely over the FS as long as they too are opposite. This leads to the *functional flow equation*

$$\frac{dV(\theta_1-\theta_3)}{dt}=-\int d\theta V(\theta_1-\theta)\,V(\theta-\theta_3). \tag{13.5}$$

By expanding in terms of angular momentum eigenfunctions we get an infinite number of flow equations for the coefficients V_m:

$$\frac{dV_m}{dt}=-V_m^2, \tag{13.6}$$

one for each coefficient. These equations tell us that if the potential in angular momentum channel m is repulsive, it will get renormalized down to zero (a result derived many years ago by Morel and Anderson [6]), while if it is attractive, it will run off, causing the BCS instability. This is the reason the V's are not a part of Landau theory, where the assumed adiabatic continuity requires that there be no phase transitions.

It was at this point that I tried to publish my results but ran into a brick wall. The main objection was that I had not computed any numbers. Finally I got a break when I was asked to contribute to Michael Fisher's 60-th birthday Festschrift in a special issue of Physica [7] in 1991. Since Michael knew of this work and had heartily encouraged me (as on many others occasions), I was delighted to publish my work and also honor him with a work he approved of.

For the next two years I traveled a lot, describing my work. I discovered and resolved many subtleties I had initially overlooked. I learned a lot from my audience and clarified their issues. I summarized my findings in the Reviews of Modern Physics issue that came out in 1994 [11]. John Wilkins insisted, as editor, that I should take all the space I needed to render the article accessible to a wide audience, especially graduate students. I needed no coaxing and jumped at that invitation. Here are some highlights from that review.

- In more than two spatial dimensions, Fermi liquids still behave like $1+1$ dimensional theories, with θ getting replaced by more angles. The RG now allows non-forward scattering amplitudes but their effect (say on lifetime) is small due to phase-space. Only the forward scattering amplitudes enter the energetics.
- I was made to realize that fixed point I had found described Landau Fermi liquid theory [8]. In particular, the non-singular fixed point coupling $F(\theta)$ was the Landau parameter, which he had arrived at by considering the ground state energy as functional of the shape of the Fermi surface. Subsequently his students [9] gave a diagrammatic proof of his results.

  Let us remember that getting the fixed point action is only half the task. One still needs to solve for the Green's functions of the cut-off theory. For these one draws Feynman diagrams with vertices F and loop momenta that run in the range $-\Lambda \leq k \leq \Lambda$. (I ignore the Cooper channel following Landau.) Here one finds a great simplification: the diagrams are given by a sum of iterated ZS bubbles that can be summed geometrically.

For example $\bar{\psi}\psi - \bar{\psi}\psi$ correlation function (with vanishing values of external frequency and momentum transfer and in the limit $\omega/q \to 0$) takes the form

$$\chi = \frac{\chi_0}{1 + F_0}, \tag{13.7}$$

where F_0 is the angular average of $F(\theta)$ and χ_0 is the answer when $F = 0$. Note that the answer is not perturbative in F. Landau got this result by working with the ground-state energy as a functional of Fermi surface deformations. We get it by summing bubbles.

The RG provides us with not just the ground-state energy as Landau theory does, but an effective Hamiltonian (operator) for all of low-energy physics. This operator problem can be solved using large N-techniques.

The bubble sums should ring a bell with readers familiar with the $1/N$ expansion (say from the Gross–Neveu model [10]). I showed in my review that the small cut-off theory is a $1/N$ theory with $N \simeq K_F/\Lambda$. By dividing the annulus into N regions of width 2Λ in the angular direction one ends up with N patches, labeled by an index that runs from 1 to N. The kinematics is such that only diagrams where the patch index is freely summed survive in the large N limit. These turn out to be the ZS diagram (which does not contribute to the flow but does contribute to the iterated bubble in the computation of Green's functions) and the Cooper diagram that gives rise to the flow.

Since in the large N limit the one-loop β-function for the fermion–fermion coupling is exact, it follows that the marginal nature of the Landau parameters F and the flow of V in Eq. (13.6) are both exact as $\Lambda \to 0$.

- If a Fermi surface is nested, that is, if there is a momentum Q that takes every point on the FS to another point on the FS, the system has infinite susceptibility to charge density wave (CDW) order at that Q. So order sets in for arbitrarily small coupling, just as the BCS instability sets in for arbitrarily weak attraction. I showed that in this case, as $\Lambda \to 0$, the kinematics allows another marginal amplitude W, besides F and V. At one loop W flows to strong values, exhibiting the CDW instability. I computed the functional flow equation, but in the absence of rotational invariance, could not separate them into decoupled flows for individual angular momentum amplitudes.

  Thus the RG seems to not only get the Fermi liquid as a fixed point, but also all its known instabilities. The only way a non-Fermi liquid can arise is if the entire diagrammatic approach fails at the outset due to strong interactions that invalidate any perturbative analysis that starts with free fermions.
- I laid out a road map for solving realistic problems by first integrating high energy modes perturbatively, getting some effective couplings and then letting the RG run its course, mainly to see if a repulsive microscopic interaction could lead to

pairing. Unlike Ken Wilson, I talked about it, but did not actually do this. Luckily this was subsequently done by many others, as I will describe momentarily.

13.2. Ken's reaction

Of all my talks, one of the most significant ones was at Ohio State where Ken was present. He showed keen interest, especially in the way the RG demystified and so naturally led to Landau theory. I also got the chance to have a long discussion after my talk on a problem that vexed me at that time. The discussion ended in a manner that seems to have been experienced and chronicled by many others. He simply said at the end of my long statement of the problem "You must integrate and see what happens". I have turned that oracular sentence over and over in my head, the way scholars of the Rg Veda or the Talmud do, in search of hidden meanings. I am not sure what he meant, but ever since, when in doubt, I have always integrated to see what happens.

13.3. Contemporaneous and subsequent developments

After this rather personal description of events I want to add a few words on related works. Benfatto and Gallavotti [12] considered this problem around the time I did. Their language is very different from mine, but the parallels (including the BCS flow) are readily visible. Joe Polchinski [13] discovered the marginal nature of the Fermi liquid couplings in his study of Effective Field Theories. Feldman *et al.* [14] attacked the same problem more formally and later independently derived the $1/N$ expansion [15].

One can ask how a non-Fermi liquid of the type Anderson [4] envisaged may emerge by asking how the above derivation could fail. One way is at strong coupling where the perturbative RG does not converge. Another option is to consider fermions coupled to gauge fields either in strongly correlated electrons (Polchinski [16]) or in the Fractional Quantum Hall problem at half-filling (Halperin, Lee and Read [17]). Earlier I had raised the question of whether a repulsive interaction can spawn superconductivity within the RG. This proposed attack has two stages: integrate high energy modes perturbatively (to one loop say) to get an effective set of interactions and then run the flow as above starting with these as initial couplings. Among others, this was done by Zanchi and Schulz [18], Halboth and Metzner [19], Wang *et al.* [20], and Honerkamp *et al.* [21]. More rigorous derivations of the ideas espoused here were derived, among others, by Salmhofer [22], though as of now, no small parameters besides $\Lambda/K_F = 1/N$ or weak coupling have emerged to justify approximations which inevitably follow in any practical calculation.

The $1/N$ expansion may be used to study problems where the interaction has frequency dependence, as in the BCS problem with retardation (Tsai S. W. *et al.* [23]).

It is possible to use the RG even if bosons are simultaneously present, provided the dynamical exponent $z = 1$ (Yamamoto and Si [24]).

The theory may be applied to nuclear matter or stars (Schwenk, Friman and Brown, [25]), as well as quarks at finite density (Rajagopal and Wilczek [26], Alford *et al.* [27]).

One can also vary the power law of the force between fermions continuously to a controllable fixed point (Nayak and Wilczek [28]).

The theory may also be extended to quantum dots to address a problem with finite size, disorder and interactions (Murthy and Shankar [29]).

If the present approach is applied to fermions in d spatial dimensions, the scaling is done only for ω and $k = K - K_F$, since the other angles (like θ) on the Fermi surface behave like internal or isospin degrees of freedom. Thus the Fermi liquid behaves like a $1+1$ dimensional field theory for all $d \geq 2$. A new generalization has been proposed (Senthil and Shankar [30]) to Fermi surfaces with more co-dimensions on which the energy depends linearly. For example in three dimensions a line Fermi surface has co-dimension 2 and behaves like a $2+1$ dimensional theory. It is also possible to vary these extra dimensions in an ε-expansion and obtain new weak coupling fixed points. This idea has been advanced further by S. S. Lee [31].

References

[1] Kadanoff L. P., Scaling laws for Ising models near T_c, *Physics*, **2**, 263 (1966). See also Kadanoff, L. P., The application of renormalization group techniques to quarks and strings, *Rev. Mod. Phys.*, **49**, 267–296 (1977).

[2] Wilson, K. G., Renormalization group and critical phenomena. I. Renormalization group and the Kadanoff scaling picture, *Phys. Rev. B*, **4** (9), 3174 (1971). See also Wilson, K. G., The renormalization group: Critical phenomena and the Kondo problem, *Rev. Mod. Phys.*, **47**, 773 (1975), and Wilson, K. G. and Kogut, J. B., *Phys. Rep.*, **12**, 7 (1974).

[3] Wilson, K. G. and Fisher, M. E., Critical exponents in 3.99 dimensions, *Phys. Rev. Lett.*, **28**, 240–243 (1972). See also Fisher, M. E., The renormalization group in the theory of critical behavior, *Rev. Mod. Phys.*, **4**, 597 (1974), and Fisher, M. E., Renormalization group theory: Its basis and formulation in statistical physics, *Rev. Mod. Phys.*, **70**, 653 (1998).

[4] Anderson, P. W., Luttinger-liquid behavior of the normal metallic state of the 2D Hubbard model, *Phys. Rev. Lett.*, **64**, 1839 (1990).

[5] Solyom, J., The Fermi gas model of one-dimensional conductors, *Advances in Physics*, **28**, 201 (1979), and Bourbonnais, C. and Caron, L. G., *Int. J. Mod. Phys. B*, **5**, 1033, (1991).

[6] Morel, P. and Anderson, P. W., Calculation of the superconducting state parameters with retarded electron-phonon interaction, *Phys. Rev.*, **125**, 1263 (1962).

[7] Shankar, R., Renormalization group for interacting fermions in $d > 1$, *Physica A*, **177**, 530 (1991).

[8] Landau, L. D., *Sov. Phys. JETP*, **3**, 920 (1956).

[9] Abrikosov, A. A., Gorkov, L. P., and Dzyaloshinski, I. E., *Methods of Quantum Field Theory in Statistical Mechanics*, Dover, New York, 1963; Pines, D. and Nozieres, P.,

The Theory of Quantum Liquids, W. A. Benjamin, 1966; Baym, G. and Pethick, C., *Landau Fermi Liquid Theory*, Wiley, New York, 1991.

[10] Gross, D. J. and Neveu, A., Dynamical symmetry breaking in asymptotically free field theories, *Phys. Rev. D*, **10**, 3235 (1974).

[11] Shankar, R., Renormalization group approach to interacting fermions, *Rev. Mod. Phys.*, **66**, 129 (1994).

[12] Benfatto, G. and Gallavotti, G., Renormalization-group approach to the theory of the Fermi surface, *Phys. Rev. B*, **42**, 9967 (1990).

[13] Polchinski, J., Effective field theory and the Fermi surface, in *Recent Directions in Particle Theory*, eds. Harvey, J. and Polchinski, J., TASI Lectures 1992, World Scientific, Singapore, 1993.

[14] Feldman, J., and Trubowitz, E., Perturbation theory for many fermion systems, *Helv. Phys. Acta*, **63**, 157 (1990).

[15] Feldman, J., Magnan, J., Rivasseau, V., and Trubowitz, E ., An intrinsic $1/N$ expansion for many fermion systems, *Europhys. Lett.*, **24**, 437 (1992).

[16] Polchinski, J., Low energy dynamics of the spinon-gauge system, *Nucl. Phys. B*, **422**, 617 (1993).

[17] Halperin, B. I., Lee, P. A., and Read, N., Theory of the half-filled Landau level, *Phys. Rev. B*, **47**, 7312 (1993).

[18] Zanchi, D. and Schulz, H. J., Weakly correlated electrons on a square lattice: Renormalization-group theory, *Phys. Rev.*, **61**, 13609 (2000).

[19] Halboth, C. J. and Metzner, W., Renormalization group analysis of the 2d Hubbard Model, *Phys. Rev. B*, **61**, 7364 (2000).

[20] Wang, F. *et al.*, Functional renormalization group study of the pairing symmetry and pairing mechanism of the FeAs based high temperature superconductors, *Phys. Rev. Lett.*, **102**, 047005 (2009).

[21] Honerkamp, C. *et al.*, Breakdown of the Landau–Fermi liquid in two dimensions due to umklapp scattering, *Phys. Rev.*, **63**, 035109 (2001); for a variation see Honerkamp, C. *et al.*, Interaction flow method for many-fermion systems, *Phys. Rev. B*, **70**, 235115 (2004).

[22] Salmhofer, M., *Renormalization: An introduction*, Springer, Heidelberg, 1998.

[23] Tsai, S. W. *et al.*, Renormalization group approach to superconductivity from weak to strong electron–phonon coupling, *Phil. Mag.*, **86**, 2631 (2006).

[24] Yamamoto, J. and Si, Q., Renormalization group for mixed boson fermion systems, *Phys. Rev. B*, **81**, 205106 (2010).

[25] Schwenk, A., Brown, G. E., and Friman, B., *Nucl. Phys. A*, **713**, 191 (2003).

[26] Rajagopal, K. and Wilczeck, F., The condensed matter physics of QCD, in *At the Frontier of Particle Physics: Handbook of QCD*, Festschrift in honor of B. L. Ioffe, ed. Shifman, M., Vol. 2, World Scientific, 2001.

[27] Alford, M. G., Schmitt, A., Rajagopal, K., and Schäfer, T., Color superconductivity in dense quark matter, *Rev. Mod. Phys.*, **80**, 1455 (2008).

[28] Nayak, C. and Wilczek, F., Non-Fermi liquid fixed points in $2+1$ dimensions, *Nucl. Phys. B*, **417**, 379 (1993).

[29] Murthy, G. and Shankar, R., Chaotic quantum dots with strongly correlated electrons, *Rev. Mod. Phys.*, **80**, 379 (2008).

[30] Senthil, T. and Shankar, R., Fermi surfaces in general codimension and a new controlled non-trivial fixed point, *Phys. Rev. Lett.*, **102**, 046406 (2009).

[31] Lee, S. S., Perturbative non-Fermi liquids from dimensional regularization, *Phys. Rev. B*, **88**, 245106 (2013).

Chapter 14

Gauge theories on the coulomb branch*

John H. Schwarz
Walter Burke Institute for Theoretical Physics
California Institute of Technology
Pasadena, CA 91125, USA
jhs@theory.caltech.edu

We construct the world-volume action of a probe D3-brane in $AdS_5 \times S^5$ with N units of flux. It has the field content, symmetries, and dualities of the $U(1)$ factor of $\mathcal{N} = 4$ $U(N+1)$ super Yang–Mills theory, spontaneously broken to $U(N) \times U(1)$ by being on the Coulomb branch, with the massive fields integrated out. This motivates the conjecture that it is the exact effective action, called a *highly effective action* (HEA). We construct an $SL(2,\mathbb{Z})$ multiplet of BPS soliton solutions of the D3-brane theory (the conjectured HEA) and show that they reproduce the electrically charged massive states that have been integrated out as well as magnetic monopoles and dyons. Their charges are uniformly spread on a spherical surface, called *a soliton bubble*, which is interpreted as a phase boundary.

14.1. Introduction

Two recent papers [1, 2] explored the construction and properties of certain actions that describe superconformal gauge theories on the Coulomb branch. The examples that were studied are ones with a lot of supersymmetry. The reason for this choice is that the large symmetry gives us more confidence in the plausibility of the conjecture that we will describe.

Even though Ref. [1] discussed several examples, this article only describes one of them, namely $\mathcal{N} = 4$ super Yang–Mills (SYM) theory. Our procedure is to begin by deriving the world-volume action of a probe D3-brane in an $AdS_5 \times S^5$ background of type IIB superstring theory. As will be explained, the probe D3-brane action involves various approximations. However, the resulting formula has a number of exact properties, especially symmetries, that make it an attractive candidate for the exact solution to a different problem. Specifically, we conjecture that it is the

*Lectures presented at the Erice 2014 International School of Subnuclear Physics.

exact effective action for $\mathcal{N} = 4$ SYM theory on the Coulomb branch, called a *highly effective action* (HEA).

More generally, we consider the world-volume action of a probe p-brane in an $AdS_{p+2} \times M_q$ background. M_q is a q-dimensional compact space with N units of q-form flux, $\int_{M_q} F_q \sim N$. In addition to a D3-brane in $AdS_5 \times S^5$, Ref. [1] also discussed

- M2-brane in $AdS_4 \times S^7/\mathbb{Z}_k$
- D2-brane in $AdS_4 \times CP^3$
- M5-brane in $AdS_7 \times S^4$

The first two of these correspond to ABJM theory in three dimensions, whereas the third one corresponds to the $(2,0)$ theory in six dimensions. As our understanding improves, we intend to extend these investigations to theories with less supersymmetry.

Section 2 provides background material in string theory and quantum field theory that is required to follow the rest of the story. People knowledgable in these subjects can skip this section. Section 3 constructs (the bosonic part of) the formula for the probe D3-brane action — the conjectured HEA — and discusses some of its properties. Section 4, based on Ref. [2], describes the construction of BPS soliton solutions of this action. We find exactly the spectrum of soliton solutions expected for the HEA. Moreover, they have an interesting structure in which the charge of the soliton resides on a spherical shell. Even though the theory is nongravitational, the solitons turn out to have unexpected analogies with black holes. This analogy suggests that it may be possible to associate an entropy to solitons with a large charge.

14.2. Background material

14.2.1. $\mathcal{N} = 4$ *super Yang–Mills theory*

When one says that a theory has $\mathcal{N}$ supersymmetries, one means that the conserved supercharges consist of $\mathcal{N}$ irreducible spinors, as appropriate for the spacetime dimension in question. $\mathcal{N} = 4$ supersymmetry is the maximal amount that is possible for an interacting nongravitational quantum field theory in four-dimensional spacetime (4d). The four spinors may be chosen to be two-component complex Weyl spinors. Thus, there are 16 real supercharges. It is believed that the only 4d nongravitational quantum field theories with this amount of supersymmetry are SYM theories. $\mathcal{N} = 4$ SYM exists for any compact gauge group, but we will focus on the case of $U(N)$, so all of the fields are $N \times N$ hermitian matrices.

$\mathcal{N} = 4$ SYM was originally derived by first constructing SYM theory in 10d [3]. It contains a ten-vector gauge field A_M and a Majorana–Weyl spinor Ψ, both in the adjoint representation of the gauge group. The 4d theory was then obtained by dimensional reduction, which simply means dropping the dependence of the fields on six of the (Euclidean) spatial dimensions. This gives a 4d theory containing a four-vector A_μ, six scalars ϕ^I, and four Weyl spinors ψ^A. The 10d SYM theory is not a consistent quantum field theory (by itself), but the 4d SYM theory obtained in this way is a consistent quantum theory. It has an $SU(4) \sim SO(6)$ global symmetry, which corresponds to rotations of the six extra dimensions. Because this symmetry also rotates the four supercharges, Q^A, it is called an R-symmetry.

$\mathcal{N} = 4$ SYM theories are conformally invariant, which implies that they are ultraviolet (UV) finite. This was the first class of UV finite 4d quantum field theories to be discovered, but many more (with less supersymmetry) are now known. Combining all of the spacetime symmetries, which are Lorentz invariance, translation invariance, supersymmetry, scale invariance, conformal invariance, R-symmetry, and conformal supersymmetry, gives the superconformal supergroup called $PSU(2,2|4)$. (A supergroup has both bosonic and fermionic generators.) Its bosonic subgroup is $SU(2,2) \times SU(4)$. The first factor is the 4d conformal group and the second factor is the R-symmetry. Anticommutators of the supersymmetry charges give the momentum operators, which are generators of spacetime translations. Similarly, anticommutators of conformal supersymmetry charges give operators that generate conformal transformations. Mixed anticommutators give the rest of the bosonic generators. Counting both the Poincaré and conformal supersymmetries, there are a total of 32 fermionic generators.

The parameters of $\mathcal{N} = 4$ SYM with gauge group $U(N)$ are a YM coupling constant g, a vacuum angle θ, and the rank of the gauge group N. For large N and fixed 't Hooft parameter

$$\lambda = g^2 N, \tag{14.1}$$

the theory has a $1/N$ expansion (for large N) with a nice topological interpretation. The leading term in this expansion (the planar approximation) has an additional symmetry called dual conformal invariance. This amount of symmetry is sufficient to make the theory completely integrable. It is not known whether this only applies to the planar approximation or whether it extends to the complete theory. Defining the complex parameter

$$\tau = \frac{\theta}{2\pi} + i\frac{4\pi}{g^2}, \tag{14.2}$$

the $U(N)$ theory has an $SL(2,\mathbb{Z})$ duality symmetry under which

$$\tau \to \frac{a\tau + b}{c\tau + d}, \tag{14.3}$$

where a, b, c, d are arbitrary integers satisfying $ad - bc = 1$. The transformation $\tau \to -1/\tau$ is called S-duality. When $\theta = 0$, this gives $g \to 4\pi/g$, which allows one to relate strong coupling to weak coupling. S-duality is an exact nonabelian electric-magnetic equivalence.

14.2.2. *Type IIB superstring theory*

Type IIB superstring theory is one of the five distinct superstring theories in 10d. The IIB theory has two Majorana–Weyl supersymmetries of the same chirality, for a total of 32 real conserved supercharges. Its massless bosonic fields are

- the 10d metric g_{MN}
- the dilaton σ
- the Neveu–Schwarz (NS-NS) two-form B_{MN}
- Ramond–Ramond (RR) zero-, two-, and four-forms C, C_{MN}, and C_{MNPQ}

The four-form C_4, constructed from C_{MNPQ}, has a self-dual field strength $F_5 = dC_4 + \ldots$.

Type IIB superstring theory has a few solutions that preserve all of the supersymmetry. The most obvious one is 10d Minkowski spacetime, *i.e.*, $g_{MN} = \eta_{MN}$. In this solution σ and C are constants and the other fields vanish. A less obvious maximally supersymmetric solution has a metric describing the geometry $AdS_5 \times S^5$ — we'll give the formula later. In this solution σ and C are again constants. However, now $F_5 \sim N[\text{vol}(AdS_5) + \text{vol}(S^5)]$. N is the number of units of five-form flux threading the five-sphere, $\int_{S^5} F_5 \sim N$, where the coefficients depend on normalization conventions. The isometry of this solution is given by the supergroup $PSU(2,2|4)$, the same supergroup as before! In particular, $SU(2,2)$ is the isometry of AdS_5 and $SU(4)$ is the isometry of S^5.

The value of $\exp(\sigma)$ is the string coupling constant g_s and the value of C is called χ. In terms of these one can form

$$\tau = \chi + i/g_s. \tag{14.4}$$

Type IIB superstring theory has an exact $SL(2,\mathbb{Z})$ symmetry under which

$$\tau \rightarrow \frac{a\tau + b}{c\tau + d}, \tag{14.5}$$

as before. More generally, the fields σ and C transform in this manner, but we are concerned with the case when they are constants.

14.2.3. *AdS/CFT duality*

The preceding sections have been presented so as to make the AdS/CFT conjecture seem obvious. However, when it is was put forward by Maldacena [4], it came as quite a surprise. Specifically, he proposed that $\mathcal{N} = 4$ SYM in 4d with $U(N)$ gauge group is exactly equivalent ("dual") to type IIB superstring theory in an $AdS_5 \times S^5$ background with N units of five-form flux. There are many other analogous AdS/CFT pairs, some of which are relevant to the other examples of probe-brane theories listed in the introduction.

The evidence for AdS/CFT duality that we have presented so far is that both the gauge theory and the string theory have $PSU(2,2|4)$ symmetry and $SL(2,\mathbb{Z})$

duality. However, by now there is much, much more evidence, which we will not describe here. Even though the evidence is overwhelming, a complete proof is not possible. The reason for saying this is that we lack a complete nonperturbative definition of type IIB superstring theory (other than the one given by AdS/CFT duality). The best one can hope to do, in the absence of such a definition, is to show that everything we know about type IIB superstring theory in an $AdS_5 \times S^5$ background with N units of five-form flux agrees with what can be deduced from $\mathcal{N} = 4$ SYM in 4d with $U(N)$ gauge group.

14.2.4. *D3-branes*

Superstring theories contain various supersymmetric (and hence stable) extended objects. Ones with p spatial dimensions are called p-branes. They carry a type of conserved current J that couples to a $(p+1)$-form gauge field A with a $(p+2)$-form field strength $F = dA$. Those p-branes on which strings can end are called Dp-branes. (D stands for Dirichlet here, since such strings satisfy Dirichlet boundary conditions in the directions orthogonal to the p-brane. They also satisfy Neumann boundary conditions in the directions parallel to the Dp-brane.)

In type IIB superstring theory supersymmetric Dp-branes exist for $p = 1, 3, 5, 7$. Only the D3-brane is invariant under $SL(2, \mathbb{Z})$ transformations. This theory has an infinite number of different kinds of strings. A (p, q) string arises as a bound state of p fundamental strings and q D-strings, provided p and q are relatively prime. These strings transform irreducibly under $SL(2, \mathbb{Z})$, and any one of them can end on a D3-brane. From the point of view of the world-volume theory of a single D3-brane, which is an abelian gauge theory, the end of the string appears to carry p units of electric charge and q units of magnetic charge.

The various p-branes also act as sources of gravitational and other fields. Maldacena was led to his conjecture by realizing that N coincident D3-branes give a "black brane," which is a higher-dimensional generalization of a black hole, whose near-horizon geometry is $AdS_5 \times S^5$ with N units of 5-form flux on the sphere. In this way the branes are replaced by a 10d geometry with a horizon, and the 10d string theory is represented "holographically" by a 4d quantum field theory.

A stack of coincident flat Dp-branes has fields that are localized on its $(p+1)$-dimensional world volume. They define a "world-volume theory," which is maximally supersymmetric when the background in which they are embedded is maximally supersymmetric. In particular, we will study the world-volume theory of a single D3-brane embedded in an $AdS_5 \times S^5$ background geometry, which may be regarded as having been created by N other coincident D3-branes.

14.2.5. *The Coulomb branch*

A stack of N coincident flat D3-branes has a world-volume theory, which is a $U(N)$ gauge theory. Consider starting with N coincident flat D3-branes and pulling them apart along one of the orthogonal axes (let's call it the x direction) into clumps $N_1 + N_2 + \ldots + N_k = N$. Then, the world-volume gauge symmetry is broken spontaneously,

$$U(N) \to U(N_1) \times U(N_2) \times \ldots \times U(N_k). \tag{14.6}$$

The world-volume fields that arise as the lowest mode of a fundamental string with one end attached to the ith D3-brane and the other end attached to the jth D3-brane acquire the mass

$$m_{ij} = |x_i - x_j| T, \tag{14.7}$$

where T is the fundamental string tension. This is similar to the Higgs mechanism. The main difference is that the scalar fields that are eaten by the gauge fields that become massive belong to the adjoint representation of the original gauge group (in contrast to the $SU(2)$ Higgs doublet in the Standard Model). To emphasize this distinction, this phase of the gauge theory is called a Coulomb branch (rather than a Higgs branch).

Let us consider $\mathcal{N} = 4$, $d = 4$ SYM theory with a $U(N)$ gauge group. For most purposes one can say that a free $U(1)$ multiplet decouples leaving an $SU(N)$ theory. However, this 'decoupled' $U(1)$ is needed to get the full $SL(2,\mathbb{Z})$ duality group rather than a subgroup [5] [6]. In the special case of $U(2)$, if we ignore the decoupled $U(1)$, the remaining $SU(2)$ theory on the Coulomb branch has unbroken $U(1)$ gauge symmetry. Let us refer to the massless supermultiplet as the "photon" supermultiplet and the two massive supermultiplets as $W^{\pm}$.

The $U(2)$ $\mathcal{N} = 4$ SYM theory on the Coulomb branch has a famous soliton solution: the 't Hooft–Polyakov monopole. This solution preserves half of the supersymmetry. One says that it is "half BPS." This monopole is part of an infinite $SL(2,\mathbb{Z})$ multiplet of half-BPS (p,q) states, with p units of electric charge and q units of magnetic charge, where p and q are coprime. The masses, determined by supersymmetry, are

$$M_{p,q} = vg|p + q\tau| = vg\sqrt{\left(p + \frac{\theta}{2\pi}q\right)^2 + \left(\frac{4\pi q}{g^2}\right)^2}$$

where $\langle\phi\rangle = v$ is the vev of a massless scalar field. The W mass is $M_{1,0} = gv$ and the monopole mass is $M_{0,1} = 4\pi v/g$ for $\theta = 0$. As discussed above, a (p,q) dyon is introduced when a (p,q) string ends on a D3-brane.

14.2.6. *The probe approximation*

Consider a D3-brane embedded in a 10d spacetime. The probe approximation involves neglecting the back reaction of the brane on the geometry and the other background fields. Since the brane is a source for one unit of flux, this requires that the background flux N is large, so that the distinction between N and $N+1$ becomes negligible. The world-volume action of a single D-brane contains a $U(1)$

field strength, $F_{\alpha\beta} = \partial_\alpha A_\beta - \partial_\beta A_\alpha$. Since derivatives of F are included in the formula, F is required to vary sufficiently slowly so that its derivatives can be neglected. A similar restriction applies to the other world-volume fields, as well.

Despite these approximations, the probe D3-brane action has some beautiful exact properties: It precisely realizes the isometry of the $AdS_5 \times S^5$ background geometry as a world-volume superconformal symmetry $PSU(2,2|4)$. Only the bosonic subgroup is taken into account in the subsequent discussion, since fermi fields are omitted. (A version of the complete formula with fermions is given in Ref. [7].) The brane action also has the duality symmetry of the background, which is $SL(2,\mathbb{Z})$ for the D3-brane example. Furthermore, the D3-brane world-volume action is most naturally formulated with local symmetries, which are general coordinate invariance and a fermionic symmetry called kappa symmetry. It is only after implementing a suitable gauge choice that one is left with a conventional nongravitational quantum field theory in Minkowski spacetime.

In principle, one can integrate out the massive fields of the Coulomb branch theory exactly, thereby producing a very complicated formula in terms of the massless photon supermultiplet only. Very schematically,

$$\exp(iS_{\text{HEA}}) = \int DW^+ DW^- \exp(iS). \tag{14.8}$$

If one could do this integral exactly, which is not possible in practice, the resulting action would capture the entire theory on the Coulomb branch, and it would be valid at all energies. This is in contrast with the more common notion of a low-energy effective action, which only includes the leading terms in a derivative expansion. We have proposed to call such an exact Coulomb branch action a *highly effective action* (HEA). Even though we cannot carry out such an exact computation, in some cases we know many of the properties that the HEA should possess. One can hope that they go a long way towards determining it.

14.3. The highly effective action

14.3.1. *General requirements for an HEA*

An HEA should have all of the unbroken and spontaneously broken global symmetries of the original Coulomb branch theory with the massive W fields included. The conformal symmetry is spontaneously broken as a consequence of assigning a vacuum expectation value (vev) to a massless scalar field that has a flat potential. In other words, this vev spontaneously breaks spacetime symmetries (though not Poincaré symmetry) as well as gauge symmetry. An HEA should have the same duality properties as the Coulomb branch theory containing explicit W fields. The global symmetry and duality groups are $PSU(2,2|4)$ and $SL(2,\mathbb{Z})$ in the example considered here.

A further requirement for an HEA is that it should have the same spectrum of supersymmetry protected (or BPS) states as the original Coulomb branch theory. In particular, the $W^{\pm}$ supermultiplets, which have been integrated out, should reappear in the HEA as solitons. By contrast, this would not be expected for a low-energy effective action. When we examine these soliton solutions explicitly, we will be led to a certain refinement of the interpretation of this requirement. Specifically, a complete understanding of the BPS soliton solutions also requires knowledge of the original conformal branch gauge theory for which the gauge symmetry and conformal symmetry are unbroken.

Reference [1] conjectured that the probe D3-brane action, in an $AdS_5 \times S^5$ background with one unit of flux ($N = 1$), is precisely the HEA for the $\mathcal{N} = 4$, 4d SYM theory with $U(2)$ gauge symmetry on the Coulomb branch despite the fact that it is only an approximate solution to a different problem. In fact, $N = 1$ is the worst possible case for the probe approximation. However, this shouldn't matter, because we are solving a different problem. Still, it is helpful to keep the 10d description in mind.

I am not certain that this conjecture is correct. On the one hand, the formula that we will obtain seems to be too simple to be the exact answer for such a complicated path integral. On the other hand, we will find that it has all of the expected properties of the HEA. If it is not the conjectured HEA, then it would seem to define a new maximally supersymmetric 4d quantum field theory, whose existence is unexpected. So either conclusion is remarkable.

14.3.2. *The AdS Poincaré patch*

The AdS_{p+2} geometry with unit radius can be described as a hypersurface embedded in a $(p+3)$-dimensional Lorentzian space of signature $(p+1, 2)$:

$$y \cdot y - u\phi = -1, \tag{14.9}$$

where $y \cdot y = -(y^0)^2 + \sum_1^p (y^i)^2$. The equation eliminates one of the two time directions leaving a manifold of signature $(p+1, 1)$. The Poincaré-patch metric of radius R is

$$ds^2 = R^2(dy \cdot dy - du d\phi). \tag{14.10}$$

Note that this preserves the $SO(p+1, 2)$ symmetry of the embedding equation. Defining $x^{\mu} = y^{\mu}/\phi$ and eliminating $u = \phi^{-1} + \phi x \cdot x$,

$$ds^2 = R^2(\phi^2 dx \cdot dx + \phi^{-2} dv^2). \tag{14.11}$$

It is more customary to write the formula in terms of a coordinate $z = \phi^{-1}$, which has dimensions of length. However, the choice ϕ is convenient for our purposes, since it will correspond to a scalar field, also called ϕ, with dimensions of inverse length in the D3-brane action.

The Poincaré patch is not geodesically complete. Rather there exists a different coordinate choice that describes a geodesically complete spacetime, called global AdS, for which the Poincaré patch is just a region. The Poincaré patch contains a horizon, which is not present in the global AdS metric. In this respect it is analogous to the horizon of Rindler spacetime, whose geodesic completion is Minkowski spacetime. The Poincaré patch of AdS is sufficient for our purposes.

The ten-dimensional $AdS_5 \times S^5$ metric $ds^2 = g_{MN}(x)dx^M dx^N$ is

$$ds^2 = R^2\left(\phi^2 dx \cdot dx + \phi^{-2} d\phi^2 + d\Omega_5^2\right) = R^2\left(\phi^2 dx \cdot dx + \phi^{-2} d\phi \cdot d\phi\right), \quad (14.12)$$

where $d\Omega_5^2$ is the metric of a round unit-radius five-sphere, and ϕ is now the length of the six-vector ϕ^I. It will be important later that ϕ cannot be negative.

14.3.3. *The D3-brane in $AdS_5 \times S^5$*

According to the AdS/CFT dictionary, the value of the radius R, expressed in string units, satisfies

$$R^4 = 4\pi g_s N l_s^4. \quad (14.13)$$

The positive integer N is the number of units of five-form flux on the five-sphere. Also, $\int_{S^5} F_5 = 2\pi N$ for an appropriate normalization, and N is the rank of the $U(N)$ gauge group of the dual CFT.

The probe D3-brane action is the sum of two terms: $S = S_1 + S_2$. S_1 is a Dirac–Born–Infeld (DBI) functional of the ten embedding functions $x^M(\sigma^\alpha)$ and a world-volume $U(1)$ gauge field $A_\beta(\sigma^\alpha)$ with field strength $F_{\alpha\beta} = \partial_\alpha A_\beta - \partial_\beta A_\alpha$. S_2 is a Chern–Simons-like (CS) term, which is linear in the RR fields. S_1 and S_2 also depend on the fermi fields that we are omitting.

The general formula for the bosonic part of the DBI term is

$$S_1 = -T_{D3} \int \sqrt{-\det\left(G_{\alpha\beta} + 2\pi\alpha' F_{\alpha\beta}\right)}\, d^4\sigma, \quad (14.14)$$

where $G_{\alpha\beta}$ is the induced 4d world-volume metric

$$G_{\alpha\beta} = g_{MN}(x)\partial_\alpha x^M \partial_\beta x^N. \quad (14.15)$$

As usual, $\alpha' = l_s^2$ and the D3-brane tension is

$$T_{D3} = \frac{2\pi}{g_s(2\pi l_s)^4}. \quad (14.16)$$

In fact, only two dimensionless combinations of parameters occur in the brane action:

$$R^4 T_{D3} = \frac{N}{2\pi^2} \quad \text{and} \quad 2\pi\alpha'/R^2 = \sqrt{\pi/g_s N}. \quad (14.17)$$

Thus, S_1 only depends on the dimensionless string coupling constant g_s and the integer N. We shall see shortly that the g_s dependence also drops out.

The nonvanishing Chern–Simons terms (aside from coefficients, which can be found in Ref. [1]) are

$$S_2 \sim \int C_4 + \chi \int F \wedge F. \tag{14.18}$$

The RR four-form potential C_4 has a self-dual field strength $F_5 = dC_4$. As mentioned previously,

$$F_5 \sim N \left(\mathrm{vol}(S^5) + \mathrm{vol}(AdS_5)\right), \tag{14.19}$$

and χ is the value of the RR 0-form C_0. It is proportional to the theta angle of the gauge theory.

S_1 contains a potential term $\int \phi^4 d^4x$, which should not appear in S, since there should be no net force acting on the brane when $\phi > 0$. In fact, this term is canceled by the $\int C_4$ term in S_2. To show this we consider the region M of AdS_5 for which the coordinate ϕ is less than the value of ϕ that specifies the position of the D3-brane. Then, using Stokes' theorem,

$$\int_{D3} C_4 = \int_M F_5 \sim \int_0^{\phi} \int_{D3'} \mathrm{vol}(AdS_5)', \tag{14.20}$$

where

$$\mathrm{vol}(AdS_5)' \sim (\phi')^3 d\phi' \wedge dx^0 \wedge dx^1 \wedge dx^2 \wedge dx^3. \tag{14.21}$$

Thus,

$$\int_{D3} C_4 \sim \int \phi^4 d^4x. \tag{14.22}$$

The coefficients work perfectly.

The general coordinate invariance of the D3-brane action allows one to fix a gauge. A convenient choice for our purposes is the static gauge in which the four world-volume coordinates σ^α are identified with the four Lorentzian coordinates x^μ, which were introduced previously,

$$x^\mu(\sigma) = \delta^\mu_\alpha \sigma^\alpha. \tag{14.23}$$

In this gauge $\phi^I(\sigma)$ and $A_\mu(\sigma)$ are the only remaining bosonic world-volume fields. Moreover, due to the gauge choice, they become functions of x^μ.

The complete D3-brane action in the static gauge (aside from fermions), expressed in terms of canonically normalized fields $\phi^I(\sigma)$ and $A_\mu(\sigma)$, is

$$S = \frac{1}{\gamma^2} \int \phi^4 \left(1 - \sqrt{-\det M_{\mu\nu}}\right) d^4x + \frac{1}{4} g_s \chi \int F \wedge F, \tag{14.24}$$

where

$$\gamma = \sqrt{N/2\pi^2} \tag{14.25}$$

and

$$M_{\mu\nu} = \eta_{\mu\nu} + \gamma \frac{F_{\mu\nu}}{\phi^2} + \gamma^2 \frac{\partial_\mu \phi^I \partial_\nu \phi^I}{\phi^4}, \tag{14.26}$$

where $\eta_{\mu\nu}$ denotes the Lorentz metric. It is an important fact that the first term in S is independent of g_s and χ and that they only appear (as a product) in the coefficient of the second term.

The scale invariance of the action (14.24) is manifest, since all terms have dimension four and all parameters are dimensionless. On the Coulomb branch there is a scale, the vev of the scalar field ϕ. However, the full conformal symmetry is realized on the action. Only the choice of vacuum breaks the symmetry (and all choices are equivalent). The D3-brane action contains inverse powers of the scalar field. However, the vev of this field ensures that these are not singular. Individual terms in the action can be arbitrarily complicated and still end up with dimension four by including an appropriate (inverse) power of ϕ.

Regarded as an HEA, Eq. (14.24) should encode all the quantum effects that arise from integrating out the $W^\pm$ supermultiplets. On the other hand, Eq. (14.24) should have its own loop expansion (or path integral) to take account of the quantum effects of the massless supermultiplet. The loop expansion is expected to be free of UV divergences, because of the (spontaneously broken) conformal symmetry. There are IR divergences, but they can be treated by standard methods. Rescaling all fields in Eq. (14.24) by γ brings the action to a form in which all the N dependence appears as an overall factor of N, so that $S(N) = NS(1)$. This shows that the loop expansion of this theory is a $1/N$ expansion.

14.3.4. *S-duality*

The $SL(2,\mathbb{Z})$ transformation $\tau \to -1/\tau$ duality is accompanied by an electric-magnetic transformation in the S-duality transformation of the gauge theory. This invariance has not been proved in the nonabelian formulation with W fields, but it was proved for the abelian D3-brane action in Ref. [1]. Recall that $\tau = \tau_1 + i\tau_2 = \chi + i/g_s$, but that the action only depends on $g_s\chi = \tau_1/\tau_2$. When $\tau \to -1/\tau$, $g_s\chi \to -g_s\chi$. To show invariance under this sign change, we must also examine the transformation that exchanges electric and magnetic fields.

The procedure is standard. It involves replacing the $U(1)$ gauge field A_μ by a new one such that the Bianchi identity

$$\partial_{[\mu}F_{\nu\rho]} = 0 \tag{14.27}$$

becomes a field equation, and the field equation

$$\partial_\mu \left(\frac{\partial \mathcal{L}}{\partial F_{\mu\nu}} \right) = 0 \tag{14.28}$$

becomes a Bianchi identity. The nontrivial fact, which was proved in Ref. [1], is that the new action, obtained by this field replacement together with the sign change $g_s\chi \to -g_s\chi$, is identical to the original one.

14.3.5. *Summary and discussion*

We have conjectured that the world-volume action of a probe D3-brane in an $AdS_5 \times S^5$ background of type IIB superstring theory, with one unit of flux ($N = 1$), can be reinterpreted as the HEA of $U(2)$ $\mathcal{N}=4$ SYM theory on the Coulomb branch. An explicit formula for the bosonic part of the action has been presented. It is likely that there is a generalization in which the formula with $N>1$ plays a role in the Coulomb branch decomposition $U(N+1)\to U(N)\times U(1)$. However, when $N>1$ certain issues still need to be clarified.

The evidence presented so far for the conjecture that the probe D3-brane action is the desired HEA is that the action incorporates all of the required symmetries and dualities: $PSU(2,2|4)$ superconformal symmetry (when fermions are included) and $SL(2,\mathbb{Z})$ duality. The next section will describe BPS soliton solutions of this action. Their properties will give further support to the conjecture. They will also lead us to refine the conjecture somewhat.

14.4. BPS soliton solutions

The previous section described the construction of the bosonic part of a world-volume action for a probe D3-brane in $AdS_5\times S^5$. We also discussed $\mathcal{N}=4$ SYM theory, with gauge group $U(N+1)$, on the Coulomb branch. A suitable scalar field vev breaks the gauge symmetry so that

$$U(N+1)\to U(1)\times U(N). \tag{14.29}$$

The $2N$ supermultiplets corresponding to the broken symmetries acquire mass gv. The probe D3-brane action was conjectured to be the HEA for the $U(1)$ factor.

In this section we will derive classical supersymmetric soliton solutions of the action (14.24). Even though they are classical, they are supposed to take account of all quantum effects due to the massive fields that have been integrated out. Furthermore, they will turn out be BPS, which implies that they are protected from additional quantum corrections. The soliton solutions will be independent of N (or γ), since N is an overall factor of the action written in terms of rescaled fields. Thus, this parameter drops out of the classical field equations.

For the purpose of constructing the soliton solutions, it is sufficient to consider a D3-brane that is localized at a fixed position on the S^5. In this case, the S^5 coordinates do not contribute to the D3-brane action. The radial position in the AdS_5 space is encoded in the nonnegative scalar field ϕ. The dependence of the action on this field and the $U(1)$ gauge field are all that are required in this section. The fermions are not required, since they vanish for these solutions.

14.4.1. *Solitons*

Let us begin by looking for spherically symmetrical static solutions, centered at $r=0$, for which the action (14.24) is stationary. We require that $\vec{E}$ and $\vec{B}$ (the electric and magnetic parts of $F_{\mu\nu}$) only have radial components, denoted E and B. Also, E, B, and ϕ are functions of the radial coordinate r only. It then follows that

$$-\det(M_{\mu\nu}) = -\det\left(\eta_{\mu\nu} + \gamma^2 \frac{\partial_\mu\phi\partial_\nu\phi}{\phi^4} + \gamma\frac{F_{\mu\nu}}{\phi^2}\right)$$
$$= \left(1 + \gamma^2 \frac{(\phi')^2 - E^2}{\phi^4}\right)\left(1 + \gamma^2 \frac{B^2}{\phi^4}\right). \tag{14.30}$$

This results in the Lagrangian density

$$\mathcal{L} = \frac{1}{\gamma^2}\phi^4\left(1 - \sqrt{\left(1 + \gamma^2\frac{(\phi')^2 - E^2}{\phi^4}\right)\left(1 + \gamma^2\frac{B^2}{\phi^4}\right)}\right) + g_s\chi BE. \tag{14.31}$$

The equation of motion for A_0 is $\frac{\partial}{\partial r}(r^2 D) = 0$, where

$$D = \frac{\partial\mathcal{L}}{\partial E} = E\sqrt{\frac{1 + \gamma^2 B^2/\phi^4}{1 + \gamma^2[(\phi')^2 - E^2]/\phi^4}} + g_s\chi B. \tag{14.32}$$

For a soliton centered at $r = 0$, with p units of electric charge g and q units of magnetic charge g_m, where $g_m = 4\pi/g$, we have

$$D = \frac{pg}{4\pi r^2} \quad \text{and} \quad B = -\frac{qg_m}{4\pi r^2}. \tag{14.33}$$

The mass of the soliton is given by the Hamiltonian. $H = \int \mathcal{H} d^3x = 4\pi \int \mathcal{H} r^2 dr$. The Hamiltonian density is $\mathcal{H} = DE - \mathcal{L}$. Eliminating E in favor of D gives

$$\mathcal{H} = \frac{1}{\gamma^2}\left(\sqrt{(\phi^4 + \gamma^2(\phi')^2)(\phi^4 + \gamma^2 X^2)} - \phi^4\right), \tag{14.34}$$

where

$$X = \sqrt{\tilde{D}^2 + B^2} = Q/r^2 \tag{14.35}$$

and

$$\tilde{D} = D - g_s\chi B. \tag{14.36}$$

Thus, $\tilde{D}^2 + B^2 = Q^2/r^4$, where

$$Q = \frac{g}{4\pi}|p + q\tau|. \tag{14.37}$$

As before,

$$\tau = \chi + i/g_s = \frac{\theta}{2\pi} + i\frac{4\pi}{g^2}. \tag{14.38}$$

We want to find functions $\phi(r)$ that give BPS extrema of H with the boundary condition $\phi \to v$ as $r \to \infty$. The BPS condition turns out to require that the two factors inside the square root in Eq. (14.34) are equal, which implies that $\mathcal{H} = (\phi')^2$. The proof goes as follows. One first writes the formula for $\mathcal{H}$ in the form

$$(\gamma^2\mathcal{H} + \phi^4)^2 = (\gamma^2 X|\phi'| + \phi^4)^2 + \gamma^2\phi^4(X - |\phi'|)^2. \tag{14.39}$$

Thus,

$$(\gamma^2\mathcal{H} + \phi^4)^2 \geq (\gamma^2 X|\phi'| + \phi^4)^2, \tag{14.40}$$

which implies $\mathcal{H} \geq X|\phi'|$. Saturation of the BPS bound is achieved for $|\phi'| = X$ and then

$$\mathcal{H} = X^2 = (\phi')^2 = Q^2/r^4. \tag{14.41}$$

The equation $(\phi')^2 = Q^2/r^4$, together with the boundary condition $\phi \to v$ as $r \to \infty$, has two BPS solutions

$$\phi_\pm(r) = v \pm Q/r, \tag{14.42}$$

where $Q = \frac{g}{4\pi}|p + q\tau|$. The ϕ_+ solution is similar to the flat-space case studied by Callan and Maldacena [8]. It describes a funnel-shaped protrusion of the D3-brane extending to the boundary of AdS at $\phi = +\infty$. This solution gives infinite mass (proportional to $\int dr/r^2$), and thus it is not the solution we are seeking.

The ϕ_- solution is different. $\phi = 0$ corresponds to the horizon of the Poincaré patch of AdS_5. Thus, since ϕ is nonnegative, the ϕ_- solution must be cut off at

$$r_0 = \frac{Q}{v}. \tag{14.43}$$

Then the masses of BPS solitons are given by

$$M = 4\pi \int_{r_0}^{\infty} \mathcal{H} r^2 dr = \frac{4\pi Q^2}{r_0} = 4\pi v^2 r_0 = vg|p + q\tau|, \tag{14.44}$$

exactly as was expected. We have obtained a complete infinite $SL(2,\mathbb{Z})$ multiplet, which includes the W particles that were integrated out, as well as monopoles and dyons.

The charge of the ϕ_- solution is uniformly spread on the sphere $r = r_0$, which we call a *soliton bubble.* The interior of the bubble should not contribute to the mass of the soliton. So, how should we think about the interior of the bubble in the QFT? The only sensible interpretation is that the gauge theory is in the ground state of the *conformal phase* of $U(N+1)$ inside the sphere. This implies that the bubble is a *phase boundary.* This interpretation has the advantage that the parameter τ is required to describe the $U(N+1)$ theory in the conformal phase. This would explain how the soliton solutions know what the values of g and θ are. The $U(1)$ action does not contain all the required information. One also needs to know the nonabelian theory in the conformal phase.

14.4.2. *Comparison with the BPS 't Hooft–Polyakov monopole*

Let us compare the monopole solution of the D3-brane action that we have obtained with the corresponding BPS 't Hooft–Polyakov monopole solution. The single monopole solution of the nonabelian $SU(2)$ gauge theory on the Coulomb branch (for $\theta = 0$) has a triplet of scalar fields ϕ^a whose internal symmetry is aligned with the spatial directions as follows[1]:

$$\phi^a(\vec{x}) = \frac{x^a}{r}\phi(y) \tag{14.45}$$

with

$$\phi(y) = v(\coth y - 1/y), \tag{14.46}$$

where $y = M_W r$ and $M_W = gv$. $\phi(y)$ is strictly positive for $y > 0$, and $\phi^a(\vec{x})$ is nonsingular at the origin. Thus, there is no sign of a soliton bubble in the nonabelian description. Both constructions give the correct mass and charge for the monopole, but the D3-brane solution gives a soliton bubble whereas the nonabelian solution does not. So, which formula more accurately describes what is happening?

Equation (14.46) differs from

$$\phi_-(r) = v(1 - 1/y), \tag{14.47}$$

the D3-brane theory result for the monopole, by a series of terms of the form $\exp(-2nM_W r)$, where n is a positive integer. In the context of the $\mathcal{N} = 4$ theory, the effect of integrating out the fields of mass M_W should be to cancel these exponential terms for $y > 1$ and to give $\phi = 0$ for $y < 1$. After all, the $U(1)$ HEA is supposed to incorporate all of the contributions due to W loops, and these exponentials are a plausible form for those contributions. Hence, we conclude that the bubble is real and that the D3-brane solution gives a more accurate description of what is happening than the usual classical solution (14.46) of the nonabelian theory.

14.4.3. *Black hole analogy*

We have found a universal formula relating the mass and radius of BPS soliton bubbles: $M = 4\pi v^2 r_0$, which is valid for all (p, q). For comparison, the radius of the horizon of a 4d extremal Reissner–Nordstrom asymptotically Lorentzian black hole in four dimensions is $r_0 = MG$, where G is Newton's constant, for all (p, q). In the latter case the charge Q should be large for the classical analysis to be valid. Thus, the relation between mass and radius is the same in both cases, with $(4\pi v^2)^{-1}$ the analog of Newton's constant. The BPS condition ensures that the analogy extends to the relation between mass and charge.

[1]The $\mathcal{N} = 4$ theory has six such triplets, but only one of them is utilized in the construction. This choice corresponds to the choice of a point on the five-sphere in the D3-brane construction.

This analogy is rather surprising, because the D3-brane theory is a nongravitational theory in 4d Minkowski spacetime. If one tries to pursue this analogy, there is a natural question: Does the Bekenstein–Hawking entropy of the black hole, which is proportional to Q^2 (for large Q), have an analog for the solitons of the D3-brane theory? For example, is there an entanglement entropy between the inside and outside of the soliton bubble with this value? If the solitons can be shown to have a well-defined entropy of this sort (for large Q), then one may be tempted to take the black-hole analogy seriously.

Even though the D3-brane theory is defined on a 4d Lorentzian spacetime, we know that the field ϕ can be interpreted as a radial coordinate in AdS_5. From this point of view the soliton solutions have a nontrivial geometry induced from their embedding in AdS_5. From the 5d (or 10d) viewpoint, the bubble is on the horizon of the Poincaré patch of AdS_5, where it intersects the boundary of global AdS. This fact may be useful for understanding the origin of the black-hole analogy.

14.4.4. *Multi-soliton solutions*

It is easy to derive the generalization of $\phi_-(r)$ to the case of n solitons of equal charge.[2] Since supersymmetry ensures that the forces between them should cancel when they are at rest, their centers can be at arbitrary positions $\vec{x}_k$, $k = 1, 2, \ldots, n$. Since ϕ satisfies Laplace's equation, the solutions can be superposed. The obvious guess, which is easy to verify, is

$$\phi(\vec{x}) = v - Q \sum_{k=1}^{n} \frac{1}{|\vec{x} - \vec{x}_k|}. \tag{14.48}$$

The surfaces of the bubbles, which are no longer spheres, are given by $\phi(\vec{x}) = 0$. The fields $\vec{D}$ and $\vec{B}$ are then proportional to $\vec{\nabla}\phi$, with coefficients determined by the charges. Their values at $\phi = 0$ determine the charge densities on the bubble surface. This is much simpler than the usual multi-monopole analysis, which involves nonlinear equations! It should also be more accurate.

14.4.5. *Previous related work*

Soliton bubbles like those found here have appeared in the literature previously. We will briefly discuss the examples that we are aware of. There may be others.

Using attractor flow equations [9], Denef [10, 11] found similar structures to our soliton bubbles in the context of supergravity solutions in which a D3-brane wraps a cycle of a Calabi–Yau manifold that vanishes at a conifold point, where the central charge modulus is zero.

Gauntlett *et al.* [12] studied soliton solutions of a probe D3-brane in an asymptotically flat black D3-brane supergravity background. This problem is closely

[2] Pairs of solitons with different charges are mutually nonlocal and therefore difficult to describe.

related to the one we have considered, since this geometry has $AdS_5 \times S^5$ as its near-horizon limit. They identified half-BPS solutions, like those found here, "in which a point charge is replaced by a perfectly conducting spherical shell." In Ref. [13] the authors examined the DBI action of a D3-brane probe in F theory. They constructed a monopole solution containing a soliton bubble that coincides with a 7-brane. The formation of soliton bubbles may also be related to the *enhançon* mechanism in Ref. [14]. This mechanism circumvents the appearance of a class of naked singularities, known as *repulsons.*

In Ref. [15], Popescu and Shapere studied the low-energy effective action of $\mathcal{N} = 2$ $SU(2)$ gauge theory without additional matter in the Coulomb phase. The Seiberg–Witten *low-energy* effective action [16] should only be valid below the mass of W bosons. However, the theory has a BPS monopole and a BPS dyon that become massless at points in the moduli space of vacua. Therefore, these particles should be obtainable as soliton solutions of the Seiberg–Witten action. Popescu and Shapere constructed these solutions and discovered that they exhibit a spherical shell of charge, just like what we have found.

There is some interesting related evidence for soliton bubbles in a nonsupersymmetric field theory context [17–20]. By considering multi-monopole solutions of large magnetic charge in the 4d $SU(2)$ gauge theory with adjoint scalars on the Coulomb branch, Bolognesi deduced the existence of "magnetic bags" with properties that are very close to those of the soliton bubbles obtained here. Bolognesi's magnetic bags do not have sharply defined surfaces, though they become sharp in the limit of large charge. Bolognesi also pointed out the analogy to black holes [19].

14.5. Conclusion

The action of a probe D3-brane in $AdS_5 \times S^5$ is a candidate for the HEA for a $U(1)$ factor of $\mathcal{N} = 4$ SYM theory on the Coulomb branch. It incorporates all the required symmetries and dualities, and it gives the expected BPS soliton solutions. Even so, it might only be an approximation to the true HEA that is sufficient for computing SUSY-protected quantities. This would also be noteworthy, since then it would be a candidate for a new interacting UV finite abelian $\mathcal{N} = 4$ quantum field theory in 4d. The existence of such a theory is unexpected. In either case, it is important to settle this question.

There is much more that remains to be explored. We need to understand the extent to which symmetry and other general considerations determine the HEA and whether the world-volume theory of a probe D3-brane should give this HEA. We should construct other analogous p-brane actions and explore their BPS soliton solutions. (BPS soliton solutions of the M5-brane example are discussed briefly in Ref. [2].) It should be illuminating to explore tree-approximation scattering amplitudes of the D3-brane theory. It seems likely that they will exhibit beautiful properties, maybe even a (spontaneously broken) Yangian symmetry. Finally, we would like to generalize the analysis to higher-rank gauge theories on the Coulomb branch, which have multiple abelian factors.

Acknowledgment

The author wishes to acknowledge discussions and communications with Hee-Joong Chung, Frank Ferrari, Abhijit Gadde, Jerome Gauntlett, Sergei Gukov, Nicholas Hunter-Jones, Elias Kiritsis, Arthur Lipstein, Nick Manton, Hirosi Ooguri, Jaemo Park, Nati Seiberg, Savdeep Sethi, Yuji Tachikawa, David Tong, and Wenbin Yan.

This work was supported in part by DOE Grant # DE-SC0011632. The author acknowledges the hospitality of the Aspen Center for Physics, where he began working on this project in the summer of 2013, and where he wrote this manuscript in the summer of 2014. The ACP is supported by the National Science Foundation Grant No. PHY-1066293.

References

[1] J. H. Schwarz, "Highly effective actions," JHEP **1401**, 088 (2014) [arXiv:1311.0305 [hep-th]].

[2] J. H. Schwarz, "BPS soliton solutions of a D3-brane action," [arXiv:1405.7444 [hep-th]].

[3] L. Brink, J. H. Schwarz and J. Scherk, "Supersymmetric Yang–Mills theories," Nucl. Phys. B **121**, 77 (1977).

[4] J. M. Maldacena, "The large-N limit of superconformal field theories and supergravity," Adv. Theor. Math. Phys. **2**, 231 (1998) [hep-th/9711200].

[5] D. Belov and G. W. Moore, "Conformal blocks for AdS(5) singletons," [hep-th/0412167].

[6] O. Aharony, N. Seiberg and Y. Tachikawa, "Reading between the lines of four-dimensional gauge theories," JHEP **1308**, 115 (2013) [arXiv:1305.0318 [hep-th]].

[7] R. R. Metsaev and A. A. Tseytlin, "Supersymmetric D3-brane action in AdS(5) × S^5," Phys. Lett. B **436**, 281 (1998) [hep-th/9806095].

[8] C. G. Callan and J. M. Maldacena, "Brane dynamics from the Born-Infeld action," Nucl. Phys. B **513**, 198 (1998) [hep-th/9708147].

[9] S. Ferrara, R. Kallosh and A. Strominger, "$N = 2$ extremal black holes," Phys. Rev. D **52**, 5412 (1995) [hep-th/9508072].

[10] F. Denef, "Attractors at weak gravity," Nucl. Phys. B **547**, 201 (1999) [hep-th/9812049].

[11] F. Denef, "Supergravity flows and D-brane stability," JHEP **0008**, 050 (2000) [hep-th/0005049].

[12] J. P. Gauntlett, C. Kohl, D. Mateos, P. K. Townsend and M. Zamaklar, "Finite energy Dirac–Born–Infeld monopoles and string junctions," Phys. Rev. D **60**, 045004 (1999) [hep-th/9903156].

[13] R. de Mello Koch, A. Paulin-Campbell and J. P. Rodrigues, "Monopole dynamics in $N = 2$ super Yang–Mills theory from a three-brane probe," Nucl. Phys. B **559**, 143 (1999) [hep-th/9903207].

[14] C. V. Johnson, A. W. Peet and J. Polchinski, "Gauge theory and the excision of repulson singularities," Phys. Rev. D **61**, 086001 (2000) [hep-th/9911161].

[15] I. A. Popescu and A. D. Shapere, "BPS equations, BPS states, and central charge of $N = 2$ supersymmetric gauge theories," JHEP **0210**, 033 (2002) [hep-th/0102169].
[16] N. Seiberg and E. Witten, "Electric–magnetic duality, monopole condensation, and confinement in $N = 2$ supersymmetric Yang–Mills theory," Nucl. Phys. B **426**, 19 (1994); Erratum, *ibid.* B **430**, 485 (1994) [hep-th/9407087].
[17] S. Bolognesi, "Multi-monopoles, magnetic bags, bions and the monopole cosmological problem," Nucl. Phys. B **752**, 93 (2006) [hep-th/0512133].
[18] K.-M. Lee and E. J. Weinberg, "BPS magnetic monopole bags," Phys. Rev. D **79**, 025013 (2009) [arXiv:0810.4962 [hep-th]].
[19] S. Bolognesi, "Magnetic bags and black holes," Nucl. Phys. B **845**, 324 (2011) [arXiv:1005.4642 [hep-th]].
[20] N. S. Manton, "Monopole planets and galaxies," Phys. Rev. D **85**, 045022 (2012) [arXiv:1111.2934 [hep-th]].

Chapter 15

Screening clouds and Majorana fermions*

Ian Affleck[†,§] and Domenico Giuliano[‡]

[†] *Department of Physics and Astronomy, University of British Columbia, Vancouver, BCV6T1Z1, Canada*
iaffleck@phas.ubc.ca

[‡] *Dipartimento di Fisica, Università della Calabria Arcavacata di Rende, 87036 Cosenza, Italy*

I.N.F.N., Gruppo collegato di Cosenza, Arcavacata di Rende 87036 Cosenza, Italy

CNR SPIN, Monte S. Angelo, via Cinthia, 80126, Napoli, Italy

Ken Wilson developed the Numerical Renormalization Group technique which greatly enhanced our understanding of the Kondo effect and other quantum impurity problems. Wilson's NRG also inspired Philippe Nozières to propose the idea of a large "Kondo screening cloud". While much theoretical evidence has accumulated for this idea it has remained somewhat controversial and has not yet been confirmed experimentally. Recently a new possibility for observing an analogous crossover length scale has emerged, involving a Majorana fermion localized at the interface between a topological superconductor quantum wire and a normal wire. We give an overview of this topic both with and without interactions included in the normal wire.

Keywords: Kondo screening cloud; topological superconductors; Majorana fermions; renormalization group.

15.1. Introduction

A rather unusual feature of Ken Wilson's contributions to theoretical physics is that they include both fundamental conceptual ideas and novel numerical techniques. The most celebrated numerical technique which he helped to develop is lattice gauge theory, now leading to a quantitative understanding of the strong

---

*For a special issue of J. Stat. Phys. in memory of Kenneth G. Wilson. This article was originally published in *J. Stat. Phys.* **157**, 666–691 (2014).

§Corresponding author.

interactions. But Wilson also played a huge role in condensed matter theory by developing the Numerical Renormalization Group (NRG) technique [1] to study the Kondo problem. This technique has since been applied to a host of other quantum impurity models and is also used in the dynamical mean field theory [2] approach to strongly correlated systems where translationally invariant models are reduced to impurity models by taking the limit of large spatial dimension.

Typical quantum impurity models which are studied using the NRG involve localized quantum mechanical degrees of freedom, such as a spin-1/2 operator, interacting with a spatially extended gapless system such as a Fermi gas of free electrons. These models can generically be reduced to one-dimensional ones, for example by using s-wave projection in cases where the impurity interactions are short ranged. Wilson's NRG eventually maps the system to a one-dimensional lattice model with the quantum impurity at one end of a chain and the hopping terms between neighbouring sites dropping off exponentially with distance from the impurity. These quantum impurity models usually exhibit a renormalization group cross-over between ultraviolet and infrared fixed points. In the case of the spin-1/2 Kondo model the ultraviolet fixed point, which describes the physics at sufficiently high energies when the bare anti-ferromagnetic Heisenberg exchange coupling of the electrons to the spin is very weak, corresponds to a free spin. The infrared fixed point, describing the physics when the renormalized exchange coupling is large, corresponds to a single electron emerging from the Fermi sea to form a spin singlet with the impurity. The remaining electrons in the Fermi sea must adjust to the presence of this spin singlet, leading to universal low energy behaviour. Wilson's NRG approach studies this crossover by focussing on the dependence on distance away from the impurity, in particular the crossover as more sites are added to the "Wilson chain". These ideas led Philippe Nozières to propose the existence of a large "Kondo screening cloud" [3–5]. A cartoon picture is that the electron which forms a singlet with the impurity has an extended wave-function of radius $\xi_K \approx v_F/T_K$. Here T_K is the Kondo temperature which is the RG crossover energy scale and v_F is the Fermi velocity. We work in units where $k_B = \hbar = 1$. The low energy degrees of freedom of the free electrons correspond to relativistic Dirac fermions with speed of light replaced by the Fermi velocity, which is thus the natural parameter to convert an energy scale, T_K into a length scale, ξ_K.

More precisely, one might expect that physical quantities depending on distance from the impurity, r, should exhibit a universal crossover at the length scale ξ_K, being described by the ultraviolet fixed point at distances $r \ll \xi_K$ and by the infrared fixed point at distances $r \gg \xi_K$. This naive expectation has been largely confirmed by a number of theoretical calculations where RG predictions were compared to NRG and other numerical results.[1] This screening cloud picture is actually quite disturbing if one notes that typical values of ξ_K in experiments may correspond

[1]For a review of work on the Kondo screening cloud see [6].

to hundreds or thousands of lattice spacings. No experiments have yet detected this large screening cloud. A partial explanation of this is that the cloud is so big that it is difficult to see. Typical correlation functions involve a product of a power law decaying factor, an oscillating factor and a slowly varying envelope function, crossing over at ξ_K. Thus the correlation functions have become very small at the distance ξ_K where the crossover occurs. One must also consider the fact that in most experiments there is a finite density of quantum impurities and their average separations are $\ll \xi_K$. Furthermore, one should consider that the underlying models are indeed just models. They leave out, for example, electron-electron interactions away from the impurity. While this may be justifiable based on Fermi liquid theory renormalization group ideas, at any finite temperature there is a finite inelastic scattering length beyond which ignoring these interactions is not justified. Thus experimental confirmation of the Kondo screening cloud remains elusive (Fn 1).

The existence of a "screening cloud size" ξ_K, i.e. a characteristic length scale for crossover between ultraviolet and infrared fixed points in quantum impurity models, is not restricted to the Kondo problem; it is expected to be quite generic. In this article, we will focus on a new type of quantum impurity model where this issue arises and new opportunities present themselves for experimental confirmation. It was pointed out by Kitaev [7] that a simple 1D p-wave superconductor defined on a finite line interval has a "Majorana mode" localized at each end of the system. This is most easily seen for the tight-binding model of spinless electrons with equal hopping and pairing amplitudes and zero chemical potential:

$$H = w \sum_{j=1}^{M-1} [-a_j^\dagger a_{j+1} + a_j a_{j+1} + h.c.] \tag{15.1.1}$$

(Here $h.c.$ stand for Hermitean conjugate and a_j annihilates an electron on lattice site j.) We can always rewrite the a_j operators in terms of their Hermitean and anti-Hermitean parts, so-called Majorana operators:

$$a_j = (\gamma_{2j-1} + i\gamma_{2j})/2, \quad (j = 1, 2, 3, \dots M) \tag{15.1.2}$$

where

$$\gamma_j^\dagger = \gamma_j, \quad \{\gamma_j, \gamma_k\} = 2\delta_{j,k}. \tag{15.1.3}$$

H becomes

$$H = \frac{iw}{2} \sum_{j=1}^{M-1} \gamma_{2j}\gamma_{2j+1}. \tag{15.1.4}$$

This model is readily diagonalized by reassembling the Majoranas into "Dirac" operators in a different way:

$$b_j \equiv (\gamma_{2j} + i\gamma_{2j+1})/2, \quad (j = 1, 2, 3, \dots M - 1), \tag{15.1.5}$$

yielding:

$$H = w \sum_{j=1}^{M-1} b_j^\dagger b_j. \tag{15.1.6}$$

We obtain $M-1$ single particle levels with energy w, which are localized on the links of the lattice. However, notice that γ_1 and γ_{2M} don't appear in the Hamiltonian at all. They are a pair of Majorana modes, one at each end of the chain, which will be the focus of our discussion. They can, of course, be combined into a Dirac operator giving a zero energy state of the chain which has equal amplitude at both ends.

While the above simple discussion was for a very special model, this phenomenon is actually quite robust, persisting for a range of unequal hopping and pairing amplitudes and for a range of chemical potential. For more general parameters the Majorana modes (MM's) are localized near the two ends of the chain, decaying exponentially towards the centre of the chain with a finite decay length. Thus for a long enough chain they are nearly decoupled, leading to an exponentially low energy level with equal amplitudes at both ends of the chain. Such a system, containing a MM at each end, is known as a topological superconductor. Even allowing general values for the parameters, this still might seem like a highly unrealistic model. However, it was shown that such a topological phase occurs in a realistic model of a quantum wire with spin-orbit interactions, adjacent to a bulk s-wave superconductor, in an applied magnetic field [8, 9]. The pairing term then arises from the proximity effect. Signs of these MM's were apparently seen in experiments on indium antimonide quantum wires proximity coupled to a niobium titanium nitride superconductor [10]. In the experiments, one end of the quantum wire extended past the edge of the superconductor, over an insulator and then eventually contacted a normal electrode. Indications of a MM localized in the quantum wire near the edge of the superconductor were obtained from current measurements. Other experiments were performed in which the two ends of the normal wire were in contact with two different superconducting electrodes with the central region of the wire resting on an insulator. Curved quantum wires with both ends in contact with the same superconducting electrode and a magnetic flux applied inside the loop produced by the wire and the superconductor have also been considered theoretically.

We discuss here the low energy behaviour of a long normal wire of length ℓ with a MM at each end,[2] corresponding to a superconductor-normal-superconductor (SNS)

[2]We assume the superconducting wires are long enough that we only need consider two MM's, one at each SN junction, which are coupled to the normal wire but not directly to each other. In the case of two distinct superconductors this will be true if the length of each superconductor is long compared to the proximity effect-induced coherence length, in which case the coupling to the MM's at the far ends of the superconductors will be exponentially small. In the case where the normal wire is contacted at both ends of the same topological superconductor, the two MM's which we consider have an additional coupling through the superconductor. However, this will be exponentially small when the superconductor is long compared to the coherence length.

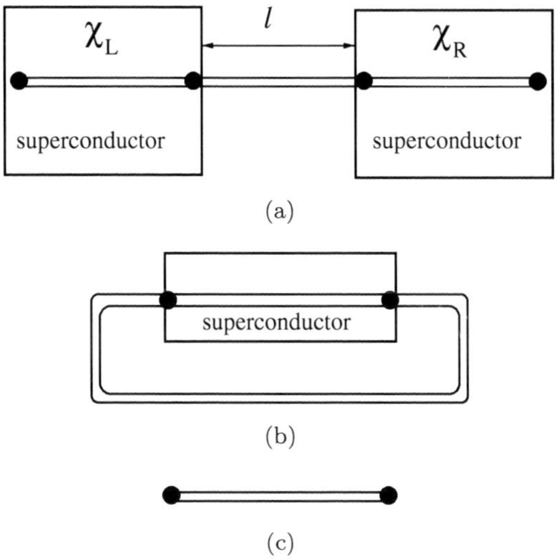

Fig. 15.1. (a) A sketch of the device considered. The portions of the quantum wire on top of the conventional bulk superconductors become topological superconductors, with different phases χ_L and χ_R of the superconducting order parameter. Majorana modes, denoted by *black circles*, exist at both ends of each topological superconductor. (b) A closed curved quantum wire, part of which is on top of a superconductor. Now the two Majorana modes of the topological superconductor couple to each end of the normal portion of the wire. (c) The effective low energy system. Only the normal wire and the two Majorana modes at the SN junctions are retained.

junction where each superconductor is topological.[3] See Fig. 15.1. In order to conveniently study the effects of electron-electron interactions inside the normal wire and to obtain universal results, we assume that the temperature and the finite size gap of the normal wire, ($\propto v_F/\ell$) are both small compared to the induced superconducting gap in the topological superconductor, Δ [Δ is given by the parameter w in the simple model of Eq. (15.1.1). Note that the superconductor is actually gapless, due to the MM. Δ represents the gap for bulk excitations. The MM exists as a single zero energy state inside this gap]. We can then safely integrate out the gapped modes in the superconductors, keeping only the MM's and the excitations of the normal wire in our low energy effective Hamiltonian, H_{eff} [13–15]. This leaves a type of quantum impurity model, similar to the Kondo model, in which the quantum impurities are the MM's and the delocalized gapless excitations are those of the normal wire. This long junction limit corresponds to $\ell \gg \xi_0$ where $\xi_0 \propto v_F/\Delta$ is the superconducting coherence length in the topological superconductor. Our assumption $v_F/\ell \ll \Delta$ implies that there are many Andreev bound states in the normal region.

We first consider each SN junction separately, appropriate for infinite ℓ. We show that an analogue of Kondo screening occurs. At low energies, the MM combines with another Majorana degree of freedom within the normal wire to form a Dirac

[3]Related finite size effects have been discussed in [11, 12].

mode localized near the SN interface. There is a finite energy cost to depopulate this localized level, analogous to the Kondo temperature. We may think of this localized Dirac mode as being extended over some finite distance into the normal wire, ξ_M, the analogue of the Kondo screening length, ξ_K. This state with the MM "screened" corresponds to a low energy strong coupling fixed point of the model. On the other hand, if the MM is weakly coupled to the normal wire, there is an unstable high energy fixed point in which the MM is decoupled. We then consider the dc Josephson current through the finite length SNS junction. This provides an excellent method for measuring the screening length, ξ_M since the current behaves very differently in the two regimes $\xi_M \ll \ell$ and $\xi_M \gg \ell$. Essentially the wire length ℓ acts as an infrared cut-off on the renormalization of the coupling between the MM and the wire. There are strong similarities between this system and the 2 impurity Kondo model where a competition occurs between screening of both impurities by the electron bath and singlet formation between the two impurities due to the Ruderman–Kittel–Kasuya–Yosida (RKKY) interaction.

Note that the MM introduces another characteristic length scale, ξ_M, in addition to ξ_0 the coherence length in the superconductor. For weak tunnelling across the SN junction, $\xi_M \gg \xi_0$, the limit we consider. This additional characteristic length scale is a special feature of topological superconductor SN junctions, which doesn't exist for an ordinary SN junction. While crossover of the Josephson current as a function of ℓ/ξ_0 in both ordinary [16] and topological [17, 18] SNS junctions has been frequently analysed, the crossover at much longer length scales in topological SNS junctions, as a function of ℓ/ξ_M, is quite novel. This very different length dependence in the topological case is discussed further in Sect. 15.2.1.

There is an important simplification of the MM impurity model compared to the Kondo model. The Kondo model exhibits very non-trivial many body effects despite the fact that the electron bath is usually treated as non-interacting. These effects are due to the spin degree of freedom of the impurity and the associated spin-flip scattering. On the other hand, the relevant coupling between the MM and the normal wire is a simple tunnelling term so that, if we ignore interactions in the normal wire, the Hamiltonian remains harmonic. Thus the case of a non-interacting normal wire provides a relatively simple example where the screening cloud physics can already be seen. This is a bit reminiscent of the non-interacting resonant level model,[4] which has some relation to the Kondo model although in that case the spin degree of freedom of the impurity is dropped. Thus we begin by discussing the non-interacting SNS junction. Simple analytic formulas can be obtained for the Josephson current in the two limits $\xi_M \ll \ell$ and $\xi_M \gg \ell$. We supplement these with numerical results which demonstrate the scaling behaviour and the cross-over between weak coupling and strong coupling fixed points. We then

[4]For a recent discussion of the crossover length scale for the non-interacting resonant level model see [19].

turn to the case of an interacting normal wire, again obtaining analytic expressions for the current at weak and strong coupling fixed points. Numerical simulation of the interacting case is more challenging, requiring for example use of the Density Matrix Renormalization Group method, and we don't attempt it here.

15.2. Non-interacting model

We begin by writing down a simple non-interacting tight-binding model. The continuum model which captures the universal low energy physics can be derived from this tight-binding model which we also find useful for numerical simulations. We decompose the Hamiltonian into a "bulk" term, H_0 describing the normal wire and a boundary term, H_b describing the coupling to the two MM's.

$$H = H_0 + H_b$$

$$H_0 = -J \sum_{j=1}^{N-2} \left\{ c_j^\dagger c_{j+1} + c_{j+1}^\dagger c_j \right\} - \mu \sum_{j=1}^{N-1} c_j^\dagger c_j \tag{15.2.1}$$

$$H_b = -it_L \gamma_L \left\{ c_1 \mathrm{e}^{i\frac{\chi_L}{2}} + c_1^\dagger \mathrm{e}^{-i\frac{\chi_L}{2}} \right\} + t_R \gamma_R \left\{ c_{N-1} \mathrm{e}^{-i\frac{\chi_R}{2}} - c_{N-1}^\dagger \mathrm{e}^{i\frac{\chi_R}{2}} \right\}. \tag{15.2.2}$$

Here γ_L and γ_R are the two MM's manufactured by the two topological superconductors at the left and right hand side of the normal wire. χ_L and χ_R are the phases of the superconducting order parameter in the left and right superconductor. The Josephson current is a function of the phase difference $\chi \equiv \chi_L - \chi_R$. For convenience, we henceforth set $\chi_L = \chi/2$ and $\chi_R = -\chi/2$. The dc Josephson current is determined by the derivative of the equilibrium free energy, F, with respect to the phase difference:

$$I[\chi] = 2e \frac{\partial F}{\partial \chi}. \tag{15.2.3}$$

At $T = 0$, which we focus on here, F becomes the ground state energy.

We now turn to a continuum field theory description. Note that this seems highly appropriate since our impurity model is only valid at energy scales below the induced superconducting gap of the topological superconductors, a scale which is expected to be much less than bandwidth, J of the normal wire. We also assume the tunnelling amplitudes to the MM, $t_{L/R} \ll J$ and that the length of the normal region, ℓ, is large compared to microscopic scales like the lattice constant. Then it is expected that the χ dependence of the ground state energy, and hence the Josephson current, depends only on universal low energy information [20]. This field theory model simplifies calculations in this section and is crucial for including interactions in the next section.

Keeping only a narrow band of wavevectors around the Fermi points, $\pm k_F$, we write:

$$c_j \approx \left[\mathrm{e}^{ik_F a j} \psi_+(aj) + \mathrm{e}^{-ik_F a j} \psi_-(aj) \right] \sqrt{a} \tag{15.2.4}$$

where $+/-$ label right and left movers respectively, a is the lattice constant and we define k_F by:

$$\mu = -2J \cos k_F a. \tag{15.2.5}$$

The fields $\psi_\pm$ vary slowly on the lattice scale. Linearizing the dispersion relation near the Fermi energy, the low energy bulk Hamiltonian becomes:

$$H_0 \approx iv_F \int_0^\ell dx \left[-\psi_+^\dagger \frac{d}{dx} \psi_+ + \psi_-^\dagger \frac{d}{dx} \psi_- \right] \tag{15.2.6}$$

where $v_F = 2J \sin k_F a$. This Hamiltonian must be supplemented by boundary conditions to be well-defined. For the lattice Hamiltonian of Eq. (15.2.1), with free ends, the boundary conditions correspond to requiring the operators to vanish on the "phantom sites", $j = 0$ and $j = N$, implying

$$0 = \psi_+(0) + \psi_-(0) = \mathrm{e}^{ik_F \ell} \psi_+(\ell) + \mathrm{e}^{-ik_F \ell} \psi_-(\ell) \tag{15.2.7}$$

where

$$\ell \equiv Na. \tag{15.2.8}$$

It is convenient to make an "unfolding" transformation, taking advantage of the boundary condition at $x = 0$ to define:

$$\psi_-(-x) \equiv -\psi_+(x), \quad (0 < x < \ell). \tag{15.2.9}$$

Then we can write the Hamiltonian in terms of left-movers only, on an interval of length 2ℓ with

$$H_0 = iv_F \int_{-\ell}^{\ell} dx \psi_-^\dagger \frac{d}{dx} \psi_- \tag{15.2.10}$$

and the boundary condition:

$$\psi_-(\ell) = \mathrm{e}^{2ik_F \ell} \psi_-(-\ell). \tag{15.2.11}$$

Letting

$$\mathrm{e}^{ik_F \ell} = -i \mathrm{e}^{i\alpha/2}, \quad (\text{with } |\alpha| < \pi) \tag{15.2.12}$$

it is convenient to define the field $\tilde{\psi}_-(x)$ by:

$$\tilde{\psi}_-(x) \equiv \mathrm{e}^{-i\alpha x/(2\ell)} \psi_-(x). \tag{15.2.13}$$

$\tilde{\psi}$ obeys the more convenient anti-periodic boundary condition:

$$\tilde{\psi}_-(\ell) = -\tilde{\psi}_-(-\ell) \tag{15.2.14}$$

at the cost of introducing a chemical potential term into H_0. Henceforth we drop the cumbersome — subscript and the tilde letting:

$$\tilde{\psi}_- \to \psi \tag{15.2.15}$$

so that

$$H_0 = v_F \int_{-\ell}^{\ell} dx \psi^\dagger \left[i\frac{d}{dx} + \frac{\alpha}{2\ell} \right] \psi. \tag{15.2.16}$$

H_b now becomes:

$$H_b \approx -\tilde{t}_L \gamma_L \left[e^{i\chi/4}\psi(0) - e^{-i\chi/4}\psi^\dagger(0) \right] - \tilde{t}_R \gamma_R \left[e^{-i\chi/4}\psi(\ell) - e^{i\chi/4}\psi^\dagger(\ell) \right] \tag{15.2.17}$$

where

$$\tilde{t}_{L/R} \equiv 2\sin(k_F a) t_{L/R} \sqrt{a}. \tag{15.2.18}$$

15.2.1. *Single S-N junction*

We start by considering the case of infinite ℓ where an incoming left-moving plane wave interacts with the Majorana γ_L as it passes the origin. The fermonic operators which diagonalize the Hamiltonian are of the form:

$$\Gamma_k = \phi_{Lk}\gamma_L + \int_{-\infty}^{\infty} dx e^{ikx} [P_k(x)\psi(x) + H_k(x)\psi^\dagger(x)]. \tag{15.2.19}$$

Requiring

$$[\Gamma_k, H] = v_F k \Gamma_k \tag{15.2.20}$$

gives the Bogoliubov-de Gennes (BdG) equations:

$$\begin{aligned} i v_F \partial_x P_k(x) &= -2\tilde{t}_L e^{i\chi/4} \phi_{Lk} \delta(x) \\ i v_F \partial_x H_k(x) &= 2\tilde{t}_L e^{-i\chi/4} \phi_{Lk} \delta(x) \\ \tilde{t}_L [e^{i\chi/4} H_k(0) - e^{-i\chi/4} P_k(0)] &= \phi_{Lk} v_F k. \end{aligned} \tag{15.2.21}$$

Clearly $P_k(x)$ and $H_k(x)$ are step functions whose form is fully determined by the S-matrix:

$$\begin{bmatrix} P_k(-\infty) \\ H_k(-\infty) \end{bmatrix} = \begin{bmatrix} S_{PP}(k) & S_{PH}(k) \\ S_{HP}(k) & S_{HH}(k) \end{bmatrix} \begin{bmatrix} P_k(\infty) \\ H_k(\infty) \end{bmatrix} \tag{15.2.22}$$

The particle-hole symmetry of the BdG equations implies

$$\begin{aligned} S_{HH}(k) &= S^*_{PP}(-k) \\ S_{HP}(k) &= S^*_{PH}(-k). \end{aligned} \tag{15.2.23}$$

Thus we only need to solve for S_{PP} and S_{HP}. From Eqs. (15.2.21) we find:

$$S_{HP}(k) = \frac{2\tilde{t}_L^2 e^{-i\chi/2}}{2\tilde{t}_L^2 + iv_F^2 k}$$

$$S_{PP}(k) = \frac{iv_F^2 k}{2\tilde{t}_L^2 + iv_F^2 k}. \qquad (15.2.24)$$

Note that at zero energy:

$$S_{HP}(0) = e^{-i\chi/2}$$
$$S_{PP}(0) = 0, \qquad (15.2.25)$$

corresponding to perfect Andreev reflection (or "Andreev transmission" in the unfolded system). On the other hand at sufficiently large $|k|$, $S_{HP}(k) \approx 0$ and $S_{PP}(k) \approx 1$, corresponding to perfect normal reflection. The crossover length scale between these two behaviours is seen from Eq. (15.2.24) to be

$$\xi_M \equiv \frac{v_F^2}{\tilde{t}_L^2}. \qquad (15.2.26)$$

We define this to be the "Majorana screening cloud length" in the non-interacting system. Beyond this length scale, the featureless nature of the S-matrix indicates that the MM has been "screened". See Sect. 15.3 for a further discussion of this intuitive picture from a different viewpoint that also applies to the interacting case.

The fact that $\xi_M \propto \tilde{t}_L^{-2}$ follows from a renormalization group scaling analysis of H_b of Eq. (15.2.17), although such sophisticated methods are not really necessary for the non-interacting model. $\psi(0)$ has dimension 1/2 while the MM is dimensionless. A boundary term in the Hamiltonian must have dimension 1, corresponding to energy. Therefore $\tilde{t}_L$ is relevant, with dimension 1/2, implying this scaling of ξ_M. Note that the situation is very different for an ordinary SN junction, with no MM. For a long junction, $\ell \gg \xi_0$, we may again integrate out the excitations of the superconductor. Now the most relevant induced interaction in the normal region is a proximity effect pairing term. For spinful fermions, analyzed in [13], this is a marginal boundary interaction $[\Delta_B \psi_\uparrow(0)\psi_\downarrow(0) + h.c.]$. For weak tunnelling across the SN junction Δ_B is again of order $\tilde{t}_L^2$, like ξ_M. However, it does not introduce a characteristic length scale, being marginal and no corresponding crossover of the Josephson current with junction length occurs [13]. Rather the current scales as $1/\ell$ at all lengths $\gg \xi_0$. Nonetheless, the current depends in a non-trivial way on Δ_B, being sinusoidal for small Δ_B and being a sawtooth for a fine-tuned large value of Δ_B [13]. As we will see, the behavior of the current is much more interesting for a topological SN junction due to the presence of this new length scale, ξ_M. For an SN junction between an ordinary superconductor and spinless electrons, the most relevant induced pairing term is $\psi(0)\partial_x\psi(0)$ which is irrelevant, of dimension 2. Now Δ_B *does* introduce a characteristic length scale, $\xi \propto \Delta_B \propto \tilde{t}_L$. Note this length scale shrinks with decreasing tunnelling, unlike ξ_M which grows. Starting with weak tunnelling, this scaling doesn't lead to very interesting behavior, since the current is sinusoidal for all junction lengths, unlike the topological case analysed here.

15.2.2. *SNS junction*

In this simple non-interacting model, it is possible to write down an explicit expression for the current, for any values of the parameters, in terms of an elementary integral which can be readily evaluated numerically. Integrating out the ψ field exactly gives the imaginary time effective action for the MM's:

$$S = \int_0^\infty \frac{d\omega}{2\pi} \left\{ \frac{i\omega}{2}\gamma_L(-\omega)\gamma_L(\omega) + \frac{i\omega}{2}\omega\gamma_R(-\omega)\gamma_R(\omega) \right.$$
$$\left. -[\gamma_L(-\omega), \gamma_R(-\omega)]\mathcal{M}(\omega) \begin{bmatrix} \gamma_L(\omega) \\ \gamma_R(\omega) \end{bmatrix} \right\} \tag{15.2.27}$$

where

$$\mathcal{M}(\omega) \equiv \begin{bmatrix} \tilde{t}_L^2[G(\omega, x=0) - G(-\omega, x=0)] & \tilde{t}_L\tilde{t}_R[e^{i\chi/2}G(\omega, x=\ell) + e^{-i\chi/2}G(-\omega, x=\ell)] \\ -\tilde{t}_L\tilde{t}_R[e^{-i\chi/2}G(\omega, x=\ell) + e^{i\chi/2}G(-\omega, x=\ell)] & \tilde{t}_R^2[G(\omega, x=0) - G(-\omega, x=0)] \end{bmatrix} \tag{15.2.28}$$

Here G is the Matsubara Green's function for the ψ fermions in the case $H_b = 0$:

$$G(\omega, x) = \frac{1}{2\ell} \sum_{n=-\infty}^{\infty} \frac{e^{i\pi(2n+1)(x/2\ell)}}{i\omega - v_F[(n+1/2)\pi + \alpha/2]/\ell}. \tag{15.2.29}$$

Note that α measures the amount of breaking of particle-hole symmetry in the spectrum of H_0:

$$E_n = \frac{v_F}{\ell}[(n+1/2)\pi + \alpha/2]. \tag{15.2.30}$$

This spectrum is plotted in Fig. 15.2. Particle-hole symmetry is broken except when $|\alpha| = 0$ or π and there is a zero mode in the spectrum when $\alpha = \pi$. We see from Eq. (15.2.29) that

$$G(\omega, x=0) = -\frac{1}{2v_F} \tan\left[\frac{i\omega\ell}{v_F} + \frac{\alpha}{2}\right]$$
$$G(\omega, x=\ell) = \frac{i}{2v_F} \sec\left[\frac{i\omega\ell}{v_F} + \frac{\alpha}{2}\right]. \tag{15.2.31}$$

Next we integrate out the MM's to express the ground state energy as

$$E_0 = \int_0^\infty \frac{d\omega}{2\pi} \ln \mathrm{Det}\left[\frac{i\omega}{2}\mathbf{I} + \mathcal{M}(\omega)\right]. \tag{15.2.32}$$

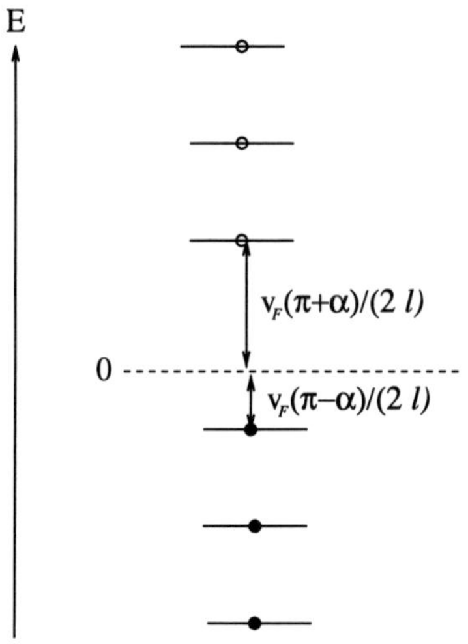

Fig. 15.2. Spectrum of Hamiltonian, with states filled in ground state marked, before coupling to MM's. Particle-hole symmetry is broken except when $\alpha = 0$ or π and there is a zero model when $\alpha = \pi$.

Differentiating with respect to χ then gives the exact formula for the current. In terms of a rescaled integration variable, $w = 2\omega\ell/v_F$:

$$
\begin{aligned}
I[\chi] = \frac{v_F(2e)}{8\pi\ell}\int_{-\infty}^{\infty} & dw\ \sin\chi[\cos\alpha + \cosh w] \\
\Bigg/ \Bigg\{ & \left[\frac{v_F^2 w}{4\ell\tilde{t}_L\tilde{t}_R}(\cos\alpha + \cosh w) + \left(\frac{\tilde{t}_L^2 + \tilde{t}_R^2}{2\tilde{t}_L\tilde{t}_R}\right)\sinh w\right]^2 \\
& - \left(\frac{\tilde{t}_L^2 - \tilde{t}_R^2}{2\tilde{t}_L\tilde{t}_R}\right)^2 \sinh^2 w + [1 + \cos\alpha\cosh w + \cos\chi(\cos\alpha + \cosh w)]\Bigg\}.
\end{aligned}
\qquad (15.2.33)
$$

Clearly $\ell I/(ev_F)$ is a scaling function of χ, α and

$$
\begin{aligned}
z &\equiv \ell/\sqrt{\xi_{ML}\xi_{MR}} = \frac{\ell\tilde{t}_L\tilde{t}_R}{v_F^2} \\
r &\equiv \frac{\xi_{ML}}{\xi_{MR}} = \frac{\tilde{t}_R^2}{\tilde{t}_L^2}.
\end{aligned}
\qquad (15.2.34)
$$

Note that the current depends on k_F in two distinct ways. There is a weak dependence via the Fermi velocity, v_F and then an additional, much stronger dependence via α, the fractional part of $k_F\ell/\pi$. In Fig. 15.3 we plot the result of a numerical integration of Eq. (15.2.33) for three values of the parameters. As shown in Fig. 15.3, in the two limits $\ell/\xi_M \gg 1$ and $\ell/\xi_M \ll 1$ the current is given by simple analytic expressions which can be obtained straightforwardly from Eq. (15.2.33). It is instructive to derive these expressions by physical arguments, given in the

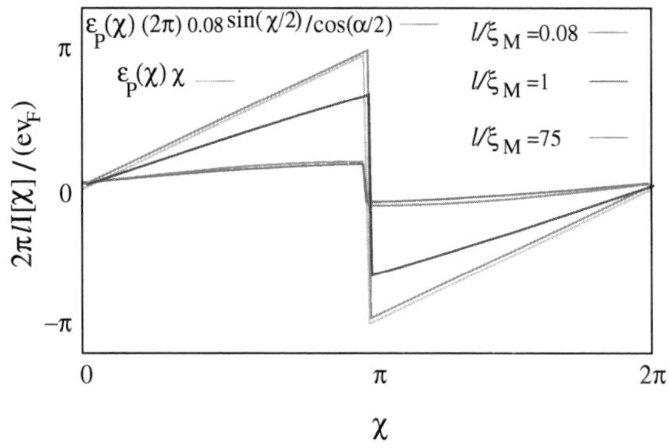

Fig. 15.3. Josephson current versus phase difference, χ from Eq. (15.2.33) for particle-hole asymmetry parameter $\alpha = \pi/10$ and a left-right symmetric junction. ξ_M is defined in Eq. (15.2.26). Fermion parity conservation is *not* imposed. Our analytic predictions are also shown, fitting the data well for $\ell \ll \xi_M$ and $\ell \gg \xi_M$. Note that steps occur for all ℓ/ξ_M.

remainder of this section. These expressions are then extended to the interacting case in Sect. 15.3.

15.2.3. $\ell \ll \xi_M$

In this case, for general values of α, it is convenient to integrate out the ψ field to obtain an effective action for the MM's as done above. Since the energy level spacing in the normal wire is v_F/ℓ we may ignore retardation effects in this short junction limit, $\tilde{t}^2/v_F \ll v_F/\ell$. Evaluating the Green's function at $\omega = 0$, the action of Eq. (15.2.27) reduces to a simple Hamiltonian:

$$H_{eff} = -2\tilde{t}_L\tilde{t}_R\cos(\chi/2)G(\omega = 0, \ell)\gamma_L\gamma_R \tag{15.2.35}$$

where $G(\omega = 0, \ell) = i/[2v_F\cos(\alpha/2)]$, from Eq. (15.2.31). Clearly this approach fails when $\alpha = \pi$. This is due to the presence of a zero mode in the spectrum of the normal region for $\tilde{t}_{L/R} = 0$, Eq. (15.2.30). A separate treatment of this special case is given in Sect. 15.5. On the other hand, the exact result of Eq. (15.2.33) can be used to calculate the current even in this case.

Once we have reduced the Hamiltonian to Eq. (15.2.35), containing only the 2 MM's, it is easy to calculate the Josephson current. We may combine the two MM's to make a single Dirac operator

$$\psi \equiv (\gamma_R + i\gamma_L)/2 \tag{15.2.36}$$

in terms of which

$$H_{eff} = 2\frac{\tilde{t}_L\tilde{t}_R}{v_F\cos\alpha/2}\cos(\chi/2)(\psi^\dagger\psi - 1/2). \tag{15.2.37}$$

We see that the ground state has the single fermionic energy level vacant for $\cos\chi/2 > 0$ and occupied for $\cos\chi/2 < 0$:

$$E_0 = -\frac{\tilde{t}_L\tilde{t}_R}{v_F\cos\alpha/2}|\cos\chi/2|. \qquad (15.2.38)$$

The corresponding Josephson current is

$$I[\chi] = 2e\frac{\partial E_0}{\partial\chi} \approx e\frac{\tilde{t}_L\tilde{t}_R}{v_F\cos\alpha/2}\sin(\chi/2)\varepsilon_P(\chi) = \frac{ev_F}{\xi_M\cos(\alpha/2)}\sin(\chi/2)\varepsilon_P(\chi), \qquad (15.2.39)$$

Here $\varepsilon_P(\chi)$ is a periodic step function with period 4π:

$$\begin{aligned}\varepsilon_P(\chi) &= 1, \quad (-\pi < \chi < \pi)\\ &= -1, \quad (\pi < \chi < 3\pi)\\ &= 1, \quad (3\pi < \chi < 5\pi)\end{aligned} \qquad (15.2.40)$$

et cetera. Note that while neither $\sin(\chi/2)$ nor $\varepsilon_P(\chi)$ has 2π periodicity, their product does. Also note that we have assumed that the dominant χ dependence of the ground state energy, for $\ell \ll \xi_M$, is given by the lowest energy level. The other energy levels are $O(v_F/\ell)$, much larger than $\frac{\tilde{t}_L\tilde{t}_R}{v_F}$ for $\ell \ll \xi_M$. They are only weakly perturbed by the coupling to the MM's and are thus insensitive to the phase of that coupling. Eq. (15.2.39) agrees well with the exact result of Eq. (15.2.33), for a small $\ell/\xi_M = .08$ plotted in Fig. 15.3.

15.2.4. $\ell \gg \xi_M$

Let's now consider the Josephson current for a long but finite length normal region with $\ell \gg \xi_M$. We may now write the operators which diagonalize the Hamiltonian as:

$$\Gamma_k = \phi_{Lk}\gamma_L + \phi_{Rk}\gamma_R + \int_{-\ell}^{\ell} dx\left[e^{i(k-\alpha/2\ell)x}P_k(x)\psi(x) + e^{i(k+\alpha/2\ell)x}H_k(x)\psi^\dagger(x)\right]. \qquad (15.2.41)$$

We can again satisfy $[H,\Gamma_k] = v_F k\Gamma_k$ by solving the BdG Eqs. (15.2.21) and the corresponding equations at $x = \ell$, yielding Eq. (15.2.33). Here we just discuss the low energy solutions with $|k| \ll 1/\xi_M$. In this limit we may approximate the transmission process at each interface as being purely Andreev. Thus a particle travelling to the left towards the interface at $x = 0$ is transmitted as a hole while picking up a phase $e^{-i\chi/2}$ as we see from Eq. (15.2.25). At the other interface at $x = \ell$ it turns back into a particle picking up a further phase of $e^{-i\chi/2}$. In addition it acquires a "transmission phase" of $e^{-2ik\ell}$. In this small k approximation, α, related to the fractional part of $2k_F\ell$, does not enter into the acquired phase due to a cancellation of particle and hole contributions. Imposing antiperiodic boundary conditions then determines the allowed wave-vectors by:

$$e^{-2ik\ell - i\chi} = -1. \tag{15.2.42}$$

On the other hand, a hole travelling towards the interface at $x = 0$ acquires a phase $e^{i\chi/2}$. Therefore the two types of low energy solutions correspond to wave-vectors

$$k_{n\pm} = \frac{1}{\ell}[\pi(n + 1/2) \pm \chi/2], \quad (n \in Z) \tag{15.2.43}$$

and corresponding energies $E = v_F k$. Remarkably, the low energy eigenvalues are independent of α for $\ell \gg 1/\xi_M$, quite unlike the opposite limit $\ell \ll \xi_M$, discussed in the previous sub-section, where there is very strong dependence on α.

Note that in this case important contributions are made to the current from many energy levels and we must sum over all of them. We now consider the entire set of energy levels in a general microscopic theory such as the tight-binding model introduced above. In general the Hamiltonian can be diagonalized in the form:

$$H = \sum_n \epsilon_n (\psi_n^\dagger \psi_n - 1/2) + C \tag{15.2.44}$$

with all $\epsilon_n \geq 0$. (The constant C is independent of χ.) Thus the ground state energy is

$$E_0 = -\frac{1}{2}\sum_n \epsilon_n + C. \tag{15.2.45}$$

The allowed wave-vectors are quite generally

$$k_{n\pm} = \frac{\pi n + \delta_{n\pm}}{\ell} \tag{15.2.46}$$

where k is now the full wave-vector, unshifted by k_F as in the field theory treatment and $\delta_{n\pm}$ are phase shifts. A general technique for doing such sums at large ℓ was developed in [13]. To order $1/\ell$, the ground state energy can be written

$$\begin{aligned} E_0 = \ell \int_0^{k_F} \frac{dk}{2\pi}\epsilon(k) + \frac{1}{2\pi}\int_{\epsilon_0}^{\epsilon_F} d\epsilon[\delta_+(\epsilon) + \delta_-(\epsilon)] \\ + \frac{\pi v_F}{4\ell}\left[\left(\frac{\delta_+(k_F)}{\pi}\right)^2 + \left(\frac{\delta_-(k_F)}{\pi}\right)^2 - \frac{1}{6}\right] + C. \end{aligned} \tag{15.2.47}$$

Here $\epsilon(k)$ is the full dispersion relation of the microscopic model, ϵ_0 is the bottom of the band and ϵ_F is the Fermi energy. The reason that the $O(1/\ell)$ term arises only at ϵ_F, not the bottom of the band, ϵ_0, is related to the vanishing of $d\epsilon/dk$ at the bottom of the band, true for general dispersion relations.

Generally, $\delta_+(\epsilon) + \delta_-(\epsilon)$ is independent of χ, so only the last term, of order $1/\ell$ contributes to the current. From Eq. (15.2.43) we see that:

$$\delta_\pm(k_F) = (\pi \pm \chi)/2 \tag{15.2.48}$$

implying:

$$E_0(\chi) = \text{constant} + \frac{v_F \chi^2}{8\pi\ell}, \quad (\text{mod } 2\pi). \tag{15.2.49}$$

and a Josephson current

$$I = \frac{ev_F}{2\pi\ell}\chi, \quad (\text{mod } 2\pi). \tag{15.2.50}$$

We have determined the current for all χ by demanding 2π periodicity. We now obtain a sawtooth with jumps at $\chi = (2n+1)\pi$. As we see from Eq. (15.2.43), these correspond to the values of χ at which an energy level passes through zero. We see from Fig. 15.3 that Eq. (15.2.50) agrees well with the exact result of Eq. (15.2.33) when $\ell \gg \xi_M$.

15.3. Interacting case

We now consider adding general interactions in the normal region. To the tight-binding model of Eq. (15.2.2) we could add nearest neighbor repulsive interactions between the electrons on sites 1 to $N-1$. To the Hamiltonian density of Eq. (15.2.6) we could add a $\psi_+^\dagger\psi_+\psi_-^\dagger\psi_-$ term. (We assume that Umklapp scattering is not relevant due to an incommensurate filling factor so the system remains gapless.) In the low energy effective field theory approach it is very convenient to use "bosonization" techniques which map interacting fermions in one dimension into non-interacting bosons. The left and right movers are represented as:

$$\psi_\mp(x) = C\Gamma \exp\{i\sqrt{\pi}[\phi(x)/\sqrt{K} \pm \sqrt{K}\theta(x)]\}. \tag{15.3.1}$$

Here Γ is a "Klein factor", another unphysical Majorana fermion introduced to obtain the correct anti-commutation relations with the physical MM's, $\gamma_{L/R}$ [21]. K is the Luttinger parameter, having the value $K = 1$ for non-interacting fermions and $K < 1$ for repulsive interactions. C is some cut-off dependent constant which we can assume is positive. The low energy Hamiltonian is:

$$H_0 = \frac{u}{2}\int_0^\ell \left[\left(\frac{d\phi}{dx}\right)^2 + \left(\frac{d\theta}{dx}\right)^2\right] \tag{15.3.2}$$

with the fields ϕ and θ obeying the canonical commutation relations:

$$[\phi(x), \theta(y)] = -\frac{i}{2}\text{sign}(x - y). \tag{15.3.3}$$

Here u is a renormalized Fermi velocity. The boundary conditions of Eq. (15.2.7) become:

$$\theta(0) = \sqrt{\pi/4K} \tag{15.3.4}$$

$$\theta(\ell) = \alpha/\sqrt{4\pi K}. \tag{15.3.5}$$

H_b thus becomes, in the low energy effective Hamiltonian:

$$H_b \approx -i\tilde{t}_L\gamma_L\Gamma C e^{i[\sqrt{\pi/K}\phi(0)+\chi/4]} - \tilde{t}_R\gamma_R\Gamma C e^{i[\sqrt{\pi/K}\phi(\ell)-\chi/4]} + h.c. \tag{15.3.6}$$

15.3.1. *Single S-N junction*

Consider first the case of an infinite normal region so that we may consider the effect of a single S-N junction, say the left one, with

$$H_b = -i2\tilde{t}_L\gamma_L\Gamma\cos\left[\sqrt{\pi/K}\phi(0) + \chi/4\right]. \tag{15.3.7}$$

Then this boundary operator has a renormalization group scaling dimension of

$$d = 1/(2K) \tag{15.3.8}$$

and $\tilde{t}_L$ is relevant for $d < 1$ which includes a large range of repulsive interaction strengths, $K > 1/2$. It is expected to renormalize to a strong couplng fixed point. It is natural that there are actually two equivalent fixed points at which $i < \gamma_L\Gamma >= \pm 1$. We may form a Dirac fermion from the physical MM γ_L of the topological superconductor on the left side and from the Klein factor Γ coming from the normal region:

$$\psi_0 = (\gamma_L + i\Gamma)/2. \tag{15.3.9}$$

Then

$$i\gamma_L\Gamma = 2\psi_0^\dagger\psi_0 - 1 \tag{15.3.10}$$

and we see that the two fixed points correspond to this energy level being occupied or empty. This level being filled corresponds to a "Schroedinger cat state" in which an electron is simultaneously localized at the end of the superconductor and delocalized in the normal region. At these two fixed points $\phi(0)$ is pinned at $-\sqrt{K/\pi}\chi/4$ or $\sqrt{\pi K} - \sqrt{K/\pi}\chi/4$ (mod $2\sqrt{\pi K}$). The physical meaning of this boundary condition on the normal region is perfect Andreev reflection as can be seen from the fact that it implies:

$$\psi_+^\dagger(0) = e^{i\chi/2}\psi_-(0). \tag{15.3.11}$$

Note that pinning $\phi(0)$ means that the dual field $\theta(0)$ must fluctuate wildly and vice versa, due to the commutation relations of Eq. (15.3.3).

A cartoon picture of this low energy fixed point is that the MM γ_L has paired with another MM from the normal wire, represented by the Klein factor, Γ, to form a Dirac mode. The remaining low energy degrees of freedom of the normal wire then simply exhibit perfect Andreev reflection. This is reminiscent of the cartoon picture of the strong coupling Kondo fixed point discussed in Sect. 15.1.

We may again estimate a "screening cloud size", ξ_M from the RG equations. Starting with a small tunnelling to γ_L, we see that the renormalized dimensionless tunnelling amplitude becomes O(1) at the length scale:

$$\xi_M \approx a\left(\frac{u}{\sqrt{a}\tilde{t}_L}\right)^{\frac{1}{1-d}}. \tag{15.3.12}$$

Here a is an ultraviolet cut-off scale such as the lattice constant in the tight binding model. Of course for the non-interacting case where $K = 1$, $d = 1/2$, we recover Eq. (15.2.26).

Again, we expect that the length of the SNS junction, ℓ, will act as an infrared cut-off on the growth of the renormalized couplings to the two MM's with distinct simple behaviours in the two limits $\ell \ll \xi_M$ and $\ell \gg \xi_M$ which we now consider.

15.3.2. $\ell \ll \xi_M$

It is convenient to decompose the boson fields into left and right movers:

$$\begin{aligned}\phi(x) &= \phi_-(x) + \phi_+(x)\\ \theta(x) &= \phi_-(x) - \phi_+(x).\end{aligned} \tag{15.3.13}$$

We may again use the boundary condition at $x = 0$ of Eq. (15.3.4) to "unfold" the system, defining

$$\phi_-(-x) \equiv \phi_+(x) + \sqrt{\pi/4K}, \quad (0 < x < \ell). \tag{15.3.14}$$

The Hamiltonian can now be written in terms of left-movers, $\phi_-(x)$ only on an interval of length 2ℓ:

$$H_0 = u\int_{-\ell}^{\ell}\left(\frac{d\phi_-}{dx}\right)^2 \tag{15.3.15}$$

with commutation relations:

$$[\phi_-(x), \phi_-(y)] = -\frac{i}{4}\mathrm{sign}(x-y) \tag{15.3.16}$$

and twisted boundary condition:

$$\phi_-(\ell) = \phi_-(-\ell) + \alpha/\sqrt{4\pi K}, \quad \left(\mathrm{mod}\sqrt{\frac{\pi}{K}}\right). \tag{15.3.17}$$

It is then convenient to define a new field obeying periodic boundary conditions:

$$\begin{aligned}\tilde{\phi}_-(x) &\equiv \phi_-(x) - \frac{\alpha x}{2\ell\sqrt{4\pi K}}\\ \tilde{\phi}_-(\ell) &= \tilde{\phi}_-(-\ell), \quad \left(\mathrm{mod}\sqrt{\frac{\pi}{K}}\right).\end{aligned} \tag{15.3.18}$$

We now have:

$$H_0 = u\int_{-\ell}^{\ell} dx \left[\left(\frac{d\tilde{\phi}_-}{dx}\right)^2 + \frac{\alpha}{\ell\sqrt{4\pi K}}\frac{d\tilde{\phi}_-}{dx}\right]. \tag{15.3.19}$$

The boundary interactions become:

$$\begin{aligned} H_b = &-\tilde{t}_L\gamma_L\Gamma C \mathrm{e}^{i(\pi/2)(1-1/K)+i\chi/4}\mathrm{e}^{i\sqrt{4\pi/K}\tilde{\phi}_-(0)} \\ &-\tilde{t}_R\gamma_R\Gamma C\mathrm{e}^{-i\chi/4}\mathrm{e}^{i\sqrt{4\pi/K}\tilde{\phi}_-(\ell)} + h.c. \end{aligned} \tag{15.3.20}$$

for a non-universal positive constant C. We may again integrate out the degrees of freedom in the central region, $\tilde{\phi}_-(x)$, leaving

$$H_{eff} = -i\tilde{t}_L\tilde{t}_R C^2\gamma_L\gamma_R \mathrm{e}^{i\chi/2-i\pi/2K}G(\omega=0,\ell) + h.c. \tag{15.3.21}$$

where

$$G(\tau,x) \equiv < \mathrm{e}^{i\sqrt{4\pi/K}\tilde{\phi}_-(0,0)}\mathrm{e}^{-i\sqrt{4\pi/K}\tilde{\phi}_-(\tau,x)} > . \tag{15.3.22}$$

This Green's function for a left-moving boson with periodic boundary conditions on a cylinder of circumference 2ℓ may be calculated from the Green's function on the infinite plane, $G \propto 1/(u\tau - ix)^{1/K}$ by the conformal transformation [22]

$$u\tau' - ix' \longrightarrow \mathrm{e}^{\pi(u\tau-ix)/2\ell}, \tag{15.3.23}$$

giving, for $\alpha = 0$, and an appropriately chosen positive constant C,

$$G(\tau,x) = \frac{\pi u}{2\ell^{1/K}\sinh[\pi(u\tau-ix)/(2\ell)]^{(1/K)}}. \tag{15.3.24}$$

Thus

$$G(\tau,\ell) = \frac{\mathrm{e}^{i\pi/2K}\pi u}{2\ell^{1/K}\cosh[\pi(u\tau)/(2\ell)]^{(1/K)}} \tag{15.3.25}$$

This gives:

$$G(\omega=0,\ell) = \frac{\mathrm{e}^{i\pi/2K}}{\ell^{1/K-1}}\int_{-\infty}^{\infty} dy\frac{1}{\cosh^{1/K}y}. \tag{15.3.26}$$

The K-dependent phase in Eq. (15.3.21) cancels and we are left with:

$$H_{eff} = -2i\tilde{t}_L\tilde{t}_R C^2\gamma_L\gamma_R\cos(\chi/2)|G(\omega=0,\ell)|. \tag{15.3.27}$$

So, we see that repulsive interactions, in the case $\ell \ll \xi_M$ have the effect of suppressing the Josephson current by an extra factor of $1/\ell^{1/K-1}$. It now has the form

$$I \propto \frac{e\tilde{t}_L\tilde{t}_R}{\ell^{1/K-1}}\sin\chi/2, \quad (\mathrm{mod}\ 2\pi). \tag{15.3.28}$$

This has qualitatively the same form for all values of the Luttinger parameter K. Note that the overall amplitude is proportional to

$$\frac{\tilde{t}_L\tilde{t}_R}{\ell^{1/K-1}} \propto \frac{1}{\ell}\left(\frac{\ell^2}{\xi_{ML}\xi_{MR}}\right)^{1-d} \tag{15.3.29}$$

where ξ_{ML} and ξ_{MR} are the sizes of the screening clouds for the left and right MM's, defined in Eq. (15.3.12), and $d = 1/K$ is the scaling dimension of the tunnelling terms.

To calculate the Green's function for non-zero α we must take into account the extra term in the Hamiltonian in Eq. (15.3.19):

$$\delta H \equiv u\frac{\alpha}{\ell\sqrt{4\pi K}}\int_{-\ell}^{\ell} dx \frac{d\tilde{\phi}_-}{dx} \tag{15.3.30}$$

which commutes with the first term in H_0 in Eq. (15.3.19). Writing:

$$G(\tau,\ell) = < 0|\mathrm{e}^{i\sqrt{4\pi/K}\tilde{\phi}_-(0)}\mathrm{e}^{H\tau}\mathrm{e}^{-i\sqrt{4\pi/K}\tilde{\phi}_-(\ell)}\mathrm{e}^{-H\tau}|0> \tag{15.3.31}$$

we see that the Green's function picks up an extra factor:

$$\exp\left\{-i\sqrt{4\pi/K}[\delta H, \tilde{\phi}_-(\ell)]\tau\right\} = \exp[-u\alpha\tau/(2K\ell)] \tag{15.3.32}$$

where the commutator of Eq. (15.3.16) was used. Thus

$$G(\omega = 0,\ell) = \frac{\mathrm{e}^{i\pi/2K}}{\ell^{1/K-1}}\int_{-\infty}^{\infty} dy \frac{\mathrm{e}^{-\alpha y/(\pi K)}}{\cosh^{1/K} y} \equiv \frac{\mathrm{e}^{i\pi/2K}}{\ell^{1/K-1}} f_K(\alpha). \tag{15.3.33}$$

implying the current:

$$I \propto \frac{f_K(\alpha)\tilde{t}_L\tilde{t}_R}{\ell^{1/K-1}}\sin\chi/2, \quad (\mathrm{mod}\ 2\pi). \tag{15.3.34}$$

Again, this is only valid for $|\alpha| < \pi$ where the integral converges. At $\alpha \to \pm\pi$ a zero mode appears in the bosonic spectrum which must be treated separately. See Sect. 15.5.

15.3.3. $\ell \gg \xi_M$

In this limit, assuming the tunnelling parameters $\tilde{t}_{L/R}$ in Eq. (15.3.6) renormalize to large values, we expect $\phi(0)$ and $\phi(\ell)$ to be pinned, implying that $\theta(0)$ and $\theta(\ell)$ are fluctuating. As discussed in Sect. 15.3.1, this corresponds to perfect Andreev scattering. In this large ℓ limit, the finite size of the normal region doesn't prevent this renormalization from occurring independently at both boundaries. We may now simply forget about the MM's, which are screened and just study the free boson system with these Andreev boundary conditions. We may again transform to left-movers, $\phi_-(x)$ on an interval of length 2ℓ with the Hamiltonian of Eq. (15.3.15) and the twisted boundary conditions, determined by Eq. (15.3.6) :

$$\phi_-(\ell) = \phi_-(-\ell) + \sqrt{\frac{K}{\pi}}\frac{\chi}{2} \quad (\mathrm{mod}\ \sqrt{\pi K}). \tag{15.3.35}$$

The field $\phi_-(x)$ may be expanded in normal modes. These are harmonic oscillator modes, vanishing at $x = -\ell$ and ℓ as well as a soliton mode of the form:

$$\sqrt{\frac{\pi}{K}}\phi_-(x) = \mathrm{constant} + \left[\frac{\chi}{2} - \pi n\right]\frac{x}{2\ell}. \tag{15.3.36}$$

[See Sect. 15.5 and in particular Eq. (15.5.12) for a more detailed discussion.] Only the soliton mode energies have any dependence on the phase difference χ of the two superconductors. Thus the χ dependence of the ground state is:

$$E_0 = \frac{uK}{2\pi\ell}\left[\frac{\chi}{2} - \pi n\right]^2 \tag{15.3.37}$$

with the integer n chosen to minimize E_0:

$$E_0 = \frac{uK\chi^2}{8\pi\ell}, \quad (\text{mod } 2\pi). \tag{15.3.38}$$

The corresponding current, $2e(dE_0/d\chi)$ is

$$I = \frac{euK}{2\pi\ell}\chi, \quad (\text{mod } 2\pi). \tag{15.3.39}$$

This is again a sawtooth, as found in the non-interacting case, for $\xi_M \ll \ell$, but now the amplitude is proportional to the product of velocity and Luttinger parameter uK.

We again expect that in general I can be expressed as a scaling function

$$I(\chi) = \frac{eu}{2\pi\ell}F\left(\frac{\ell}{\sqrt{\xi_{ML}\xi_{MR}}}, \frac{\xi_{ML}}{\xi_{MR}}, \chi, \alpha, K\right). \tag{15.3.40}$$

The above results are consistent with this, giving:

$$\begin{aligned} F(z, r, \chi, \alpha, K) &\to K\chi, \quad (z \to \infty) \\ &\propto z^{1-1/K} f_K(\alpha) \sin\chi, \quad (z \to 0) \end{aligned} \tag{15.3.41}$$

where the function $f_K(\alpha)$ is defined in Eq. (15.3.33).

15.4. Fermion parity

Our treatment of the topological SNS junction so far has implicitly assumed that it is in diffusive equilibrium with a source of electrons at chemical potential μ, which determines the parameter α by Eq. (15.2.12). Whether this is a valid model will depend on experimental details. If the only materials in contact with the normal region are superconductors and insulators then there might be a bath of Cooper pairs present but not of single electrons. In that case, as the phase difference χ (and other parameters) are varied the *parity* of the total number of electrons in the normal region would stay fixed, the number only changing in steps of two. In that case the calculation of the Josephson current still proceeds from Eq. (15.2.3) but the ground state energy as a function of chemical potential and other parameters must be calculated for a fixed fermion parity. We again expect the Josephson current to be a universal scaling function of $\ell/\sqrt{\xi_{ML}\xi_{MR}}$ and other parameters, but it clearly must be a different function that was calculated above without enforcing fermion parity conservation. In this section we extend our calculation of the current to the fermion parity symmetric case, in the two simple limits $\ell \ll \xi_M$ and $\ell \gg \xi_M$ and present numerical results for intermediate values of this ratio.

The case of weak coupling, $\ell \ll \xi_M$ is easily dealt with since the low energy effective Hamiltonian of Eq. (15.2.37) only contains a single Dirac energy level. Without fermion parity conservation, a cusp occurs in $E(\chi)$ at $|\chi| = \pi$ because the energy of this Dirac level passes through zero at those points, with the Dirac level switching from being occupied to empty in the ground state, leading to a step in the current. However, with fermion parity enforced, this Dirac level must stay empty (or occupied) for all χ, eliminating the step in the current. So, in this case we obtain simply:

$$I = \pm \frac{e\tilde{t}_L \tilde{t}_R}{v_F \cos \alpha/2} \sin \chi/2, \tag{15.4.1}$$

where the sign depends on whether the total number of fermions is even or odd. Precisely the same argument applies to the interacting case for $\ell \ll \xi_M$ so that the current becomes:

$$I \propto \frac{e\tilde{t}_L \tilde{t}_R}{\ell^{1/K-1}} \sin(\chi/2), \tag{15.4.2}$$

with no steps.

A similar argument can be easily constructed in the opposite limit $\ell \gg \xi_M$, in which Eq. (15.3.37) gives the χ-dependence of the ground state energy, including interactions. The integer n, which is the number of electrons present (measured from a convenient zero point), jumps by ± 1 at $|\chi| = \pi$ in the ground state, leading to steps in the current. With fermion parity conservation enforced, n may only change in units of ± 2. If the value of the fermion parity is such that n must be even then

$$I = \frac{euK}{2\pi\ell} \chi, \quad (\text{mod } 4\pi) \tag{15.4.3}$$

On the other hand, if n must be odd,

$$I = \frac{euK}{2\pi\ell} (\chi - 2\pi), \quad (\text{mod } 4\pi). \tag{15.4.4}$$

Steps occur at $\chi = 4\pi(n + 1/2)$ for one value of the fermion parity quantum number and at $\chi = 4\pi n$ for the other.

We see that the period of $I(\chi)$ is doubled from 2π to 4π when fermion parity is enforced, a phenomena which holds for any value of ξ_M/ℓ. On the other hand, while the steps in the current are eliminated by imposing fermion parity at $\ell \ll \xi$, they are still present at $\ell \gg \xi$, occurring at only half as many values of χ but with twice the height. In Fig. 15.4 we plot the current calculated in the tight-binding model at large N with fermion parity enforced, corresponding to $\alpha = 0$ and $\xi_{ML} = \xi_{MR}$, comparing to our analytic predictions (see the Appendix for a discussion of the tight-binding model results).

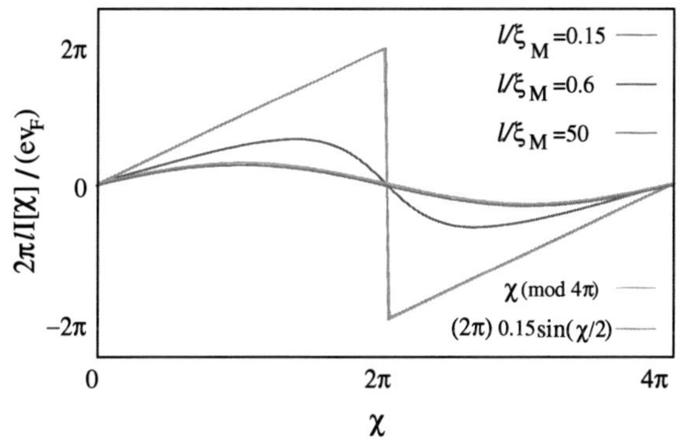

Fig. 15.4. Josephson current versus phase difference, χ from a numerical solution of the tight binding model for particle-hole and left-right symmetric junction, $\alpha = 0$, $\xi_{ML} = \xi_{MR}$ [defined in Eq. (15.2.26)] with fermion parity conservation imposed. Our analytic predictions are also shown, fitting the data well for $\ell \ll \xi_M$ and $\ell \gg \xi_M$. Note that the period is now 4π and that a step only occurs at $\ell/\xi_M \to \infty$.

15.5. Zero mode case

In the special case $|\alpha| = \pi$, the spectrum of the non-interacting normal region, Eq. (15.2.30) contains a zero energy state, before it is coupled to the MM's. While our analysis of $\ell \gg \xi_M$ is unaffected, our simple analysis of weak coupling to the MM's, $\ell \ll \xi_M$, then requires modification. We may simply project the bulk fermion operators $\psi(x)$ appearing in H_b onto the zero mode:

$$\psi(0), \psi(\ell) \to \frac{1}{\sqrt{2\ell}} c \tag{15.5.1}$$

where c is the zero mode annihilation operator and the normalization of the zero-mode wave-function has been taken into account. Integrating out the other, finite energy, modes, leads to no contribution to the low energy effective Hamiltonian to order $\tilde{t}_L \tilde{t}_R$. This can be seen from the Green's function of Eq. (15.2.29) at $\alpha = -\pi$ with the zero mode omitted:

$$G'(\omega = 0, \ell) = -\frac{1}{2\pi v_F} \sum_{n \neq 0} \frac{(-1)^n}{n} = 0. \tag{15.5.2}$$

The projected boundary Hamiltonian is:

$$H_b \to -\frac{\tilde{t}_L}{\sqrt{2\ell}} \gamma_L [e^{i\chi/4} c - e^{-i\chi/4} c^\dagger] - \frac{\tilde{t}_R}{\sqrt{2\ell}} \gamma_R [e^{-i\chi/4} c - e^{i\chi/4} c^\dagger]. \tag{15.5.3}$$

This is a simple model of 2 MM's, γ_L, γ_R, coupled to one Dirac mode, c, $c^\dagger$, which can be easily diagonalized in terms of 2 Dirac mode, $\psi_\pm$, giving:

$$H = \sum_{\pm} E_\pm(\chi) [\psi_\pm^\dagger \psi_\pm - 1/2] \tag{15.5.4}$$

where the two positive energy eigenvalues are:

$$E_\pm(\chi) \equiv \frac{1}{\sqrt{2\ell}}\left[\tilde{t}_L^2+\tilde{t}_R^2 \pm \sqrt{(\tilde{t}_L^2+\tilde{t}_R^2)^2-4\tilde{t}_L^2\tilde{t}_R^2\cos^2\chi/2}\right]^{1/2}. \quad (15.5.5)$$

Thus the ground state energy is:

$$E_0 = -\frac{1}{2}\sum_\pm E_\pm \quad (15.5.6)$$

and the current is:

$$I = \frac{e\tilde{t}_L^2\tilde{t}_R^2\sin\chi}{4\sqrt{2\ell}\sqrt{(\tilde{t}_L^2+\tilde{t}_R^2)^2-4\tilde{t}_L^2\tilde{t}_R^2\cos^2\chi/2}}\sum_\pm$$

$$\pm\left[\tilde{t}_L^2+\tilde{t}_R^2 \mp \sqrt{(\tilde{t}_L^2+\tilde{t}_R^2)^2-4\tilde{t}_L^2\tilde{t}_R^2\cos^2\chi/2}\right]^{-1/2}. \quad (15.5.7)$$

We see that this can be written:

$$I = \frac{2ev_F}{\ell}\left(\frac{\ell}{\sqrt{\xi_{ML}\xi_{MR}}}\right)^{1/2} f(\xi_{ML}/\xi_{MR},\chi), \quad (15.5.8)$$

consistent with scaling. I can again be seen to have steps at $|\chi| = \pi$ due to E_- vanishing there.

Naturally, the same result can be obtained from our exact result, Eq. (15.2.33), for the current in the non-interacting case, by taking the limit $\ell \ll \xi_M$. In this limit the integral is dominated by small ω. Setting $\alpha = \pi$ and Taylor expanding the cosh and sinh functions, Eq. (15.2.33) gives:

$$I = \frac{e}{\pi\ell^2}\tilde{t}_L^2\tilde{t}_R^2\sin\chi\int_{-\infty}^{\infty} d\omega\frac{1}{[\omega^2+(2E_+)^2][\omega^2+(2E_-)^2]} \quad (15.5.9)$$

where $E_\pm$ are given in Eq. (15.5.5). Doing this elementary integral gives back Eq. (15.5.7). Note that I is larger by a factor of $\sqrt{(\xi_{ML}\xi_{MR})^{1/2}/\ell}$, for $\ell \ll \sqrt{\xi_{ML}\xi_{MR}}$, in the case $\alpha = \pi$ with a bulk zero mode than for generic α. The crossover behaviour for α near π could also be calculated by considering the case of a low energy bulk state whose energy is not precisely zero. The result can be read off from Eq. (15.2.33).

This result may be straightforwardly extended to the case with fermion parity conservation. The fermion parity of the ground state of this 2-state system reverses at $\chi = \pi$ where $E_- = 0$. Thus the lowest energy state of fixed fermion parity has energy:

$$E_0^{FP} = -\frac{1}{2}[E_+ \pm \varepsilon_P(\chi)E_-] \quad (15.5.10)$$

where $\varepsilon_P(\chi)$ is the period 4π step function defined in Eq. (15.2.40). the plus or minus sign in Eq. (15.5.10) is determined by whether the number of fermions is maintained at even or odd parity. Then the current becomes:

$$I = \frac{e\tilde{t}_L^2\tilde{t}_R^2 \sin\chi}{4\sqrt{2}\ell\sqrt{(\tilde{t}_L^2+\tilde{t}_R^2)^2 - 4\tilde{t}_L^2\tilde{t}_R^2\cos^2\chi/2}}$$
$$\times \left\{ - \left[\tilde{t}_L^2+\tilde{t}_R^2+\sqrt{(\tilde{t}_L^2+\tilde{t}_R^2)^2 - 4\tilde{t}_L^2\tilde{t}_R^2\cos^2\chi/2}\right]^{-1/2}\right.$$
$$\left. \pm\, \varepsilon_P(\chi)\left[\tilde{t}_L^2+\tilde{t}_R^2-\sqrt{(\tilde{t}_L^2+\tilde{t}_R^2)^2 - 4\tilde{t}_L^2\tilde{t}_R^2\cos^2\chi/2}\right]^{-1/2}\right\} \quad (15.5.11)$$

which is continuous (no steps) and has period 4π. [The $\pm$ sign in Eq. (15.5.11) is determined by whether the fermion number is maintained at even or odd values.]

Finally, we may extend our analysis to the interacting case. At $\alpha = \pi$ there is again a zero energy bulk state now represented in bosonized language. In general, the mode expansion for the left-moving bosonic field is:

$$\phi_-(x) = \phi_0 + \frac{x}{4\ell}\left[Q - \sqrt{\pi/K}\alpha/\pi\right] + \sum_{n=1}^{\infty}\frac{1}{\sqrt{4\pi n}}\left[\mathrm{e}^{-i\pi n x/\ell}a_n + h.c.\right] \quad (15.5.12)$$

The first term represents the solitonic mode of the boson field. The operator Q has eigenvalues $Q = \sqrt{4\pi/K}n$ for integers n and is canonically conjugate to ϕ_0:

$$[Q, \phi_0] = -i, \quad (15.5.13)$$

necessary for $\phi_-(x)$ to obey the canonical commutation relations of Eq. (15.3.16). The last term in Eq. (15.5.12) contains the harmonic oscillator modes, with corresponding annihilation operators a_n. For $\alpha = \pi$ we see that the $Q = 0$ and $Q = \sqrt{4\pi/K}$ soliton states become degenerate ground states. For $\ell \ll \xi_M$ we project the boundary Hamiltonian of Eq. (15.3.6) onto these ground states of the solitonic mode. The operators $\mathrm{e}^{\pm i\sqrt{4\pi/K}\phi_0}$ are raising and lowering operators between these two ground states:

$$\mathrm{e}^{i\sqrt{4\pi/K}\phi_0}|- >\propto |+ >, \quad \mathrm{e}^{-i4\sqrt{\pi/K}\phi_0}|+ >\propto |- > \quad (15.5.14)$$

We may represent them as raising and lowering operators for an effective s=1/2 spin:

$$\mathrm{e}^{\pm i\sqrt{4\pi/K}\phi_0} \propto S^{\pm}. \quad (15.5.15)$$

To first order in $\tilde{t}_{L/R}$ we can integrate out the oscillator degrees of freedom by simply taking the ground state expectation value of the exponential operator in H_b:

$$< 0|\mathrm{e}^{i\sqrt{4\pi/K}\phi(0)}|0 >\propto S^+ < 0|\exp\left[i\sqrt{4\pi/K}\sum_{n=1}^{\infty}\frac{1}{\sqrt{4\pi n}}[a_n + a_n^\dagger]\right]|0 >$$
$$= S^+ \exp\left[\frac{-1}{2K}\sum_{n=1}^{\ell/a}\frac{1}{n}\right]. \quad (15.5.16)$$

Note that the ground state matrix element is taken in the harmonic oscillator space but the solitonic factor remains as an operator. We have inserted an ultraviolet cut off on the number of oscillator modes, restricting the wave-vector to $k < \pi/a$, where a is for example the lattice constant in the tight binding model, so the maximum n is $O(\ell/a)$.

Thus

$$< 0|e^{i\sqrt{4\pi/K}\phi(0)}|0 >\propto S^+ \left(\frac{a}{\ell}\right)^{1/(2K)}. \qquad (15.5.17)$$

The exponent $1/(2K)$ appearing here is the RG scaling dimension of the operator $e^{i\sqrt{4\pi/K}\phi(0)}$. We obtain the same result for $e^{i\sqrt{4\pi/K}\phi(\ell)}$. Then the boundary Hamiltonian, projected onto the two degenerate solitonic ground states, with the harmonic oscillator modes integrated out takes the form:

$$H_b \propto -\frac{\tilde{t}_L}{\ell^{1/2K}}\gamma_L\Gamma[e^{i\chi/4}S^- - e^{-i\chi/4}S^+] - \frac{\tilde{t}_R}{\ell^{1/2K}}\gamma_R\Gamma[e^{-i\chi/4}S^- - e^{i\chi/4}S^+]. \qquad (15.5.18)$$

Finally, we note that we can make a Jordan-Wigner-type transformation and represent the product of the Klein factor Γ and the spin raising and lowering operators $S^{\pm}$ by a Dirac pair c, $c^{\dagger}$:

$$\begin{aligned} \Gamma S^- &= c \\ \Gamma S^+ &= c^{\dagger}. \end{aligned} \qquad (15.5.19)$$

We thus recover the boundary Hamiltonian of the non-interacting case, Eq. (15.5.3) apart from the different power of $1/\ell$. Thus the current is again given by Eq. (15.5.7) with this different power of $1/\ell$. The current therefore has the scaling form:

$$I = \frac{eu}{2\pi\ell}\sqrt{\tilde{t}_L\tilde{t}_R}\ell^{1-1/2K}f(\tilde{t}_L/\tilde{t}_R,\chi) = \frac{eu}{2\pi\ell}\left(\ell/\sqrt{\xi_{ML}\xi_{MR}}\right)^{1-d}g(\xi_{ML}/\xi_{MR},\chi) \qquad (15.5.20)$$

where Eqs. (15.3.8) and (15.3.12) were used. Comparing to the case $\ell \ll \xi_M$ with generic α, Eq. (15.3.28), Eq. (15.3.29), we see that the current is enhanced by a factor of $(\sqrt{\xi_{ML}\xi_{MR}}/\ell)^{1-d}$ at $\alpha = \pi$ due to the presence of the bulk zero mode. The case with fermion parity conservation can be analysed as above for the non-interacting limit.

15.6. Conclusions

Quantum impurity models quite generally exhibit a characteristic crossover length scale, or screening cloud size, an idea which may have first emerged from Ken Wilson's work on the Kondo problem. We have studied this crossover in the dc Josephson current through a long normal wire between two topological superconductors, which is very generally a scaling function of the ratio of the screening cloud size associated with the Majorana mode at each SN junction to the length of the normal region. Analytic formulas were derived when this ratio is large or small, based on simple physical pictures. The crossover, in the non-interacting limit, was studied numerically and confirmed these simple pictures. Experimental measurements on such long topological SNS junctions could provide evidence for the existence of Majorana modes *and* of a Kondo-like screening cloud.

Acknowledgments

We would like to thank Y. Komijani and A. Tagliacozzo for helpful discussions. This research was supported in part by NSERC of Canada and CIfAR.

Appendix A: Current in tight binding model

To diagonalize the Hamiltonian of Eq. (15.2.2) we make a BdG transformation:

$$\Gamma_E = \sum_{j=1}^{N-1} \{u_j c_j + v_j c_j^\dagger\} + w_L \gamma_L + w_R \gamma_R. \tag{A.1}$$

On requiring that $[\Gamma_E, H] = E\Gamma_E$ we obtain the BdG equations

$$\begin{aligned} Eu_j &= -J\{u_{j+1} + u_{j-1}\} - \mu u_j \\ Ev_j &= J\{v_{j+1} + v_{j-1}\} + \mu v_j, \end{aligned} \tag{A.2}$$

for $2 < j < N-2$, supplemented by the boundary conditions

$$\begin{aligned} Eu_1 &= -Ju_2 - \mu u_1 - it_L w_L \mathrm{e}^{i\frac{\chi}{4}} \\ Ev_1 &= Jv_2 + \mu v_1 - it_L w_L \mathrm{e}^{-i\frac{\chi}{4}} \\ Ew_L &= 2t_L i\{u_1 \mathrm{e}^{-i\frac{\chi}{4}} + v_1 \mathrm{e}^{i\frac{\chi}{4}}\}, \end{aligned} \tag{A.3}$$

at the left-hand boundary, and

$$\begin{aligned} Eu_{N-1} &= -Ju_{N-2} - \mu u_{N-1} + t_R w_R \mathrm{e}^{-i\frac{\chi}{4}} \\ Ev_{N-1} &= Jv_{N-2} + \mu v_{N-1} - t_R w_R \mathrm{e}^{i\frac{\chi}{4}} \\ Ew_R &= 2t_R\{u_{N-1} \mathrm{e}^{i\frac{\chi}{4}} - v_{N-1} \mathrm{e}^{-i\frac{\chi}{4}}\}, \end{aligned} \tag{A.4}$$

at the right-hand boundary. In view of the "bulk" equations in Eq. (A.2), we expect that a generic solution to the BdG equations is given by

$$\begin{bmatrix} u_j \\ v_j \end{bmatrix} = \begin{bmatrix} A\mathrm{e}^{ikaj} + B\mathrm{e}^{-ikaj} \\ C\mathrm{e}^{-ik'aj} + D\mathrm{e}^{ik'aj} \end{bmatrix}, \tag{A.5}$$

with

$$E = -2J\cos(ka) - \mu = 2J\cos(k'a) + \mu. \tag{A.6}$$

In view of the explicit form of the solution in Eq. (A.5), Eq. (A.3) simplify to

$$0 = Ju_0 - it_L w_L e^{i\frac{\chi}{4}} \tag{A.7}$$
$$0 = Jv_0 + it_L w_L e^{-i\frac{\chi}{4}} \tag{A.8}$$
$$Ew_L = 2t_L i\{u_1 e^{-i\frac{\chi}{4}} + v_1 e^{i\frac{\chi}{4}}\}, \tag{A.9}$$

and, similarly, Eq. (A.4) simplify to

$$0 = Ju_N + t_R w_R e^{-i\frac{\chi}{4}} \tag{A.10}$$
$$0 = Jv_N + t_R w_R e^{i\frac{\chi}{4}} \tag{A.11}$$
$$Ew_R = 2t_R\{u_{N-1} e^{i\frac{\chi}{4}} - v_{N-1} e^{-i\frac{\chi}{4}}\}. \tag{A.12}$$

Substituting Eqs. (A.5) into Eqs. (A.7), (A.8), (A.10) and (A.11) we can solve for A, B, C and D in terms of w_L and w_R. Equations (A.9) and (A.12) then yield:

$$\frac{1}{J}\begin{pmatrix} 2t_L^2\left(-\frac{\sin k(\ell-a)}{\sin k\ell} + \frac{\sin k'(\ell-a)}{\sin k'\ell}\right) & 2it_L t_R\left(-\frac{\sin k}{\sin k\ell}e^{-i\chi/2} - \frac{\sin k'}{\sin k'\ell}e^{i\chi/2}\right) \\ 2it_L t_R\left(\frac{\sin k}{\sin k\ell}e^{i\chi/2} + \frac{\sin k'}{\sin k'\ell}e^{-i\chi/2}\right) & 2t_R^2\left(-\frac{\sin k(\ell-a)}{\sin k\ell} + \frac{\sin k'(\ell-a)}{\sin k'\ell}\right) \end{pmatrix}$$
$$\times \begin{pmatrix} w_L \\ w_R \end{pmatrix}$$
$$= E\begin{pmatrix} w_L \\ w_R \end{pmatrix}. \tag{A.13}$$

The allowed values of k are thus given by the solutions of:

$$\begin{aligned} 0 = &\left\{\left[EJ + 2t_L^2\left(\frac{\sin k(\ell-a)}{\sin k\ell} - \frac{\sin k'(\ell-a)}{\sin k'\ell}\right)\right]\right. \\ &\times\left.\left[EJ + 2t_R^2\left(\frac{\sin k(\ell-a)}{\sin k\ell} - \frac{\sin k'(\ell-a)}{\sin k'\ell}\right)\right]\right\} \\ &- 4t_L^2 t_R^2\left(\frac{\sin ka}{\sin k\ell}e^{-i\chi/2} + \frac{\sin k'a}{\sin k'\ell}e^{i\chi/2}\right)\left(\frac{\sin ka}{\sin k\ell}e^{i\chi/2} + \frac{\sin k'a}{\sin k'\ell}e^{-i\chi/2}\right) \end{aligned} \tag{A.14}$$

with k and k' related by Eq. (A.6). A simple special case occurs at half-filling, $\mu = 0$ when $k' = \pi - k$. In this case, Eq. (15.2.12) determines $\alpha = 0$ for odd N and $\alpha = \pi$ for even N. To get a simple solution for $\alpha = 0$, we choose N odd and also assume $t_L = t_R \equiv t$. Eqs. (A.14) simplifies to

$$EJ\sin k\ell + 4t^2\sin k(\ell - a) = \pm 2t^2 \sin ka \cos\chi/2. \tag{A.15}$$

We have solved Eq. (A.15) for for all allowed values of k for odd N up to 149 finding agreement with the universal field theory predictions as $N \to \infty$ and small $t_{L/R}$. We may then calculate the ground state energy:

$$E_0(\chi) = -J\sum_n \cos[k_n(\chi)a] \tag{A.16}$$

where the sum is over all solutions of Eq. (A.15) with $|k| < \pi/2$. The Josephson current is then given by $I = 2e(dE_0/d\chi)$. To enforce fermion parity conservation, we sum over an even number of values of k_n, which means omitting the smallest negative term for a range of χ. This result was used to produce Fig. 15.4.

References

1. Wilson, K.G.: The renormalization group: critical phenomena and the Kondo problem. Rev. Mod. Phys. **47**, 773–840 (1975)
2. For a review see Georges, A., Kotliar, G., Krauth, W., Rozenberg, M.J.: Dynamical mean-field theory of strongly correlated fermion systems and the limit of infinite dimensions. Rev. Mod. Phys. **68**, 13–125 (1996)
3. Nozières, P.: A "fermi-liquid" description of the Kondo problem at low temperatures. J. Low Temp. Phys. **17**, 31–42 (1974)
4. Nozières, P.: Kondo effect for spin 1/2 impurity a minimal effort scaling approach. J. Phys. **39**, 1117–1124 (1978)
5. Nozières, P.: In: Krusius, M., Vuorio, M. (eds.) Proceedings of 14th International Conference on Low Temperature Physics, vol. 5. North Holland, Amsterdam (1975)
6. Affleck, I.: The Kondo screening cloud: what is it and how to observe it. In: Perspectives of Mesoscopic Physics: Dedicated to Yoseph Imry's 70th Birthday, pp. 1–44. World Scientific, Singapore (2010). http://arxiv.org/abs/0911.2209 arXiv:0911.2209.
7. Yu, A.: Kitaev, unpaired Majorana fermions in quantum wires. Phys. Usp. **44**, 131 (2001)
8. Lutchyn, R.M., Sau, J.D., Das Sarma, S.: Majorana fermions and a topological phase transition in semiconductor–superconductor heterostructures. Phys. Rev. Lett. **105**, 077001 (2010)
9. Oreg, Y., Refael, G., von Oppen, F.: Helical liquids and Majorana bound states in quantum wires. Phys. Rev. Lett. **105**, 177002 (2010)
10. Mourik, V., Zuo, K., Frolov, S.M., Plissard, S.R., Bakkers, E.P.A.M., Kowenhoven, L.P.: Signatures of Majorana fermions in hybrid superconductor–semiconductor nanowire devices. Science **336**, 1003–1007 (2012)
11. Klinovaja, J., Loss, D.: Related finite size effects have been discussed in composite Majorana fermion wave functions in nanowires. Phys. Rev. **B86**, 085408 (2012)
12. J. Klinovaja and D. Loss, Spin and Majorana Polarization in Topological Superconducting Wires, D. Sticlet, C. Bena and P. Simon, Phys. Rev. Lett. **108**, 096802 (2012)
13. Affleck, I., Caux, J.S., Zagoskin, A.: Andreev scattering and Josephson current in a one-dimensional electron liquid. Phys. Rev. B **62**, 1433–1445 (2000)
14. Fidowski, L., Alicea, J., Lindner, N.H., Lutchyn, R.M., Fisher, M.P.A.: Universal transport signatures of Majorana fermions in superconductor-Luttinger liquid junctions. Phys. Rev. **B85**, 245121 (2012)
15. Affleck, I., Giuliano, D.: Topological superconductor-Luttinger liquid junctions. J. Stat. Mech. P06011 (2013)
16. Svidzinsky, A.V., Antsygina, T.N., Bratus, E.N.: Concerning the theory of the Josephson effect in pure SNS junctions. J. Low Temp. Phys. **10**, 131–136 (1973)
17. Beenakker, C.W.J., Pikulin, D.I., Hyart, T., Schomerus, H., Dahlhaus, J.P.: Fermion-Parity Anomaly of the Critical Supercurrent in the Quantum Spin-Hall Effect. Phys. Rev. Lett. **110**, 017003 (2013)

18. Zhang, S., Zhu, W., Sun, Q.: Josephson junction on one edge of a two dimensional topological insulator affected by magnetic impurity. J. Phys. Condens. Mat. **25**, 295301 (2013)
19. Ghosh, S., Ribeiro, P., Haque, M.: For a recent discussion of the crossover length scale for the non-interacting resonant level model see Real-space structure of the impurity screening cloud in the resonant level model. J. Stat. Mech. P04011 (2014)
20. Giuliano, D., Affleck, I.: The Josephson current through a long quantum wire. J. Stat. Mech. P02034 (2013)
21. Béri, B.: Majorana-Klein hybridization in topological superconductor junctions. Phys. Rev. Lett. **110**, 216803 (2013)
22. Cardy, J.L.: In: Brézin, E., Zinn-Justin, J. (eds.) Conformal invariance and statistical mechanics, fields, strings and critical phenomena, Les Houches 1988. North-Holland, Amsterdam (1990)

Chapter 16

Kenneth Wilson and lattice QCD*

Akira Ukawa
RIKEN Advanced Institute for Computational Science,
7-1-26 Minami-machi, Minatojima,
Chuoku, Kobe 650-0047, Japan
akira.ukawa@riken.jp

We discuss the physics and computation of lattice QCD, a space-time lattice formulation of quantum chromodynamics, and Kenneth Wilson's seminal role in its development. We start with the fundamental issue of confinement of quarks in the theory of the strong interactions, and discuss how lattice QCD provides a framework for understanding this phenomenon. A conceptual issue with lattice QCD is a conflict of space-time lattice with chiral symmetry of quarks. We discuss how this problem is resolved. Since lattice QCD is a non-linear quantum dynamical system with infinite degrees of freedom, quantities which are analytically calculable are limited. On the other hand, it provides an ideal case of massively parallel numerical computations. We review the long and distinguished history of parallel-architecture supercomputers designed and built for lattice QCD. We discuss algorithmic developments, in particular the difficulties posed by the fermionic nature of quarks, and their resolution. The triad of efforts toward better understanding of physics, better algorithms, and more powerful supercomputers have produced major breakthroughs in our understanding of the strong interactions. We review the salient results of this effort in understanding the hadron spectrum, the Cabibbo–Kobayashi–Maskawa matrix elements and CP violation, and quark-gluon plasma at high temperatures. We conclude with a brief summary and a future perspective.

Keywords: Lattice QCD; strong interactions; Standard Model; high performance computing; parallel supercomputers.

16.1. Introduction

In early 1974, Kenneth Wilson circulated a preprint entitled "Confinement of Quarks". The paper was received by Physical Review D on 12 June 1974, and

*This article was originally published in *J. Stat. Phys.*, doi:10.1007/s10955-015-1197-x.

was published in the 15 October issue of that journal [1]. In this paper, Wilson formulated gauge theories on a space-time lattice.[1] Using an expansion in inverse powers of the bare gauge coupling constant, Wilson demonstrated that lattice gauge theories at strong coupling confine charged states. He also argued that the absence of Lorentz invariance (or Euclidean invariance for the imaginary time used for the lattice formulation) is not a hindrance if there is a second order phase transition at some value of the gauge coupling constant, for in the vicinity of such a phase transition the lattice spacing can be taken to zero while fixing the physical correlation length at a finite value.

This paper laid the conceptual foundation for understanding the quark confinement phenomenon. It showed that large quantum fluctuations of gauge fields at strong coupling can generate a force between charged states which stays constant at arbitrary distances. This is a novel type of force, essentially different from the Yukawa force due to exchange of particles which tends to zero at large distances.

Initially, Wilson's idea did not catch on rapidly. Techniques were hard to come by which allowed calculations of physical quantities such as hadron masses and connect them to physical predictions in the continuum space-time. The situation changed dramatically around 1979–1980 when Creutz *et al.* [4], and Wilson himself [5], showed the possibility of numerically calculating the observables on a computer. Particularly dramatic was the calculation of the static quark-antiquark potential by Creutz for SU(2) gauge group in 1979 [6], and the calculations in 1981 of hadron masses by Weingarten [7] and by Hamber and Parisi [8].

Traditional quantum field theory until that time could only deal with weakly coupled bound states such as hydrogen. The possibility of a technique which enables calculation of the properties of relativistic and strongly coupled bound states such as pion and proton was entirely new.

The timing was also perfect from computational point of view. The CRAY-1 supercomputer which appeared in 1976 had revolutionized scientific computing, and lattice QCD computation could quickly exploit vector supercomputers in the 80s. Perhaps more important in retrospect, rapid development of microprocessors in the 70s stimulated more than a few groups of particle theorists around the world to start developing parallel computers for lattice QCD.

The development of lattice QCD has been continuous since then. With large-scale numerical simulations on parallel supercomputers, understanding of physics of lattice QCD progressed, which in turn led to better algorithms for computation. These algorithms allowed a better exploitation of the next generation of more powerful computers, which brought even more progress of physics.

---

[1]Contemporary research toward lattice gauge theory was described by Wilson in his plenary talk at 1983 Lepton Photon Symposium [2], and in more detail in his historical talk at 2004 International Symposium on Lattice Field Theory at FNAL [3].

In the four decades of progress, lattice QCD has brought a deep understanding on the physics of the strong interactions. It has matured to the point where one can make calculations with the physical values of quark masses, on lattices with sufficiently large sizes, at lattice spacings small enough so that a continuum limit can be carried out with confidence.

In this article we review lattice QCD from four perspectives in the following four sections. In Sect. 16.2, we discuss the foundation of lattice QCD, touching upon the connection between quantum field theory in Euclidean space time and statistical mechanics which was consciously exploited by Wilson. We show how it led to a conceptual breakthrough in the understanding of confinement. We also explain the issues related with chiral symmetry. In Sect. 16.3, we discuss the computational aspects. We review how lattice QCD embodies an ideal case of massively parallel computation, and how this led to the development of parallel supercomputers for lattice QCD, which impacted seriously on the history of supercomputers up to the present time. Also discussed is a special computational difficulty posed by the fermionic nature of quark fields, and how overcoming that difficulty has led to the algorithm in use today. In Sect. 16.4 we discuss some major physics results achieved so far in lattice QCD. The themes include the hadron mass spectrum, the determination of the Cabibbo–Kobayashi–Maskawa matrix and CP violation in the Standard Model, and the properties of quark gluon plasma at high temperatures and densities. Finally, in Sect. 16.5 we collect some thoughts on how Kenneth Wilson's thinking and vision helped develop the subject.

The tone of this article is partly historical, describing the development of lattice QCD and the role Kenneth Wilson played in it. It also reviews the achievements made in the four decades since its inception in 1974.

16.2. Quantum chromodynamics on a space-time lattice

16.2.1. *Hadrons, quarks, quantum chromodynamics*

If one looks up the Reviews of Particle Physics web page [9], one finds there an entire list of particles and their properties experimentally discovered to date. In addition to "Gauge and Higgs Bosons", "Leptons" and "Quarks", there are two lists named "Mesons" and "Baryons", each of which contains hundreds of particles. Protons and neutrons, which make up atomic nuclei, are two representative particles belonging to the family of baryons. Pions and K mesons are less familiar, but important particles for binding protons and neutrons into nuclei, and they belong to the family of mesons. The mesons and baryons are collectively called "hadrons". Their chief characteristic is that they participate in the strong interactions in addition to the electromagnetic and weak interactions, while leptons participate only in the latter two interactions.

Many of hadrons were discovered in the accelerator experiments in the 50s and 60s. In 1964 Gell-Mann and Zweig proposed that hadrons are composed of more fundamental particles, which Gell-Mann named quarks. Quarks were predicted to have unusual properties such as fractional charge in units of electron charge. Evidence has gradually built up, however, that quarks are real entities. Yet experimental efforts for detecting them in isolation have been unsuccessful. This situation is often called "quark confinement".

Quantum chromodynamics (QCD) proposes to explain the constitution of hadrons from quarks and their interactions. It is a quantum field theory with local SU(3) gauge invariance in which the quark field $q(x)$, with x the space-time coordinate, transforms under the fundamental **3** representation of SU(3). The SU(3) quantum numbers are called color since the basic premise of the theory is that only color neutral states, trivial under SU(3), carry finite energy and hence exist as physical states. This is a general statement of "quark confinement".

Local gauge invariance is a requirement that the frame of reference of an internal symmetry may be freely rotated at each space-time point without altering the content of the theory. Thus QCD as a gauge theory is to remain invariant under the transformation $q(x) \to V(x)q(x)$ where $V(x) \in \mathrm{SU}(3)$ may vary from point to point in space-time. This invariance requires the existence of a vector gluon field $A_\mu(x)$ with values in the Lie algebra of SU(3) which tells how the local frame at a point x and at a different point y are related.

QCD as field theory thus contains gluon as well as quark fields. It is defined by the Lagrangian density given by

$$\mathcal{L}_{\mathrm{QCD}} = \frac{1}{2g^2}\mathrm{Tr}\left(F_{\mu\nu}(x)^2\right) + \overline{q}(x)\left(i\gamma_\mu D_\mu + m\right)q(x). \tag{16.1}$$

Here, $q(x) = (q_1(x), \ldots, q_{N_f}(x))$ is a spin 1/2 Dirac spinor field for quarks, N_f denotes the number of quark flavors with $m = (m_1, \ldots, m_{N_f})$ the quark mass matrix, and $D_\mu = \partial_\mu - iA_\mu(x)$ is the covariant derivative with $F_{\mu\nu} = \partial_\mu A_\nu - \partial_\nu A_\mu + [A_\mu, A_\nu]$ the gluon field strength, and g is the QCD coupling constant.

The discovery of asymptotic freedom [10, 11] in 1974 showed that the coupling strength of non-Abelian gauge theories decreases toward zero at large momenta. Since deep inelastic electron nucleon scattering experiments carried out in the late 60s indicated just such a behavior called scaling, the discovery boosted QCD to the leading candidate of the theory of strong interactions.

A beautiful prediction of asymptotic freedom is the existence of logarithmic violation of scaling which can be quantitatively calculated via renormalization group methods. The prediction was later confirmed by experiments, thus establishing the validity of QCD beyond doubt at high energies.

Since asymptotic freedom at high energies means that the coupling strength increases in the opposite limit of low energies, it was natural to speculate that QCD also provided a solution to the long standing puzzle that quarks had never been observed in experiments. However, quantum field theory at the time, though quite sophisticated, did not possess means to analyze the behavior of QCD for large coupling constant expected at low energies.

16.2.2. *Formulation of QCD on a space-time lattice*

An essential ingredient for a strong coupling analysis is a mathematically well-defined formulation of QCD in which ultraviolet divergences are controlled. Kenneth Wilson approached this problem with three key ideas: (i) use Euclidean space-time with imaginary time rather than Minkowski space-time with physical time, (ii) replace Euclidean space-time continuum by a discrete 4-dimensional lattice to control ultraviolet divergence, and (iii) maintain local gauge invariance as the guiding principle to construct field variables and action on the lattice.

The use of Euclidean space-time brings out a beautiful and powerful connection between quantum field theory and statistical mechanics. This connection was realized in the 60s by Kurt Symanzik and others from the viewpoint of rigorously defining quantum field theory [12]. The explosive development in the theory of critical phenomena due to Leo Kadanoff, Michael Fisher, Wilson himself and others in the late 60s and early 70s brought the importance of the concept to the foreground [13]. A formal proof that quantum field theory in real time can be recovered from that in imaginary time under a set of axioms was given in the first half of 70s [14,15]. This connection, then, was a hot topic at the time and Wilson consciously exploited it in his work on lattice QCD.

Let us consider a simple cubic 4-dimensional lattice in Euclidean space-time. The lattice points, called sites, are labeled by an integer component 4-vector $n = (n_1, n_2, n_3, n_4) \in \mathbf{Z}^4$, and have a physical coordinate $x = na$ with a the lattice spacing. A pair of neighboring sites at n and $n + \hat{\mu}$ with $\hat{\mu}$ the unit vector in the direction $\mu = 1, 2, 3, 4$ are connected by the link between the two sites which may be denoted as $\ell = (n, \mu), n \in \mathbf{Z}^4, \mu = 1, 2, 3, 4$. An elementary square, or plaquette, on the lattice is then labelled by $P = (n, \mu, \nu)$ with n the lowest corner and the two directions spanning the square denoted by μ and ν. The lattice construction is illustrated in Fig. 16.1.

It is natural to place a quark field $q(x)$ on each site $q(x) = q_n$ with $x = na$. If one replaces the derivative $\partial_\mu q(x)$ by a finite difference $(q_{n+\hat{\mu}} - q_{n-\hat{\mu}})/2a$, the quark Lagrangian on a lattice will contain a bilocal term $\overline{q}_n q_{n\pm\hat{\mu}}$.

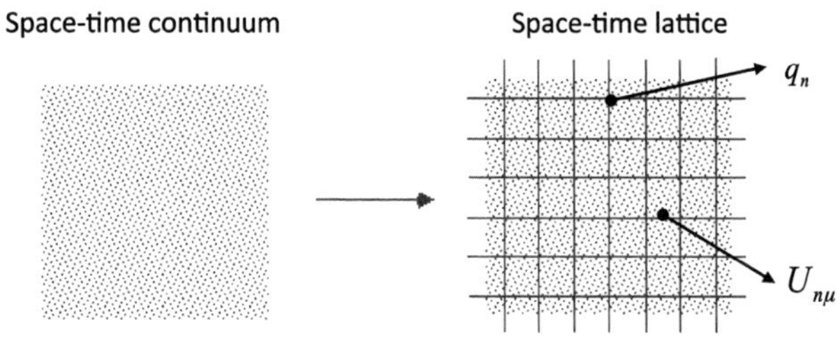

Fig. 16.1. Replacing continuum space-time by a space-time lattice. Quark fields are placed at lattice sites, and gluon fields on links.

In the continuum space-time, bilocal quantities such as $\overline{q}(y)q(x)$ are rendered gauge invariant by inserting a path ordered phase factor $U(y,x) = P\exp(i\int_x^y A_\mu(x)dx_\mu)$, which transforms as $U(y,x) \to V^\dagger(y)U(y,x)V(x)$ under the local gauge transformation. Wilson proposed to employ the phase factor along the lattice link $U_{n\mu} \sim P\exp(i\int_{na}^{(n+\hat{\mu})a} A_\mu(x)dx_\mu)$ as the fundamental gluon field variable on the lattice. The lattice quark action can then be made invariant by inserting appropriate $U_{n\mu}$'s to the bilocal terms. One thus finds for the quark action on the lattice,

$$S_{\text{quark}} = \frac{a^3}{2}\sum_{n\mu}\left(\overline{q}_n\gamma_\mu U_{n\mu}q_{n+\hat{\mu}} - \overline{q}_n\gamma_\mu U^\dagger_{n-\hat{\mu}\,\mu}q_{n-\hat{\mu}}\right) + a^4\sum_n \overline{q}_n m_0 q_n, \quad (16.2)$$

with m_0 the bare quark mass matrix.

A classical action for gluons which reduces to the continuum action in the limit $a \to 0$ can be constructed by taking the product of $U_{n\mu}$'s around the boundary of a plaquette P [1]:

$$S_{\text{gluon}} = \frac{1}{g_0^2}\sum_P \text{Tr}\left[\prod_{n\mu\in\partial P} U_{n\mu}\right], \quad (16.3)$$

where Tr stands for trace over SU(3) indices. Here g_0 is the bare gauge coupling constant at the energy scale of the lattice cutoff $1/a$.

Lattice QCD as quantum theory is defined by the Feynman path integral. If $O(U,q,\overline{q})$ is an operator corresponding to some physical quantity, the vacuum expectation value over the quantum average is given by

$$\langle O(U,q,\overline{q})\rangle = \frac{1}{Z_{\text{QCD}}}\int\prod_{n\mu}dU_{n\mu}\int\prod_n d\overline{q}_n dq_n O(U,q,\overline{q})\exp(S_{\text{QCD}}) \quad (16.4)$$

with

$$Z_{\text{QCD}} = \int\prod_{n\mu}dU_{n\mu}\int\prod_n d\overline{q}_n dq_n \exp(S_{\text{QCD}}), \quad (16.5)$$

where

$$S_{\text{QCD}} = S_{\text{gluon}} + S_{\text{quark}}. \quad (16.6)$$

The gluon link variable $U_{n\mu}$ takes values in the group SU(3) rather than the Lie algebra for the vector gluon field $A_\mu(x)$. The integration over $U_{n\mu}$'s should be defined as invariant integration over the group SU(3). Since SU(3) has a finite volume under this integration, the lattice Feynman path integral is well-defined without gauge fixing.

The integration over the quark fields also requires some care. Since quarks are fermions, the path integral has to be defined in terms of Grassmann numbers which anticommute under exchange, i.e., $q_n q_m = -q_m q_n$. These points are clearly spelled out in the Wilson's original paper in 1974 [1].

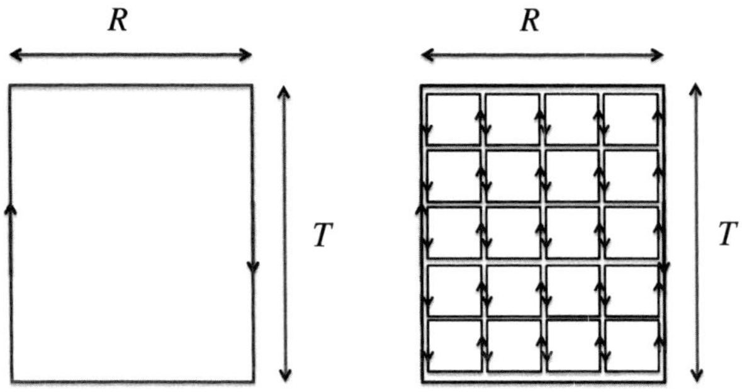

Fig. 16.2. *Left panel*: Wilson loop for a temporal size T and a spatial size R measures the potential energy for a pair of static quark and antiquark. *Right panel*: The leading order of the strong coupling expansion is obtained by tiling the Wilson loop by plaquettes.

16.2.3. *Confinement of color*

Whether quarks are confined in QCD can be examined if one knows how to calculate the energy of an isolated quark in interaction with the gluon fields: an isolated quark would exist only if the energy of that state is finite. An elegant method invented by Wilson is to consider a pair of static quark and antiquark, which is created at some point in space-time, then separated to some distance, stays in that configuration for some time, and finally is brought together to a point and annihilated. Geometrically, the space-time trajectory of the pair forms an oriented closed loop C. Since the color charge of the quark interacts with gluon fields, the creation and annihilation of the pair inserts a phase factor

$$W\left[C\right] = P\exp\left(i\oint_C dx_\mu A_\mu(x)\right) \tag{16.7}$$

in the path integral where P indicates ordering along the path. This is the Wilson loop operator.

If the loop has the shape of a rectangle of a width R in the spatial direction and an extent T in the temporal direction as shown in the left panel of Fig. 16.2, the quantum average of the Wilson loop operator measures the energy $E(R)$ of the quak-antiquark pair relative to the vacuum over a temporal length T so that

$$\langle W\left[R\times T\right]\rangle \propto \exp\left(-E(R)T\right), \quad T\to\infty. \tag{16.8}$$

Wilson speculated that for large values of the gauge coupling, fluctuations of gluon fields would be large, leading to a significant cancellation among the contributions to the Wilson loop. This would mean that the Wilson loop average rapidly decreases for larger loops, or equivalently the probability decreases that the quark and antiquark are found in a well-separated configuration. This would mean confinement. Another equivalent statement would be that the energy $E(R)$ grows for larger separations R, so that separating quark from antiquark is not possible.

An amazingly simple calculation suffices to verify this picture if one employs lattice QCD. Let us assume, following Wilson, that quark fields do not play an essential role. When the bare coupling g_0 becomes large, the gluon path integral in (16.4) can be calculated by an expansion in inverse powers of g_0^2. The leading term is obtained when one tiles the $R \times T$ surface of the Wilson loop by a set of plaquettes from the expansion of the gluon part of the weight $\exp(S_{\text{gluon}})$, as shown in the right panel of Fig 16.2. We then find that

$$\langle W[R \times T] \rangle = N_c \left(\frac{1}{N_c g_0^2} \right)^{RT/a^2} \left(1 + O(g_0^{-8})\right), \quad g_0^2 \to \infty, \tag{16.9}$$

with $N_c = 3$ for SU(3) so that

$$E(R) \to \sigma R, \quad R \to \infty \quad \text{with} \quad \sigma = \frac{1}{a^2} \left(\log\left(N_c g_0^2\right) + O(g_0^{-8}) \right), \tag{16.10}$$

namely, a static pair of quark and antiquark are bound by a potential linearly rising with the separation. Hence they cannot be separated to infinite distance with any finite amount of energy.

A closed loop C has two geometrical characteristics, the length of the loop $L[C]$ and the area of the minimal surface $A[C]$ spanned by the loop. The confinement behavior corresponds to an area decay for large loop;

$$\langle W[C] \rangle \propto \exp\left(-\sigma A[C]\right). \tag{16.11}$$

If, on the other hand, the Wilson loop expectation value decays with the loop length,

$$\langle W[C] \rangle \propto \exp\left(-\mu L[C]\right), \tag{16.12}$$

the energy of a quark-antiquark pair saturates to a constant $E(R) \to 2\mu$ for large separation $R \to \infty$. Hence there is no longer confinement. The confining and non-confining phases of gauge theories are thus distinguished by the behavior of the Wilson loop expectation value.

The possibility that there can be both confining and non-confining phases raises an interesting question whether a confining phase can turn into a non-confining phase if some parameters of the theory are varied. Temperature is such an important parameter. As we discuss in more detail in Sect. 16.4.4, the confining property becomes lost through a phase transition when the temperature is raised sufficiently.

16.2.4. *Continuum limit*

Let us go back to the strong coupling result in (16.10). If one uses the phenomenological value $\sqrt{\sigma} \approx 440\,\text{MeV}$, and if one assumes that a value $\log\left(N_c g_0^2\right) \sim O(1)$ is sufficient for the strong coupling expansion to converge, the corresponding value of the lattice spacing equals $a \sim 0.5$ fm with 1 fm $= 10^{-15}$ m. This value is comparable to a typical length scale of the strong interactions, e.g., the charge radius of proton $R_p \sim 0.9$ fm. We certainly know that space-time is continuous well below such length scales. Thus one needs to know if confinement holds for smaller lattice spacings.

In Fig. 16.3 we show the static potential $E(R)$ calculated by Monte Carlo simulation of the pure gluon theory [16]. We observe a linearly rising potential $E(R) \sim \sigma R$

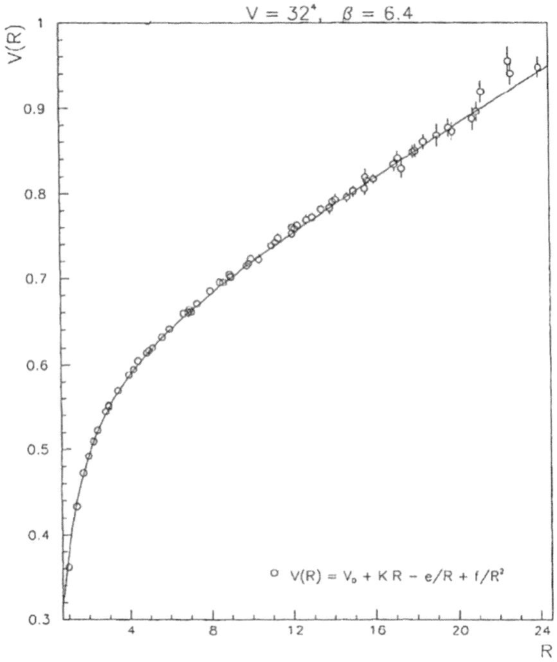

Fig. 16.3. Monte Carlo result for the static potential energy $E(R)$ calculated in pure gluon theory showing a linearly rising term at long distances and a Coulomb term at short distances. From Ref. [16].

at large distances. There is also a Coulomb behavior $V(R) \sim \alpha/R$ at short distances, which is consistent with a perturbative one gluon exchange expected from asymptotic freedom. The lattice spacing estimated from the phenomenological value $\sqrt{\sigma} \approx 440$ MeV is $a \approx 0.0544(4)$ fm. This is one of many evidences that the property of confinement holds not just at strong coupling but also toward weak couplings with small lattice spacings.

The question then is whether the confinement property really persists as the lattice spacing is taken toward the limit of continuous space-time $a \to 0$. For Wilson, who elucidated critical phenomena with his renormalization group ideas, this was not a conceptually difficult issue. The strong coupling calculation demonstrates that the system of gluons without quarks is in a confining phase. When one decreases the coupling constant g_0, this property persists as long as one does not encounter a phase transition to a non-confining phase characterized by perimeter decay (16.12). Suppose that there is such a second-order phase transition at $g_0^2 = g_c^2$. Close to the critical point, the correlation length measured in lattice units diverges $\xi(g_0)/a \to \infty$ as $g_0 \to g_c$. This means that one should be able to hold the physical correlation length $\xi(g_0)$ fixed to a physical value observed in experiment while sending the lattice spacing to zero $a \to 0$, recovering physics in the space-time continuum.

One can define lattice gauge theory for a variety of groups and space-time dimensions. As with the case for spin systems, the existence and position of second-order phase transition points would depend on them. For the group SU(3) in 4 dimensions, a particularly attractive possibility is $g_c = 0$. In fact, this is the only possibility if confinement at low energies and asymptotic freedom at high energies are to coexist in the same phase.

The renormalization group allows a more concrete discussion. The change of gauge coupling $g_0 \to g_0 + dg_0$ under a change of cutoff scale $a \to a + da$ while fixing a physical scale constant defines a renormalization group β function

$$a\frac{dg_0(a)}{da} = -\beta(g_0), \tag{16.13}$$

where the minus sign is inserted to be compatible with the conventional definition using momentum slicing. The strong coupling result (16.10) shows that $\beta(g_0)$ is large and negative for large g_0. For small values of g_0, one can employ perturbation theory to confirm the asymptotic freedom result,

$$\beta(g_0) = -b_0 g_0^3 - b_1 g_0^5 + O(g_0^7) \tag{16.14}$$

with

$$b_0 = \frac{1}{(4\pi)^2}\left(\frac{11}{3}N_c - \frac{2}{3}N_f\right), \tag{16.15}$$

$$b_1 = \frac{1}{(4\pi)^4}\left(\frac{34}{3}N_c^2 - \frac{N_c^2-1}{N_c}N_f - \frac{10}{3}N_c N_f\right), \tag{16.16}$$

for $SU(N_c)$ gauge group and N_f flavors of fermions in the fundamental representation. The first two coefficients are negative for our world with $N_c = 3$ and $N_f = 6$. A natural supposition is that the beta function is negative for all values of the coupling, with the only zero residing at $g_c = 0$.

Wilson started a numerical Monte Carlo calculation to check this by a block spin renormalization group for gauge group SU(2) [5]. This attempt was followed by several serious calculations with SU(3) gauge group in the 80s. The results, though mostly restricted to the pure gluon theory and had fairly large errors, showed that the beta function stayed negative and became consistent with the two-loop result (16.14) toward weak coupling (see, e.g., Refs. [17, 18]).

A different, and perhaps a more elegant, approach was developed in the 90s by the Alpha Collaboration [19]. Their method of step scaling function defines the renormalized coupling constant at a scale $1/L$ through the Schrödinger functional on a lattice of a finite size $L^3 \times T$ with a prescribed boundary condition in time, and follows the evolution of the coupling under a change of scale by a factor 2. The continuum limit is systematically taken in this process. Therefore, the end result is the evolution of the renormalized coupling from low to high energies in the continuum theory.

In Fig. 16.4 we show the result for the pure gluon theory (solid circles) and a comparison with perturbation theory (dashed and dotted lines). The full evolution

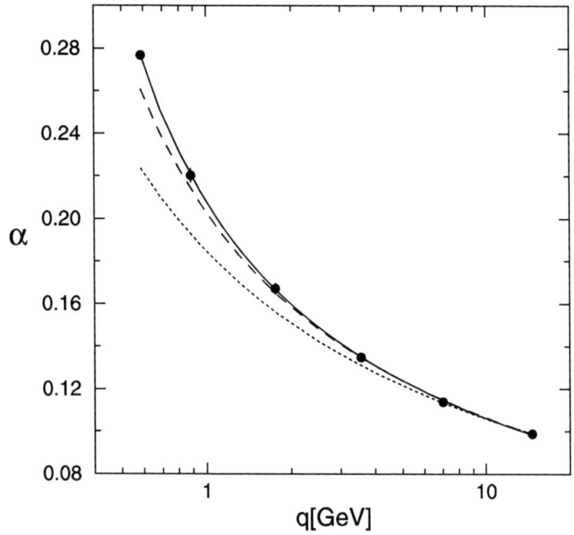

Fig. 16.4. Running of renormalized coupling constant $\alpha(q^2) = g^2(q^2)/4\pi$ in the Schrödinger functional scheme as a function of momenta scale q. *Dashed* and *dotted lines* are two- and one-loop evolution starting from the right-most point. The physical scale is fixed by the Sommer scale $r_0 = 0.5$ fm. From Ref. [19].

runs from 10 GeV down to a few hundred MeV, where the system is in a strong coupling non-perturbative regime. Thus the confining behavior at large distance is continuously connected with the asymptotically free behavior at short distances. The full evolution runs almost parallel to the two-loop evolution. Similar results have been obtained in full QCD with $N_f = 2, 3, 4$ dynamical quarks [20–22].

Based on the results described above, one can state, though a mathematically rigorous proof is yet lacking, that QCD is in the confining phase over the entire range of coupling from zero to infinity.

16.2.5. *Quarks and chiral symmetry*

16.2.5.1. *Chiral symmetry*

An important feature of the strong interaction which was recognized in the late 50s and early 60s is chiral symmetry. Studies in this period led to the concept of spontaneous breakdown of symmetry, and a realization that it is accompanied by the emergence of massless bosons, called Nambu–Goldstone bosons today.

In QCD language, chiral symmetry is invariance under the global transformation $q(x) \rightarrow Wq(x)$, where $W = \exp(i\alpha\gamma_5) \in \mathrm{SU(N_f)}$ acts on the Dirac-flavor indices and rotates left and right handed chiral components of $q(x)$ in the opposite direction. This symmetry is explicitly broken by the quark mass term. Hence it holds approximately for three light quarks, up, down and strange. It is not very relevant for the heavier quarks, charm, bottom and top. The octet of pseudo scalar mesons, π, K, η, are identified as the Nambu–Goldstone bosons corresponding to spontaneous breakdown of chiral symmetry. With a strongly interacting dynamics at large distances, QCD should dynamically explain spontaneously broken chiral symmetry and its physical consequences.

16.2.5.2. *Chiral symmetry on a space-time lattice*

Introduction of a space-time lattice affects bosons and fermions in different ways. This is most easily seen by looking at the kinetic term in the free field case. For a boson field ϕ_n, the second-order derivative $-\partial_\mu^2\phi(x)$ is discretized as $-(\phi_{n+\hat{\mu}} + \phi_{n-\hat{\mu}} - 2\phi_n)/2a^2$, whereas for a fermion field $q(x)$, the first order derivative $\gamma_\mu\partial_\mu q(x)$ is discretized as $\gamma_\mu(q_{n+\hat{\mu}} - q_{n-\hat{\mu}})/2a$. The momentum space expression then becomes $2(1-\cos(p_\mu a))/a^2$ and $i\gamma_\mu \sin(p_\mu a)$, respectively. There is a crucial difference in the location of the zeros in the Brillouin zone: the boson case has a zero only at $p_\mu = 0$ for all μ whereas the fermion case has a set of zeros for $p_\mu = 0$ or π/a for each μ. Since the zeros gives rise to poles in the propagator, a naive fermion discretization leads to multiple copies of the state at $p_\mu = 0$. This is the fermion species doubling problem.

Building upon a pioneering work by Karsten and Smit [23], Nielesen and Ninomiya [24] proved that essentially the same conclusion holds under a set of rather general conditions. The theorem states that if the fermion action satisfies (i) chiral symmetry $q_n \to \exp(i\alpha\gamma_5)q_n$, (ii) invariance under unit translation on the lattice, (iii) Hermiticity, and (iv) locality, then the spectrum of fermions contains even number of particles, half of them left handed and the other half right handed.

An elegant topological proof runs as follows [25]. If we write a Dirac fermion action on a lattice in a general form, $S_F = \sum_{n,m} \bar{q}_n D_{n,m} q_m$, the assumptions (i) and (ii) imply that $D_{n,m} = \sum_\mu \gamma_\mu F_\mu(n-m)$ where $F_\mu(n)$ is a vector function which, by (iii) satisfies $F_\mu^\dagger(n) = F_\mu(-n)$, and by (iv) rapidly decreases as $|n|$ becomes large. The Fourier transform $\tilde{F}_\mu(p)$ is, therefore, a well-defined real vector field over the Brillouin zone in momentum space. Let $p = p^{(i)}, i = 1, 2, \ldots$ be the zeros of the vector field $\tilde{F}_\mu(p) = f_{\mu\nu}(p_\nu - p_\nu^{(i)}) + O((p - p^{(i)})^2)$. They correspond to the poles of the propagator $D^{-1}(p)$, and hence to the particle states. The relative chirality of these states are determined by the index, $\mathrm{sign}(\det f_{\mu\nu})$. Since the sum of index of a vector field over the 4-dimension torus vanishes by the Poincaré-Hopf theorem, there has to be an equal number of fermion states with opposite chirality.

The Nielsen–Ninomiya theorem indicates that one either has to abandon chiral symmetry or one has to allow for the presence of species doubling. Wilson's choice, which he actually wrote down [26] prior to the publication of the theorem, was to add a term which softly breaks the chiral symmetry of the naive lattice action in (16.2):

$$\delta S_{\mathrm{W}} = \frac{a^3}{2} \sum_{n\mu} \left(2\bar{q}_n q_n - \bar{q}_n U_{n\mu} q_{n+\hat{\mu}} - \bar{q}_n U^\dagger_{n-\hat{\mu},\mu} q_{n-\hat{\mu}} \right). \tag{16.17}$$

For the free field case, this term adds $\sum_\mu (1 - \cos p_\mu a)/a$ to the kinetic term $i\sum_\mu \gamma_\mu \sin p_\mu a$, and hence removes the zeros at $p_\mu = \pi/a$.

Let us add that toward the continuum limit $a \to 0$ the Wilson's added term becomes of form $a\overline{q}(x)D_\mu^2 q(x)$ relative to the original term $\overline{q}(x)i\gamma_\mu D_\mu q(x)$ in (16.2), i.e., higher order in a. Apart from removing the doublers, the effect of the added term disappears in the continuum limit.

Chiral breaking effects of the Wilson's added term can be analyzed by the method of Ward identities [27]. In particular, a definition of quark mass can be given that satisfies the PCAC relation [27, 28]. A detailed analysis of the phase structure on the (g_0, m_0) plane was made [29], and the existence of a massless pion, in spite of chiral symmetry breaking, was explained as due to spontaneous breakdown of parity Z(2) symmetry [30]. With these analytical developments, Wilson's formulation provides a quantitative framework for the computation of physical observables, and is used extensively in Monte Carlo studies.

A variant of Wilson's formulation is to consider two flavors of quarks as a pair and add a twisted mass term to the naive action (16.2) [31]:

$$\delta S_{\rm tm} = a^4 \sum_n \overline{q}_n i\mu_q \gamma_5 \tau_3 q_n, \tag{16.18}$$

where τ_3 acts on the flavor index. This formulation has an attractive feature that one can twist the angle $\alpha = \arctan \mu_q/m_0$ in such a way that $O(a)$ lattice artifacts are absent in physical observables [32, 33]. Large-scale simulations are being made using such "maximally" twisted QCD.

There is a different method, called the staggered formulation [34], which retains a U(1) chiral symmetry at the cost of a four fold species doubling. In the 4-dimensional Euclidean formulation [35], it starts with a single component fermion field χ_n at each site, and reconstructs four species of Dirac fields $\psi = (\psi_1, \ldots, \psi_4)$ from 16 χ's on the 16 vertices of a 4-dimensional unit hypercube [36]. The original action is invariant under an even-odd U(1) symmetry $\chi_n \to \exp(i(-1)^{|n|}\alpha)\chi_n$, which translates into an axial U(1) transformation on the four Dirac fields of form $\psi \to \exp(i\alpha\gamma_5 \otimes \xi_5)\psi$ where ξ_5 acts on the species index α of the Dirac field ψ_α, $\alpha = 1, \ldots, 4$.

It is generally believed that lattice QCD with the staggered fermion formalism converges to continuum QCD with $N_f = 4$ degenerate flavors of quarks [37]. An elaborate effective theory description has been developed to control the breaking of the full 4 species chiral symmetry down to the U(1) subgroup at finite lattice spacing [38]. On these theoretical bases, the staggered formalism is also extensively used in Monte Carlo simulations.

16.2.5.3. *Lattice fermion action with chiral symmetry*

In 1981, a year after Nielsen and Ninomiya presented their theorem, Wilson revisited the issue of chiral symmetry from a different perspective with Ginsparg [39]. He asked what would be the relation satisfied by a lattice Dirac operator $D_{n,m}$ if it were derived by a (chiral symmetry breaking) block spin transformation from a chiral invariant theory. The answer turned out to be remarkably simple; it is given by

$$\gamma_5 D_{n,m} + D_{n,m}\gamma_5 = a \sum_k D_{n,k}\gamma_5 D_{k,m}. \tag{16.19}$$

Since D does not anticommute with γ_5, assumption (i) of the Nielsen–Ninomiya theorem is not satisfied, and so the conclusion of the theorem does not hold. In fact, there is no species doubling. Furthermore, the axial vector current for this action has the correct U(1) anomaly.

One can rewrite (16.19) in terms of the propagator $G = D^{-1}$ as

$$\gamma_5 G_{n,m} + G_{n,m}\gamma_5 = a\delta_{nm}\gamma_5. \tag{16.20}$$

Hence the breaking of chiral symmetry is a local effect. One may expect then that a modified form of symmetry may exist. This was found almost 20 years later [40]. Fermion actions that satisfy the Ginzparg–Wilson relation (16.19) are invariant under an infinitesimal transformation given by

$$\delta\overline{\psi} = \overline{\psi}\left(1 - \frac{1}{2}aD\right)\gamma_5, \quad \delta\psi = \gamma_5\left(1 - \frac{1}{2}aD\right)\psi. \tag{16.21}$$

Several forms of fermion action which satisfy the Ginsparg–Wilson relation were discovered in the 90s. One form is a domain-wall formalism [41, 42] in which the 4-dimensional fermion field is constructed as the zero mode of a 5-dimensional theory generated by a mass defect at the boundary in a fictitious fifth dimension. Another form is given by the overlap formalism [43, 44]. In this case an explicit form of the operator D is given by

$$D = \frac{1}{a}\left(1 + \gamma_5 H (H^\dagger H)^{-1/2}\right), \quad H = \gamma_5(aD_W - 1) \tag{16.22}$$

with $aD_W + m_0$ the operator for the Wilson fermion action with a negative mass $m_0 = -1$. The two forms are equivalent in the limit of infinite fifth dimension [45,46].

Yet another form is the perfect action [47], so named because it is defined as the fixed point of a block spin transformation of renormalization group for QCD. This form follows from the line of reasoning of Ginsparg and Wilson, but it was pursued and arrived at independently almost two decades later.

All forms of action, particularly the domain wall and overlap actions, have come to be used extensively in the last decade. The domain wall formalism has been exploited by RBC-UKQCD Collaborations (see, e.g., [48]), and the overlap formalism by JLQCD (see [49] for a recent review). The situation with the perfect action is reported in [50].

16.2.5.4. *Spontaneous breakdown of chiral symmetry*

Spontaneous breakdown of chiral symmetry is best studied by examining the behavior of the order parameter of chiral symmetry. This is given by the quark bilinear

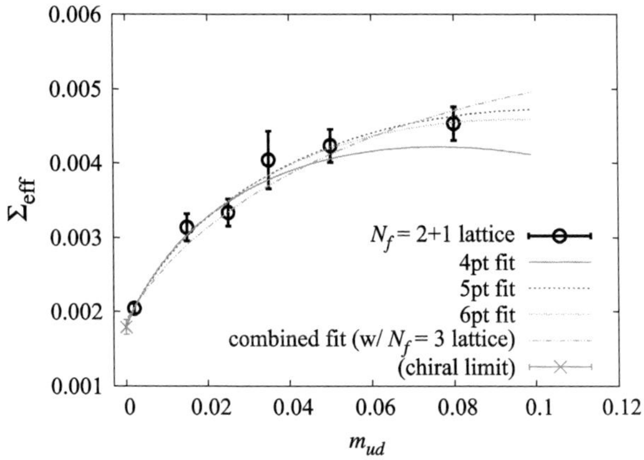

Fig. 16.5. (Color online) Chiral order parameter $\Sigma = \langle \overline{q}_n q_n \rangle$ as a function of degenerate up-down quark mass in lattice units in $N_f = 2 + 1$ flavor lattice QCD with overlap fermion formalism. The strange quark mass is held fixed at a value close to experiment. From Ref. [49].

operator $\Sigma = \langle \overline{q}_n q_n \rangle$. If $\Sigma \neq 0$ after sending the spatial volume to infinity $V \to \infty$ followed by the limit of quark mass to zero $m_q \to 0$, then chiral symmetry is spontaneously broken.

In Fig. 16.5 we show the result for Σ as a function of the degenerate up-down quark mass m_{ud} in lattice units obtained with the overlap formulation for $N_f = 2+1$ lattice QCD. We choose these data since the overlap formalism is the cleanest regarding the chiral aspect among lattice fermion formulations. The terminology $N_f = 2 + 1$ refers to the fact that the up and down quark masses are taken degenerate, while the strange quark mass has a separate value. Strictly speaking, the values shown in Fig. 16.5 are obtained from the eigenvalue distribution of the Dirac operator and not by calculating the condensate directly.

The concave curvature as a function of m_{ud} is consistent with the presence of a logarithm term predicted by chiral perturbation theory. Extrapolating to $m_{ud} = 0$ including the effect of the logarithm yields a non-zero value, supporting spontaneous breakdown of chiral symmetry. The dependence of the pion mass is consistent with $m_\pi^2 \propto m_{ud}$ (up to logarithmic corrections), as follows from the Nambu–Goldstone theorem.

Similar results have been obtained for the staggered fermion formalism which has U(1) chiral symmetry. The analysis is more complicated for the Wilson fermion formulation, since one needs to carry out a subtractive renormalization for both chiral condensate and quark mass due to the soft chiral symmetry breaking induced by the Wilson term (16.17). Once these subtractions are done, one finds the signatures expected for spontaneous breakdown of chiral symmetry, such as the relation $m_\pi^2 \propto m_{ud}$ for an appropriately defined quark mass [27, 28].

16.2.6. *Heavy quarks on a lattice*

The three light quark quantum numbers, i.e., up, down and strange, have been known from 1930s and 1950s. In contrast, heavier quarks were discovered more recently, i.e., charm quark in 1974, bottom quark in 1977, and top quark in 1995. These heavy quarks occupied the central place in the experimental and theoretical studies toward establishing the Standard Model, particularly with the construction of B factories in the 90s and experiments with them in the 2000s.

Studying heavy quarks with lattice QCD poses the problem that for a large value of a heavy quark mass m_h, the dimensionless combination $m_h a$ is also large, leading to an amplification of lattice discretization errors, especially if $m_h a \gg 1$. This situation applies to the bottom quark with a mass $m_b \approx 5$ GeV, since lattice QCD simulations to date have been made with inverse lattice spacings in the range $a^{-1} \approx 2 - 4$ GeV.

Several methods have been formulated and employed to deal with this problem. The static approximation [51] is an expansion in $1/m_h$, NRQCD [52] is a reformulation of QCD for non-relativistic quarks with an expansion in powers of the quark velocity v, and a relativistic formalism [53–55] modifies the Wilson quark action so as to systematically reduce the effects of large $m_h a$.

All three methods have been extensively used to calculate physical quantities involving charm or bottom quark. In particular, matrix elements such as the pseudoscalar decay constants and form factors calculated with these methods have been playing an important role in constraining the Cabibbo–Kobayashi–Maskawa matrix elements including the CP violation phase.

It should be mentioned that with increasingly smaller lattice spacings becoming accessible with progress of algorithms and computer power, direct simulations are replacing calculations with effective heavy quark theories. This has already occurred for charm quark, and it may not be too far into the future that bottom quark becomes treated in a similar way.

16.3. Lattice QCD as computation

16.3.1. *Numerical simulation and lattice QCD*

Lattice QCD offered a framework for conceptually understanding the dynamics of non-Abelian gauge fields. In particular, it elucidated the mechanism of confinement in a way which would have been impossible in the perturbative framework of field theory. Nonetheless, calculation methods to obtain physical results did not really exist; for example, higher order strong coupling expansions were very cumbersome and hard to extrapolate to the continuum limit expected at $g_0 \to 0$.

Monte Carlo simulations offered a new approach to solve this impasse. Of course Monte Carlo methods had been known since the pioneering era of electronic computers in the late 40s and early 50s. The Metropolis algorithm to handle multi-dimensional integrations for statistical mechanical systems was formulated in 1953 [56]. Applications to spin systems in statistical mechanics started to appear in the 60s and was pursued increasingly in the 70s.

It is in this context that Creutz, Jacobs and Rebbi carried out a Monte Carlo study of Ising gauge theory with Z(2) gauge group in 4 dimensions in 1979, finding a first-order phase transition [4]. Creutz extended the application to SU(2) gauge group [6], and extracted the string tension σ in front of the area decay of the Wilson loop. The dependence of σ on the gauge coupling g_0 turned out to be consistent with the scaling law predicted by the renormalization group. This suggested the possibility that a continuum limit could be successfully taken, giving rise to a hope that the confinement problem could be solved in a numerical way.

Wilson's interest in numerical analyses and computer applications started early in his career [3]. He often used numerical methods to carry out his analyses.[2] In his 1979 lecture at the Cargése Summer School [5], Wilson reported a block-spin renormalization group analysis for SU(2) gauge group using Monte Carlo methods to evaluate the necessary path integral averages.

These attempts introduced a method hitherto unknown in field theory. The method looked very promising, and immediately attracted attention of particle theorists. In particular, it was applied to calculate masses of hadrons directly from quarks and gluons. The method was based on the observation that if $O_{\rm H}(t)$ is an operator at a time slice t corresponding to a hadron H, e.g., $\pi^+ \sim \overline{u}\gamma_5 d$ for pion or $p \sim ({}^t u C\gamma_5 d)u$ for proton, then the 2-point Green's function for this operator behaves for large times as

$$\left\langle O_{\rm H}(t)O_{\rm H}^\dagger(0)\right\rangle \to Z\exp\left(-m_{\rm H}at\right), \quad t\to\infty, \tag{16.23}$$

where $m_{\rm H}$ denotes the mass of hadron H. Therefore, calculating the two-point function by a Monte Carlo simulation and extracting the slope of the exponential decay would yield the mass of hadron H.

This program was carried out by Weingarten [7], and by Hamber and Parisi [8] in 1981. The lattice size employed was 12^4 for the Icosahedral subgroup of SU(2) for the former, and $6^3 \times 12$ for SU(3) for the latter work. Albeit very modest in today's standards, their studies clearly demonstrated the feasibility of their approach, which accelerated an explosive development of Monte Carlo simulations in lattice QCD.

[2]Probably his most famous numerical work is a renormalization group solution of the Kondo problem [57]. The numerical rigor he maintained for this work has become legendary. For the universal ratio of the two temperatures T_K and T_0 characterizing the high and low temperature scales, he obtained $W/(4\pi) = (4\pi)^{-1}T_K/T_0 = 0.1032(5)$. Six years later, an exact solution by the Bethe ansatz yielded $W/(4\pi) = 0.102676\ldots$ [58], verifying the Wilson's number to the fourth digit within the estimated error!

16.3.2. *Massive parallelism and lattice QCD*

Monte Carlo simulation of lattice QCD is based on the method of importance sampling. Let ϕ_n be a field variable at a site n on a lattice Λ with V lattice points, and $O(\phi)$ an operator. In field theory, one wishes to calculate the average value of $O(\phi)$ weighted with the action $S[\phi]$ according to,

$$\langle O(\phi)\rangle = \frac{1}{Z}\int \prod_n d\phi_n O(\phi)\exp\left(S[\phi]\right), \tag{16.24}$$

where

$$Z = \int \prod_n d\phi_n \exp\left(S[\phi]\right). \tag{16.25}$$

Let us define a configuration C as a point $C = (\phi_1, \ldots, \phi_V)$ in the integration space of all fields on the lattice Λ. Monte Carlo evaluation of the integral proceeds by setting up a stochastic chain of configurations $C_1 \to C_2 \to \cdots \to C_N$ such that the distribution of configurations converges to the weight of the integral:

$$\rho_N(C) = \frac{1}{N}\sum_{i=1}^{N}\delta(C - C_i) \to \frac{\exp\left(S[C]\right)}{Z}, \quad N \to \infty. \tag{16.26}$$

In the well-known Metropolis algorithm, the convergence is guaranteed by accepting or rejecting a new trial configuration C_{trial} according to the probability $P_{\text{acc}} = \text{Min}\left(1, \exp(S[C_{\text{trial}}] - S[C_{\text{old}}])\right)$ where C_{old} is the latest configuration of the chain.

A very important property in Monte Carlo calculations of field theoretical systems is locality. A field variable ϕ_n at a site n interacts only with those variables ϕ_m's in a limited neighborhood of the site n. In creating a trial configuration C_{trial} from an old configuration C_{old}, and in calculating the difference of the action $S[C_{\text{trial}}] - S[C_{\text{old}}]$, the numerical operations at a site n can be carried out independently of those at a site m unless the pair is within the limited neighborhood of each other.

As shown in Fig. 16.6, let us divide the entire lattice Λ into a set of sub-lattices $\Lambda_\alpha, \alpha = 1, 2, \ldots, N_P$, and assign each sub-lattice Λ_α to a processor P_α. Because of

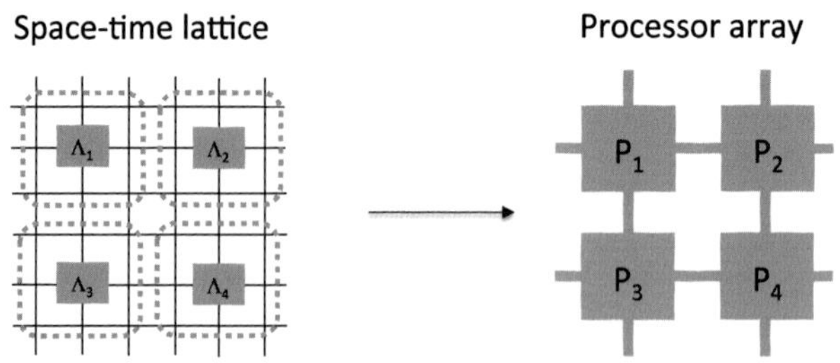

Fig. 16.6. (Color online) Mapping of space-time lattice $\Lambda = \cup_{\alpha=1}^{N_P}\Lambda_\alpha$ to an array of processors $P_\alpha, \alpha = 1, \ldots, N_P$ interconnected according to the space-time lattice topology.

Table 16.1 Representative microprocessors developed in the 70s.

Year	Name	Bit	Price/chip
1971	Intel 4004	4 bit	$60
1972	Intel 8008	8 bit	$120
1974	Intel 8080	8 bit	$360
1974	Motorola 6800	8 bit	$360
1978	Intel 8086	16 bit	$320
1979	Motorola 68000	32 bit	

Price/chip is from Wikipedia.

locality, calculations by the processor P_α can be carried out independently of those by the other processors, except that the processors with overlapping boundaries have to exchange values of ϕ_k's in the boundaries before and/or after the calculations in each sub lattice. This means that, for a fixed lattice size, the computation time can be reduced by a factor N_P, and for a fixed sub-lattice size, one can enlarge the total lattice size proportionately to the number of processors N_P without increasing the computation time.

This is an ideal case of the parallel computation paradigm. The locality property, which is one of the fundamental premises of field theoretic description of our universe, allows a mapping of calculations on a space-time lattice to a parallel array of processors interconnected with each other according to the connection of space-time sub-lattices.

16.3.3. *Parallel computers for lattice QCD*

Immediately after Monte Carlo calculations started in lattice QCD, several groups started to plan the building of a parallel computer for lattice QCD calculations. A crucial factor which helped push such an activity was a rapid development of microprocessors in the 70s. As shown in Table 16.1, starting with a 4 bit Intel 4004 in 1971, increasingly more powerful microprocessors were developed and were put out on the market at prices affordable by academic scientific projects.

In Table 16.2 we list the parallel computers developed by physicists for lattice QCD in the 80s. Typically, these machines employed a commercial microprocessor such as those listed in Table 16.1 as the control processor, and combined it with a floating point unit to enhance numerical computation capabilities.

The famous CRAY-1 vector supercomputer came on the market in 1976. Vector supercomputers developed rapidly, and dominated the market in the 70s and 80s. However, the progress of parallel computers was even faster. By the end of the 80s, parallel computers caught up and even overtook vector computers in speed. We can observe this trend by tracking blue and green symbols (vector computers) and red and violet symbols (parallel computers) from the 80s to early 90s in Fig. 16.7.

In Table 16.3 we list the parallel computers developed in the 90s and later for lattice QCD. Big success continued with CP-PACS in Japan, which occupied the

Table 16.2 Parallel computers dedicated to lattice QCD developed in the 80s.

Name	Year	Authors	CPU	FPU	Peak speed
PAX-32[a]	1980	Hoshino–Kawai	M6800	AM9511	0.5 MFlops
Columbia	1984	Christ–Terrano	PDP11	TRW	—
Columbia-16	1985	Christ *et al.*	Intel 80286	TRW	0.25 GFlops
APE1	1988	Cabibbo–Parisi	3081/E	Weitek	1 GFlops
Columbia-64	1987	Christ *et al.*	Intel 80286	Weitek	1 GFlops
Columbia-256	1989	Christ *et al.*	M68020	Weitek	16 GFlops
ACPMAPS	1991	Mackenzie *et al.*	micro VAX	Weitek	5 Gflops
QCDPAX	1991	Iwasaki–Hoshino	M68020	LSI-logic	14 GFlops
GF11	1992	Weingarten	PC/AT	Weitek	11 GFlops

[a]Not for lattice QCD.
"Year" marks the completion date.

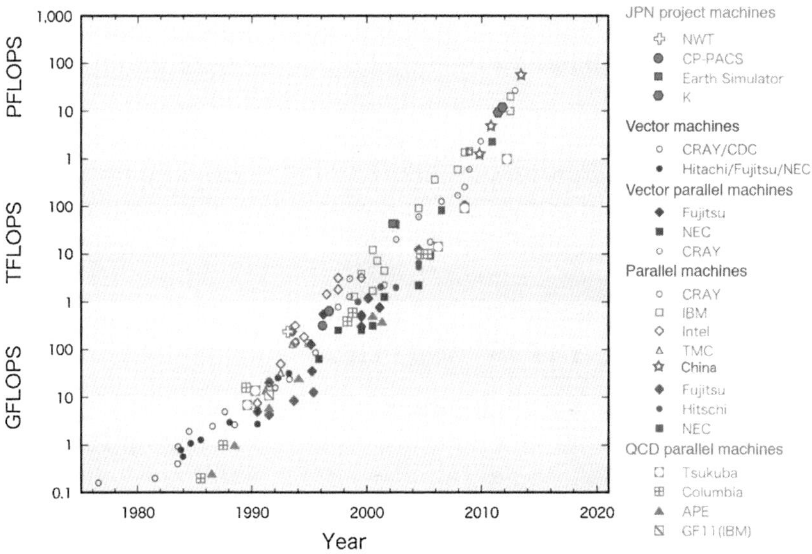

Fig. 16.7. (Color online) Peak speed of supercomputers. The *circle* at the *bottom left* is CRAY-1. *Red fancy symbols* show parallel machines developed for QCD.

top position in the Top 500 list of supercomputers in November 1996, and QCDSP in USA. We observe an increasing involvement of major vendors such as Hitachi and IBM. This was necessary to secure advanced technology and computer building knowhow to develop those fast computers. Probably the most well-known of this trend is the QCDOC project, which gave rise to the IBM BlueGene/L, and the QCDCQ project which ran parallel with the IBM BlueGene/Q development. In this way lattice QCD has seriously influenced the development of parallel supercomputers for scientific computing. Another trend one can observe in Table 16.3 is the use of commodity processors such as Intel Xeon for a quick machine building. This was the option adopted by PACS-CS and QPACE.

Table 16.3 Parallel computers developed for lattice QCD in the 90s and later.

Name	Year	Authors	CPU	Vendor	Peak speed
APE100	1994	APE Collab.	Custom	—	0.1Tflops
CP-PACS	1996	Iwasaki *et al.*	PA-RISC-based	Hitachi(SR2201)	0.6TFlops
QCDSP	1998	Christ *et al.*	TI DSP	—	0.6TFlops
APEmille	2000	APE Collab.	Custom	—	0.8Tflops
QCDOC	2005	Christ *et al.*	PPC-based	IBM(BG/L)	10TFlops
PACS-CS	2006	Ukawa *et al.*	Intel Xeon	Hitachi	14TFlops
QCDCQ	2011	Christ *et al.*	PPC-based	IBM(BG/Q)	500TFlops
QPACE	2012	Wettig *et al.*	PowerXCell	—	200TFlops

"Year" marks the completion date.

Today, the fastest supercomputers have reached the peak speed of $O(50)$ Pflops [59]. This has been achieved in two ways. The K computer in Japan and Sequoia (BlueGene/Q) in USA connected $O(10^5)$ multi-core processors running at $O(100)$ Gflops. Tianhe-2 in China and Titan in USA boosted the computing power by adding multiples of GPGPU's running at $O(1)$ Tflops to each node. Further increase of computing speed faces a serious issue that the memory cannot supply data fast enough to the processing units and the power consumption is becoming too large for large systems. Serious effort is already going on, however, to overcome these problems.

16.3.4. *Fermion problem and hybrid Monte Carlo method*

Monte Carlo calculation for the gluon fields, though somewhat complicated by the SU(3) nature of the field variable, is straightforward. Calculations for the quark fields, on the other hand, cannot be directly put on a computer since quark fields are represented by anticommuting Grassmann numbers. Instead one uses the identity[3]

$$\int \prod_n d\overline{q}_n dq_n \exp\left(\sum_{n,m} \overline{q}_n D_{n,m}(U) q_m\right) \tag{16.27}$$

$$= \int \prod_n d\phi_n^\dagger d\phi_n \exp\left(-\sum_{n,m} \phi_n^\dagger D_{n,m}^{-1}(U)\phi_m\right), \tag{16.28}$$

where D represents the lattice Dirac operator and ϕ_n represents a bosonic field with 4 Dirac and 3 color indices to rewrite

$$Z_{\text{QCD}} = \int \prod_{n\mu} dU_{n\mu} \int \prod_n d\phi_n^\dagger d\phi_n \exp\left(S_{\text{gluon}} - \sum_{n,m} \phi_n^\dagger D_{n,m}^{-1}(U)\phi_m\right). \tag{16.29}$$

with which the fundamental variables $U_{n\mu}$'s and ϕ_n's are all bosonic.

[3]Strictly speaking, this identity requires positivity of the Hermitian part of matrix D. We shall not go into this technical detail and mention only that this can be guaranteed.

The inverse $D^{-1}_{n,m}(U)$ is a non-local quantity. A change of a ϕ_n spreads across the lattice through the inverse. Therefore preparing a trial configuration whose acceptance can be controlled is not straightforward. A number of methods were developed in the 80s including the micro canonical [60, 61] and Langevin [62] methods. The latter was also explored by the group at Cornell including Wilson [63]. The standard method has settled on the hybrid Monte Carlo (HMC) method proposed in 1987 [64], which we now discuss.

The first step of HMC is to introduce an SU(3)-algebra valued momentum $P_{n\mu}$ conjugate to $U_{n\mu}$, and rewrite the path integral (16.29) of full QCD as a partition function of a fictitious classical system of $U_{n\mu}$'s and $P_{n\mu}$'s as

$$Z_{\rm QCD} = \int \prod_{n\mu} dP_{n\mu} \prod_{n\mu} dU_{n\mu} \int \prod_{n} d\phi_n^\dagger d\phi_n \exp\left(-H\right), \tag{16.30}$$

where H is a fictitious Hamiltonian defined by

$$H = \frac{1}{2}\sum_{n\mu} {\rm Tr}\left(P_{n\mu}^2\right) - S_{\rm gluon}(U) + \sum_{n,m} \phi_n^\dagger D^{-1}_{n,m}(U)\phi_m. \tag{16.31}$$

We now wish to generate a set of configurations distributed according to the weight $\exp(-H)/Z_{\rm QCD}$ by a Monte Carlo procedure. For this purpose, one introduces a fictitious time τ conjugate to the Hamiltonian H. Starting with a given configuration of $U_{n\mu}$ and ϕ_n at $\tau = 0$, we solve Hamilton's equations for the canonical pair,

$$\frac{d}{d\tau}U_{n_\mu} = iU_{n\mu}P_{n\mu}, \tag{16.32}$$

$$\begin{aligned} \frac{d}{d\tau}P_{n\mu} &= \frac{\partial}{\partial U_{n\mu}} S_{gluon}(U) \\ &+ \sum_{n,k,l,m} \phi_n^\dagger D^{-1}_{n,k}(U) \frac{\partial}{\partial U_{n\mu}} D_{k,l}(U) D^{-1}_{l,m}(U)\phi_m \end{aligned} \tag{16.33}$$

over some interval of τ. The configuration at a final time τ is used as a new configuration in the Monte Carlo procedure.

In numerical implementations, a continuous fictitious time evolution is discretized with a finite step size $\delta\tau$. Since the Hamiltonian is no longer conserved, the configuration generated after a number of steps $\tau_n = n\delta\tau$ suffers from a bias. This is corrected by accepting or rejecting the generated configuration according to the Metropolis probability $P_{\rm acc} = \min(1, \exp(-\delta H))$ where $\delta H = H(P(\tau_n), U(\tau_n)) - H(P(0), U(0))$ is the difference of Hamiltonian over the trajectory of length $\tau_n = n\delta\tau$.

Hybrid Monte Carlo is an elegant method. It is (i) exact, (ii) allows control of acceptance, i.e., the probability that a trial configuration is accepted, through the magnitude of $\delta\tau$, and (iii) solving Hamltion's equations can be executed in a parallel fashion. However, at every discretized step $\delta\tau$ in updating the fields according to the Hamilton's equation, it requires the inverse of the lattice Dirac operator of form $x_n = \sum_m D^{-1}_{n,m}(U)\phi_m$ for given $U_{n\mu}$ and ϕ_n.

The inverse can be obtained by solving the lattice Dirac equation,

$$\sum_m D_{n,m}(U)x_m = \phi_n. \tag{16.34}$$

This is a linear equation for a large but sparse matrix $D_{n,m}$, which can be obtained by iterative algorithms such as the conjugate gradient method. The number of iterations $N_{\rm iter}$ needed to reach an approximate solution is controlled by the condition number $\kappa(D)$ of the matrix D. It is given by $\kappa(D) = \lambda_{\rm max}/\lambda_{\rm min}$ where $\lambda_{\rm max}$ and $\lambda_{\rm min}$ are the maximum and minimum value of the eigenvalues of D. Typically, the iteration number $N_{\rm iter}$ is inversely proportional to the condition number. Since the minimum eigenvalue for the lattice Dirac operator is of the order of quark mass $m_q a$ in lattice units, and the maximum eigenvalue is $O(1)$, one finds $N_{\rm iter} \propto 1/m_q a$.

Among the six types of quarks known experimentally, the lightest up and down quarks have masses of the order of a few MeV. This is three orders of magnitude smaller than the typical hadronic scale of 1 GeV. The condition number of the lattice Dirac operator for these quarks is large, and therefore, hybrid Monte Carlo simulation slows down considerably as one tries to approach the physical values of quark masses.

16.3.5. *Physical point calculation*

Lattice QCD simulations including quark effects through hybrid Monte Carlo algorithm developed rapidly toward the end of the 90s. By the turn of the century, there was enough experience accumulated to empirically estimate how much computing is needed to calculate observables with a quotable error for a given quark mass. In Fig. 16.8 is shown a typical plot [65], taking the case of $N_f = 2$ flavor simulations for a lattice box of a size 3 fm, a minimum value to contain a hadron such as a nucleon within the box. The vertical axis shows the amount of computing in units of Tflops $\times$ year, i.e., 1 year of running a computer executing 1 Tflop $= 10^{12}$ floating point operations per second. The horizontal axis is the ratio of π meson to ρ meson masses, which varies with up and down quark masses and equals $m_\pi/m_\rho = 0.18$ in experiment. The three curves correspond to the inverse lattice spacing $1/a = 1, 2, 3$ GeV, with $1/a = 2$ GeV or larger being needed for a reliable continuum extrapolation. We observe a sharp rise of the curves toward the physical value of $m_\pi/m_\rho = 0.18$. This is a critical slowing down of the hybrid Monte Carlo algorithm, primarily arising from the slowdown of the Dirac solver toward $m_q a \to 0$.

The rapid increase presented a major problem for lattice QCD simulations. Without overcoming this problem, one could only compute at quark masses much heavier than the physical values. The results had to be extrapolated to the physical point, but such an extrapolation involved large systematic errors because of the

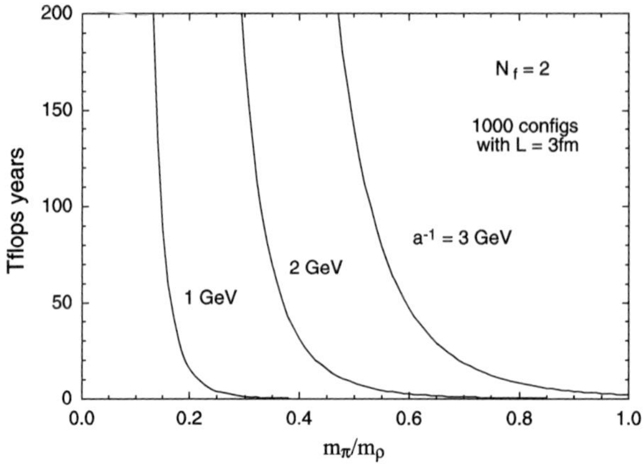

Fig. 16.8. Computational cost for generating 1000 independent configurations for $N_f = 2$ flavor full QCD as a function of m_π/m_ρ whose physical value equals 0.18. The three curves correspond to the lattice spacing $a^{-1} = 1, 2, 3$ GeV. Units is Tflops × years, i.e., 1 year of time with the average speed of 1 Tflops. From Ref. [65].

existence of logarithmic terms of the form $\sim m_q \log m_q/\Lambda$ expected at vanishing quark mass.

The difficulty was resolved in the middle of the 2000s [66]. Since the force in the Hamilton's equation (16.33) involves $D^{-1}_{nm}(U)$, there are contributions coming from the short-distance neighborhood of the link $n\mu$ being updated and from those further away. It is possible to rewrite the quark determinant $\det D(U)$ such that these two types of contributions are separated. It was shown in [66] that, decomposing the lattice Λ into a set of sub-lattices $\Lambda_i, i = 1, 2, \ldots$, one can write

$$\det D(U) = \prod_i \det D_i(U) \cdot \det R(U), \tag{16.35}$$

where $D_i(U)$ is the Dirac operator restricted to a sub-lattice Λ_i and R couples sites belonging to different sub-lattices. Rewriting each determinant factor on the right hand side as a bosonic integral, one can write the force term as

$$F_{n\mu} = F^{\text{gluon}}_{n\mu} + F^{\text{UV}}_{n\mu} + F^{\text{IR}}_{n\mu}, \tag{16.36}$$

where $F^{\text{gluon}}_{n\mu}$ is the term coming from the gluon action, $F^{\text{UV}}_{n\mu}$ those arising from $D_i(U)$'s, and hence represents short-distance contributions, and $F^{\text{IR}}_{n\mu}$ is the term from $R(U)$ giving long-distance contributions.

In Fig. 16.9 we show the magnitude of the three terms observed in a hybrid Monte Carlo run [66]. There is a clear separation of magnitude for the gluon, and UV and IR quark contributions. Therefore, the step size $\delta\tau_{\text{IR}}$ for the IR part $F^{\text{IR}}_{n\mu}$ can be taken much larger than $\delta\tau_{\text{UV}}$ for the UV part, which in turn can be taken much larger than $\delta\tau_{\text{gluon}}$ for the gluon part. An evolution with different step sizes for different parts of the force can be realized by a multi-time step integrator originally developed in [67].

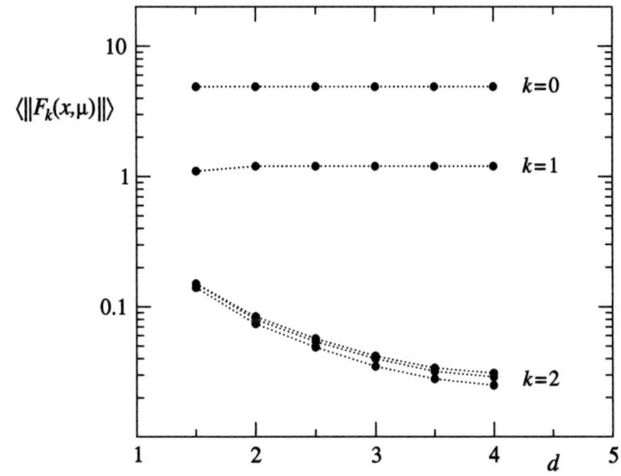

Fig. 16.9. Magnitude of the force term $F^{\rm gluon}_{n\mu}(k=0)$, $F^{\rm UV}_{n\mu}(k=1)$, and $F^{\rm IR}_{n\mu}(k=2)$ as a function of distance from sub lattice boundary. From Ref. [66].

Since inverting the long-distance part of the operator R is the computationally most intensive, using different step sizes for the three parts of the force lead to acceleration of the evolution by a large factor of order $\delta\tau_{\rm IR}/\delta\tau_{\rm gluon} \approx 10 - 50$. Simulations incorporating such an acceleration technique were first carried out in the late 2000s [68], and reached the physical quark masses for the up and down quarks.

The separation of UV and IR parts of the quark force can be realized in a different manner [69] by rewriting the Dirac operator for a quark mass m_0 as a product of ratios of successively heavier masses,

$$D(U, m_0) = \prod_{i=0}^{N} \left(D(U, m_i) D^{-1}(U, m_{i+1})\right) \cdot D(U, m_N), \tag{16.37}$$

where m_{i+1} is taken larger than m_i. Qualitatively speaking, one is separating the contributions to the force according to the eigenvalues in the range $[m_i, m_{i+1}]$ with m_{N+1} equal to the largest eigenvalue.

A somewhat different method to accelerate the hybrid Monte Carlo algorithm is provided by the rational hybrid Monte Carlo algorithm [70]. One writes

$$\det D = \left[\det D^{1/n}\right]^n = \int \prod_{i=1}^{n} d\phi^{i\dagger} d\phi^i \exp\left[-\sum_{i=1}^{n} \phi^{i\dagger} D^{-1/n} \phi^i\right] \tag{16.38}$$

for some positive integer n, and applies a rational approximation to the fractional power $x^{-1/n}$. If κ is the condition number of D, the nth root $D^{1/n}$ will have a smaller condition number $\kappa^{1/n}$. The magnitude of the force, compared with that of the hybrid Monte carlo method, will be reduced by a factor $\kappa^{1-1/n}$. Hence the step size can be increased by the corresponding factor, leading to an acceleration of the molecular dynamics evolution.

A final comment concerns the iterative inversion of the lattice Dirac operator (16.34). Since the increase of the number of iterations toward small quark masses is the cause of the slowdown of the hybrid Monte Carlo algorithm, an ultimate optimization of the algorithm would be realized if one could remove the critical slowdown. This has recently been achieved by understanding the modes corresponding to the small eigenvalues of the lattice Dirac operator. One can either construct these modes explicitly and "deflate" (i.e., remove) them from the inversion [71], or employ multi-grid techniques to adaptively generate and incorporate these modes in the inversion [72, 73].

Over the years, the improvements as described here plus the development of an increasingly powerful computers have made it possible to carry out lattice QCD calculations at the physical quark masses. Such calculations are now routinely done. This is an impressive achievement. It also carries an aesthetic appeal; with the physical quark masses, one is no longer simulating, but rather calculating, the physical processes of the strong interactions as they are taking place in our universe.

16.4. Physics results

16.4.1. *Light hadron spectrum*

16.4.1.1. *Mass spectrum of light Mesons and Baryons*

Since the masses of hadrons are dynamical quantities, whether lattice QCD can quantitatively explain the experimentally known mass spectrum provides a stringent test of the validity of QCD at low energies. Furthermore, the success of such a calculation forms the basis on which the reliability of lattice QCD predictions of other physical quantities are to be built. For these reasons, the calculation of the hadron mass spectrum, in particular those of the ground states of light mesons and baryons composed of up, down, and strange quarks, has been pursued since the beginning of lattice QCD and numerical simulations.

A precise determination of the mass spectrum has to control a number of sources of errors. They are (i) statistical error due to the Monte Carlo nature of the calculation, (ii) systematic error due to extrapolation to the physical quark masses, (iii) systematic error due to finite lattice sizes, (iv) systematic error due to finite lattice spacings. In addition, until the late 90s, most calculations were carried out with the so-called quenched approximation in which effects of quark-antiquark pair creation and annihilation are entirely ignored by dropping the quark determinant $\det D$ from the path integral. This was due to a high cost of including the quark effects into the simulations.

The first systematic attempt at a precision calculation of the light hadron spectrum was carried out by Weingarten and his colleagues within the quenched approximation in 1993 [74]. The calculation took 1 year of dedicated computer time on GF11 computer developed by him at IBM. This was a landmark calculation in

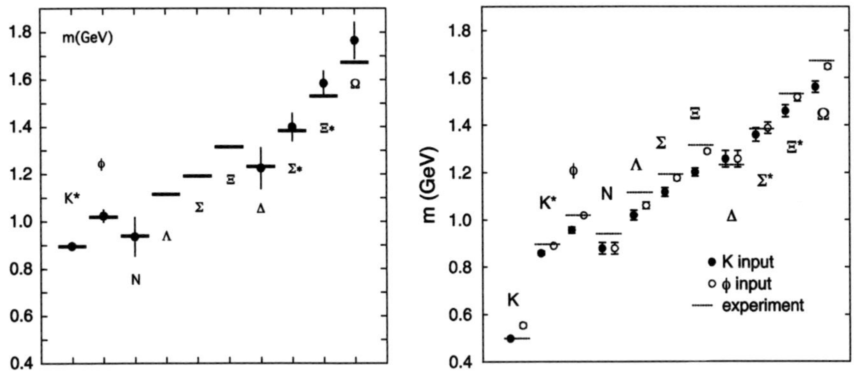

Fig. 16.10. The *left panel* shows the result of the first systematic quenched light hadron mass spectrum calculation [74] published in 1993. The *right panel* shows the definitive quenched result [75] reported in 2000.

that a controlled extrapolation to the physical quark masses, to infinite volume and to zero lattice spacing, was systematically attempted. The masses of π, ρ and K mesons were employed to fix degenerate up and down, and strange quark masses. The masses of K^* and ϕ for the meson octet, nucleon N for the baryon octet, and Δ, Σ^*, Ξ^* and Ω for the baryon decuplet were obtained as predictions. They are plotted by filled circles in the left panel of Fig. 16.10. The masses of Λ, Σ and Ξ of the baryon octet were not reported. The horizontal bars show the experimental values. We observe that the calculated values are consistent with experiment within one standard deviation. However, for baryons, a sizable magnitude of the errors of O(10%) make it difficult to conclude if there is a precise agreement.

A definitive calculation of the hadron spectrum within the quenched calculation followed in 2000 [75]. This work took half a year of CP-PACS computer which was 55 times faster than GF11. The results for the light meson and baryon ground states are shown in the right panel of Fig. 16.10. As one clearly sees there, the quenched spectrum systematically deviates from the experimental spectrum. If one uses the K meson mass m_K as input to fix the strange quark mass (filled data points), the vector meson masses m_{K^*} and m_ϕ are smaller by 4% (4σ) and 6% (5σ), the octet baryon masses are smaller by 6–9% ($4 - 7\sigma$), and the decuplet mass splittings are smaller by 30% on average. Alternatively, if the ϕ meson mass m_ϕ is employed (open circles), m_{K^*} agrees with experiment within 0.8% (2σ) and the discrepancies for baryon masses are much reduced. However, m_K is larger by 11% (6σ). In other words, there is no way to match the entire spectrum beyond 5 to 10% precision in quenched QCD.

The CP-PACS calculation heralded the end of the era of quenched calculations. Efforts toward full QCD simulations, which had already been taking shape in the late 90s, intensified. A first systematic calculation in full QCD was made by the CP-PACS Collaboration with dynamical up and down quarks ($N_f = 2$) [76].

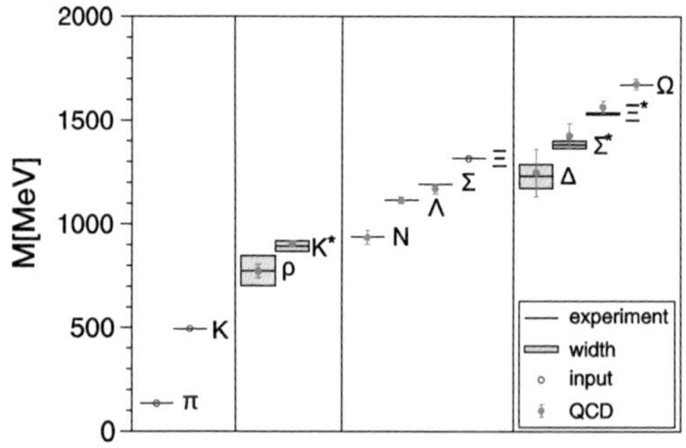

Fig. 16.11. (Color online) Light hadron mass spectrum in $N_f = 2+1$ flavor QCD [77] reported in 2008. Masses of π, K and Ξ are used to fix the degenerate up-down and strange quark masses as well as scale. *Boxes* of shaded *grey show* resonance masses and widths.

With the algorithmic development in the middle of the 2000s, which we described in Sect. 16.3.5, calculations around the physical quark masses became possible. The PACS-CS Collaboration carried out $N_f = 2+1$ flavor calculations in which the strange quark mass was taken close to the physical value and the pion mass was decreased down to $m_\pi = 155$ MeV as compared to the physical value of 135 MeV [68].

Finally, the BMW Collaboration carried out an infinite volume and continuum limit extrapolated $N_f = 2+1$ calculation in 2008 [77]. We reproduce their results in Fig. 16.11. There is good agreement with the experimental spectrum within the errors of 2–5% except for Δ for which the error is 13%.

Some of hadrons, including ρ and Δ, are resonances which decay for physical quark masses and infinite volume. A general relation between energy levels of two body states at finite volume and scattering phase shift, from which resonance parameters can be extracted, was established by Lüscher some time ago [78, 79]. The BMW Collaboration assumed the Breit-Wigner form for the phase shift, and used this relation to correct the measured masses for the effects of decay coupling.

A more rigorous approach in which the phase shift is first extracted from measured energy levels and Lüscher's formula, and resonance masses and widths are subsequently determined from the phase shift, was pioneered in [80, 81]. Because of severe computational cost, it will take more time before physical point calculations can treat resonances numerically precisely in this way.

16.4.1.2. *Isospin splittings of light hadron masses*

The isospin multiplets of hadrons exhibit small mass differences of a few MeV. These tiny effects are nonetheless very important to understand our universe. For example, neutron with a mass $m_n = 939.565379(21)$ MeV is heavier than proton of a mass $m_p = 938.272046(21)$ MeV by $m_n - m_p = 1.2933322(4)$ MeV. Because of this difference, neutron undergoes a β decay with a mean life of 880.3(1.1) s, which has important consequences in nucleosynthesis after the Big Bang, and the composition of nuclei as we see them today.

Since isospin mass splittings arise from the mass difference of up and down quarks, $N_f = 1+1+1$ simulations with different masses for up, down, and strange quarks become necessary. Once we take the quark mass difference into account, one also likes to consider the electromagnetic effects, since the magnitude of the effect, expected at the order of a few MeV, is similar to those arising from the up-down quark mass difference.

A pioneering work on isospin breaking effects was carried out in the mid-90s [82]. The RBC Collaboration rekindled interest by pointing out its importance [83]. Both quenched QED [83] and full QED [84, 85] have been attempted. The most recent calculation has been reported by the BMW Collaboration [86]. They carried out $N_f = 1+1+1+1$ QCD and QED simulations with independent up, down, strange, and charm quark masses and three values of QED coupling for a variety of lattice sizes and spacings. A careful analysis of finite size effects due to the infinite range of photon was made. The infinite volume and continuum limit extrapolation was carried out. The final result is

$$m_n - m_p = +1.51(28)\ \text{MeV} \tag{16.39}$$

in good agreement with experiment. Treating the hadron mass differences to first order in $m_u - m_d$ and QED coupling α, they could separate QCD and QED contributions with the result,

$$m_n - m_p = +2.52(30)_{\text{QCD}} - 1.00(16)_{\text{QED}}\ \text{MeV}. \tag{16.40}$$

There is a delicate cancellation between the QCD and QED effects before the final number settles on the experimental value.

16.4.2. *Fundamental constants of the strong interaction*

16.4.2.1. *Quark masses*

The determination of quark masses is a very important consequence of hadron mass spectrum calculations with lattice QCD. Since quarks are confined unlike leptons, lattice QCD provides the only reliable way for finding the values of this set of fundamental constants of our universe.

In Table 16.4 we list representative lattice results as well as estimates by Particle Data Group over the years. Even as late as 1998, the Review of Particle Physics listed quite a wide band of values as shown in this table. This exemplifies how uncertain quark masses were only a decade and a half ago. Two years later, the CP-PACS quenched spectrum calculation narrowed the range considerably. However, the limitation of quenched QCD is manifest in a large discrepancy of the strange quark mass depending on the input.

The $N_f = 2$ full QCD calculation [76] including dynamical up and down quarks but treating strange quark in the quenched approximation showed that (i) quark

Table 16.4 Light quark masses in MeV units in the $\overline{\mathrm{MS}}$ scheme at $\mu = 2$ GeV.

Year		Action	$\frac{m_u+m_d}{2}$	m_u	m_d	m_s
Quenched QCD					$^{1)}K$ meson mass as input	$^{2)}\phi$ meson mass as input
2000	CP-PACS [75]	Wilson	4.57(18)	—	—	$115.6(2.3)^{1)}$ $143.7(5.8)^{2)}$
$N_f = 2$ QCD					$^{1)}K$ meson mass as input	$^{2)}\phi$ meson mass as input
2000	CP-PACS [76]	Wilson	$3.45^{+0.14}_{-0.20}$	—	—	89^{+3}_{-6} $^{1)}$ 90^{+5}_{-10} $^{2)}$
$N_f = 2+1$ QCD						
2010	MILC [87]	Staggered	3.19(18)	1.96(14)	4.53(32)	89.0(4.8)
2010	BMW [88]	Wilson	3.469(67)	2.15(11)	4.79(14)	95.5(1.9)
2012	RBC/ UKQCD [48]	Domain wall	3.37(12)	—	—	92.3(2.3)
Reviews of Particle Physics						
1998	PDG [89]		2–6	1.5–5	3–9	60–170
2012	PDG [90]		$3.5^{+0.9}_{-0.2}$	$2.3^{+0.7}_{-0.5}$	$4.8^{+0.5}_{-0.3}$	95 ± 5

Table 16.5 Heavy quark masses in GeV units in the $\overline{\text{MS}}$ scheme at $\mu = m_h$.

Year		Action	m_c	m_b	m_t
$N_f = 2 + 1$ QCD					
2010	HPQCD [91]	Staggered	1.273(6)	4.164(27)	—
1998	PDG [89]		1.1–1.4	4.1–4.4	173.8(5.2)
2012	PDG [90]		1.275(25)	4.18(3)	173.07(89)

masses are significantly smaller than indicated by the quenched results, and (ii) the discrepancy of strange quark mass depending on the input almost disappears.

The recent results from $N_f = 2 + 1$ full QCD listed in Table 16.4 covers 3 types of fermion actions. All three calculations carry out infinite volume and continuum extrapolations, albeit the degree of sophistication differs among the three. The separate values of up and down quark masses are estimated using additional input on isospin breaking such as the $K^0 - K^+$ mass difference and estimations on electromagnetic effects. The three sets of results are reasonably consistent. The 2012 Review of Particle Physics values reflect this progress.

In Table 16.5 we list the masses of charm and bottom quarks as determined by lattice QCD, together with those in Reviews of Particle Physics over the years. We also list the value for the top quark for completeness.

Lattice QCD is rapidly moving into the era when all three light quarks are treated independently, and dynamical effects of charm quark is also included. It will be soon that such calculations yield a direct calculation of each of the quark masses with a few % error.

16.4.2.2. *Strong coupling constant*

The value of the strong coupling constant $\alpha_s(\mu) = g^2(\mu)/4\pi$ defined in terms of the QCD coupling $g(\mu)$ at some prescribed scale μ is another fundamental constant of our universe, on a par with the fine structure constant $\alpha = e^2/4\pi$ for electromagnetism with the famous value $\alpha^{-1} = 137.035999074(44)$.

In Fig. 16.12 we plot the determinations obtained from scattering and decay data combined with perturbative QCD as of Fall 2013 [92], and compare them with lattice QCD values from recent $N_f = 2 + 1$ calculations. By convention, the scale is taken to be the Z boson mass $m_Z = 91.1876(21)$ GeV, and five flavors of quarks excluding top is incorporated in the running of the coupling. Lattice QCD determinations employ experimental quantities at low energies such as hadron masses and decay constants for fixing the scale. The methods employed range from step scaling function using the Schrödinger functional by PACS-CS [68], current 2-point function and Wilson loops by HPQCD [91], to static quark antiquark potential by Bazavov *et al.* [93]. The vertical band corresponds to the average over the four experimental determinations [92], $\alpha_s^{(5)}(M_Z) = 0.1183(12)$.

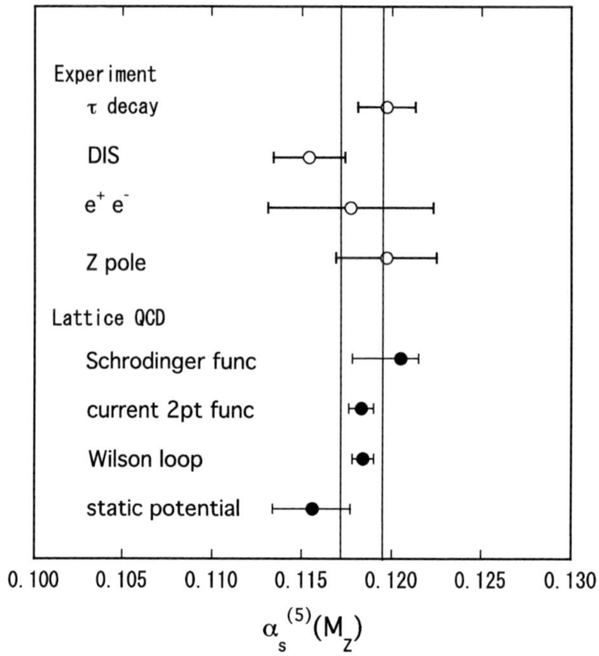

Fig. 16.12. Strong coupling constant $\alpha_s^{(5)}(M_Z)$ obtained from experiment and from lattice QCD. Here experiment means determinations from scattering and decay data combined with perturbative QCD. For further explanations, see text.

The determinations with perturbative QCD still show some scatter and relatively large errors, and so do lattice QCD determinations depending on the method, the ones from current 2-point function and Wilson loops having the smallest error. There is consistency with experiment at 1% level, and the precision of lattice determinations will steadily improve.

16.4.3. *CP violation in the standard model*

16.4.3.1. *CKM matrix elements*

The connection between six types of quarks and CP violation is a salient feature of the Standard Model. The complex phase characterizing CP violation is encoded in the Cabibbo–Kobayashi–Maskawa matrix V which appears in the weak interaction Lagrangian as

$$L_{int} = ig\left(\overline{u}, \overline{c}, \overline{t}\right)_L \gamma_\mu V \begin{pmatrix} d \\ s \\ b \end{pmatrix}_L W_\mu^- + \text{h.c.} \tag{16.41}$$

In the Wolfenstein parametrization, the matrix takes the form,

$$V = \begin{pmatrix} V_{ud}, V_{us}, V_{ub} \\ V_{cd}, V_{cs}, V_{cb} \\ V_{td}, V_{ts}, V_{tb} \end{pmatrix} = \begin{pmatrix} 1 - \lambda^2/2, \lambda, A\lambda^3(\rho - i\eta) \\ -\lambda, 1 - \lambda^2/2, A\lambda^2 \\ A\lambda^3(1 - \rho - i\eta), -A\lambda^2, 1 \end{pmatrix} + O(\lambda^4), \quad (16.42)$$

with three real parameters A, λ, ρ and an imaginary term $i\eta$ characterizing CP violation.

Constraining the CKM matrix requires matching experimental data on weak processes of hadrons to the theoretical expressions for the transition amplitudes which follow from the interaction Lagrangian above. Since quarks interact strongly according to QCD, the weak transition amplitudes are dressed by QCD corrections. These corrections can be calculated by lattice QCD.

One of weak processes which plays an important role is the CP violating mixing of neutral K^0 and $\overline{K}^0$ mesons. The experimentally measured state mixing amplitude ϵ can be written as

$$\epsilon = C\hat{B}_K \left\{ \eta_1 S(x_c)\mathrm{Im}(\lambda_c^2) + \eta_2 S(x_t)\mathrm{Im}(\lambda_t^2) + 2\eta_3 S(x_c, x_t)\mathrm{Im}(\lambda_c\lambda_t) \right\}. \quad (16.43)$$

Here C, $\eta_{1,2,3}$, $S(x,y)$, $S(x) = S(x,x)$ are known constants and functions, $x_q = m_q^2/m_W^2$ with m_q and m_W the mass of quark q and W boson, and $\lambda_q = V_{qs}^* V_{qd}$ is the product of CKM matrix elements. The K meson bag parameter B_K is defined as

$$B_K = \frac{\left\langle \overline{K}^0 \right| (\overline{s}\gamma_\mu d)_L (\overline{s}\gamma_\mu d)_L \left| K^0 \right\rangle}{\frac{8}{3} f_K^2 m_K^2}, \quad (16.44)$$

where the expectation value is taken for QCD, and $\hat{B}_K$ is its renormalization group invariant value. We see then that the precision with which we can compute B_K directly reflects in the precision in the determination of the CKM matrix elements through the products $\lambda_q = V_{qs}^* V_{qd}$.

Similarly, for the mixing of B_q^0 and $\overline{B}_q^0$ mesons with $q = d$ or s, the oscillation frequency Δm_q is proportional to the B_q meson decay constant f_{B_q} and the bag parameter B_{B_q} defined by

$$B_{B_q} = \frac{\langle \overline{B}^0 | (\overline{b}\gamma_\mu q)_L (\overline{b}\gamma_\mu q)_L | B^0 \rangle}{\frac{8}{3} f_{B_q}^2 m_{B_q}^2}. \quad (16.45)$$

In addition, the decays of B mesons in the leptonic ($B \to \ell\nu$) and semi-leptonic (such as $B \to \pi\ell\nu$ and $B \to D^*\ell\nu$) channels are used to constrain the matrix elements $|V_{ub}|$ and $|V_{cb}|$.

The constraint on the CKM matrix is usually written on the complex $(\overline{\rho}, \overline{\eta})$ plane defined by

$$\overline{\rho} + i\overline{\eta} = \left(1 - \frac{\lambda^2}{2}\right)(\rho + i\eta). \quad (16.46)$$

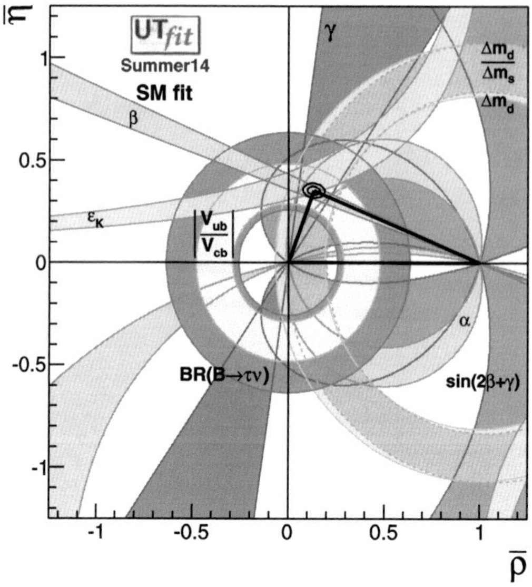

Fig. 16.13. (Color online) Latest constraints on the CKM matrix element expressed in the $(\overline{\rho}, \overline{\eta})$ plane. From Ref. [95].

Table 16.6 CP violating observables and relevant matrix elements.

Observable	PDG [9]/ HFAG [96]		Matrix Element	FLAG [97]	
ϵ	$2.228(11) \times 10^{-3}$	0.49%	$\hat{B}_K$	0.7661(99)	1.3%
$\Delta m_s/\Delta m_d$	34.8(2)	0.60%	$\frac{f_{B_s}\sqrt{B_{B_s}}}{f_{B_d}\sqrt{B_{B_d}}}$	1.268(63) [98]	5.0%
$\mathcal{B}(B \to \tau\nu)$	$1.14(27) \times 10^{-4}$	24%	f_B(MeV)	190.5(4.2)	2.2%
$\mathcal{B}(B \to \pi\ell\nu)^{1)}$	$0.38(2) \times 10^{-4}$	5%	$\Delta\zeta^{B\to\pi}$	2.16(50) [99, 100]	23%
$B \to D^*\ell\nu$	$35.85(45) \times 10^{-3}$	1.3%	$F^{B\to D^*1)}$	0.906(13) [101]	1.4%

Percentage values of errors are also listed. See text for explanations.
1) Integral over $16\text{GeV}^2 \leq q^2 \leq q^2_{\text{max}}$.

In Fig. 16.13 we show the latest result compiled by the UTfit Group [94]. The result by the CKMfitter group [95] is similar. The inside of various regions are allowed from each of the constraints. There has to be a common overlap region if the Standard Model is to be consistent with experiment. Fitting data to the Standard Model, UTfit finds

$$\overline{\rho} = 0.132(23), \quad \overline{\eta} = 0.351(13). \tag{16.47}$$

For each region in Fig. 16.13 the width of the allowed band represents experimental uncertainties as well as those of lattice QCD determinations. In the two left columns of Table 16.6, we list the experimental inputs, and their percentage errors, used for constraining the allowed regions. The quantities and percentage errors listed in the two right columns are the QCD matrix elements which are needed to convert the experimental inputs to constraints in the $(\overline{\rho}, \overline{\eta})$ plane.

The experimental values are taken from Reviews of Particle Physics [9] and a compilation of heavy flavor data by the Heavy Flavor Averaging Group [96]. The lattice QCD values for hadronic matrix elements are taken from a recent compilation by the Flavor Lattice Averaging Group (FLAG) [97]. We select the values quoted for $N_f = 2+1$ calculations that passes the FLAG criteria for the control of systematics errors including the continuum extrapolation. In some cases, only a few calculations are available, in which cases we quote the original references.

We observe that B meson related quantities still has a significant margin for improvement, both on the experimental and lattice QCD sides. The SuperKEKB experiment, which will start in a few years, is expected to reduce the experimental error by another order of magnitude. Improvements in lattice QCD values should keep pace. We should also remark that the large error in the B decay measurement affects the determination of V_{cb}, which in turn broadens the band for ϵ. The determination of B_K itself is already quite precise.

16.4.3.2. *CP violation in the two pion decays of K meson*

Historically, CP violation was discovered through an observation of K mesons decaying into two pions in 1964. The strength of direct CP violation relative to that in the state mixing is measured by the ratio

$$\frac{\epsilon'}{\epsilon} = \frac{\omega}{\sqrt{2}|\epsilon|}\left[\frac{\mathrm{Im}A_2}{\mathrm{Re}A_2} - \frac{\mathrm{Im}A_0}{\mathrm{Re}A_0}\right], \tag{16.48}$$

where A_I denotes the decay amplitude with isospin I in the final state, and

$$\omega^{-1} = \frac{\mathrm{Re}A_0}{\mathrm{Re}A_2}. \tag{16.49}$$

Two heroic experiments, NA48 at CERN [102] and KTEV at FNAL [103], spanning two decades from the 80s to early 2000s, measured ϵ'/ϵ with the result,

$$\frac{\epsilon'}{\epsilon} = \begin{cases} 18.5(7.3) \times 10^{-4}, & \text{NA48} \\ 20.7(2.8) \times 10^{-4}, & \text{KTeV} \end{cases} \tag{16.50}$$

It has also been a long standing puzzle that the amplitude for the $I = 0$ final state is sizably larger than that for $I = 2$, namely $\omega^{-1} \approx 22$. This is called the $\Delta I = 1/2$ rule. Whether the Standard Model can successfully explain these features of $K \to 2\pi$ decay has been a major problem in particle physics. Since the issue boils down to calculating the strong interaction corrections to the effective weak interaction Hamiltonian, this has been an important challenge to lattice QCD since the 80s.

There are three obstacles to a successful calculation of the $K \to \pi\pi$ amplitudes in lattice QCD. The first obstacle is chiral symmetry. Chirality plays an essential role in weak interactions. The effective 4-quark weak interaction Hamiltonian obtained at a lattice cutoff scale of a few GeV starting from a much higher weak interaction scale has a definite chiral structure [104, 105]. Thus one has to employ a lattice fermion formulation which has chiral symmetry. If, on the other hand, one uses non-chiral formulations such as Wilson's, one has to carefully control chiral symmetry violation effects under renormalization. The former option was not available until the late 90s when domain-wall and overlap formulations were proposed. The latter problem was successfully resolved for the Wilson fermion action only recently, with an enticing conclusion that the renormalization structure is the same as in the continuum for $K \to \pi\pi$ decay [106, 107].

The second obstacle stems from the fact that, in the Euclidean Green's function for the $K \to \pi\pi$ transition by the weak Hamiltonian, the two-pion state with zero relative momentum, being the state with lowest energy in this channel, dominates for large times. This contradicts the physical kinematics of the decay; the two pions decaying from a K meson at rest should have an equal and opposite momentum $p = \sqrt{m_K^2/4 - m_\pi^2}$. Thus a naive calculation does not yield physical results. A resolution was found in 2001 [108]. One makes a calculation for a finite lattice volume chosen such that the energy of the two pion state matches the energy of the K meson. The physical amplitude for infinite volume can then be obtained by the following formula,

$$|A_{phys}(K \to \pi\pi)|^2 = 8\pi \frac{E_{\pi\pi}}{p} \left\{ p \frac{d\delta_{\pi\pi}(p)}{dp} + q \frac{d\phi(q)}{dq} \right\} |\langle K|H_W|\pi\pi\rangle_{lat}|^2. \tag{16.51}$$

Here p is the pion momentum, $\delta_{\pi\pi}(p)$ the elastic $\pi\pi$ phase shift at momentum p, $q = pL/(2\pi)$, and $\phi(q)$ is defined by

$$\tan\phi(q) = -\frac{q\pi^{3/2}}{Z_{00}(1;q^2)}, \quad Z_{00}(1;q^2) = \frac{1}{\sqrt{4\pi}} \sum_{\vec{n}\in Z^3} \frac{1}{\vec{n}^2 - q^2}. \tag{16.52}$$

The third issue is specific to the $I = 0$ channel and computational in character. In this channel, there are diagrams with disconnected quark loops, e.g., quarks from the pions annihilate themselves. In addition, the so-called Penguin diagrams in which a pair of quarks from the weak Hamiltonian forms a loop are also present. These diagrams suffer from large statistical fluctuations, rendering the statistical average difficult. This problem is gradually being overcome with efficient algorithms for computing disconnected and Penguin contributions, and with increase of computing power which allows a large number of Monte Carlo ensembles.

Recently there has been significant progress assembling these developments together. The RBC Collaboration has developed applications of the domain-wall formulation of QCD having chiral symmetry. It has succeeded in calculating the $I = 2$ amplitude for the physical pion mass [109, 110]. Their result, obtained at a lattice spacing of $a \approx 0.14$ fm, is given by

$$\mathrm{Re}A_2 = +1.381(46)_{\mathrm{stat}}(258)_{\mathrm{syst}} \times 10^{-8}\,\mathrm{GeV}, \quad (16.53)$$

$$\mathrm{Im}A_2 = -6.54(46)_{\mathrm{stat}}(120)_{\mathrm{syst}} \times 10^{-13}\,\mathrm{GeV}. \quad (16.54)$$

The real part is in good agreement with experiment: $\mathrm{Re}A_2^{\mathrm{exp}} = 1.479(4)\times 10^{-8}$ GeV.

The RBC Collaboration has been attacking the much more difficult $I = 0$ channel, using a special G-periodic boundary condition to force the pions to carry momentum so that an energy matching is realized between the K meson and 2 pions. Preliminary results have been presented at Lattice 2014 Conference this year.

Another group at Tsukuba has been pursuing the problem using the Wilson fermion formulation [107]. The renormalization structure for the operator relevant for $K \to \pi\pi$ decay turned out to be the same as in the continuum except for the mixing with dimension 3 operator $\bar{s}\gamma_5 d$, which is subtracted non-perturbatively. Their calculation so far does not achieve physical kinematics. The π meson mass is artificially taken large so that $2m_\pi = m_K$ is satisfied. Nonetheless they reported an encouraging first result for $I = 0$ channel as well as for $I = 2$ at Lattice 2014 Conference.

It is hoped that progress in the near future brings definitive results on the $I = 0$ amplitude A_0. This will put an independent constraint on the CP violating phase $\bar{\eta}$ in Fig. 16.13 as the ratio ϵ'/ϵ is proportional to $\bar{\eta}$.

16.4.4. *Quark-gluon matter at high temperature and density*

Confinement of quarks and spontaneous breakdown of chiral symmetry are both dynamical consequences of QCD. A very interesting question then is how these properties may or may not change if parameters (or dials) external to QCD are varied. One of important dials is temperature, which increases toward the Big Bang in the Early Universe. Another dial is baryon number density, which has a large value in extreme conditions such as in the core of neutron stars.

In both cases one is interested in an aggregate of hadrons, rather than individual hadrons, and wants to understand how its properties change at extremely high temperature or density. As we discuss below, a general conclusion from lattice studies is that a gas of hadrons turns into a different state in which quark and gluon degrees of freedom becomes manifest. The novel state of matter is often called quark-gluon plasma (QGP).

16.4.4.1. *Phase diagram of QCD at finite temperature: analytical considerations*

The Euclidean formulation adopted for lattice QCD is well suited for studies of its properties at finite temperatures. If one considers a lattice with N_t sites in the temporal direction, and imposes periodic and antiperiodic boundary condition for gluon and quark fields, respectively, the lattice path integral (16.5) is equal to the

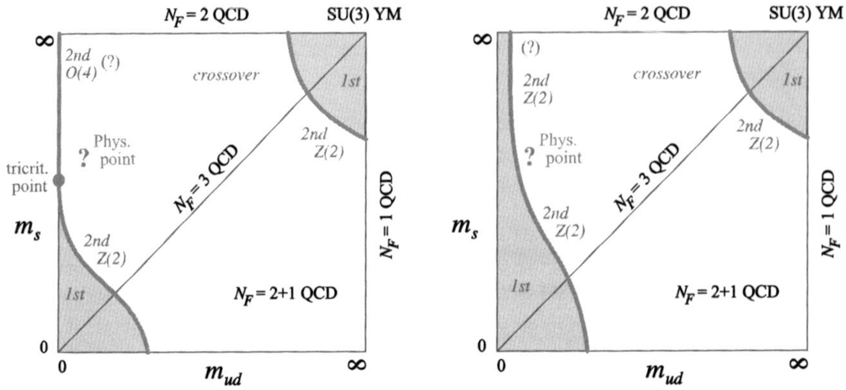

Fig. 16.14. (Color online) Phase digram in 2+1 flavor QCD as a function of the degenerate u and d quark mass m_{ud} and the s quark mass m_s. *Left panel* shows the case for a second order chiral transition for two-flavor QCD. *Right panel* shows the case for a first order two-flavor chiral transition. From Ref. [112].

canonical partition function

$$Z_{\rm QCD} = {\rm Tr}\left[\exp\left(-H_{\rm QCD}/T\right)\right], \tag{16.55}$$

at a physical temperature

$$T = \frac{1}{N_t a}. \tag{16.56}$$

This connection shows that methods developed for zero temperature studies, including Monte Carlo calculations, are readily applicable to the finite temperature case.

In order to discuss what happens as the temperature is raised from zero, it is helpful to consider a two-dimensional plane of the average up-down quark mass m_{ud} and the strange quark mass m_s, often dubbed Columbia Plot, as depicted in Fig. 16.14 [111]. One can examine various limiting cases as follows.

(i) Pure gluon theory: The top right corner in Fig. 16.14 for $m_{ud} = m_s = \infty$ corresponds to the pure gluon theory without quark degrees of freedom. This case is important, nonetheless, since confinement is a dynamical consequence of the gluon fields.

The pure gluon theory possesses a center $Z(3)$ symmetry defined by the transformation

$$U_{n4} \to \zeta U_{n4}, \quad \zeta \in Z(3), \tag{16.57}$$

for the sites n on some fixed time slice t. The corresponding order parameter is the Wilson loop winding around the space-time in the time direction at a fixed spatial site $\vec{n}$ defined by

$$\Omega(\vec{n}) = {\rm Tr}\left(\prod_{n_t=1}^{N_t} U_{(\vec{n},n_t)4}\right). \tag{16.58}$$

This operator, often called Polyakov loop, transforms under the center Z(3) symmetry as

$$\Omega(\vec{n}) \to \zeta\Omega(\vec{n}). \tag{16.59}$$

Polyakov [113] and Susskind [114] argued that the center Z(3) symmetry is intact at low temperatures with $\langle\Omega(\vec{n})\rangle = 0$, but becomes spontaneously broken beyond a certain temperature $T = T_c$ with a non-zero expectation value $\langle\Omega(\vec{n})\rangle \neq 0$.

The connection with deconfinement is most easily understood in the following way. The 2-point correlation function of Polyakov loops defined by $\langle\Omega(\vec{n})\Omega^\dagger(\vec{m})\rangle$ is connected with the free energy $F(|\vec{n}-\vec{m}|)$ of a static quark at site $\vec{n}$ and an antiquark at site $\vec{m}$ by

$$\exp\left(-F(r)/T\right) = \langle\Omega(\vec{n})\Omega^\dagger(\vec{m})\rangle, \quad r = |\vec{n}-\vec{m}|. \tag{16.60}$$

Assuming the presence of a mass gap μ, the 2-point function on the right hand side is expected to behave for large r as

$$\langle\Omega(\vec{n})\Omega^\dagger(\vec{m})\rangle \to \langle\Omega(\vec{n})\rangle \cdot \langle\Omega^\dagger(\vec{m})\rangle + O\left(\exp\left(-\mu r\right)\right), \quad r \to \infty. \tag{16.61}$$

Depending on the expectation value $\langle\Omega\rangle$, one finds for $r \to \infty$ that

$$F(r) \to \begin{cases} \sigma r, & \langle\Omega\rangle = 0, \quad T < T_c \\ c + O(\exp\left(-\mu_D r\right)), & \langle\Omega\rangle \neq 0, \quad T > T_c \end{cases} \tag{16.62}$$

where $\sigma = T\mu$ for $T < T_c$, and c is a constant and $\mu_D = \mu$ is a color electric Debye screening mass for $T > T_c$.

Analytical considerations indicate that this deconfinement phase transition should be of first order [115]. If this is the case, the first-order transition will extend into the region of large but finite quark masses [116], terminating at a line of second-order transition as depicted in Fig. 16.14.

(ii) $N_f = 2$ and $N_f = 3$ massless QCD: The top left corner of Fig. 16.14 corresponds to $N_f = 2$ massless QCD with $m_{ud} = 0$ but $m_s = \infty$, and the bottom left corner to $N_f = 3$ massless QCD with $m_{ud} = 0$ and $m_s = 0$.

For N_f flavors of massless quarks, QCD is invariant under a global $SU(N_f) \otimes SU(N_f)$ chiral symmetry, which is spontaneously broken down to vector $SU(N_f)$ symmetry with a non-zero vacuum expectation value of the order parameter $\Sigma = \langle\overline{q}_n q_n\rangle \neq 0$. The $N_f^2 - 1$-plet of pseudo scalar mesons are the corresponding Nambu–Goldstone bosons.

When one raises the temperature, thermal fluctuations tend to destabilize the chiral order parameter Σ. Thus one expects a phase transition restoring chiral symmetry at some temperature. A more detailed examination is possible using an effective non-linear sigma model for the order parameter field $\Phi_n^{ij} = \overline{q}_n^i q_n^j$ with $i, j = 1, \ldots, N_f$ and renormalization group methods. The result [117] indicates that (i) for $N_f = 3$ or larger, chiral symmetry is restored through a first order phase transition, while (ii) for $N_f = 2$ the phase transition is of first or of second order depending on whether flavor singlet $U(1)$ axial symmetry is effectively restored or not at $T > T_c$.

(iii) Connecting $N_f = 2$ and $N_f = 3$ massless QCD: One can interpolate between the $N_f = 2$ and 3 cases by changing the strange quark mass m_s from ∞ to 0. If the $N_f = 2$ transition is of second order, the first order transition for $N_f = 3$ at $m_s = 0$ has to change to a second order transition at a tricritical point at some $m_s = m_s^c$.

(iv) $N_f = 3$ symmetric QCD: In general, a first order phase transition is stable under symmetry breaking perturbations up to a critical value where it terminates with a second order phase transition. For the second order case, the phase transition immediately disappears if symmetry breaking perturbations are turned on. Thus one expects a sheet of first order transition extending from the bottom left corner corresponding to $N_f = 3$ massless QCD to the interior of the phase diagram. The sheet should terminate at a line of second order phase transitions, which is expected to belong to the Ising universality class [118].

If the $N_f = 2$ transition at $m_{ud} = 0$ is of second order, this line of second order transitions will hit the vertical axis at $m_s = m_s^c$ with a power law $m_{ud} \propto (m_s^c - m_s)^{5/2}$ [119], as depicted in the left panel of Fig. 16.14. If, on the other hand, the $N_f = 2$ transition is of first order, the line will go up to the top horizontal line (the right panel of Fig. 16.14).

16.4.4.2. *Monte Carlo study of the finite-temperature phase diagram*

Lattice QCD simulations are carried out for a fixed temporal lattice size N_t. One then regulates the temperature indirectly by varying the bare gauge coupling constant g_0^2; since the continuum limit $a = 0$ is located at $g_0^2 = 0$, weaker couplings correspond to smaller lattice spacing, and hence to higher temperatures, and *vice versa* for stronger couplings and to lower temperatures.

The basic tool for studying the phase diagram is finite size scaling theory developed in the late 60s and 70s [120]. This theory helps to analyze how singularities marking phase transitions develop from numerical data obtained at finite volumes.

(i) Pure gluon theory: The first instance where finite size scaling method was crucially effective was the deconfinment transition of the pure gluon theory. Early simulations quickly found that the Polyakov loop expectation value exhibited a rapid increase from $\langle\Omega\rangle \approx 0$ to non-zero values over a narrow range of temperature as expected. A subtle issue is if a rapid increase actually signifies a phase transition, and if so, whether it is a first order phase transition with a discontinuous jump of $\langle\Omega\rangle$ at T_c or a second order phase transition with a continuous $\langle\Omega\rangle$ but having a singular derivative.

The susceptibility of Polyakov loop is defined by

$$\chi_\Omega = \frac{1}{L^d} \sum_{\vec{n},\vec{m}} \langle \Omega(\vec{n}) \Omega^\dagger(\vec{m}) \rangle \tag{16.63}$$

with L the linear extent of the system, and d the space dimension. For a finite volume, the susceptibility exhibits a peak. The infinite behavior of the peak height $\chi_\Omega^{\max}$ distinguishes the order of the phase transition. According to finite size scaling theory, if one parametrizes the volume dependence by a power law,

$$\chi_\Omega^{\max}(L) \propto L^\alpha, \quad L \to \infty, \tag{16.64}$$

the value $\alpha = d$, i.e., the dimensionality of space, signals a first order transition, while values $\alpha < d$ characterizes a second order transition in which case α is related to the critical exponents. Similar criterion holds for the volume dependence of the width of the peak.

These criteria were utilized to establish the first order nature of the deconfinement transition in the pure gluon theory [121,122], in agreement with the analytical considerations.

(ii) Chiral transition with quarks: The chiral phase transition in the presence of quarks has also be studied extensively. Ideally one would like to use a fermion formulation preserving chiral symmetry such as the domain-wall or overlap. They became available only in the early 2000s, and are computationally very costly, however. For these reasons, most of the calculations have utilized, and still use, the staggered fermion formulation and, to a lesser extent, the Wilson formulation.

Broadly speaking, results accumulated to date are as follows. For $N_f = 3$ degenerate flavors, the phase transition is consistent with being of first order for small quark masses. Increasing the quark mass, the transition weakens and terminates at a second order transition whose exponents are consistent with the Ising universality class (see Refs. [123, 124] for early representative work and references). These calculations are made at most for $N_t = 6$ lattices, and the continuum limit is yet to be taken [125].

For $N_f = 2$, the situation is more complicated. Early simulations were consistent with a second order phase transition both with the staggered [126] and with the Wilson [127] fermion formulations. For the staggered case, however, the critical exponents did not come out consistent with either $O(4)$ nor $O(2)$ values. There have also been simulations suggesting consistency with a first order transition [128].

As already pointed out in Ref. [117], the order of the $N_f = 2$ chiral transition is connected with the $U_A(1)$ anomaly. Recently, a theoretical argument has been put forward [129] that the anomaly effects disappear for certain sets of correlation functions if chiral symmetry is restored. At the moment, the order of the $N_f = 2$ transition is not settled completely.

16.4.4.3. *Thermodynamics with physical quark masses*

Physically, the crucial issue is where the point with physical quark masses lies on the phase diagram in Fig. 16.14. The staggered results are unanimous that it lies beyond the line of critical end points. Hence there is only a continuous crossover and no phase transition with a singular behavior. The basis for this conclusion includes

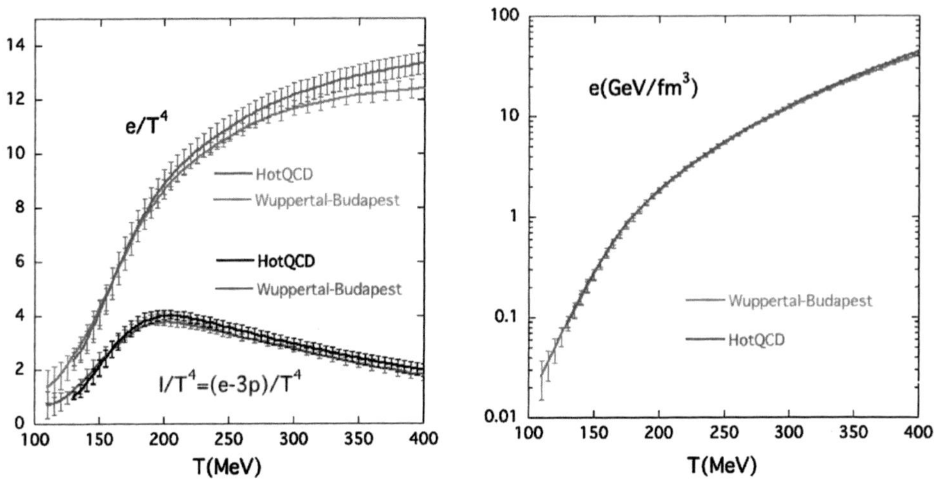

Fig. 16.15. (Color online) Thermodynamic quantities in the continuum limit in 2+1 flavor QCD as a function of temperature. *Left panel* shows the energy density e/T^4 and interaction measure $I/T^4 = (e-3p)/T^4$ in units of T^4, and *right panel* shows energy density e in units of GeV/fm^3. From Refs. [134, 135].

an extensive study at physical quark masses with infinite volume and continuum extrapolations using susceptibilities of various physical observable [130].

Since the transition is a continuous crossover, the transition temperature is not uniquely determined but depends on the quantity used. For example [131, 132], one finds $T_c = 147(2)(3)$ MeV if one uses the susceptibility peak of chiral order parameter, and $T_c = 157(4)(5)$ MeV from the inflection point of the energy density. An independent calculation [133] reported $T_c = 154(9)$ MeV from $O(4)$ scaling analysis of chiral susceptibility. The results are consistent, and altogether indicate $T \approx 150 - 160$ MeV as the temperature range across which the physical chiral transition takes place.

In the left panel of Fig. 16.15 we show the energy density e/T^4 and interaction measure $I/T^4 = (e - 3p)/T^4$, with p the pressure, in units of T^4 calculated in the staggered quark formalism by two groups [134, 135]. They are obtained at the physical quark masses, and infinite volume and continuum extrapolations are made. We observe very good agreement up to about 200 MeV beyond which a one standard deviation difference appears. Physically important is the feature that the Stefan-Boltzmann value for free gluons and quarks, $e_{\rm SB}/T^4 = \pi^2(8+21N_f/4)/15 = 15.62\ldots$ for $N_f = 3$, is reached only slowly, with significant deviations remaining at $T/T_c \sim 2-3$. This indicates that the quark-gluon matter is strongly interacting at these temperature ranges. The right panel shows the energy density in units of GeV/fm^3.

Experimental effort toward detection of quark-gluon plasma through heavy ion collisions in accelerators has been going on for a long time. It started with the

Bevalac at Berkeley in the 70s, the AGS at BNL and the SPS at CERN in the 80s, with RHIC at BNL since 2000 and with the LHC at CERN most recently.

Let us see how lattice QCD predictions compare with heavy ion experiments at RHIC and the LHC. Phenomenological estimates on the energy density reached at the initial stage of the collision range from $e \approx 5.6$ GeV/fm^3 for Au–Au collisions at RHIC with $\sqrt{s_{NN}} = 200$ GeV [136], to $e \approx 15$ GeV/fm^3 for Pb–Pb collisions at the LHC with $\sqrt{s_{NN}} = 2.76$ TeV [137]. Looking at the right panel of Fig. 16.15 we read out a temperature of order $T \approx 200$ and 300 MeV, respectively, for these energy densities which are high enough for the collision product to be in the high temperature phase. For comparison, an experimental estimate of the temperature using the low p_T excess of direct photons that are supposed to come from the initial thermalized state gave $T = 221(27)$ MeV at RHIC [138] and $T = 304(51)$ MeV at the LHC [139].

The initial fireball rapidly cools as it expands, and hadrons are formed once the temperature falls below the transition temperature T_c. It has been known that the yield of various hadrons from the collision could be well fitted with statistical thermal distribution parametrized by a chemical freeze out temperature T_f. This temperature increases with the collision energy and saturates at $T_f \approx 160$ MeV at RHIC energies [140], though slightly decreasing at LHC energies.

Furthermore, the azimuthal anisotropy in the transverse hadron yields, quantified in terms of the elliptic flow v_2 and higher moments where $v_n = \langle \cos n\phi \rangle$, is well described by hydrodynamical models with no or small viscosity and using the equation of state from lattice QCD.

Thus, while small viscosity was a surprise, the experimental data are broadly consistent with the production of strongly interacting high temperature quark-gluon matter with the properties as predicted by lattice QCD. It is interesting to note that small shear viscosity close to the quantum limit $\eta/s = 1/4\pi$ [141] is observed by lattice QCD calculations of transport coefficients in the pure gluon case [142, 143].

From a heavy-ion collision point of view, the transition temperature and equation of state are only indirectly reflected in the characteristics of the final hadronic state. The moments of conserved charges such as electric charge Q, strangeness S, and baryon number B provide interesting quantities which are calculable in lattice QCD and are directly observable in experiments. Thus they have been attracting much interest recently [144–148].

16.4.4.4. *Dynamics at finite baryon number density*

Theoretical expectations on the phases of QCD at finite temperature T and finite baryon number density ρ_B is depicted in Fig. 16.16. The chiral crossover transition at $T_c \approx 150-160$ MeV continues into the finite ρ_B region, and hits a second order critical point at some (T_E, ρ_E) beyond which the transition turns into a first order phase transition. For sufficiently large baryon number density, one expects novel phases such as a color superconductor.

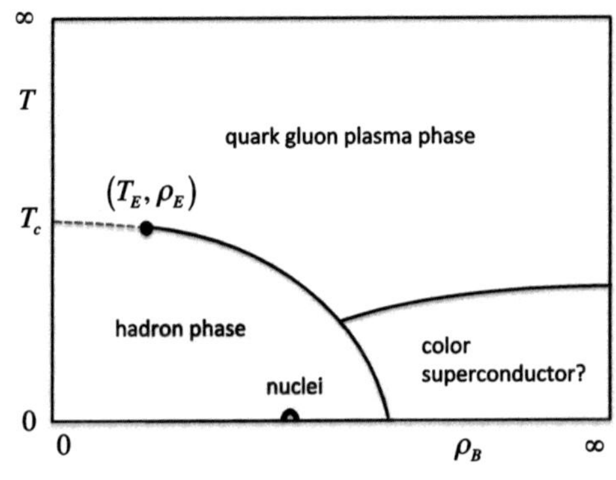

Fig. 16.16. Schematic phase diagram of QCD on the temperature T-baryon density ρ_B plane.

Finite baryon number density can be introduced by adding a quark chemical potential μ to the Hamiltonian. On a lattice, this is achieved by multiplying the positive temporal hopping term of the quark action by $\exp(\mu a)$, and the negative hopping term by $\exp(-\mu a)$.

Solid quantitative results are still not available on the phase diagram. The main reason is that for non-zero quark chemical potential, the quark Dirac determinant $\det D(U, \mu)$ is complex, so that the Monte Carlo methods based on a probability interpretation of the weight $\det D \cdot \exp(-S_G(U))$ no longer work in general.

One can see the difficulty by defining the phase of the determinant by

$$\theta(U, \mu) = -i \log \frac{\det D(U, \mu)}{|\det D(U, .\mu)|}, \tag{16.65}$$

and rewriting expectation values in the following way:

$$\langle O \rangle = \frac{\langle O \exp(i\theta(U, \mu)) \rangle_{||}}{\langle \exp(i\theta(U, \mu)) \rangle_{||}}, \tag{16.66}$$

where $\langle \cdot \rangle_{||}$ means the average with respect to the phase quenched determinant $|\det D(U, \mu)|$. The quenched average of the phase factor in the denominator is a ratio of the two partition functions:

$$\langle \exp(i\theta(U, \mu)) \rangle_{||} = \frac{Z(\mu)}{Z_{||}(\mu)}, \tag{16.67}$$

$$Z_{||}(\mu) = \int \prod_{n\mu} dU_{n\mu} |\det D(U, \mu)| \exp(S_{\text{gluon}}(U)). \tag{16.68}$$

If the quenched average defines a statistical system with a free energy density $f_{||}(\mu)$, one can write

$$\langle \exp(i\theta(U, \mu)) \rangle_{||} = \exp\left(-\frac{V(f(\mu) - f_{||}(\mu))}{T} \right), \tag{16.69}$$

with $f(\mu)$ the free energy density of the original system. The right hand side vanishes exponentially fast for large spatial volumes $V \to \infty$. This is one way of explaining the sign problem.

A variety of methods have been devised and explored to overcome this problem. Besides the phase quenched calculation described above, they include reweighting of the determinant [149,150], analytic continuation from imaginary chemical potential [151, 152], Taylor expansion in powers of μ/T [153, 154], canonical ensemble simulation [155, 156], complex Langevin simulation [157–159], and others. These methods have yielded some success for not too large values of μ/T. The main problem, however, has been the difficulty of controlling errors in the results. The precise location of the critical point E and physics of finite density QCD for larger baryon number density is still largely open at present.

16.5. Conclusions

Lattice QCD is a major contribution of Kenneth Wilson to physics. In our view, it is on a par in significance with his renormalization group theory of critical phenomena which won him Nobel Prize in Physics in 1982. Conceptually, it clarified how quantum fluctuations of gauge fields give rise to a confining force which is essentially distinct in its origin from the forces arising from exchange of a particle such as the electromagnetic force. At the same time, coupled with supercomputers, it opened a way to calculate the physical predictions from its first principles, making possible detailed comparisons with experiment.

In 2004, Kenneth Wilson delivered a talk entitled “The Origins of Lattice Gauge Theory” at International Symposium on Lattice Field Theory held at Fermi National Accelerator Laboratory [3]. Looking back on the development of lattice QCD, he said that *The lattice gauge theory was a discovery waiting to happen, once asymptotic freedom was established.*, and went on to describe works of his contemporaries who could have preceded or shared the discovery. Nonetheless it is clear that, because of his previous studies leading to his encounter with statistical mechanics and phase transitions, he was in a unique position to be the first to grasp the deep significance of strongly coupled gauge dynamics in relation to confinement.

In the same article, he commended on the vast progress in lattice QCD in the thirty years since 1974, the year of his seminal paper [1], due *in part to improved algorithms, in part to increased computing power, and in part to the increased scale of the research effort underway today*, but urged on that *this does not mean that the present state of lattice gauge computations is fully satisfactory. The knowledge and further advances that will likely accumulate over the next thirty years should be just as profound.*

As if corroborating his view, in just a decade since then, the physical point computation was realized, and many beautiful physics results ensued, as summarized in Sect. 16.4.

The past decade, then, was a turning point of lattice QCD in our view; prior to this event, despite the premise that it provided a first-principle calculation, it remained a method of uncertain reliability, requiring chiral extrapolations which were difficult to control. We are now able to calculate and understand many properties of hadron states including masses and matrix elements, at the physical quark masses and controlling errors of calculations. The precision of the calculation has now reached the level of a few % or better in many quantities. Consequently, important constraints are now available on the CKM matrix elements and CP violation in the Standard Model.

This is not to say that progress has been uniform in all fronts. In thermodynamics of QCD, while some basic quantities such as the transition temperature and equation of state have been calculated, the region of finite baryon number density is still largely unexplored. A major methodological breakthrough will be needed to extend our understanding to the entire phase diagram of QCD.

One area which did not even exist at the time of Wilson's talk in 2004 is nuclear physics based on lattice QCD. In fact, while there were pioneering studies on H dibaryon in the late 80s [160,161] and nucleon scattering lengths in 1994 [162,163], it is in 2007 that the two nucleon potential was first extracted from lattice QCD [164], and in 2009 that the binding energy of Helium was calculated directly from Helium correlation function [165]. This is a challenging area in terms of physics as well as in calculational techniques; one has to deal with a small energy scale of 0(10) MeV, the number of Wick contractions for quark fields increases factorially fast with the mass number, and so does the statistical error of nuclei propagators for large time separations.

In theoretical physics, theory and computing go hand in hand. Calculations are indispensable in order to confirm the validity of theory, and even more so to explore the consequences of theory which help us better understand why our world works the way it does. Lattice QCD is a prime example of such a relationship between theory and computing. Kenneth Wilson clearly foresaw the importance and future potential of supercomputing in this connection, and many of his thoughts and vision in the early 80s [166] came to be realized since then. Looking toward future, as he prophesied [2], our understanding of the strong interactions will become more profound as the computing power increases toward exascale and possibly beyond in the decades to come.

Personal reminiscences

My first encounter with lattice QCD was in the spring of 1974 in Tokyo, Japan, when I was a graduate student under the supervision of Kazuhiko Nishijima at University of Tokyo. Kenneth Wilson's preprint was introduced at a seminar organized by graduate students of the particle theory group. I was struck by the novelty of the idea, which I vividly remember, but I do not think I understood the full impact of that preprint at that time.

I had a good fortune to spend two years from September 1975 to August 1977 as a postdoctoral fellow of Toichiro Kinoshita at Laboratory of Nuclear Studies of Cornell University. Kenneth Wilson had his office one floor below. John Kogut's office was on the same floor across the corridor, and there were a number of graduate students, all of them very active, including Gyan Bhanot, Paul Mackenzie, Michael Peskin, Serge Rudaz, Steve Shenker, and Junko Shigemitsu.

At Cornell I worked with Tom Kinoshita on mass singularities and, in later terminology, perturbative QCD. I was of course aware of the work of Ken Wilson, operator product expansions, renormalization group, lattice QCD, Kondo problem, but somehow he was a somewhat distant figure for me throughout my stay at Cornell. But I do remember him very well. A lean and taciturn man, often with his shirt tail hanging out in the back, and he always wore a slight kindly smile on his face, which deepened from time to time when something apparently amused him.

My serious involvement with lattice QCD started after I left Cornell, first with analytic studies on Z(n) duality at Princeton, and then, after I came back to Japan, with numerical simulations and even supercomputer development. The 40 years of lattice QCD since 1974 overlap with my scientific career. It is an honor for me to write this article on Kenneth Wilson and lattice QCD.

Acknowledgments

I would like to thank Sinya Aoki, Norman Christ, Carleton De Tar, Zoltan Fodor, Shoji Hashimoto, Yoichi Iwasaki, Kazuyuki Kanaya, Frithjof Karsch, Andreas Kronfeld, Martin Lüscher, and Paul Mackenzie for valuable comments on the manuscript.

References

1. Wilson, K.G.: Confinement of Quarks, Cornell preprint CLNS-262 (Feb 1974), published in Phys. Rev. D **10**, 2445 (1974)
2. Wilson, K.G.: Future directions in particle theory. In: Proceedings of the 1983 Lepton Photon Symposium at High Energies, p. 812. Cornell University Press, Ithaca (1983)
3. Wilson, K.G.: The origins of lattice gauge theory. Nucl. Phys. B (Proc. Suppl.) **140**, 3–19 (2005)
4. Creutz, M., Jacobs, L., Rebbi, C.: Experiments with a Gauge invariant Ising system. Phys. Rev. Lett. **42**, 1390 (1979)
5. Wilson, K.G.: Monte Carlo calculations for the lattice gauge theory. In: Proceedings of the 1979 Cargese Summer Institute. NATO Sci. Ser. **B59**, 363 (1980)
6. Creutz, M.: Solving quantized SU(2) Gauge theory, Brookhaven National Laboratory Print-79-0919 (Sep 1979): Monte Carlo Study of Quantized SU(2) Gauge Theory. Phys. Rev. D **21**, 2308 (1980)
7. Weingarten, D.: Monte Carlo evaluation of Hadron Masses in lattice gauge theories with fermions. Phys. Lett. B **109**, 57 (1982)

8. Hamber, H., Parisi, G.: Numerical estimates of Hadronic Masses in a pure SU(3) gauge theory. Phys. Rev. Lett. **47**, 1792 (1981)
9. Particle Data Group web page `http://pdg.lbl.gov`
10. Gross, D.J., Wilczek, F.: Ultraviolet behavior of non-Abelian gauge theories. Phys. Rev. Lett. **30**, 1343 (1974)
11. Politzer, H.D.: Reliable perturbative results for strong interactions? Phys. Rev. Lett. **30**, 1346 (1974)
12. Symanzik, K.: Euclidean quantum field theory. In: Jost, R. (eds.) Local Quantum Field Theory, p. 152. Academic Press, New York (1969)
13. Wilson, K.G., Kogut, J.: The renormalization group and the ϵ expansion. Phys. Rep. **12**, 75 (1974)
14. Osterwalder, K., Schrader, R.: Axioms for Euclidean green's functions. Commun. Math. Phys. **31**, 83 (1973)
15. Osterwalder, K., Schrader, R.: Axioms for Euclidean green's functions II. Commun. Math. Phys. **42**, 281 (1975)
16. Bali, G.S., Schilling, K.: Running coupling and the lambda parameter from SU(3) Lattice simulations. Phys. Rev. D **47**, 661 (1993)
17. Bowler, K.C., Hasenfratz, A., Hasenfratz, P., Heller, U.M., Karsch, F., Kenway, R.D., Pawley, G.S., Wallace, D.J.: The SU(3) beta function at large beta. Phys. Lett. B **179**, 375 (1986)
18. Gupta, R., Kilcup, G.W., Patel, A., Sharpe, S.R.: The beta function for pure gauge SU(3). Phys. Lett. B **211**, 132 (1988)
19. Lüscher, M., Sommer, R., Weisz, P., Wolff, U.: A precise determination of the running coupling in the SU(3) Yang–Mills theory. Nucl. Phys. **B413**, 481 (1994)
20. ALPHA Collaboration, Della Morte, M., Frezzotti, R., Heitger, J., Rolf, J., Sommer, R., Wolff, U.: Computation of the strong coupling in QCD with two dynamical flavours. Nucl. Phys. **B713**, 378 (2005)
21. PACS-CS Collaboration, Aoki, S., Ishikawa, K.I., Ishizuka, N., Izubuchi, T., Kadoh, D., Kanaya, K., Kuramashi, Y., Murano, K., Namekawa, Y., Okawa, M., Taniguchi, Y., Ukawa, A., Ukita, N., Yoshié, T.: Precise determination of the strong coupling constant in $N_f = 2+1$ Lattice QCD with the Schrödinger Functional Scheme, JHEP 0910 (2009) 053
22. Sommer, R., Tekin, F., Wolff, U.: Running of the SF-coupling with four massless flavours. PoS(Lattice 2010) (2010) 241
23. Karsten, L.H., Smit, J.: Lattice fermions, species dubling, chiral invariance and the triangle anomaly. Nucl. Phys. B **183**, 103 (1981)
24. Nielesen, H.B., Ninomiya, M.: Absence of neutrinos on a lattice. 1. Proof by Homotopy Theory. Nucl. Phys. **B185**, 20 (1981): Erratum-ibid. B195 (1982) 541
25. Karsten, L.H.: Lattice fermions in euclidean space-time. Phys. Lett. B **104**, 315 (1981)
26. Wilson, K.G.: Quarks and strings on a lattice. In: Zichichi, A. (ed.) Proceedings of the 14th Course of the International School of Subnuclear Physics, Erice, 1975. Plenum, New York (1977)
27. Bochicchio, M., Maiani, L., Martinelli, G., Rossi, G.C., Testal, M.: Chiral symmetry on the lattice with Wilson Fermions. Nucl. Phys. B **262**, 331 (1985)
28. Itoh, S., Iwasaki, Y., Oyanagi, Y., Yoshié, T.: Hadron spectrum in quenched QCD on a $12^3 \times 24$ lattice with renormalization group improved lattice SU(3) Gauge action. Nucl. Phys. B **274**, 33 (1986)
29. Kawamoto, N.: Towards the phase structure of Euclidean Lattice Gauge theories with Fermions. Nucl. Phys. B **190**, 617 (1981)

30. Aoki, S.: New phase structure for lattice QCD with Wilson Fermions. Phys. Rev. D **30**, 2653 (1984)
31. Frezzotti, R., Grassi, P., Sint, S., Weisz, P.: Lattice QCD with a chirally twisted mass term. JHEP **0108**, 058 (2001)
32. Frezzotti, R., Rossi, G.C.: Chirally improving Wilson Fermions—I. O(a) Improvement, JHEP 08 (2004) 007
33. Aoki, S., Baer, O.: Automatic $O(a)$ improvement for twisted mass QCD in the presence of spontaneous symmetry breaking. Phys. Rev. D **74**, 034511 (2006)
34. Susskind, L.: Lattice Fermions. Phys. Rev. D **16**, 3031 (1977)
35. Sharatchandra, H.S., Thus, H.J., Weisz, P.: Susskind Fermions on a Euclidean lattice. Nucl. Phys. B **192**, 205 (1981)
36. Kluberg-Stern, H., Morel, A., Napoly, O., Petersson, B.: Flavors of lagrangian susskind fermions. Nucl. Phys. B **220**, 447 (1983)
37. Sharpe, S.R.: Rooted staggered fermions: good, bad or ugly?. PoS (Lattice 2006) 022 (2006)
38. Lee, W.-J., Sharpe, S.R.: Partial flavor symmetry restoration for chiral staggered fermions. Phys. Rev. D **60**, 114503 (1999)
39. Ginsparg, P.H., Wilson, K.G.: A remnant of chiral symmetry on the lattice. Phys. Rev. D **25**, 2649 (1982)
40. Lüscher, M.: Exact chiral symmetry on the lattice and the Ginsparg–Wilson relation. Phys. Lett. B **428**, 342 (1998)
41. Kaplan, D.B.: A method for simulating chiral fermions on the lattice. Phys. Lett. B **288**, 342 (1992)
42. Furman, V., Shamir, Y.: Axial symmetries in Lattice QCD with Kaplan fermions. Nucl. Phys. B **439**, 54 (1995)
43. Narayanan, R., Neuberger, H.: A construction of lattice chiral gauge theories. Nucl. Phys. B **443**, 305 (1995)
44. Neuberger, H.: Exactly massless quarks on the Lattice. Phys. Lett. B **417**, 141 (1998)
45. Neuberger, H.: Vector-like gauge theories with almost massless fermions on the lattice. Phys. Rev. D **57**, 5417 (1998)
46. Borici, A.: Truncated overlap fermions. Nucl. Phys. B (Proc. Suppl.) **83**, 771 (2000)
47. Hasenfratz, P., Laliena, V., Niedermayer, F.: The index theorem in QCD with a finite cut-off. Phys. Lett. B **427**, 125 (1998)
48. Arthur, R., Blum, T., Boyle, P.A., Christ, N.H., Garron, N., Hudspith, R.J., Izubuchi, T., Jung, C., Kelly, C., Lytle, A.T., Mawhinney, R.D., Murphy, D., Ohta, S., Sachrajda, C.T., Soni, A., Yu, J., Zanotti, J.M.: (RBC and UKQCD Collaborations), Domain Wall QCD with near-physical pions. Phys. Rev. D **87**, 094514 (2013)
49. Aoki, S., Chiu, T.-W., Cossu, G., Feng, X., Fukaya, H., Hashimoto, S., Hsieh, T.-H., Kaneko, T., Matsufuru, H., Noaki, J.: Simulation of quantum chromodynamics on the lattice with exactly chiral lattice fermions. PTEP **2012**, 01A106 (2012)
50. Hasenfratz, A., Hasenfratz, P., Niedermayer, F.: Simulating full QCD with the fixed point action. Phys. Rev. D **72**, 114508 (2005)
51. Eichten, E.: Heavy quarks on the lattice. Nucl. Phys. Proc. Suppl. **4**, 170 (1988)
52. Lepage, G.P., Thacker, B.A.: Effective Lagrangians for simulating heavy quark systems. Nucl. Phys. Proc. Suppl. **4**, 199 (1988)
53. El-Khadra, A.X., Kronfeld, A.S., Mackenzie, P.B.: Massive fermions in lattice gauge theory. Phys. Rev. D **55**, 3933 (1997)
54. Aoki, S., Kuramashi, Y., Tominaga, S.: Relativistic Heavy Quarks on the Lattice. Prog. Theor. Phys. **109**, 383 (2003)

55. Christ, N.H., Li, M., Lin, H.-W.: Relativistic heavy quark effective action. Phys. Rev. D **76**, 074505 (2007)
56. Metropolis, N., Rosenbluth, A.W., Rosenbluth, M.N., Teller, A.H., Teller, E.: Equation of state calculations by fast computing machines. J. Chem. Phys. **21**, 1087 (1953)
57. Wilson, K.G.: The renormalization group: critical phenomena and the Kondo problem. Rev. Mod. Phys. **47**, 773 (1975)
58. Andrei, N., Lowenstein, J.H.: Scales and scaling in the Kondo model. Phys. Rev. Lett. **46**, 356 (1981)
59. http://www.top500.org/lists/2014/11/
60. Callaway, D.J.E., Rahman, A.: Microcanonical ensemble formulation of lattice gauge theory. Phys. Rev. Lett. **49**, 613 (1982)
61. Polonyi, J., Wyld, H.W.: Microcanonical simulation of fermionic systems. Phys. Rev. Lett. **51**, 2257 (1983)
62. Ukawa, A., Fukugita, M.: Langevin simulation including dynamical quark loops. Phys. Rev. Lett. **55**, 1854 (1985)
63. Batrouni, G.G., Katz, G.R., Kronfeld, A.S., Lepage, G.P., Svetitsky, B., Wilson, K.G.: Langevin simulations of lattice field theories. Phys. Rev. D **32**, 2736 (1985)
64. Duane, S., Kennedy, A.D., Pendleton, B.J., Roweth, D.: Hybrid Monte Carlo. Phys. Lett. B **195**, 216 (1987)
65. Ukawa, A. for CP-PACS and JLQCD Collaborations. Computational cost of full QCD simulations experienced by CP-PACS and JLQCD Collaborations. Nucl. Phys. B (Proc. Suppl.) **106** 195 (2002)
66. Lüscher, M.: Schwarz-preconditioned HMC algorithm for two-flavour lattice QCD. Comput. Phys. Commun. **165**, 199 (2005)
67. Sexton, J.C., Weingarten, D.H.: Hamiltonian evolution for the hybrid Monte Carlo Algorithm. Nucl. Phys. B **380**, 665 (1992)
68. Aoki, S., Ishikawa, K.-I., Ishizuka, N., Izubuchi, T., Kadoh, D., Kanaya, K., Kuramashi, Y., Namekawa, Y., Okawa, M., Taniguchi, Y., Ukawa, A., Ukita, N., Yoshié, T.: (PACS-CS Collaboration), $2+1$ flavor lattice QCD toward the physical point. Phys. Rev. D **79**, 034503 (2009)
69. Hasenbusch, M.: Speeding up the hybrid Monte Carlo algorithm for dynamical fermions. Phys. Lett. B **519**, 177 (2001)
70. Horvath, I., Kennedy, A.D., Sint, S.: A new exact method for dynamical fermion computations with non-local actions. Nucl. Phys. B (Proc. Suppl.) **73**, 834 (1999)
71. Lüscher, M.: Local coherence and deflation of the low quark modes in lattice QCD. JHEP **0707**, 081 (2007)
72. Babich, R., Brannick, J., Brower, R.C., Clark, M.A., Manteuffel, T.A., McCormick, S.F., Osborn, J.C., Rebbi, C.: Adaptive multigrid algorithm for the lattice Wilson-dirac operator. Phys. Rev. Lett. **105**, 201602 (2010)
73. Frommer, A., Kahl, K., Krieg, S., Leder, B., B. Rottmann, B.: Adaptive Aggregation based domain decomposition multigrid for the Lattice Wilson Dirac Operator. e-Print: http://arxiv.org/abs/1303.1377 arXiv:1303.1377 [hep-lat]
74. Butler, F., Chen, H., Sexton, J., Vaccarino, A., Weingarten, D.: Hadron mass predictions of the valence approximation to lattice QCD. Phys. Rev. Lett. **70**, 2849 (1993)
75. CP-PACS Collaboration, Aoki, S., Boyd, G., Burkhalter, R., Ejiri, S., Fukugita, M., Hashimoto, S., Iwasaki, Y., Kanaya, K., Kaneko, T., Kuramashi, Y., Nagai, K., Okawa, M., Shanahan, H.P., Ukawa, A., Yoshié, T.: Quenched light hadron spectrum. Phys. Rev. Lett. **84**, 238 (2000)

76. CP-PACS Collaboration: Ali Khan, A., Aoki, S., Boyd, G., Burkhalter, R., Ejiri, S., Fukugita, M., Hashimoto, S., Ishizuka, N., Iwasaki, Y., Kanaya, K., Kaneko, T., Kuramashi, Y., Manke, T., Nagai, K., Okawa, M., Shanahan, H.P., Ukawa, A., Yoshié, T.: Dynamical quark effects on light quark masses. Phys. Rev. Lett. **85**, 4674 (2000) ; Erratum-ibid. 90 (2003) 029902
77. Durr, S., Fodor, Z., Frison, J., Hoelbling, C., Hoffmann, R., Katz, S.D., Krieg, S., Kurth, T., Lellouch, L., Lippert, T., Szabo, K.K., Vulvert, G.: Ab-initio determination of light hadron masses. Science **322**, 1224 (2008)
78. Lüscher, M.: States, two particle, on a torus and their relation to the scattering matrix. Nucl. Phys. B **354**, 531 (1991)
79. Lüscher, M.: Signatures of unstable particles in finite volume. Nucl. Phys. B **364**, 237 (1991)
80. Aoki, S., Fukugita, M., Ishikawa, K.-I., Ishizuka, N., Kanaya, K., Kuramashi, Y., Namekawa, Y., Okawa, M., Sasaki, K., Ukawa, A., Yoshié, T.: Lattice QCD calculation of the ρ Meson decay width. Phys. Rev. D **76**, 094506 (2007)
81. Aoki, S., Fukugita, M., Ishikawa, K.-I., Ishizuka, N., Kanaya, K., Kuramashi, Y., Namekawa, Y., Okawa, M., Sasaki, K., Ukawa, A., Yoshié, T.: ρ Meson decay in $2+1$ flavor lattice QCD. Phys. Rev. D **84**, 094505 (2011)
82. Duncan, A., Eichten, E., Thacker, H.: Electromagnetic splittings and light quark masses in lattice QCD. Phys. Rev. Lett. **76**, 3894 (1996)
83. Blum, T., Zhou, R., Doi, T., Hayakawa, M., Izubuchi, T., Uno, S., Yamada, N.: Electromagnetic mass splittings of the low lying hadrons and quark masses from $2+1$ flavor lattice QCD+QED. Phys. Rev. D **82**, 094508 (2010)
84. Ishikawa, T., Blum, T., Hayakawa, M., Izubuchi, T., Jung, C., Zhou, R.: Full QED + QCD low-energy constants through reweighting. Phys. Rev. Lett. **109**, 072002 (2012)
85. Aoki, S., Ishikawa, K.-I., Ishizuka, N., Kanaya, K., Kuramashi, Y., Nakamura, Y., Namekawa, Y., Okawa, M., Taniguchi, Y., Ukawa, A., Ukita, N., Yoshié, T.: (PACS-CS Collaboration), $1+1+1$ flavor QCD + QED Simulation at the Physical Point. Phys. Rev. D **86**, 034507 (2012)
86. Borsanyi, S., Durr, S., Fodor, Z., Hoelbling, C., Katz, S.D., Krieg, S., Lellouch, L., Lippert, T., Portelli, A., Szabo, K.K.: Ab initio calculation of the neutron–proton mass difference, http://arxiv.org/abs/1406.4088 arXiv:1406.4088 (2014)
87. Bazavov, A., Bernard, C., DeTar, C., Du, X., Freeman, W., Gottlieb, S., Heller, U.M., Hetrick, J.E., Laiho, J., Levkova, L., Oktay, M.B., Osborn, J., Sugar, R., Toussaint, D., Van de Water, R. S.: (The MILC Collaboration), MILC Results for Light Pseudo Scalars, PoS (CD09) 007 (2009)
88. Durr, S., Fodor, Z., Hoelbling, C., Katz, S., Krieg, S., et al.: Lattice QCD at the physical point: light quark masses. Phys. Lett. B **701**, 265 (2011)
89. Particle Data Group Collaboration: (C. Caso et al.), Review of particle physics. Eur. Phys. J. **C3**, 1 (1998)
90. Particle Data Group Collaboration: (J. Beringer et al.) Review of particle physics. Phys. Rev. D **86**, 010001 (2012)
91. McNeile, C., Davies, C.T.H., Follana, E., Hornbostel, K., Lepage, G.P.: High-precision c and b masses and QCD coupling from current-current correlators in lattice and continuum QCD. Phys. Rev. D **82**, 034512 (2010)
92. Quantum chromodyamics section in `http://pdg.lbl.gov/2013/reviews/`
93. Bazavov, A., Brambilla, N., Garcia i Tormo, X., Petreczky, P., Soto, J., Vairo, A.: Determination of α_s from the QCD static energy. Phys. Rev. D **86** 114031 (2012)

94. http://www.utfit.org/UTfit/
95. http://ckmfitter.in2p3.fr
96. http://www.slac.stanford.edu/xorg/hfag/
97. FLAG Working Group, Aoki, S., Aoki, Y., Bernard, C., Blum, T., Colangelo, G., Della Morte, M., Durr, S., El-Khadra, A.X., Fukaya, H., Horsley, R., Juttner, A., Kaneko, T., Laiho, J., Lellouch, L., Leutwyler, H., Lubicz, V., Lunghi, E., Necco, S., Onogi. T., Pena, C., Sachrajda, C.T., Sharpe, S.R., Simula, S., Sommer, R., Van de Water, R.S., Vladikas, A., Wenger, U., Wittig, H.: Review of lattice results concerning low energy particle physics, http://arxiv.org/abs/1310.8555 arXiv:1310.8555 (August 2014)
98. Bazavov, A., Bernard, C., Bouchard, C.M., DeTar, C., Di Pierro, M., El-Khadra, A.X., Evans, R.T., Freeland, E.D., Gamiz, E., Gottlieb, S., Heller, U.M., Hetrick, J.E., Jain, R., Kronfeld, A.S., Laiho, J., Levkova, L., Mackenzie, P.B., Neil, E.T., Oktay, M.B., Simone, J.N., Sugar, R., Toussaint, D., Van de Water, R.S.: Neutral B-meson mixing from three-flavor lattice QCD: determination of the SU(3)-breaking ratio ξ. Phys. Rev. D **86**, 034503 (2012)
99. Gulez, E., Gray, A., Wingate, M., Davies, C.T.H., Lepage, G.P., Shigemitsu, J.: B Meson semileptonic form factors from unquenched lattice QCD. Phys. Rev. D **73**, 074502 (2006)
100. Bailey, J.A., Bernard, C., DeTar, C., Di Pierro, M., El-Khadra, A.X., Evans, R.T., Freeland, E.D., Gamiz, E., Gottlieb, S., Heller, U.M., Hetrick, J.E., Kronfeld, A.S., Laiho, J., Levkova, L., Mackenzie, P.B., Okamoto, M., Simone, J.N., Sugar, R., Toussaint, D., Van de Water, R.S.: The $B \to \pi\ell\nu$ semi-leptonic form factor from three-flavor lattice QCD: a model-independent determination of $|V_{ub}|$. Phys. Rev. D **79**, 054507 (2009)
101. Bailey, J.A., Bazavov, A., Bernard, C., Bouchard, C.M., DeTar, C., Du, D., El-Khadra, A.X., Foley, J., Freeland, E.D., Gámiz, E., Gottlieb, S., Heller, U.M., Kronfeld, A.S., Laiho, J., Levkova, L., Mackenzie, P.B., Neil, E.T., Qiu, S.-W., Simone, J., Sugar, R., Toussaint, D., Van de Water, R.S., Zhou, R.: Update of $|V_{cb}|$ from the $\overline{B} \to D^*\ell\overline{\nu}$ form factor at zero recoil with three-flavor lattice QCD. Phys. Rev. D **89**, 114504 (2014)
102. Fanti, V., et al.: A new measurement of direct CP violation in two pion decays of the neutral Kaon. Phys. Lett. B **465**, 335 (1999)
103. Arabi-Harati, A., et al.: Measurements of direct CP violation, CPT symmetry, and other parameters in the neutral kaon system. Phys. Rev. D **67**, 012005 (2003)
104. Buchalla, G., Buras, A.J., Harlander, M.K.: The Anatomy of ϵ'/ϵ in the Standard Model, Nucl. Phys. B **337**, 313 (1990)
105. Buchalla, G., Buras, A.J., Lautenbacher, M.E.: Weak decays beyond leading logarithms. Rev. Mod. Phys. **68**, 1125 (1996)
106. Donini, A., Gimenez, V., Martinelli, G., Talevi, M., Vladikas, A.: Non-perturbative renormalization of lattice four-fermion operators without power subtractions. Eur. Phys. J. C **10**, 121 (1999)
107. Ishizuka, N., Ishikawa, K.I., Ukawa, A., Yoshié, T.: Calculation of $K \to \pi\pi$ decay amplitudes with improved Wilson Fermion, http://arxiv.org/abs/1311.0958 arXiv:1311.0958 (2013)
108. Lellouch, L., Lüscher, M.: Weak transition matrix elements from finite volume correlation functions. Commun. Math. Phys. **219**, 31 (2001)
109. Blum, T., Boyle, P.A., Christ, N.H., Garron, N., Goode, E., Izubuchi, T., Jung, C., Kelly, C., Lehner, C., Lightman, M.: The $K \to (\pi\pi)_{I=2}$ decay amplitude from lattice QCD. Phys. Rev. Lett. **108**(2012), 141601 (2012)

110. Blum, T., Boyle, P.A., Christ, N.H., Garron, N., Goode, E., Izubuchi, T., Jung, C., Kelly, C., Lehner, C., Lightman, M.: The $K \to (\pi\pi)_{I=2}$ Decay Amplitude from Lattice QCD. Phys. Rev. D **86**, 074513 (2012)
111. Brown, F.R., Butler, F.P., Chen, H., Christ, N.H., Dong, Z.-H., Schaffer, W., Unger, L.I., Vaccarino, A.: On the existence of a phase transition for QCD with three light quarks. Phys. Rev. Lett. **65**, 2491 (1990)
112. Kanaya, K: Finite temperature QCD on the lattice — Status 2010. PoS (Lattice 2010) (2010) 012
113. Polyakov, A.M.: Thermal properties of gauge fields and quark liberation. Phys. Lett. **72B**, 477 (1978)
114. Susskind, L.: Hot quark soup. Phys. Rev. D **20**, 2610 (1979)
115. Yaffe, L.G., Svetitsky, B.: First order phase transition in the SU(3) gauge theory at finite temperature. Phys. Rev. D **26**, 963 (1982)
116. Banks, T., Ukawa, A.: Deconfining and chiral phase transitions in quantum chromodynamics at finite temperature. Nucl. Phys. B **225**, 145 (1983)
117. Pisarski, R.D., Wilczek, F.: Remarks on the chiral phase transition in chromodynamics. Phys. Rev. D **29**, 338 (1984)
118. Gavin, S., Gocksch, A., Pisarski, R.D.: QCD and the chiral critical point. Phys. Rev. D **49**, 3079 (1994)
119. Rajagopal, K.: The chiral phase transition in QCD: critical phenomena and long wavelength pion oscillations, quark-gluon plasma, vol. 2 (1995) (arXiv: http://arxiv.org/abs/hep-ph/9504310 hep-ph/9504310)
120. Barber, M.N.: Finite size scaling, phase transitions and critical phenomena, Vol. 8, Lewobitz, C. Academic Press, Domb and J (1973)
121. Fukugita, M., Okawa, M., Ukawa, A.: Order of the deconfining phase transition in SU(3) lattice gauge theory. Phys. Rev. Lett. **63**, 1768 (1989)
122. Fukugita, M., Okawa, M., Ukawa, A.: Finite size scaling study of the deconfining phase transition in pure SU(3) Lattice Gauge Theory. Nucl. Phys. B **337**, 181 (1990)
123. JLQCD Collaboration: Aoki, S., Fukugita, M., Hashimoto, S., Ishikawa, K-I., Ishizuka, N., Iwasaki, Y., Kanaya, K., Kaneda, T., Kaya, S., Kuramashi, Y., Okawa, M., Onogi, T., Tominaga, S., Tsutsui, N., Ukawa, A., Yamada, N., Yoshié, T.: Phase structure of lattice QCD at finite temperature for $2+1$ flavors of Kogut-Susskind quarks. Nucl. Phys. B (Proc.Suppl.) **73**, 459 (1999)
124. Karsch, F., Laermann, E., Schmidt, C.: The chiral critical point in three-flavor QCD. Phys. Lett. B **520**, 41 (2001)
125. Ding, H.-T., Bazavov, A., Hegde, P., Karsch, F., Mukherjee, S., Petreczky, P.: Exploring phase diagram of $N_f = 3$ QCD at $\mu = 0$ with HISQ Fermions. PoS (Lattice 2011) 191 (2011)
126. Aoki, S., Fukugita, M., Hashimoto, S., Ishizuka, N., Iwasaki, Y., Kanaya, K., Kuramashi, Y., Mino, H., Okawa, M., Ukawa, A., Yoshié, T.: (JLQCD Collaboration), scaling study of the two flavor chiral phase transition with the kogut-susskind quark action in lattice QCD. Phys. Rev. D **57**, 3910 (1998)
127. Ali Khan, A., Aoki, S., Burkhalter, R., Ejiri, S., Fukugita, M., Hashimoto, S., Ishizuka, N., Iwasaki, Y., Kanaya, K., Kaneko, T., Kuramashi, Y., Manke, T., Nagai, K., Okamoto, M., Okawa, M., Ukawa, A., and Yoshié, T., (CP-PACS Collaboration)): Phase structure and critical temperature of two flavor QCD with renormalization group improved gauge action and clover improved Wilson quark action. Phys. Rev. D **63**, 034502 (2001)
128. Bonati, C., Cossu, G., D'Elia, M., Di Giacomo, A., Pica, C.: The order of the QCD transition with two light flavors. Nucl. Phys. A **820**, 243C (2009)

129. Aoki, S., Fukaya, H., Taniguchi, Y.: Chiral symmetry restoration, the eigenvalue density of the dirac operator, and the axial U(1) anomaly at finite temperature. Phys. Rev. D **86**, 114512 (2012)
130. Aoki, Y., Endrodi, G., Fodor, Z., Katz, S.D., Szabo, K.K.: The order of the quantum chromodynamics transition predicted by the standard model of particle physics. Nature **443**, 675 (2006)
131. Aoki, Y., Fodor, Z., Katz, S.D., Szabo, K.K.: The QCD Transition Temperature: Results with Physical Masses in the Continuum Limit. Phys. Lett. B **643**, 46 (2006)
132. Wuppertal-Budapest Collaboration, Borsanyi, S., Fodor, Z., Hoelbling, C., Katz, S.D., Krieg, S., Ratti, C., Szabo, K.K.: Is there still any T_c Mystery in Lattice QCD? results with physical masses in the continuum limit III, JHEP 1009 (2010) 073
133. Bazavov, A., Bhattacharya, T., Cheng, M., DeTar, C., Ding, H.-T., Gottlieb, S., Gupta, R., Hegde, P., Heller, U.M., Karsch, F., Laermann, E., Levkova, L., Mukherjee, S., Petreczky, P., Schmidt, C., Soltz, R.A., Soeldner, W., Sugar, R., Toussaint, D., Unger, W., Vranas, P.: (HotQCD Collaboration), Chiral and deconfinement aspects of the QCD transition. Phys. Rev. D **85**, 054503 (2012)
134. Borsanyi, S., Fodor, Z., Hoelbling, C., Katz, S.D., Krieg, S., Szabo, K.K.: Full result for the QCD equation of state with $2+1$ flavors. Phys. Lett. **B730**, 99 (2014)
135. HotQCD Collaboration, Bazavov, A., Bhattacharya, T., DeTar, C., Ding, H.-T., Gottlieb, S., Gupta, R., Hegde, P., Heller, U.M., Karsch, F., Laermann, E., Levkova, L., Mukherjee, S., Petreczky, P., Schmidt, C., Schroeder, C., Soltz R.A., Soeldner, W., Sugar, R., Wagner, M., Vranas, P.: The equation of state in 2+1-Flavor QCD, e-Print: http://arxiv.org/abs/1407.6387 arXiv:1407.6387 (2014)
136. PHENIX Collaboration, Adler, S.S. *et al.*: Systematic studies of the centrality and $\sqrt{s_{NN}}$ dependence of the $dE_T/d\eta$ and $dN_{ch}/d\eta$ in heavy ion collisions at mid-rapidity. Phys. Rev. C **71**, 034908 (2005)
137. CMS Collaboration, Chatrchyan, S. *et al.*: Measurement of the pseudorapidity and centrality dependence of the transverse energy density in Pb-Pb collisions at $\sqrt{s_{NN}} =$ 2.76 TeV. Phys. Rev. Lett. **109**, 152303 (2012)
138. PHENIX Collaboration, Enhanced production of direct photons in Au + Au collisions at $\sqrt{s_{NN}} = 200$ GeV and implications for the initial temperature. Phys. Rev. Lett. 104, 132301 (2010)
139. Wilde, M., for ALICE Collaboration. Measurement of direct photons in pp and Pb–Pb collisions with ALICE. http://arxiv.org/abs/1210.5958 arXiv.1210.5958 (2012)
140. Andronica, A., Braun-Munzinger, P., Stachel, J.: Thermal hadron production in relativistic nuclear collisions: the hadron mass spectrum, the horn, and the QCD phase transition. Phys. Lett. B **673**, 142 (2009)
141. Kovtun, P., Son, D.T., Starinets, A.O.: Viscosity in strongly interacting quantum field theories from black hole physics. Phys. Rev. Lett. **94**, 111601 (2005)
142. Nakamura, A., Sakai, S.: Lattice study of gluon viscosities: a step towards RHIC physics. Acta Phys. Polon. B **37**, 3371 (2006)
143. Meyer, H.B.: A calculation of the shear viscosity in SU(3) gluodynamics. Phys. Rev. D **76**, 101701(R) (2007)
144. Stephanov, M., Rajagopal, K., Shyuriak, E.: Event-by-event fluctuations in heavy ion collisions and the QCD critical point. Phys. Rev. D **60**, 114028 (1999)
145. Ejiri, S., Karsch, F., Redlich, K.: Hadronic fluctuations at the QCD phase transition. Phys. Lett. B **633**, 275 (2006)
146. Karsch, F.: Determination of freeze-out conditions from lattice QCD calculations. Central Eur. J. Phys. **10**, 1234 (2012)

147. Bazavov, A., Ding, H.-T., Hegde, P., Kaczmarek, O., Karsch, F., Laermann, E., Mukherjee, S., Petreczky, P., Schmidt, C., Smith, D., Soeldner, W., Wagner, M.: Freeze-out conditions in heavy ion collisions from QCD thermodynamics. Phys. Rev. Lett. **109**, 192302 (2012)
148. Borsanyi, S., Fodor, Z., Katz, S.D., Krieg, S., Ratti, C., Szabo, K.K.: Freeze-out parameters from electric charge and baryon number fluctuations: is there consistency? Phys. Rev. Lett. **113**, 052301 (2014)
149. Fodor, Z., Katz, S.D.: A new method to study lattice QCD at finite temperature and chemical potential. Phys. Lett. B **534**, 87 (2002)
150. Fodor, Z., Katz, S.D.: Lattice determination of the critical point of QCD at finite T and μ. JHEP **0203**, 014 (2002)
151. Roberge, A., Weiss, N.: Gauge theories with imaginary chemical potential and the phases of QCD. Nucl. Phys. B **275**, 734 (1986)
152. de Forcrand, P., Philipsen, O.: The QCD phase diagram for small densities from imaginary chemical potential. Nucl. Phys. B **642**, 290 (2002)
153. Allton, C.R., Ejiri, S., Hands, S.J., Kaczmarek, O., Karsch, F., Laermann, E., Schmidt, C.: The equation of state for two flavor QCD at non-zero chemical potential. Phys. Rev. D **68**, 014507 (2003)
154. Ejiri, S., Kanaya, K., Umeda, T.: Ab initio study of the thermodynamics of quantum chromodynamics on the lattice at zero and finite densities. PTEP **2012**, 01A104 (2012)
155. Li, A., Alexandru, A., Liu, K.-F., Meng, X.: Finite density phase transition of QCD with $N_f = 4$ and $N_f = 2$ using canonical ensemble method. Phys. Rev. D **84**, 071503 (2011)
156. Li, A., Alexandrou, A., Liu, K.F.: Critical point of Nf = 3 QCD from lattice simulations in the canonical ensemble. Phys. Rev. D **84**, 071503 (2011)
157. Aarts, G., Stamatescu, I.-O.: Stochastic quantization at finite chemical potential. JHEP **0809**, 018 (2008)
158. Aarts, G., Seiler, E., Stamatescu, I.O.: The complex langevin method: when can It be trusted? Phys. Rev. D **81**, 054508 (2010)
159. Sexty, D.: Simulating full QCD at non-zero density using the complex Langevin equation. Phys. Lett. B **729**, 108 (2014)
160. Mackenzie, P.B., Thacker, H.B.: Evidence against a stable Dibaryon from lattice QCD. Phys. Rev. Lett. **55**, 2539 (1985)
161. Iwasaki, Y., Yoshié, T., Tsuboi, Y.: The H Dibaryon in lattice QCD. Phys. Rev. Lett. **60**, 1371 (1988)
162. Fukugita, M., Kuramashi, Y., Mino, H., Okawa, M., Ukawa, A.: An exploratory study of nucleon-nucleon scattering lengths in lattice QCD. Phys. Rev. Lett. **73**, 2176 (1994)
163. Fukugita, M., Kuramashi, Y., Okawa, M., Mino, H., Ukawa, A.: Hadron scattering lengths in lattice QCD. Phys. Rev. D **52**, 3003 (1995)
164. Ishii, N., Aoki, S., Hatsuda, T.: The nuclear force from lattice QCD. Phys. Rev. Lett. **99**, 022001 (2007)
165. Yamazaki, T., Kuramashi, Y., Ukawa, A.: (PACS-CS Collaboration), Helium nuclei in quenched lattice QCD. Phys. Rev. D **81**, 111504 (2010)
166. Lax, P.D.: Chairman, report of the panel on Large Scale Computing in Science and Engineering, 1982 (scanned pdf file is found at `http://www.pnl.gov/scales/docs/laxreport1982.pdf`)

Chapter 17

Skeleton graph expansion of critical exponents in "cultural revolution" years*

Bailin Hao

Department of Physics, Fudan University,
220 Handan Road, Shanghai 200433, P. R. China
Institute of Theoretical Physics, Academia Sinica,
55 East Zhongguancun Street, Beijing 100190, P. R. China
hao@mail.itp.ac.cn

Kenneth Wilson's Nobel Prize winning breakthrough in the renormalization group theory of phase transition and critical phenomena almost overlapped with the violent "cultural revolution" years (1966–1976) in China. An unexpected chance in 1972 brought the author of these lines close to the Wilson–Fisher ϵ-expansion of critical exponents and eventually led to a joint paper with Lu Yu published entirely in Chinese without any English title and abstract. Even the original acknowledgment was deleted because of mentioning foreign names like Kenneth Wilson and Kerson Huang. In this article I will tell the 40-year old story as a much belated tribute to Kenneth Wilson and to reproduce the essence of our work in English. At the end, I give an elementary derivation of the Callan–Symanzik equation without referring to field theory.

Keywords: Kenneth Wilson; renormalization group; ϵ-expansion of critical exponents; skeleton graph expansion; Callan–Symansik equation.

PACS numbers: 01.65.+g, 05.70.Fh, 11.10.Hi

17.1. Introduction

There has never been a tower of ivory for scientists working in a developing country. In his report to the South Commission[1] Abdus Salam reproduced a figure by Dadison Frame. The figure showed the annual publication of scientific and technological papers versus GNP for the year 1973. Most points fell around two straight lines, that

*This article was originally published in *Int. J. Mod. Phys. B* **28**, 1430008 (2014).

for developed and developing countries, respectively. There was, however, a striking lonely outlier far below many least-developed countries. It represented China.

In the summer of 1972, an unthinkable opportunity threw four Chinese physicists into the Canadian Congress of Physicists held in Edmonton, Alberta. Michael Fisher talked about his joint work with Kenneth Wilson on ϵ-expansion of the critical exponents.[2] With my poor English at that time, I could only appreciate the importance of the renormalization group approach. Upon return to Beijing, I immediately read the two 1971 papers of Wilson.[3,4]

I must explain how could I get access to scientific literature in a time when all libraries were closed. At that time, the Institute of Physics was led by representatives of the People's Liberation Army (PLA). In order to prepare for "science reform" they appointed a group of scientists to do investigations on physics research in China and abroad. This allowed me to get into the closed libraries. Xerox machine was something unheard of. We had to make notes by hand.

The first two papers of Wilson were quite hard to grasp. Early 1973, Professor Kerson Huang paid a visit to Beijing. In a discussion, he mentioned that Kenneth Wilson had given a series of lectures in Princeton and promised to write to Kenneth. Soon, Wilson sent us a Cornell preprint[5] which later appeared in *Physics Reports.*[6] The "later" here meant at least half a year delay or more as the libraries had only surface mail subscriptions.

After digesting the available information, our goal was clear: calculate the ϵ expansion to high powers of ϵ in order to compare with experimental measurements and to check the scaling relations among critical exponents. An approach by comparing skeleton graph with scaling relation[7,8] came to my attention. Skeleton graphs were closer to my heart as more than 10 years ago in one of the Landau seminars in Moscow I listened to K. A. Ter-Martirosyan talking about applying skeleton graph expansion to meson scattering. The authors of Refs. 7, 8 considered the special case of $n = 2$ Bose systems and their result could not be extended to general n and higher-orders. Especially, the analysis at the critical point was lacking. I undertook to study the general n case.

There was a theoretical division in the Institute of Physics, of which I was the Deputy Head. The division was created in 1959 upon reflection of the "Great Leap Forward". In 1969, it was disbanded by the PLA officers as a "typical example of isolation of theory from practice". In 1973 I succeeded in restoring a small theoretical group in the laboratory of magnetism. Another important fact that allowed me to do some science consisted in that I was nailed to the bed by lumbar disc rupture. Our group member Lu Yu came and worked at my bedside and then reported to the group. Eventually, Yu was the only one who could catch up with the work.

The official journal of the Chinese Physical Society *Acta Physica Sinica* stopped publication for more than seven years, from the fall of 1966 to the end of 1973. It was decided to restore publication from January 1974. Our paper[9] arrived at the editorial office on 5 December 1973 and appeared in print only in May 1975. The paper did not have an English title and abstract. Originally, we thanked Kerson Huang and Kenneth Wilson for helping with the Princeton lecture notes. However, we were asked for the political attitude of these foreigners and we decided to delete the whole acknowledgment. In the spring of 1977, a group of solid state physicists from the Chinese Academy of Sciences visited France and Germany for the first time after many years of isolation from the outside world. After my talk at Orsay on closed-form approximation for the three-dimensional Ising model, I had a discussion with Eduard Brézin by pointing to formulas in our Chinese reprint. This was the start of our many-decade friendship with Eduard Brézin.

17.2. Critical exponents and scaling relations

In continuous phase transitions, thermodynamic functions and their first derivatives are continuous, but high-order derivatives may be singular at the transition point T_c. The behavior of thermodynamic quantities near critical point is described by various critical exponents. For example, the singularity of specific heat near a critical point is described by the exponent α:

$$c_v \sim (T - T_c)^{-\alpha} \quad (T \geq T_c) \tag{17.1}$$

The behavior of spin or density correlation function near the critical point is better expressed via their Fourier transform as

$$G(P \to 0, T = T_c) \sim P^{-2+\eta}, \tag{17.2}$$

which involves another exponent η.

Another limit of the same correlation function is associated with the initial magnetic susceptibility (or isothermal compressibility) χ_γ:

$$\chi_\gamma \sim G(P = 0, T \to T_c + 0) \sim (T - T_c)^{-\gamma}, \tag{17.3}$$

where a third exponent γ is introduced.

Historically, various phase transition analyses were unified in the Landau mean field theory which yields the same exponents: $\alpha = 0$ (finite discontinuity), $\gamma = 1$, and $\eta = 0$.

The situation was more or less satisfactory until the mid 1960s when precise experimental measurements and exact statistical models all showed that there were definite deviations of critical exponent values from the mean field theory. Nevertheless, some relations between the critical exponents turned out to be holding. Leo Kadanoff[10] and Michael Fisher[11] called attention of the statistical physics community to this challenge. Unfortunately, their well-known reviews[10,11] were overlooked by Chinese physicists as the "cultural revolution" broke out on 1st June 1966.

17.3. Classical field theory representation of statistical problem near critical point

We adopt the model of Wilson.[4] Consider the interaction of classical spins with n components in a d-dimensional lattice. Here, "spin" is nothing but the order parameter in theory of continuous phase transition. In the calculation of statistical partition function, the summation goes over all states $\{S\}$ from nearest-neighbor lattice points

$$Z = \sum_{\{S\}} \exp\left(-\frac{E}{kT}\right) = \sum_{\{S\}} \exp\left(\frac{K}{2}\sum_{m,i} S_m^\alpha S_{m+i}^\alpha\right), \quad K = \frac{J}{kT}, \tag{17.4}$$

where J denotes the exchange integral, m represents the position vector of the cell, i represents the relative position vector of the nearest neighbors (in what follows we do not use boldface for vectors). The repeated spin superscript α performs summation from 1 to n. By introducing a convergent factor $-\frac{b}{2}S_m^\alpha S_m^\alpha$ and treating S_m^α as a variable taking continuous values, the summation in Eq. (17.4) may be replaced by integration. Furthermore, changing the lattice point function S_m to a function of continuous medium $S(x)$ and absorbing K by rescaling S, we get

$$Z = \left(\int \mathcal{D}S\right) \exp\left[-\frac{1}{2}\int_x ((\nabla S)^2 + r_0 S^2)\right], \quad r_0 = \frac{b}{K} - 2d. \tag{17.5}$$

From now on, functional integration over infinite function system as well as ordinary d-dimensional integration will be represented by shorthand notations:

$$\left(\int \mathcal{D}S\right) \equiv \lim_{m\to\infty}\left(\prod_m \int_{-\infty}^{\infty} dS_m\right), \quad \int_x \equiv \int d^d x, \int_q \equiv \int \frac{d^d q}{(2\pi)^d}. \tag{17.6}$$

The gradient term $(\nabla S)^2$ comes from nearest neighbor interaction. Performing Fourier transformation $S(x) = \int_q e^{iqx}\sigma_q$ and neglecting the common factor which appears during function substitution in the functional integral, we have

$$Z = \left(\int \mathcal{D}\sigma\right) \exp(H_0) \tag{17.7}$$

and

$$H_0 = -\frac{1}{2}\int_q (q^2 + r_0)\sigma_q^\alpha \sigma_{-q}^\alpha. \tag{17.8}$$

H_0 is the Hamilton function of a free classical field. It is the simple Gaussian model in statistics. All statistical averages are Gaussian averages. For example, spin correlation function is nothing but the propagator of the classical field:

$$\langle \sigma_1^\alpha \sigma_2^\beta \rangle = G_0(q_1, r_0)\delta(1+2)\delta_{\alpha\beta} = \frac{\delta(1+2)}{q_1^2 + r_0}\delta_{\alpha\beta}, \tag{17.9}$$

where 1, 2 are shorthand for the momenta q_1, q_2. We know the temperature dependence near critical point from Eq. (17.3):

$$\chi_\gamma \sim G_0(0, r_0) = \frac{1}{r_0}, \quad r_0 \sim (T - T_c)^\gamma. \tag{17.10}$$

In the Gaussian model, all high even-order spin correlation functions decompose into combinations of second-order correlation functions, and odd-order correlation functions vanish. This decomposition corresponds to Wick theorem in field theory.

Wilson observed that critical exponents of Gaussian model coincide with that of mean-field theory. Spin distribution function in Gaussian model has a maximum at $S = 0$, far departing from general statistical models. However, if one inserts a fourth-order term into the convergence factor of Eq. (17.5), one may go beyond the mean field theory. Then, the Hamilton function reads:

$$H = H_0 + H_I = H_0 - \frac{u_0}{3}\Delta(\alpha\beta\gamma\delta) \int \sigma_1^\alpha \sigma_2^\beta \sigma_3^\gamma \sigma_4^\delta \delta(1 + 2 + 3 + 4), \tag{17.11}$$

where u_0 corresponds to the coupling constant in field theory with ϕ^4 interaction, $\Delta(\alpha\beta\gamma\delta)$ is a fully symmetric unit tensor:

$$\Delta(\alpha\beta\gamma\delta) = \delta_{\alpha\beta}\delta_{\gamma\delta} + \delta_{\alpha\gamma}\delta_{\beta\delta} + \delta_{\alpha\delta}\delta_{\beta\gamma}. \tag{17.12}$$

Various products and contractions of $\Delta\alpha\beta\gamma\delta$ (i.e., summations over repeated indices) frequently appear in the calculation of high-order terms. We postpone these monotonic yet useful technicalities to Sec. 17.6.2.

The average of any product A of field functions

$$\langle A \rangle = \frac{\langle A \exp(H_I) \rangle_0}{\langle \exp(H_I) \rangle_0}$$

may be decomposed and one can prove a connected-graph expansion theorem. Detailed enumeration of diagram coefficients and calculations of integrals will be given in Secs. 17.6 and 17.7.

In this model, propagators contain temperature, but bare interactions do not depend on temperature. High-order vertexes depend on temperature by way of propagators. Consequently, the momentum-independent part in self-energy diagram leads to a shift of critical point. In order to take into account this point, it is better to include this part of contribution from the self-energy diagram into r_0. This "mass renormalization" process may be realized by way of a cancelation term well-known in field theory, i.e., rewriting Eq. (17.11) to

$$H = -\frac{1}{2}\int (q^2 + r)\sigma_1^\alpha \sigma_{-1}^\alpha - \frac{1}{2}\int (r_0 - r)\sigma_1^\alpha \sigma_{-1}^\alpha - \frac{u_0}{3}\Delta(\alpha\beta\gamma\delta) \int \sigma_1^\alpha \sigma_2^\beta \sigma_3^\gamma \sigma_{-1-2-3}^\delta \tag{17.13}$$

and requiring that the exact propagator $G(q, r)$ satisfies a condition similar to Eq. (17.10) at $q \to 0$:

$$\chi_\gamma \sim G(0, r) = \frac{1}{r} \sim (T - T_c)^{-\gamma}. \tag{17.14}$$

Momemtun-independent part in self-energy diagram cancels out with the second term in Eq. (17.13), the cancelation equation defines r. When calculating a complex diagram containing self-energy part, one must introduce a subtraction term as explained in details in Eq. (17.54) in Sec. 17.6. In this paper, we use the same notation for the exact propagator and for the free propagator $G(q,r) = (q^2+r)^{-1}$ after "mass renormalization", their difference is clear from the context. Skeleton graphs are those composed of exact propagators.

17.4. Skeleton graph analysis of four-point vertex

The four-point vertex $\Gamma^{\alpha\beta\gamma\delta}(1234)$ discussed in this section is four-spin correlation function excluding disconnected diagrams and amputating single-particle external lines. Usually the r dependence is not written out explicitly. Since the σ's are commutative classical quantities, Γ is fully symmetric with respect to both spin indices and momentum. We bind together the spin superscript and momentum, i.e., binding $1 \leftrightarrow \alpha$, $2 \leftrightarrow \beta$, etc. There are 4! permutations. In fact, they should be symmetrized separately, leading to $(4!)^2$ permutations. Later on, when discussing the forward scattering amplitude at $r = 0$, we have to symmetrize the spin superscripts separately.

The 4! diagrams obtained from a four-point vertex by permutating external lines are divided into three groups: (12; 34), (13; 24) and (14; 23), corresponding to the s, t and u channels in field theory. Some diagrams are reducible in one channel, i.e., becoming disconnected parts by cutting two internal lines; some are entirely irreducible. Diagrams reducible in one channel are irreducible in the other two channels. A complex diagram may be reducible in one channel, its subdiagrams may be reducible in other channels. If all subdiagrams are reducible in one or another channel, then the complex diagram is called a parquet diagram. The total of diagrams, reducible in one channel, is denoted as, e.g., $\Gamma^{\alpha\beta;\gamma\delta}(12;34)$. The total of diagrams, irreducible in one channel, is denoted as $I^{\alpha\beta;\gamma\delta}(12;34)$, represented by squares in Fig. 17.1. $\Gamma^{\alpha\beta;\gamma\delta}$ may be obtained from $I^{\alpha\beta;\gamma\delta}$ by iteration, as shown in Fig. 17.1.

If taking out the leftmost square and summing over the remaining I, we get a fully symmetrized total vertex $\Gamma^{\alpha\beta\gamma\delta}(-5-634)$:

$$\Gamma^{\alpha\beta;\gamma\delta}(12;34) = -36\int I^{\alpha\beta;\mu\nu}(12;56)G(5)G(6)\Gamma^{\mu\nu\gamma\delta}(-5-634). \qquad (17.15)$$

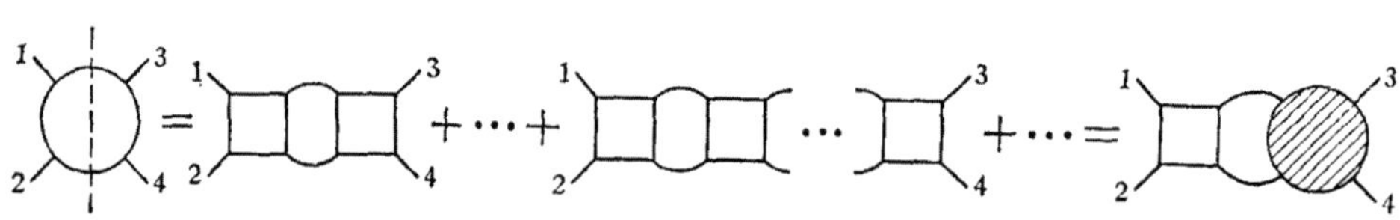

Fig. 17.1.

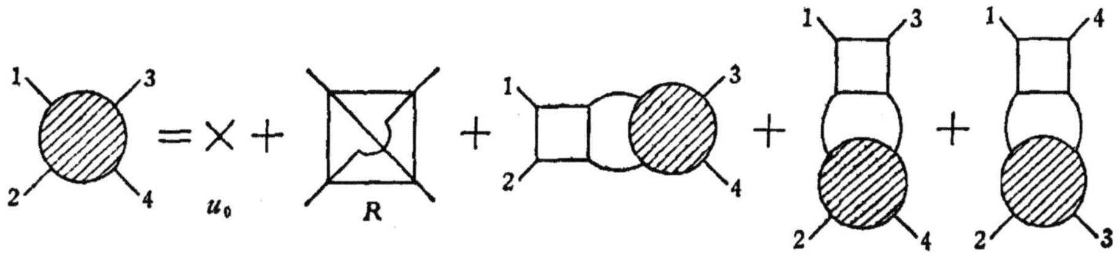

Fig. 17.2.

δ-functions ensuring momentum conservation are not written out explicitly in the above formula. The calculation of the numerical coefficient is given in Sec. 17.6. The total vertex is expressed via the sum R of vertexes reducible in some channel and diagrams irreducible in all three channels, the lowest order diagram of the latter is the 4-point diagram in Fig. 17.3(h):

$$\begin{aligned}\Gamma^{\alpha\beta\gamma\delta}(1234) &= \frac{1}{3}(u_0+R)\Delta(\alpha\beta\gamma\delta) + \frac{1}{3}\Gamma^{\alpha\beta;\gamma\delta}(12;34)\\ &\quad + \frac{1}{3}\Gamma^{\alpha\gamma;\beta\delta}(13;24) + \frac{1}{3}\Gamma^{\alpha\delta;\beta\gamma}(14;23).\end{aligned} \tag{17.16}$$

Inserting Eq. (17.15) into the above equation, we get Fig. 17.2.

$$\begin{aligned}\Gamma^{\alpha\beta\gamma\delta}(1234) &= \frac{1}{3}(u_0+R)\Delta(\alpha\beta\gamma\delta) - 12\int I^{\alpha\beta;\mu\nu}(12;56)G(5)G(6)\Gamma^{\mu\nu\gamma\delta}(-5-634)\\ &\quad - 12\int I^{\alpha\gamma;\mu\nu}(13;56)G(5)G(6)\Gamma^{\mu\nu\beta\delta}(-5-624)\\ &\quad - 12\int I^{\alpha\delta;\mu\nu}(14;56)G(5)G(6)\Gamma^{\mu\nu\beta\gamma}(-5-623).\end{aligned} \tag{17.17}$$

If separating out a part, irreducible in one channel:

$$\begin{aligned}I^{\alpha\beta;\gamma\delta}(12;34) &= \frac{1}{3}(u_0+R)\Delta(\alpha\beta\gamma\delta) - 12\int I^{\alpha\gamma;\mu\nu}(13;56)G(5)G(6)\Gamma^{\mu\nu\beta\delta}(-5-624)\\ &\quad - 12\int I^{\alpha\beta;\mu\nu}(14;56)G(5)G(6)\Gamma^{\mu\nu\beta\delta}(-5-623),\end{aligned} \tag{17.18}$$

then Eq. (17.17) may be written as an ordinary Bethe–Salpeter equation:

$$\Gamma^{\alpha\beta\gamma\delta}(1234) = I^{\alpha\beta;\gamma\delta}(12;34) - 12\int I^{\alpha\beta;\mu\nu}(12;56)G(5)G(6)\Gamma^{\mu\nu\gamma\delta}(-5-634). \tag{17.19}$$

One can write down similar equations using I irreducible in the other two channels.

In the theory of critical phenomena, we have to discuss two limits of the total four-point vertex:

(1) The long wave length limit $P = 0$ which becomes important due to divergence of the correlation length. We need the vertex U_R near the critical point $r \to 0$:

$$\Gamma^{\alpha\beta\gamma\delta}(0000; r) \equiv \frac{1}{3} U_R \Delta(\alpha\beta\gamma\delta). \tag{17.20}$$

(2) The forward scattering amplitude $\Gamma(P)$ at the critical point $r = 0$ in the $P \to 0$ limit:

$$\Gamma^{\alpha\beta\gamma\delta}\left(\frac{P}{2}\frac{P}{2}; -\frac{P}{2} - \frac{P}{2}\right) = \frac{1}{3}\Gamma(P)\Delta(\alpha\beta\gamma\delta) \tag{17.21}$$

By a skeleton graph analysis we can obtain $\frac{\partial U_R}{\partial r}$ and $\frac{\partial \Gamma(P)}{\partial P}$ from the above definitions. Further comparison with the scaling relation[6]

$$\frac{\partial U_R}{\partial r} / U_R = \frac{4 - d - 2\eta}{2 - \eta} \frac{1}{r} \tag{17.22}$$

and

$$\frac{\partial \Gamma(P)}{\partial P} / \Gamma(P) = (4 - d - 2\eta)/P, \tag{17.23}$$

one determines entirely U_R and $\Gamma(P)$ contained in the expression. A pivotal point here consists in that u_0 is not required to be a small quantity beforehand, but U_R and $\Gamma(P)$ are indeed small when the physical system approaches four dimension (small ϵ) or the internal degree of freedom n is great.

We first consider the skeleton graph expansion of U_R.

For the sake of clarity, we first keep terms up to U_R^3. Putting external momentum to zero and performing "graphical differentiation": i.e., first take derivative of propagators in between the irreducible parts, due to the arbitrariness of their positions the infinite series on both sides again sum up to Γ; then take derivatives of I at the two ends and in the middle. In this way, we obtain

$$\begin{aligned}
&\frac{\partial}{\partial r}\Gamma^{\alpha\beta\gamma\delta}(0) \\
&= -36 \int \Gamma^{\alpha\beta\mu\nu}(00q-q)\frac{\partial}{\partial r}\left(G^2(q)\right)\Gamma^{\mu\nu\gamma\delta}(-qq00) \\
&\quad - 36 \int \Gamma^{\alpha\beta\mu\nu}(00q-q)G^2(q)\frac{\partial}{\partial r}I^{\mu\nu\gamma\delta}(-qq;00) \\
&\quad - 36 \int \frac{\partial}{\partial r}\left(I^{\alpha\beta;\mu\nu}(00;q-q)\right)G^2(q)\Gamma^{\mu\nu\gamma\delta}(-qq00) \\
&\quad + 432 \int \Gamma^{\alpha\beta\gamma\delta}(00q-q)G^2(q) \\
&\quad \times \frac{\partial}{\partial r}\left(I^{\mu\nu;\rho\tau}(-qq;k-k)\right)G^2(k)\Gamma^{\rho\tau\gamma\delta}(-kk00).
\end{aligned} \tag{17.24}$$

It is clear from Eq. (17.17) that, up to terms of order U_R^2, the difference between $\Gamma(00q-q)$ and $\Gamma(0)$ consists only in the momentum transfer between vertexes. Therefore, we have

$$\Gamma^{\alpha\beta\mu\nu}(00q-q) \approx \frac{1}{3}U_R\Delta(\alpha\beta\mu\nu) \\ -\frac{4}{3}U_R^2(I(q,r)-I)\left(\Delta_2(\alpha\mu;\beta\nu)+\Delta_2(\alpha\nu;\beta\mu)\right), \quad (17.25)$$

where we have made use of the tensor contraction formulas and integration symbol given in Secs. 17.6 and 17.7. As long as the required order is reached, one may replace a vertex by U_R and take it off the integral; the contraction of spin indices in the vertexes leads to the corresponding symmetric tensors. In order to calculate higher order expansions, one iterates until the corresponding skeleton graph appears and then replaces it by U_R and then contract the spin indices. All the calculations below are done in this way. According to Eq. (17.18), the derivative of I is

$$\frac{\partial}{\partial r}I^{\alpha\beta;\mu\nu}(00;q-q) = -\frac{4}{3}U_R^2\frac{\partial}{\partial r}(I(q,r))\left(\Delta_2(\alpha\mu;\beta\nu)+\Delta_2(\alpha\nu;\beta\mu)\right). \quad (17.26)$$

Inserting the above two formulas into Eq. (17.24) and neglecting the last term of order U_R^4 in Eq. (17.24), we obtain

$$\frac{\partial}{\partial r}\Gamma^{\alpha\beta;\gamma\delta}(0) = -4\Delta_2(\alpha\beta;\gamma\delta)U_R^2 I' + 64\tau_3(\alpha\beta;\gamma\delta)U_R^3(I_b'-II').$$

Combining similar formulas in three channels is equivalent to carrying out symmetrization. The result reads

$$\frac{U_R'}{U_R} = -4(n+8)U_R I' + 64(5n+12)U_R^2(I_b'-II'). \quad (17.27)$$

Calculating the integrals at dimension $d=4-\epsilon$ and comparing with the scaling relation Eq. (17.22), we obtain the renormalized vertex:

$$U_R = \frac{2\pi^2}{n+8}\epsilon\left[1+\frac{\epsilon}{2}\left(\frac{6(3n+14)}{(n+8)^2}+C-\ln\frac{4\pi}{r}\right)\right]. \quad (17.28)$$

In calculating the above formula we have used the critical exponent η up to the order ϵ^2, easily obtainable from $\Gamma(P)$ to the order ϵ. C in Eq. (17.28) is the Euler constant. If we do not assume the smallness of ϵ, U_R is still inversely proportional to $(n+8)$; therefore, U_R remains a first-order small quantity as long as the internal freedom n is great.

In order to get the next order terms, we have to keep U_R^4 terms and take into account the contribution of irreducible skeleton graph in Fig. 17.3(h).

The result reads

$$U_R' = -4(n+8)U_R^2 I' + 64(5n+22)U_R^3 \int \left[\frac{\partial}{\partial r} G^2(q)(I(q,r)-I) + G_2(q)\frac{\partial}{\partial r} I(q,r)\right]$$
$$- 256(n^2+20n+60)U_R^4$$
$$\times \int \left[\frac{\partial}{\partial r} G_2(q)(I(q,r)-I)^2 + 2G(q)^2(I(q,r)-I)\frac{\partial}{\partial r} I(q,r)\right]$$
$$- 256(3n^2+22n+56)U_R^4$$
$$\times \int \left[\frac{\partial}{\partial r} G^2(q)(I(q,r)-I)^2 + 2G^2(q)(I(q,r)-I)\frac{\partial}{\partial r} I(q,r)\right]$$
$$- 1024(n^2+20n+60)U_R^4 \int \left\{\frac{\partial}{\partial r} G^2(q)(I(q,r)-I)G(k)(G(k+q)-G(k))\right.$$
$$\left. + G^2(q)\left[\frac{\partial}{\partial r}(G(k)G(k+q))(I(q,r)-I) + G(k)G(k+q)\frac{\partial}{\partial r} I(k,r)\right]\right\}$$
$$- 128(3n^2+22n+56)U_R^4 \int \left[2\frac{\partial}{\partial r} G^2(q)(I(k+g,r)-I(k,r))G^2(k)\right.$$
$$\left. + G^2(q)G^2(k)\frac{\partial}{\partial r} I(k+q,r)\right] - 1536(5n+22)U_R^4 I_h'. \quad (17.29)$$

The terms in the above formula correspond to subgraphs (a), (b), (c_1), (c_2), (f), (g) and (h) in Fig. 17.3. Comparing the graph and integral term by term, we can summarize the rule of differentiation of skeleton graphs as follows: take derivatives of the double propagators one by one and from the propagators that are not being differentiated subtract a term which equals the zero-momentum term of that being differentiated. For example, the second term corresponds to subgraph Fig. 17.3(b), when differentiating the first pair of reducible lines replace the right-side ring by $I(q,r) - I$.

Merging and putting in order the integrals in Eq. (17.29), we introduce the following symbols: $A = rI'$, $B = r(I_b' - II')$, $E = r(I_e' - 2II_b' + I^2 I')$, $F = r(I_f' - (II_b)' + I^2 I')$. $G = r(I_g' - 2I'I_b)$ and $H = rI_h'$. Integrals in the parentheses here may be divergent logarithmically. However, each symbol is a combination independent of the momentum cut-off. Therefore, U_R and all quantities derived from it are explicitly independent of the cut-off. Using these symbols, we may write down the next order expansion that includes Eq. (17.27) as:

$$\begin{aligned}\frac{rU_R'}{U_r} &= \frac{(4-d-2\eta)}{2-\eta}\\ &= -4(n+8)U_R A + 64(5n+22)U_R^2 B\\ &\quad - 512(2n^2+21n+58)U_R^3 E - 1024(n^2+20n+60)U_R^3 E\\ &\quad - 128(3n^2+22n+56)U_R^3 G - 1536(5n+22)U_R^3 H.\end{aligned} \quad (17.30)$$

Now we consider the skeleton graph expansion of $\Gamma(P)$.

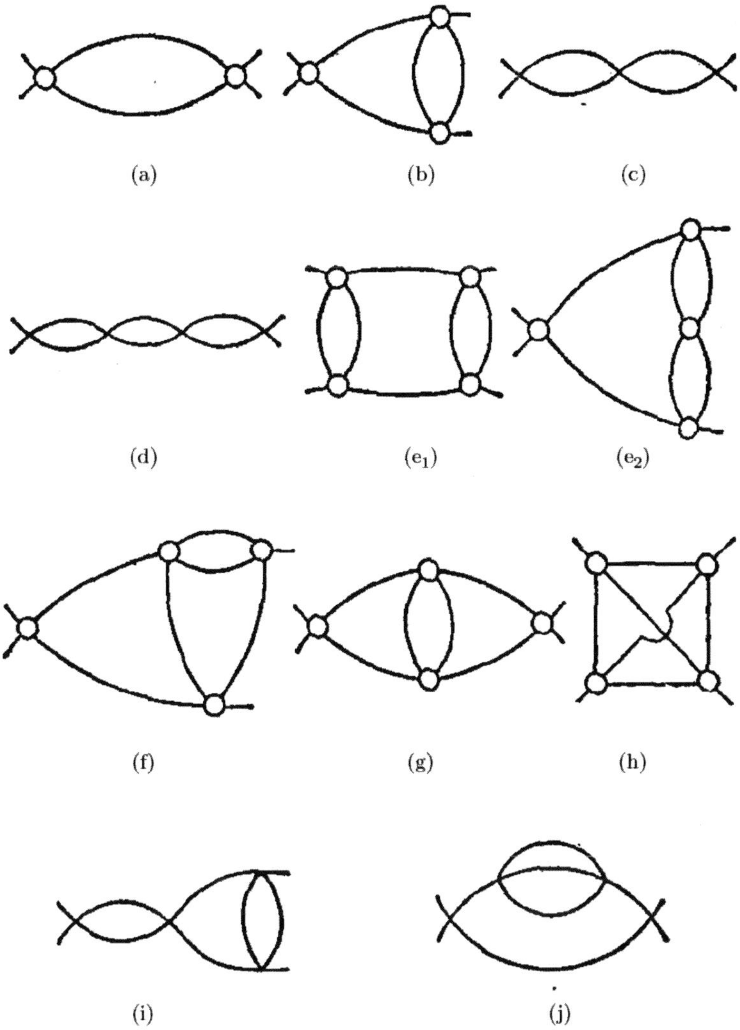

Fig. 17.3.

The analysis of $\Gamma(P)$ is similar to that for U_R. However, one must keep in mind that in terms of momentum only one of the three channels describes forward scattering and symmetrization must be carried out with respect to spin indices. Therefore, in the definition Eq. (17.21) momentum and spin are not permutated together: there is only one channel for momentum, but for spin indices there are three combinations, i.e.,

$$\Gamma^{\alpha\beta\gamma\delta}\left(\frac{P}{2}\frac{P}{2};-\frac{P}{2}-\frac{P}{2}\right)=\frac{1}{3}(u_0+R)\Delta(\alpha\beta\gamma\delta)+\frac{1}{3}\left[\Gamma^{\alpha\beta;\gamma\delta}\left(\frac{P}{2}\frac{P}{2};-\frac{P}{2}-\frac{P}{2}\right)\right.$$
$$\left.+\,\Gamma^{\alpha\gamma;\beta\delta}\left(\frac{P}{2}\frac{P}{2};-\frac{P}{2}-\frac{P}{2}\right)+\Gamma^{\alpha\delta;\beta\gamma}\left(\frac{P}{2}\frac{P}{2};-\frac{P}{2}-\frac{P}{2}\right)\right]. \tag{17.31}$$

The rule of graphical differentiation of P is similar to that for r. Then, using expansions similar to Eqs. (17.25) and (17.26) and comparing with the scaling relation Eq. (17.23), we get

$$\frac{P\Gamma'(P)}{\Gamma(P)} = 4 - d - 2\eta = -4(n+8)\Gamma(P)A_P + 64(5n+22)\Gamma^2(P)B_P, \qquad (17.32)$$

where $A_P = PI'(P)$, $B_P = P(I_b' - I(P)I'(P))$. Carrying out ϵ expansion to ϵ^2, we have

$$\Gamma(P) = \frac{2\pi^2\epsilon}{n+8}\left[1 + \frac{\epsilon}{2}\left(\frac{6(3n+14)}{(n+8)^2} + C - 2 - \ln\frac{4\pi}{P^2}\right)\right]. \qquad (17.33)$$

The last formula should be compared with Eq. (17.28).

So far in the derivation of Eqs. (17.30) and (17.32) no assumption has been made concerning the spatial dimension d and the physical nature of smallness of U_R and $\Gamma(P)$, and no property of the bare coupling constant u_0 has been used. If taking u_0 as a small quantity we can directly write down perturbation expansions for U_R and $\Gamma(P)$, then insert u_0 back as expansion of U_R (or $\Gamma(P)$), the results would be identical to Eqs. (17.30) and (17.32). Due to the logical weakness of the necessity to require the smallness of u_0 and the lack of a natural way to get integral combinations that are independent of momentum cut-off, we prefer skeleton graph expansions to the "perturbation expansions" mentioned here.

17.5. Calculation of critical exponents

Using the renormalized vertex U_R at $p = 0$, one may calculate the critical exponents γ and α, from the forward scattering amplitude $\Gamma(P)$ at $r = 0$ one may get the exponent η.

First, consider the calculation of γ from the definition Eq. (17.14) of r after the "mass renormalization"

$$r = G^{-1}(0,r) = r_0 - \Sigma(0,r) \sim (r_0 - r_{0c})^\gamma \sim (T - T_c)^\gamma,$$

it follows that if we introduce a "three-point vertex":

$$\Lambda_0 \equiv \Lambda_0(0,r) \equiv \frac{dr}{dr_0} \sim \gamma(r_0 - r_{0c})^{\gamma-1} = \gamma r^{\frac{\gamma-1}{\gamma}},$$

then on the one hand we have

$$\frac{\Lambda_0'}{\Lambda_0} = \frac{1}{r}\left(1 - \frac{1}{\gamma}\right),$$

on the other hand we get

$$\Lambda_g = 1 - \Lambda_0\frac{\partial}{\partial r}\Sigma(0,r),$$

where $\Sigma(0,r)$ is the momentum-independent contribution from the self-energy part made of exact propagators. Differentiating the self-energy part, or equivalently,

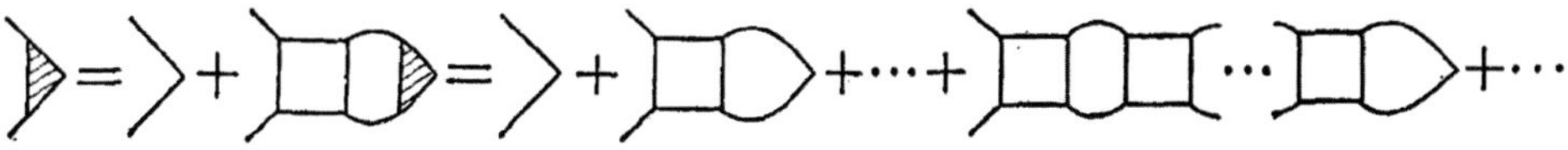

Fig. 17.4.

replacing every propagator G by $-G^2$ one after another, i.e., attach to each line a "photon" line with zero-momentum exchange. The three-point vertex comprises of the sum of all these diagrams and satisfies the Bethe–Salpeter equation (see Fig. 17.4):

$$\Lambda_0^{\alpha\beta}(p,r) = \delta_{\alpha\beta} - 12\int I^{\alpha\beta\mu\nu}(p-p;k-k)G^2(k)\Lambda_0^{\mu\nu}(k,r), \tag{17.34}$$

where $\Lambda_0^{\alpha\beta} = \Lambda_0(p,r)\delta_{\alpha\beta}$. Performing graphical differentiation in the same way as we did in the previous section:

$$\begin{aligned}\frac{\partial}{\partial r}\Lambda_0^{\alpha\beta}(0,r) = &-12\int \frac{\partial}{\partial r}(I^{\alpha\beta;\mu\nu}(00;k-k))G^2(k)\Lambda_0^{\mu\nu}(k,r)\\ &-12\int \Gamma^{\alpha\beta\mu\nu}(00;k-k)\frac{\partial}{\partial r}G^2(k)\Lambda_0^{\mu\nu}(k,r)\\ &+144\int \Gamma^{\alpha\beta\mu\nu}(00;k-k)G^2(k)\frac{\partial}{\partial r}(I^{\mu\nu;\gamma\delta}(k-k;q-q)G^2(q)\Lambda_0^{\gamma\delta}(q,r)).\end{aligned}$$

Inserting Eqs. (17.25), (17.26), etc, into the above formula, iterating $\Lambda_0(k,r)$ once in Eq. (17.34), and neglecting Λ_0 and k dependence in higher-order terms, we combine integrals and contract indices to get

$$\begin{aligned}\frac{r\Lambda_0'}{\Lambda_0} = 1 - \frac{1}{\gamma} = &-4(n+2)U_RA + 96(n+2)U_R^2B - 128(n+2)(n+8)U_R^3E\\ &- 512(n+2)(n+8)U_R^3F - 384(n+2)^2U_R^3G.\end{aligned} \tag{17.35}$$

This formula maybe obtained by using the purely "perturbation theory" method mentioned at the end of the previous section. Multiplying Eq. (17.30) by $(n+2)/(n+8)$ and subtracting Eq. (17.35), and inserting the U_R obtained by solving Eq. (17.30), we get

$$\begin{aligned}1 - \frac{1}{\gamma} = &\frac{n+2}{n+8}b + \frac{2(n+2)(7n+20)}{(n+8)^3}\frac{Bb^2}{A^2} - \frac{2(n+2)b^3}{(n+8)^4A^3}\Big[(7n^2+68n+168)E\\ &+4(n^2+24n+56)F - 8(n-1)G + 12(5n+22)H\\ &-\frac{8(5n+22)(7n+20)}{n+8}\frac{B^2}{A}\Big],\end{aligned} \tag{17.36}$$

where $b = (4-d-2\eta)/(2-\eta)$. So far we have only used the fact that U_R is a small quantity without digging into its origin. Therefore, Eq. (17.36) has a wider application than the ϵ expansion. For the ϵ expansion, just insert the integrals and using the expression (17.43) for η, we get

$$1-\frac{1}{\gamma}=\frac{n+2}{2(n+8)}\epsilon+\frac{3(n+2)(n+3)}{(n+8)^3}\epsilon^2$$
$$+(n+2)\left[\frac{55n^2+268n+424}{2(n+8)^5}-\frac{18\zeta(3)(5n+22)}{(n+8)^4}\right]\epsilon^3. \quad (17.37)$$

$\zeta(3)$ in the above formula is the Riemann ζ function. It is worth mentioning that Eq. (17.37) is an expansion both in ϵ and in $\frac{1}{n}$. In Eq. (17.36) the coefficients of b^2 and b^3 terms do not contain zeroth order of n. After calculating the integrals explicitly, the $\frac{\epsilon^3}{n}$ terms in Eq. (17.37) cancel out. Whether the cancelation of the corresponding powers of $\frac{1}{n}$ continues in high orders requires further study. When the above results were obtained in 1973 there appeared ϵ^3 terms of γ calculated by other methods[13] without revealing the details.

Using the three-point vertex, one can define the polarized ring which is proportional to specific heat[14]:

$$c_v\sim(r_0-r_{0c})^{-\alpha}\sim r^{-\frac{\alpha}{\gamma}}\sim\Pi(r)=n\int G^2(p)\Lambda_0(p,r). \quad (17.38)$$

The right-hand side of the above formula may be calculated from skeleton graph expansion

$$\Lambda_0(p,r)=\Lambda_0(0,r)+96(n+2)U_R^2\int G^2(q)(I(p+q,r)-I(q,r))\Lambda_0(q,r).$$

In fact, this formula can easily be derived from Eq. (17.34). Then compare the result with the consequence of the left-hand side

$$\frac{\Pi''}{\Pi'}=-\frac{1}{r}\left(1+\frac{\alpha}{\gamma}\right), \quad (17.39)$$

we obtain

$$1+\frac{\alpha}{\gamma}=-\frac{rI''}{I'}+8(n+2)U_RA-96(n+2)U_R^2r\left[\frac{I_g''}{I'}-\frac{I_g'I''}{(I')^2}-2II'\right]. \quad (17.40)$$

When calculating the first term one should pay attention to the contribution of Fig. 17.3(a) and (j). In the ϵ expansion there appears two momentum-independent combination of integrals. The final result reads

$$\frac{\alpha}{\gamma}=\frac{4-n}{2(n+8)}\epsilon+\frac{n+2}{(n+8)^2}\left[\frac{5}{2}-\frac{3(3n+14)}{n+8}\right]\epsilon^2. \quad (17.41)$$

Similar to Eq. (17.37), this formula is an expansion both in ϵ and $\dfrac{1}{n}$.

In order to calculate η, let $r=0$ in the Dyson equation $G^{-1}=p^2+r-\Sigma(p,r)$ and compare with Eq. (17.2), we get $p^{2-\eta}=p^2-\Sigma(p,0)$. For four-point interaction the irreducible self-energy Σ may be expressed via the four-point vertex,[12] see

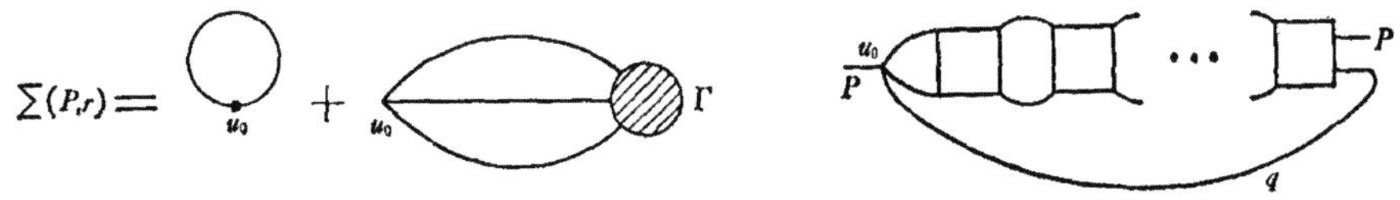

(a) (b)

Fig. 17.5.

Fig. 17.5(a). Writing down the spin indices explicitly and ignoring terms independent of p, the Dyson equation may be rewritten as

$$p^{2-\eta}\delta_{\alpha\beta} = p^2\delta_{\alpha\beta} - 32u_0\Delta(\alpha\gamma\delta\rho)\int G(p-k)G(k-q)G(q)\Gamma^{\beta\gamma\delta\rho}(p-k,k-q,q,-p).$$

Now we perform graphical differentiation in the same way as in the previous section. Since the rightmost term is not I but the bare vertex u_0, see Fig. 17.5(b), there is no term corresponding to the third term in Eq. (17.24). Furthermore, the difference between I and u_0 should be deducted from the first term. From Eq. (17.18) we have

$$\begin{aligned} I^{\alpha\gamma;\delta\rho}(p,-q;k-p,q-k) = \frac{u_0}{3}\Delta(\alpha\gamma\delta\rho) - \frac{4}{3}\Gamma^2(p-q)[I(k)\Delta_2(\alpha\delta;\gamma\rho) \\ +I(p+q-k)\Delta_2(\alpha\rho;\gamma\delta)]. \end{aligned}$$

Unlike the $r \neq 0$ and $p = 0$ limit, $\Gamma(p-q)$ must be kept within the integral. Putting in order and performing the contraction, we obtain

$$\begin{aligned} (2-\eta)p^{1-\eta} &= 2p - 32(n+2)\int G(q)\frac{\partial}{\partial p}I(p-q)\Gamma^2(p-q) \\ &\quad + \frac{128(n+2)(n+8)}{3}\int G(q)\Gamma^3(p-q)\frac{\partial}{\partial p}[G(p-q)G(k-q) \\ &\quad \times (I(k)+I(p+q-k)) - 2I^2(p-q)]. \end{aligned} \tag{17.42}$$

If the first term in the expansion of $\Gamma(p)$ does not depend on p (this is the case in the ϵ expansion, see Eq. (17.33)), the differentiation with respect to p in the last term may be taken out of the integral. Performing variable substitution, the two terms in the square brackets cancel out and Eq. (17.42) looks formally as an expansion to ϵ^2. Inserting the expansion Eq. (17.33) and performing the integration, then compare with

$$\frac{[G^{-1}(p)]''}{[G^{-1}(p)]'} = \frac{1-\eta}{p},$$

which follows from Eq. (17.2), we get

$$\eta = \frac{n+2}{2(n+8)^2}\epsilon^2 + \frac{n+2}{2(n+8)^2}\left[\frac{6(3n+14)}{(n+8)^2} - \frac{1}{4}\right]\epsilon^3. \tag{17.43}$$

In the process of calculating η, we did not define a "vector three-point vertex" $\Lambda_i = \frac{\partial}{\partial p}G^{-1}$ analogous to Λ_0, because it does not satisfy a Bethe–Salpeter equation similar to Eq. (17.34).

17.6. The coefficients of diagrams

The calculation of diagram coefficients includes $n = 1$ and $n > 1$ cases, the latter may be obtained from the former. For the parquet diagrams which are the basis of skeleton graphs we develop a method to calculate the coefficients of complex graphs.

17.6.1. *Coefficients of $n = 1$ diagrams*

The coefficient $C(1)$ of a kth-order diagram at $n = 1$ is the total number of diagrams with k vertexes and the corresponding internal and external lines. Its value is

$$C(1) = \frac{N_1 N_2}{k!}, \tag{17.44}$$

where combinatorial factor N_1 is the number of diagrams at a fixed labeling of vertexes, topological factor N_2 is the number of nonequivalent diagrams obtained by permutation of the labels. In order to calculate N_1 we introduce a matrix representation of diagrams, suggested by our colleague Dr. Fuque Pu. Each kth-order diagram corresponds to a symmetric nonnegative real matrix, whose element a_{ij} denotes the number of internal lines between vertexes i and j, the diagonal element a_{ii} is twice the number of loops at the vertex i. A connected diagram corresponds to an irreducible matrix. Suppose vertex i comes from a W_i-point vertex, then the number of external lines is given by

$$f_i = W_i - \sum_{j=1}^{k} a_{ij} \geq 0,$$

Consequently,

$$N_1 = \frac{1}{a_{11}!a_{12}!\cdots a_{1k}!a_{22}!\cdots a_{2k}!\cdots a_{kk}!} \prod_{j=1}^{k} \frac{W_i!(a_{ii}-1)!!}{f_i!}. \tag{17.45}$$

The topological factor is obtained from a case by case analysis and no general algorithm has been formulated. For example, $N_1 = 72$, $N_2 = 1$, $C_a(1) = 36$ for Fig. 17.3(a); $N_1 = 3456$, $N_2 = 3$, $C_b(1) = 1728$ for Fig. 17.3(b); $N_1 = (4!)^4$, $N_2 = 3$, $C_h(1) = 41472$ for the nonparquet diagram Fig. 17.3(h).

17.6.2. *Products and contractions of fully and partially symmetric unit tensors*

In Eq. (17.11), we have used the fully symmetric unit tensor $\Delta^{\alpha\beta\gamma\delta}$ defined in Eq. (17.12). The products and contractions (i.e., summation over repeated indices) of $\Delta(\alpha\beta\gamma\delta)$ appear frequently in the $n > 1$ high-order diagrams. The symmetry property of any four-point diagram with fixed spin indices on external lines may be expressed as $A\delta_{\alpha\beta}\delta_{\gamma\delta} + B\delta_{\alpha\gamma}\delta_{\beta\delta} + C\delta_{\alpha\delta}\delta_{\beta\gamma}$, corresponding to the three channels mentioned before. There are in total three possibilities of indices permutation in

Table 17.1 Symmetric tensors and their contractions.

Tensor	In Fig. 17.3	A	B	C	$P(n)$
$\Delta \equiv \Delta_1(\alpha\beta\gamma\delta)$	(h)	1	1	1	3
$\tau_2 = \Delta_2(\alpha\beta;\gamma\delta)$	(d), (j)	$n+4$	2	2	$n+8$
$\Delta_3(\alpha\beta;\gamma\delta)$	(c)	$n^2+6n+12$	4	4	$n^2+6n+20$
$\Delta_4(\alpha\beta;\gamma\delta)$	(d)	$n^3+8n^2+24n+32$	8	8	$n^3+8n^2+24n+48$
$\tau_3(\alpha\beta;\gamma\delta)$	(b)	$3n+10$	$n+6$	$n+6$	$5n+22$
$\tau_4(\alpha\beta;\gamma\delta)$	(e_2)	$n^2+10n+24$	$n^2+6n+16$	$n^2+6n+16$	$3n^2+22n+56$
$\rho_4(\alpha,\beta;\gamma,\delta)$	(e_1)	$8n+24$	$n^2+8n+20$	$4n+16$	$n^2+20n+60$
$\omega_4(\alpha\beta;\gamma\delta)$	(g), (i)	$3n^2+18n+32$	$2n+12$	$2n+12$	$3n^2+22n+56$
$\chi_4(\alpha\beta;\gamma\delta)$	(f)	$n^2+12n+28$	$4n+16$	$4n+16$	$n^2+20n+60$

this relation: $A = B = C$, fully symmetric, denoted as $(\alpha\beta\gamma\delta)$; $A \neq B = C$ (or similar cases), invariant with respect to indices permutation within each group or to exchange of two groups, denoted as $(\alpha\beta;\gamma\delta)$; $A \neq B \neq C \neq A$, denoted as $(\alpha,\beta;\gamma,\delta)$, invariant with respect to simultaneous permutation of indices within each group and ordered permutation of the two groups. All vertexes in the previous sections as well as symbols used below are written by using these notations. All products and contractions may easily be calculated. For example,

$$\begin{aligned}
\Delta_k(\alpha\beta;\gamma\delta) &= \Delta(\alpha\beta;\mu\nu)\Delta_{k-1}(\mu\nu;\gamma\delta) = \Delta_1(\alpha\beta;\mu\nu)\Delta_{k-1}(\mu\nu;\gamma\delta) \\
&= A_k\delta_{\alpha\beta}\delta_{\gamma\delta} + 2^{k-1}(\delta_{\alpha\gamma}\delta_{\beta\delta} + \delta_{\alpha\delta}\delta_{\beta\gamma}), \\
\Delta_k(\alpha\gamma;\beta\gamma) &= \left[A_k + 2^{k-1}(n+1)\right]\delta_{\alpha\beta},
\end{aligned}$$

where $A_k = [(n+2)^k - 2^k]/n$, which may be derived by using induction. Skipping various general expressions, we list those used in previous sections in Table 17.1.

$P(n)$ in the last column of Table 17.1 comes from symmetrization. Summing over all possible permutations of indices and dividing by the number of permutations, the result is proportional to $\Delta(\alpha\beta\gamma\delta)/3$ with coefficient $P(n)$ being a polynomial of n. It is easy to see that when $n > 1$ $P(n)$ enters into the coefficients of the corresponding diagrams. For example, Fig. 17.3(a) is obtained from the contraction of two bare vertexes

$$\Delta(\alpha\beta\mu\nu)\Delta(\mu\nu\gamma\delta) = \Delta_2(\alpha\beta;\gamma\delta),$$

and it corresponds to one channel, another channel obtained by permutation appears in Fig. 17.3(b):

$$\Delta(\alpha\beta\mu\nu)\Delta_2(\mu\gamma;\nu\beta) = \tau_3(\alpha\beta;\gamma\delta).$$

In order to obtain the total coefficient of all three channels one carries out symmetrization and produces the corresponding $P(n)$. For a kth-order diagram $P(1) = 3^k$, because each bare vertex at $n > 1$ has three ways of spin propagation, corresponding to the three channels in Eq. (17.12). Every kth order diagram

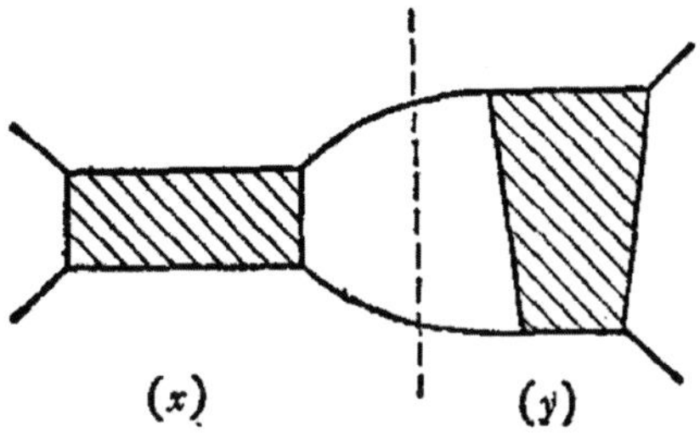

Fig. 17.6.

becomes 3^k diagrams and each spin loop brings about $\delta_{\alpha\alpha} = n$. Therefore, the coefficient of an $n > 1$ diagram is

$$C(n) = C(1)P(n)/3^k. \tag{17.46}$$

Examples: $C_a(n) = 4(n+8)$; $C_b(n) = 64(5n+22)$. The coefficient of nonparquet diagram Fig. 17.3(h) must be calculated directly and it is

$$\Delta(\alpha\mu\nu\rho)\Delta(\beta\rho\omega\sigma)\Delta(\gamma\mu\tau\omega)\Delta(\delta\tau\nu\sigma) = 3(5n+22)\Delta(\alpha\beta\gamma\delta).$$

Therefore, $C_h(n) = 1536(5n+22)$.

17.6.3. *Parquetry rules*

We first consider the case $n = 1$. Suppose that a complex diagram decomposes into two subdiagrams (x) and (y) according to the reducible channels, as given in Fig. 17.6 with symmetry property of the diagram shown symbolically. If the coefficients of the subdiagrams are known to be $C_x(1)$ and $C_y(1)$, then

$$C_{xy} = [D_x C_x(1)] * [D_y C_y(1)] * (C_2^4)^2 T. \tag{17.47}$$

Here, the coefficient D reflects the weight of the subgraph channel. Figure 17.6(x) has a weight $D = 1/3$ in the horizontal channel, and $D = 2/3$ in the vertical channel. Figure 17.6(y) has weight $D = 2/3$ in the vertical channel, but in the horizontal channel due to left–right asymmetry each parquet has weight 1/6. A fully symmetric diagram has weight $D = 1$. If the original diagram has left–right symmetry then $T = 1$, otherwise $T = 2$. The factor $C_2^4 = 6$ is the number of combinations when picking up two external lines in a four-point subgraph. Take, for example, Fig. 17.3(i), it is the result of combining two Fig. 17.3(a), therefore

$$C_i(1) = \left[\frac{1}{3}C_a(1)\right] * \left[\frac{2}{3}C_a(1)\right] \times 72 = 20736.$$

When $n > 1$, we first consider a case with fixed spin indices. The coefficient of the original graph is still given by Eq. (17.47), but when changing $C(1)$ to $C(n)$ in Eq. (17.46), $P(n)$ must be taken as that before the symmetrization. Only the final result is subject to symmetrization. When the weight in the vertical channel is $D = 2/3$, one may write down explicitly the two combinations of the spin indices to be summed over and use $D = 1/3$ instead. Take again Fig. 17.3(i) as example,

$$\left[\frac{1}{3}C_a(1)\frac{\Delta_2(\alpha\beta;\mu\nu)}{9}\right]\left[\frac{1}{3}C_a(1)\frac{\Delta_2(\mu\gamma;\nu\delta)+\Delta_2(\mu\delta;\nu\gamma)}{9}\right]\times 72 = 256\omega_4(\alpha\beta;\gamma\delta),$$

yielding $C_i(n) = 256(3n^2 + 22n + 56)$ after symmetrization. Since the definition of vertex part includes the symmetric tensor and the factor $1/3$, calculation of the coefficients of skeleton graphs reduces to that of the perturbation diagram coefficient $C(1)$. For example, the coefficients in Eq. (17.24) are $C_a(1) = 36$ and $C_c(1) = 432$.

17.7. Some integrals

We list the integrals used in this work. These integrals are calculated in a d-dimensional spherical coordinate system, expanded to the required power of ϵ. In diverging integrals we introduce a momentum cut-off Λ and retain the nonvanishing terms at $\Lambda \to 0$. The subscripts of the following integrals are the same as the labels in the corresponding diagrams. Usually the derivatives of some integrals are easier to calculate than the integrals themselves. In the following formulas, we have $L = \Lambda/\sqrt{r}$, the Euler constant $C = 0.577216$, the Riemann ζ function $\zeta(3) = 1.202057$, and an integral

$$C_1 = \int_1^\infty \frac{\ln x}{x^2 - x + 1}\frac{dx}{x+1} = 1.171954.$$

$$\begin{aligned} I(p,r) &\equiv I_a(p,r) \\ &\equiv \int G(q)G(p+q) = \frac{1}{(4\pi)^2}\left[1 + 2\ln L + \frac{\sqrt{p^2+4r}}{p}\ln\frac{\sqrt{p^2+4r}-p}{\sqrt{p^2+4r}+p}\right], \end{aligned}$$

$$\begin{aligned} I &\equiv I_a(0,r) \\ &= \frac{1}{(4\pi)^2}\left\{2\ln L - 1 + \frac{\epsilon}{2}\left[(2\ln L - 1)\left(1 - C + \ln\frac{4\pi}{r}\right) + \frac{\pi^2}{6} - 2(\ln L)^2\right]\right\}, \end{aligned} \tag{17.48}$$

$$\begin{aligned} I(p) \equiv I_a(p,0) = \frac{1}{(4\pi)^2}&\left\{(2\ln\frac{\Lambda}{p} + 1)[1 + \frac{\epsilon}{2}(\ln 4\pi - C)]\right. \\ &\left. + \epsilon\left[\frac{3}{2} - 2\ln p + \ln\Lambda + (\ln p)^2 - (\ln\Lambda)^2\right]\right\}, \end{aligned}$$

$$I_b(p,r) \equiv \int_q G\left(q - \frac{p}{2}\right)G\left(q + \frac{p}{2}\right)I(q,r),$$

$$I_b \equiv I_b(0,r) = -\frac{1}{(4\pi)^4}[1 + C_1 - 2(\ln L)^2], \tag{17.49}$$

$$I_b' \equiv \frac{d}{dr}I_b(0,r)$$
$$= \frac{1}{(4\pi)^4 r}\left\{-2\ln L + \epsilon\left[C_1 - \frac{\pi^2}{12} + 2\left(C - \ln\frac{4\pi}{r} - \frac{1}{2}\right)\ln L + (\ln L)^2\right]\right\},$$

$$I_e' = \frac{1}{(4\pi)^6 r}[1 + 2C_1 - 4(\ln L)^2], \tag{17.50}$$

$$I_f' = \frac{1}{(4\pi)^6 r}[C_1 - 2\ln L - 2(\ln L)^2], \tag{17.51}$$

$$I_g' = \frac{1}{(4\pi)^6 r}[2 - 4(\ln L)^2], \tag{17.52}$$

$$I_h' = -\frac{6}{(4\pi)^6 r}\zeta(3), \tag{17.53}$$

$$I_i \equiv \int G^3(k)I(q,r)[G(q+k) - G(q)],$$
$$I_f' = \frac{1}{(4\pi)^6 r}\left(\frac{1}{2} + \ln L\right). \tag{17.54}$$

17.8. Generalized homogeneous functions and the callan–symanzik equation

Consider a general function of, say, three variable $f(x_1, x_2, x_3)$. If under a scale change in all dimensions $x_i \to \lambda x_i$, the function remains the same except for multiplying by a numerical factor λ^n:

$$f(\lambda x_1, \lambda x_2, \lambda x_3) = \lambda^n f(x_1, x_2, x_3), \tag{17.55}$$

then f is a homogeneous function of order n. Differentiating both sides of Eq. (17.55) with respect to λ and letting $\lambda = 1$, we get a partial differential equation, namely, the Euler equation for homogeneous functions:

$$\left[x_1\frac{\partial}{\partial x_1} + x_2\frac{\partial}{\partial x_2} + x_3\frac{\partial}{\partial x_3}\right] f(x_1, x_2, x_3) = nf(x_1, x_2, x_3). \tag{17.56}$$

The expression in square brackets $\sum_i x_i \frac{\partial}{\partial x_i}$ is called a dilation operator.

If the scale change is performed differently in different dimensions $x_1 \to \lambda^{\alpha_1} x_1$, $x_2 \to \lambda^{\alpha_2} x_2$, $x_3 \to \lambda^{\alpha_3} x_3$ and the function remains the same up to a common factor λ^n:

$$f(\lambda^{\alpha_1}x_1, \lambda^{\alpha_2}x_2, \lambda^{\alpha_3}x_3) = \lambda^n f(x_1, x_2, x_3), \tag{17.57}$$

then f is a generalized homogeneous function. It satisfies a generalized Euler equation

$$\left[\alpha_1 x_1 \frac{\partial}{\partial x_1} + \alpha_2 x_2 \frac{\partial}{\partial x_2} + \alpha_3 x_3 \frac{\partial}{\partial x_3}\right] f(x_1, x_2, x_3) = \lambda^n f(x_1, x_2, x_3). \tag{17.58}$$

The dilation operator becomes

$$\alpha_1 x_1 \frac{\partial}{\partial x_1} + \alpha_2 x_2 \frac{\partial}{\partial x_2} + \alpha_3 x_3 \frac{\partial}{\partial x_3}.$$

Many scaling relations may be derived if one assumes that thermodynamic functions near critical points are generalized homogeneous functions.[15]

If, in addition, the function contains a parameter R which depends on the factor λ during the scale change, i.e., $R \to R(\lambda)$. Then there appears a term in the generalized Euler equation as well as in the dilation operator:

$$\left[\sum_i \alpha_i x_i \frac{\partial}{\partial x_i} + \beta \frac{\partial}{\partial R}\right] f(x_1, x_2, x_3) = \lambda^n f(x_1, x_2, x_3), \tag{17.59}$$

where a coefficient β is introduced:

$$\beta = \left.\frac{dR}{d\lambda}\right|_{\lambda \to 0}. \tag{17.60}$$

This is an elementary derivation of the Callan–Symanzik equation without making use of any knowledge of field theory. At critical point, the correlation length diverges and the scale change does not affect the generalized function at all. In other words, at the critical point $\beta = 0$. The critical exponents may be calculated from the zero of the coefficient function β in the Callan–Symansik equation. Had we been aware of this elementary derivation of Callan–Symansik equation 40 years ago we could have explained the relation of our skeleton graph expansion to other field theory calculation of critical exponents then.

Note: In the Reference list, we included only those available when writing paper of Ref. 9 except for Ref. 1.

References

1. A. Salam, *Notes on Science, Thechnology and Science Education in the Development of the South* (Global Publishing Co., Teaneck, NJ, 1991), p. 20 (in Chinese); The English original was published by The Third World Academy of Science, Triests, May 1991.
2. K. G. Wilson and M. E. Fisher, *Phys. Rev. Lett.* **28**(4), 240 (1972).
3. K. G. Wilson, *Phys. Rev. B* **4**, 3174 (1971).
4. K. G. Wilson, *Phys. Rev. B* **4**, 3184 (1971).
5. K. G. Wilson and J. Kogut, The renormalization group and ϵ expansion, Cornell Preprint COO 2220 (1972).

6. K. G. Wilson and J. Kogut, *Phys. Rep.* **12**, 75 (1974).
7. T. Tsuneto and E. Abrahams, *Phys. Rev. Lett.* **30**, 217 (1973).
8. M. J. Stephen and E. Abrahams, *Phys. Lett. A* **44**, 85 (1973).
9. L. Yu and B. Hao, *Acta Phys. Sinica.* **24**(3), 187 (1975) (in Chinese).
10. L. P. Kadanoff *et al.*, *Rev. Mod. Phys.* **39**, 395 (1967).
11. M. E. Fisher, *Rep. Progr. Phys.* **30**, 615 (1967).
12. A. A. Abrikosov, L. P. Gorkov and I. E. Dzialoshinskii, *Method of Quantum Field Theory in Statistical Physics*, translation from Russian by B. Hao (Science Press, Beijing, 1963), p. 90 (in Chinese).
13. J. Hubbard, *Phys. Lett. A* **39**, 365 (1972).
14. A. Z. Patashinskii and V. L. Pokrovskii, *JETP* **46**, 994 (1964) (Russian); English translation, **19**(3), 677 (1964).
15. H. E. Stanley, *Introduction to Phase Transitions and Critical Phenomena* (Oxford University Press, Oxford, 1971).

Chapter 18

Personal reflections on Kenneth Wilson at Princeton and Edinburgh*

Ken Bowler†, Richard Kenway†, Stuart Pawley† and David Wallace‡,§

†*School of Physics and Astronomy, The University of Edinburgh, James Clerk Maxwell Building, Mayfield Road, Edinburgh EH9 3JZ, UK*

‡*Churchill College, Cambridge CB3 0DS, UK*

david.wallace@chu.cam.ac.uk

This article contains recollections of the authors when they interacted with Ken Wilson at Princeton in 1971, when he gave his lectures on the renormalization group and the ε-expansion, and at Edinburgh in the early 1980s, working on Monte Carlo renormalization group calculations of critical behaviour in the three-dimensional Ising model. Edouard Brézin was a collaborator in the former work, and Robert Swendsen in the latter. Ken's vision for scientific computing impacted formatively on the development of Edinburgh Parallel Computing Centre. Reference is also made to an earlier visit to Scotland by Ken, to lecture at the Scottish Universities Summer School in Physics in 1973.

Keywords: Kenneth Wilson; theoretical physics; scientific computation.

18.1. The lectures at Princeton, 1972

18.1.1. *David Wallace writes*

At the end of his first lecture on the renormalization group and the ε-expansion at Princeton in spring 1972, Ken asked if anyone was interested in writing up the lectures with him; John Kogut instantly volunteered, and the result was the review in Physics Reports with that title [1].

At the time I was fortunate to be a Harkness Fellow in the Department. When Edouard Brézin, then also visiting, asked if anyone was interested in checking with him what Ken had been saying, I jumped at the chance. Our collaboration started for real after Ken's lecture on the ε-expansion, probably the eighth in the course.

*This article was originally published in *J. Stat. Phys.* **157**, 639–643 (2014).

§Corresponding author.

That afternoon, we worked through the one-loop calculation for the scaling of the 4-point function in φ^4 theory in 4-ε dimensions, calculating the coefficient of the logarithm when the φ^4 coupling had its special value (the renormalization group fixed point value at order ε, although I didn't really understand it that day). The predicted scaling worked; it seemed like magic. That evening, I repeated the calculation for the 2n-point function. It worked — even more magic that something primarily related to dimensional analysis should link to the multiplicities of Feynman graphs.

Edouard and I sought Ken's advice about what we might do, and he suggested calculating the equation of state for a ferromagnet, relating the external field, H, the magnetisation $M = \langle\varphi\rangle$ and the deviation from critical temperature, $T-T_c$. The Ising-like case of the uniaxial magnet (1-component field φ) to $O(\varepsilon^2)$ was straightforward. Less obvious was how to test the calculation, and I recall inconclusive efforts by Edouard and me to fit to real data. It was Edouard I think who did some numerical work to compare the results with the Gaunt and Domb estimates using Padé approximants of series expansions of the three-dimensional lattice Ising model. Ken had done much wider reading, and suggested we look at it in the context of the parametric model developed by Schofield and others. It turned out that the equation of state to $O(\varepsilon^2)$ was beautifully described by the linear parametric model. The paper was published in Physical Review Letters [2].

Edouard and I then dutifully worked out the equation of state to $O(\varepsilon^2)$ for the case of N-component field with O(N) symmetry, and took the results to Ken. He pointed out that we couldn't possibly be right, as we weren't distinguishing the longitudinal and transverse modes in the presence of an external field: this was mortifyingly embarrassing for me, given that Peter Higgs had been my PhD supervisor, and I did know about Goldstone modes... We got it right, and it was published in Phys Rev B [3]. Ken also told us about his unpublished work on the 1/N expansion, and Edouard and I were able to do the 1/N expansion for the equation of state [4].

All of this was a wonderfully enriching time. Ken had insights which were incomprehensible to me, and (a massive understatement) working with him and Edouard was inspiring and made my scientific career. At that time I was a very inexperienced 26 year-old, and the opportunity to work with scientists of such quality and experience was both great fun, and life-transforming. The intensity of the collaboration is reflected in the dates these three papers were received in Physical Review Letters and in Physical Review B: 16 June, 14 July and 31 July, 1972.

In all our interactions with him, Ken was quite undemonstrative, always willing, and never the prima donna. I suspect that he was working *very* hard. He shared with us his detailed calculation showing that at $O(\varepsilon^2)$ the exponent η of the correlation function at criticality was independent of the smoothing function adopted in the momentum-shell approach to the renormalization group (it wasn't an earth-shattering result, except that the theory would have been scuppered if it hadn't been the case). My memory is of some 30 pages of neat hand-written calculation, with no material corrections, ending in success, with the curt comment 'Whew'.

The experience and confidence that this collaboration gave me helped me to get a tenured lectureship in Physics at Southampton in 1972, and, in 1979, an interview for the Tait Chair of Mathematical Physics at Edinburgh. A vignette from the interview sets the stage for Ken's interactions with us there. I was asked if I thought computers would be any use for the kind of theoretical physics that I did. I replied with complete confidence 'No, because they will never be powerful enough'. This was the wrong answer, but no-one else on the committee knew any better and I got the job. How wrong I was began to crystallise within a few months of my return to Edinburgh in October 1979.

18.2. Parallel computing collaboration at Edinburgh

18.2.1. *Ken Bowler, Richard Kenway, Stuart Pawley and David Wallace Write*

A bit of history might be interesting, to set the context. In the late 1970s, the UK company ICL (subsequently taken over by Fujitsu) developed the Distributed Array Processor (DAP). This was an SIMD 64 × 64 array of bit-serial processors, with nearest neighbour connections and good row, column and global buses. It had a *total* of 2 Mbytes of memory — 500 bytes per processor. It was programmed in a natural and easy-to-use extension of Fortran — provided that the application mapped easily on to (and into!) the array. In many respects, programming challenges were not unlike those faced now for General Purpose Graphics Processing Units (GP GPUs). The similarity stopped there: the DAP occupied seven large cabinets, and weighed probably more than a ton. A national facility was commissioned in 1980 at Queen Mary, University of London.

Unknown to the rest of us, Stuart had started using this machine from Edinburgh for molecular dynamics simulations. Printed output results came back in the post in brown paper envelopes. Despite the challenges of using the machine, it rapidly became clear to him that this was the way of the future for scientific computation.

Independently, David learned of the machine's existence from the string theorist Michael Green, who was then at Queen Mary, and (despite his scepticism about the value of scientific computation) naively asked "Have we a DAP in Edinburgh?". Discussions triggered the possibility of lattice gauge theory calculations and we persuaded Stuart to come to a meeting at Cosener's House, Abingdon, 27–29 March 1981, which brought together theoreticians with computational interest in the field.

Ken Wilson was the main attraction at the meeting. His talk covered his parallel computing work on the Floating Point Systems Array Processor for which he had been doing heroic programming, and included a discussion of simulation of the Ising model. His enthusiasm for parallel computing made a big impact on the audience. Stuart and he talked for long periods, attracting infuriated comments from other participants: "Who is this guy monopolising Ken's time?" (not a complete monopoly, as Stuart with Ken Bowler also figured out at the meeting the DAP code layout for four-dimensional SU(3) gauge theory).

During a visit to IBM Zurich, Ken also had extensive discussions with Bob Swendsen. Bob comments: "a terrific experience and a big influence on my career". The outcome of his discussions with Bob and with Stuart was a collaboration involving Ken, Bob, Stuart and David, on Monte-Carlo renormalization group calculations of critical behaviour in the 3-dimensional Ising model. Because of its architecture, at that time the DAP outperformed other machines by some margin for Ising-model like calculations; it delivered (depending crucially of course on the random number generator used) Monte Carlo update rates of up to several 100 million/s, at a time when Cray supercomputers were managing a few tens of Mflops in practice.

By this time Edinburgh had obtained research Grants for the installation of a dedicated DAP; a second subsequently arrived for free, courtesy of the difficulty that the British National Oil Corporation found in using it. As a result, the scale of the computation possible was probably unprecedented for thirty years ago. Seven different lattice couplings were used in simulations on 8^3, 16^3, 32^3 and 64^3 lattices and a great deal of attention was paid to systematic as well as statistical errors — although we probably didn't fully satisfy Ken's legendary requirements. Our main results [5] were estimates for the critical coupling $K_1^c = 0.221654(6)$, and exponents $\nu = 0.629(4)$ and $\eta = 0.031(5)$; a review and analysis [6] of some 70 papers published up to 2002 concluded $\nu = 0.6301(4)$ and $\eta = 0.0364(5)$.

This one paper is of course only a microcosm of Ken's influence and impact on high performance computing at Edinburgh, as elsewhere. The DAPs were great work horses and the genesis of Edinburgh Parallel Computing Centre, which subsequently acquired a range of parallel machines. Within ten years the EPCC annual report for 1992–1993 summarised more than 100 projects across a wide range of the sciences and engineering, including a dozen industrial collaborations, and two Grand Challenge projects, a concept which Ken promoted, and which is still relevant today.

18.3. Scottish universities summer school in physics

As a Postscript, we record that Ken came to Scotland also in 1973, to lecture at the Scottish Universities Summer School on Physics. Because it was referred to by many other lecturers, the proceedings [7] include his previously unpublished paper *Some Experiments on Multiple Production* [Cornell preprint CLNS-131 1970].

Forty years on, and in the context of his then imminent contribution to lattice gauge theories and confinement in QCD, it is interesting to extract:

"First, though, one must say what one means by a "real theory of strong interactions". My view is that there are four essential requirements for a real theory

A. It must be derived from a few fundamental principles comprehensible to both experimentalists and theorists .

B. Any free parameters in the theory must appear explicitly and obviously as a consequence of the fundamental principles (just as s and h are explicit and obvious in ordinary quantum mechanics, and c is explicit and obvious in relativity) and there should be no arbitrary functions in the theory.
C. The fundamental principles should imply a set of equations containing the fundamental parameters whose solution will describe all aspects of strong interactions including the complete S matrix (even the S matrix for n particles going to m particles for any n and m) and all matrix elements of the weak and electromagnetic currents. For a given set of values of the parameters the equations should have one and only one solution, if this cannot be proven there should at least be plausible physical arguments suggesting it.
D. One should be able to determine qualitative features of the solution of the equations from qualitative features of the equations, or better from qualitative statements of the fundamental principles without using the equations at all."

18.4. Concluding remark

Ken was a great friend who unselfishly showed so many of us how to be better scientists.

Acknowledgements

We are very grateful to John Kogut and Bob Swendsen for comments on a draft of this paper.

References

1. Wilson, K.G., Kogut, J.: The renormalization group and the ε-expansion. Phys. Rep. **12**, 75–199 (1974)
2. Brézin, E., Wallace, D.J., Wilson, K.G.: Feynman-graph expansion for the equation of state near the critical point (Ising-like case). Phys. Rev. Lett. **29**, 591–4 (1972)
3. Brézin, E., Wallace, D.J., Wilson, K.G.: Feynman-graph expansion for the equation of state near the critical point. Phys. Rev. **B7**, 232–9 (1973)
4. Brézin, E., Wallace, D.J.: Critical behavior of a classical Heisenberg ferromagnet with many degrees of freedom. Phys. Rev. **B7**, 1967–74 (1973)
5. Pawley, G.S., Swendsen, R.H., Wallace, D.J., Wilson, K.G.: Monte Carlo renormalization-group calculations of critical behavior in the simple-cubic Ising model. Phys. Rev. **B29**, 4030–40 (1984)
6. Pelissetto, A., Vicari, E.: Critical phenomena and renormalization-group theory. Phys. Rep. **368**, 549–727 (2002)
7. Wilson, K.G.: Some experiments on multiple production. In: Crawford, R.L., Jennings, R. (eds.) Phenomenology of Particles at High Energies. Proceedings of the Fourteenth Scottish Universities Summer School in Physics, 1973, pp. 704–724. Academic Press, London and New York (1974)

Chapter 19

Ken Wilson — A Tribute: Some recollections and a few thoughts on education

H. R. Krishnamurthy
Department of Physics, Indian Institute of Science, Bangalore 560012, India

I had the marvelous good fortune to be Ken Wilson's graduate student at the Physics Department, Cornell University, from 1972–1976. In this article, I present some recollections of how this came about, my interactions with Ken, and Cornell during this period; and acknowledge my debt to Ken, and to John Wilkins and Michael Fisher, who I was privileged to have as my main mentors at Cornell. I end with some thoughts on the challenges of reforming education, a subject that was one of Ken's major preoccupations in the second half of his professional life.

I joined Cornell as a graduate student in physics in the fall of 1972, after completing an M.Sc. degree in physics from the Indian Institute of Technology (IIT), Kanpur.[a] I was pretty much sold on becoming a theoretical physicist, but wasn't very clear what area I would work in. I remember that among the graduate programs I had gained admission to, I was able to eliminate all but Caltech and Cornell relatively easily, but found the final selection between these two a tough one to make. Caltech was the more acclaimed (Feynman and Gell-Mann were among the physics faculty), and had offered me a fellowship; whereas Cornell had offered me only a teaching assistantship. I finally decided to go to Cornell nevertheless, because of the perception that this allowed me to keep my options open for doing either high energy physics (then more commonly referred to as particle physics) or condensed

[a]The first-rate education in physics I received at IIT Kanpur was a key enabler for me to get admission to Cornell. IIT Kanpur was established in 1959 with the assistance of a consortium of nine leading US research universities (see http://en.wikipedia.org/wiki/Indian_Institute_of_Technology_Kanpur), and soon became the premier institution for science and engineering education in India. A remarkably large number of well-known physicists of Indian origin are IIT Kanpur alumni.

matter physics, as Cornell was strong in both.[b] Among the flyers I received from the physics department at Cornell, there was one with a list of recent publications from their faculty, and Ken Wilson's name figured prominently at the end, with a string of 4 papers published in 1971, three of them with "renormalization group" in their titles. I looked up the papers in the library, but didn't understand much of what I read, except that the papers seemed very important. By the time I arrived in Cornell, "epsilon expansion" had been discovered [1], and there seemed to be a clear consensus that Ken had achieved a profound breakthrough.

The physics department at Cornell those days had a system that a committee of "four wise men" was designated every year to advise the entering graduate students. Ken was one of these four the year I joined, and by a fantastic stroke of luck, I was assigned to him. I met him soon after my arrival at Ithaca, and asked him whether I could skip the first year graduate courses in Quantum Mechanics (QM) as I felt I knew the course material well. He said he would give me a written exam, and advise me based on my performance. The exam consisted of a bunch of QM problems, and I had no difficulty solving them. When I met him later, after he had a chance to look at my solutions, to my great joy he suggested that I skip *all* the first year graduate courses, and instead credit the second year courses, including a special topic course that he was teaching that fall term on the renormalization group (RG) and epsilon expansion (from notes that were later published as the seminal Wilson–Kogut *Physics Reports* article [2]). I did as he suggested, but partly because of my limited exposure to advanced statistical physics, field theory and critical phenomena, and partly because of limitations in my approach to learning (I was very reluctant to plunge in and learn something without having mastered what I considered the "prerequisites"), found his course rather difficult to cope with. I eventually dropped out of crediting the course (though I probably continued to sit in on the lectures), and focused on the other courses I was taking, on Solid State Physics[c] and Quantum Field Theory. I don't think I had much interaction with Ken the rest of that academic year, except for occasional meetings for him to sign papers as my adviser.

During my second semester at Cornell, and during the summer that followed, I tried my hand a bit at experimental physics, by doing a couple of projects with Bob Buhrman, which also helped me to fulfill the experimental physics course requirements that were mandatory for all physics graduate students at Cornell. Bob was a wonderful person to work with, but by the end of this period it was clear to me that I should stick to theory. I was also clear that I wanted to work on problems related

[b]As an amusing counterpoint, Deepak Dhar, my classmate from IIT Kanpur, joined Caltech, but wrote his thesis on statistical mechanics problems, and was none the worse off! He returned to India to join the Tata Institute of Fundamental Research (TIFR) in Mumbai, and has produced outstanding research work (e.g., see his Google Scholar page and his home page at TIFR website.)
[c]Taught from notes that were later published as the celebrated Ashcroft–Mermin text book (N. W. Ashcroft and N. D. Mermin, *Solid State Physics*, Harcourt, 1976).

to RG and critical phenomena. I met Ken and asked him whether I could work with him. He basically said yes, but added that he had switched his interest to lattice gauge theory,[d] and asked whether I would be interested in working in that area. I told him that I was keener to work in condensed matter theory, and he suggested that I should talk to Michael Fisher. So I went and met Michael, and joined his group, moving to a desk in Baker Hall. By then David Nelson, a whiz-kid who was at that time in his 4th year of Cornell's famous "6 year Ph.D program",[e] had already started working with Michael on projects in RG. Michael was perhaps a bit hesitant to start me on similar problems, and asked me to look at a few other problems that were of interest to him, in more traditional areas of statistical mechanics. I did that, but couldn't get myself interested in those problems, and was puttering around for a bit, unclear where I was headed.

Then, one momentous day, to my utter amazement, the trio of Ken Wilson, Michael Fisher and John Wilkins sought me out, I think after a departmental colloquium. John was away on a sabbatical my first year at Cornell, so I had actually not met him until that moment, but had only heard of him. The first thing he said to me after being introduced was, "So you are the sucker"! It turned out that they had in mind a proposal for a research project for me to take on. A little prior to that time, Ken had completed the invention of a remarkable new technique, the numerical renormalization group (NRG), and used that to achieve yet another breakthrough — the solution of the celebrated "Kondo Problem" connected with magnetic impurities in metals.[f] The solution had been announced in a Nobel Symposium in June 1973 in Göteborg, Sweden [3], but the details were not yet published. The proposal was to have me explore the extension of the technique to the more microscopic Anderson Impurity Model (AIM).[f] I took a bit of time to do some browsing of the literature on Quantum Impurity Problems, and of Ken's notes on the NRG (the notes were later expanded and included in Ken's second famous review article on the RG, in *Reviews of Modern Physics* (*RMP*) [4]), found myself hooked, and signed on to the project.[g] Ken and John became the co-advisers of my thesis research, and Ken, John, Michael and Bob the members of my special committee.

---

[d] E.g., see Ken's anecdotal account of the origins of lattice gauge theory published in *Nucl. Phys. Proc. Suppl.* **140**, 3–19, 2005. (Also available in http://arxiv.org/abs/hep-lat/0412043v2)

[e] A 6 year accelerated Ph.D program at Cornell, right after high school, for a select few; see http://www.cs.fsu.edu/~baker/phuds/ and the other links therein.

[f] For a review, and references to the literature, see, *e.g.*, A.C. Hewson, *The Kondo Problem to Heavy Fermions*, Cambridge University Press, Cambridge (1997).

[g] I don't know the details of how the proposal got generated, but I feel amazed and grateful that three such luminaries as Ken, Michael and John put their heads together to come up with it. Michael continued to maintain a keen interest in my progress despite the fact that I was no longer working with him. The first seminar I ever gave at Cornell, on the Kondo Problem, was given in his group. After the seminar, he gave me a two-page long, meticulously composed list of "do"s and "don't"s, which I did my best to follow thereafter... I was certainly the beneficiary of extraordinary mentorship by these three.

My next two years were the most intense and wonderful period of learning I have ever experienced. I had the privilege of learning the intricacies of the renormalization group, in particular, the numerical renormalization group, from its creator, Ken. In fact, Ken gave a whole new set of lectures, essentially on the material that later went into his *RMP* article [4], and this time, I was ready; I found his lectures to be amazingly clear and insightful, and soaked them up. John was very generous in sharing his expertise and time to help me learn quantum impurity physics. I learnt a whole lot about RG and critical phenomena from Michael and his group, especially his group seminars. There were very many other exciting things happening at Cornell as well during this period, on which there were lectures and seminars I could learn from.[h] In addition to all this, I honed my computing skills developing my own NRG code for solving the AIM, learning by example from Ken, a master programmer.

Actually, the additional coding required to extend the NRG to the Anderson Impurity Model compared to the Kondo problem is relatively minor. So Ken suggested initially that I modify the NRG program he had written for the Kondo problem and use that for the AIM project, and gave me a copy of his program. I was flabbergasted when I saw it — it had well over a thousand lines of code, pretty much as one single program (except for calls to a matrix diagonalization subroutine), and there was not a single comment statement in it! Many important variable names were chosen in ways I could not fathom; there was an XXXX and a YYYY! I had to go through the code line by line, annotating it along the way, which took me a while; then I understood and appreciated how tightly and intricately knit it was. All available symmetries of the Hamiltonian had been used to reduce the sizes of matrices to be diagonalized to the minimum possible, and storage of arrays had been maximally optimized to reduce memory requirements. I have always wondered how Ken kept track of what was what in the program, and how he debugged it. Knowing how awesome he was as a programmer (he was one of the very few physicists I have come across who knew how to write machine code, and would use it to optimize the innermost computations inside 'do loops'), I am inclined to believe that he had such algorithmic clarity that he wrote code that needed very little iteration and debugging; and that he had prodigious memory which helped him keep track of obscure variable names! I wish I had preserved a copy of Ken's program for posterity, but unfortunately I did not have the foresight to do so.

In any case, being more of a mortal, I was terrified by the thought of modifying his program, having it bomb, and being unable to debug it. So I decided to write my own program starting from scratch. I probably spent too much time writing it in

[h]For example, the story of superfluid He-3 was unfolding before our eyes. The field of lattice gauge theories was emerging. de Gennes visited and gave a fascinating course of lectures on Liquid Crystals. John Kosterlitz was a post-doc, and was developing further the seminal ideas he and Thouless had put forth on the 2-d XY model... For an aspiring young condensed matter physicist, Cornell during this period was a fabulous and inspiring place to be in.

a modular fashion, putting in lots of comments, choosing variable names carefully so that they corresponded as closely as possible to the physical quantities they represented, and so on; but it was a great learning experience, and Ken and John were generous in allowing me this leeway. After benchmarking my program to ensure that it reproduced Ken's results for the Kondo problem, I did eventually start producing exciting new results for the AIM, and had great fun analyzing them and thinking about the physics that they represented. The work on the symmetric Anderson model, plus a comprehensive review of the literature, constituted my thesis.[i] The process of completion of the other "formalities" connected with my getting the Ph.D. degree was incredible for its informality. I left for a post-doc at the University of Illinois at Urbana-Champaign (UIUC) in August 1976, after having deposited the handwritten manuscript of my thesis with the venerable Velma Ray, who was Hans Bethe's secretary. She was legendary for her skills in typesetting theses with lots of equations, and I certainly needed her to typeset mine. The understanding was that I would come back by the end of that year for my thesis defense, but the process was actually completed only in Jan 1978![j] I have forgotten the details about how this came about — perhaps everybody involved, including me, forgot that my thesis defense was not yet a done deal, until more than a year had gone by!

My periodic interactions with Ken regarding my thesis research were invariably rather brief, but pleasant and rewarding. Ken was very informal — I never had to make an appointment to see him, and would walk into his office whenever I wanted to, which was typically when I had some progress to report, or to seek help when I faced some obstacles in my work. There I would generally find him, mostly in his signature grey pants and white shirt, often with his feet up on the table, and deep in thought. But he never seemed to be perturbed by the interruption, and would turn to me with the twinkle in his eye that used to be a ubiquitous feature of his demeanor, as can be seen in so many of his photographs.[k] When I reported to him the newer aspects of the physics of the AIM as they began to emerge from my NRG calculations, which I thought I was the first to discover, he would most often just nod in agreement, and it was somewhat disconcerting to find that they seemed obvious to him! When I ran into an obstacle, and broached it to him, there were only two possible outcomes — if he had a hunch as to how one might be able to get around the obstacle, he would say it succinctly, and his hunches almost always helped out;

[i]The details of my thesis research on the symmetric Anderson Model, and follow up work I did on the asymmetric Anderson Model at UIUC, eventually appeared as back to back papers in *Phys. Rev. B* **21**, 1003, 1044 (1980).

[j]After the passing of Ken, Jeevak Parpia, current chair of the Physics Department at Cornell, who overlapped with me as a co-graduate student at Cornell, sent me a most memorable surprise gift — a framed report declaring my passing of the thesis exam, carrying Ken's signature — for which I am ever so thankful to him. Jeevak told me they looked for the reports for all of Ken's students, and I was one of the lucky ones.

[k]E.g., see the pictures of Ken in Chapter 20.

otherwise, he would say he hadn't thought about the issue, and did not have any comments that might be of help. So it was somewhat difficult for me to hang around in his office for long — he never seemed to be one for "small talk". He was also probably too absorbed in his work on lattice gauge theories during this period to be very actively involved in what I was doing. I remember though, that sometime after we published the first NRG results on the symmetric Anderson model [5], he asked me how it was being received by the condensed matter community. When I said I thought it was being well received,[l] he did seem pleased. I don't know whether he kept track of the fact that the NRG has continued to thrive, especially as a solver for quantum impurity problems that arise in the context of quantum dots and Dynamical Mean Field Theory[m]; if he did, his pleasure would have been even greater.

At Cornell I also had some opportunities to observe closely Ken's approach to teaching. I was his teaching assistant for an undergraduate course on electromagnetism that he taught one of the terms, and I sat in on most of his lectures. Purcell's book was the text. My memories of the details of this experience are a bit hazy, but the impression I have retained is that Ken was not very particular about sticking to the textbook material or of covering a pre-planned set of topics, but would spend time on what he thought was interesting and useful. Even in a well-worn subject like electromagnetism, I found many of his comments and observations very original and insightful — but they might have been lost on many of the students if they did not have prior exposure to the material. He tended especially to emphasize numerical techniques of solving electromagnetics problems. For example, he taught the students the numerical technique for solving boundary value problems involving the Laplace equation for the electric potential, by choosing a square or cubic grid of points, and iteratively updating the potential at each site to the average over all its nearest neighbor sites, in great detail. This was very much in keeping with his abiding interest in computers and computing,[n] and his vision and foresight that computers were going to play a major role in the future of physics, and it was important that students get an early exposure to numerical techniques. Ken followed up on his vision in many ways. He was an active campaigner for improving computer resources for research.[o] He headed an initiative that he christened the

[l]In particular, Phil Anderson, the creator of the AIM, was very gracious and encouraging about our work, even to the extent of including a figure from the above PRL paper in his Nobel lecture in 1977: http://www.nobelprize.org/nobel_prizes/physics/laureates/1977/anderson-lecture.pdf.

[m]E.g., see R. Bulla, T. Costi, and T. Pruschke, *Rev. Mod. Phys.* **80**, 395 (2008) and references therein.

[n]For some autobiographical comments on Ken's interest in computing and how it shaped his research, as well as for insights into his views on various issues and his values, see his interview available at Caltech library (http://authors.library.caltech.edu/5456/1/hrst.mit.edu/hrs/renormalization/Wilson/index.htm).

[o]For some more comments and links to source material on Ken's role in promoting the use of computers, especially supercomputers, in research, see Chapter 20 and references therein.

"Gibbs project",[p] which attempted to create what he thought would be the ideal computing environment for physicists. What he was visualizing was *one resource* that combined the best features we have come to see in MAPLE, MATLAB, Mathematica and program libraries such as LAPACK, etc., and was user-friendly enough not to require the learning of a new language. I think we are still rather far away from having anything like what he visualized.

Unfortunately, I did not keep in regular touch with Ken after my return to the Indian Institute of Science in Bangalore, India, in 1978. This was difficult to do in the initial years in any case, as it could have been done only by snail mail, and I was a bit hesitant to write letters to him and expect him to reply. I was overjoyed by the award of the Physics Nobel Prize to him in 1982, of course, did write him a congratulatory note, and I think I did get a brief response. After the advent of e-mail, I would send him Christmas and New Year greetings occasionally, and he would not always respond. But I got to meet him briefly now and then, during visits I made to Cornell, and later to the Ohio State University (OSU) at Columbus, Ohio, where both John Wilkins and he moved in 1988.

As I look back, I am astonished as to how many of Ken's values[n] I seem to have imbibed, some consciously, and many subconsciously, that have heavily influenced me in my career and life. For example, in the context of his work on the Kondo Problem, Ken had carried out some extremely tedious 4th order (Rayleigh–Schrodinger) perturbation theory calculations for the energy levels of the Kondo NRG Hamiltonian — pages after pages of neat algebra, meticulously listing out the various terms that arise. In the course of my own research work, if I felt discouraged by some tedious and daunting algebra that needed to be done for me to make further progress, I could draw strength from his example — if a genius like Ken could sit down and carry out tedious algebra, surely I had to discipline myself to 'just do it'! I also found his commitment to societal reforms inspiring. The most prominent, and well known, are of course his role in making supercomputing widely available to the research community (see Chapter 20), and his involvement in reforming education.[q]

Reforming education was the major concern and enduring passion of Ken in the second half of his career, and perhaps the main factor responsible for his move from Cornell to OSU in 1988. A major landmark in his work on education was the publication of the book "Redesigning Education" [6]. A summary of his contributions and leadership role in reforming education in the US, and references to some of the articles he wrote on education, can be found in the obituary issued by the OSU.[q]

Most of the articles in this memorial volume on Ken are related to his research contributions in physics. The impact of his work in physics is certainly more widely recognized than his work in education, due in no small measure, of course, to the award of his Nobel Prize. Indeed there might be many who feel that he wasted

---

[p] Unfortunately I have not been able to locate any easily available references for the Gibbs Project.
[q] See the news article on OSU website (http://artsandsciences.osu.edu/news/remembering-theoretical-physicist-and-nobel-laureate-kenneth-g-wilson).

his talents working on such a complex, perhaps insoluble, problem as education. However, in the context of how Ken himself saw his role in physics research vis-à-vis his role in education, I remember him saying something along the following lines (unfortunately I have not been able to locate a precise quotation that I can reference), which I have always found heartening. While in physics the importance of a contribution often depends on a problem being *solved*, in case of societal issues such as education, even if one is able to contribute to a 1% improvement by some measure (for example, the fraction of students getting education better than some threshold), it can make a huge difference to a very large number of people! Hence I thought it might be fitting to end my tribute to Ken by airing some thoughts on the challenges confronting educational reform, especially in India.

The huge challenges that confront India in particular, and the developing and developed world in general, as regards educational reform at all levels, i.e., school, college and vocational, are of course well known. India is now entering a period of "demographic dividend" — the period when the growth rate of the working age population well exceeds the growth rate of the overall population [7]. However, as eloquently expressed by Nandan Nilekani [8], preeminent cofounder of the Indian IT company Infosys, for this to be truly a "dividend", the working age population needs to be productive. Hence, providing them education, at the least vocational education to impart to them productive skills, is an obvious necessity. If this is not available, we will instead be confronted by a "demographic disaster". The numbers are staggering — around 64% of India's population (of well over a billion) is expected to be in the age bracket of 15–59 years by 2026 [7].

National agencies in most countries are very much engaged in confronting the challenges of education, and have set many commendable goals [9]. However, numerous widely recognized obstacles block their way forward in achieving these.

A major one of these, especially in India, is the dreadful shortage of trained, high quality teachers at all levels. There is also a shortage of teaching material and laboratory and other resources. What is worse, even when these are available, the 'education' imparted is not designed to suit the people in need of the education. Conventional methods of imparting education, based on lectures, rigid curricula, undue emphasis on performance in centralized examinations rather than on learning, and their use for filtering students at various levels, governed by the "one size fits all" paradigm, are widely prevalent in India. Their limitations are well known and well documented; and they are woefully inadequate for the challenges at hand given the diversity of the student population, leaving large sections of them poorly educated and disheartened. Many voluntary groups and organizations have made laudable efforts addressing these among small batches of students all over India, but the major challenge is to come up with solutions that are *scalable to the huge numbers* of people that need the education.

Current breakthroughs in computer and communication technology, especially mobile devices, and the widespread and rapid increase in their accessibility in all parts of India, open up immense new possibilities, and perhaps a new paradigm, both for "*tailor making*" education to suit individual and societal needs, and for addressing the problem of *scale*. I discuss below some key ingredients which I believe are urgently needed for this, in the context of three clearly distinguishable aspects of education: (1) identification and publicizing of what needs to be taught or learnt, (2) teaching and learning and (3) testing and evaluation of proficiency.

Imagine that groups of people come together, perhaps supported by philanthropic foundations, and perhaps in cooperation with governments, to take the lead in the creation of a novel, model "learning tree": a framework of creatively designed, hierarchical but interconnected modules of learning content, covering all levels, from primary through college, and all aspects of human endeavor, including the learning of life skills. This model content is to be designed so as to make the best use of the wide variety of formats and types of learning material and learning methods that are viable, such as video lectures and presentations, audio lectures, movies, multimedia presentations, computer simulations, lab work, field work, computer games, mini projects, etc., in such a way that it can be flexibly adapted to educate students with varied backgrounds, levels and requirements. The primary design constraint is that students should be able to learn the material, either individually or in small groups, by working through them at their own pace, assisted by teachers who act as mentors or coaches rather than as lecturers and graders. The creation of such learning modules can surely be accomplished by teams of highly qualified and committed educationists with expertise in the different areas and people with skills in multimedia content creation working together.

Imagine that, in synergy with such a learning tree, a "testing tree", of hierarchical and interconnected testing modules, is created. Each testing module is to be associated with one or more learning modules, and consist of very large banks of carefully designed questions, problems and other testing methods, of graded difficulty levels, and covering all the well-recognized objectives of learning [10]. The question banks need to be made large enough that random selections from them can be used by the students to evaluate their own progress *while they are learning*, as well as for the purpose of final testing and certification of the extent to which they have mastered the content of a learning module. The design goal, and the challenge, is to ensure that the *evaluation can largely be done by computers*, with an end result that is nevertheless an objective, unambiguous, tamper-proof evaluation of the proficiency level the students have attained in that particular learning module.[r]

[r]Centralized, computer-evaluated examinations, such as the SAT, ACT and GRE in the US, and the IIT-JEE and several other such examinations in India, are routinely and widely used as acceptable measures of learning. What I am envisaging are vastly more extensive, creatively crafted, modular versions of these that are also available in the public domain to students as self-evaluation tools *while* they are learning.

In recent years there has been a phenomenal growth in the availability of open source learning content of various types and levels, and of steadily rising quality.[s] But a large fraction of it consists of conventional lecture courses, slide presentations and texts. We have a long way to go before anything along the lines envisioned above becomes available. Picture a gigantic network, with model modules of learning content at every node, each connected hierarchically with numerous other nodes, which students can traverse along their own paths, and at their own pace... I believe that such a framework still needs to be designed and built, and the existing and upcoming open source learning content can then be hyper-linked to the model learning modules.

As regards evaluation content, my impression is that what is currently available in open source is very limited, both in quantity and quality. Some proprietary learning and evaluation content of fairly high quality is probably available, but the majority of students in India, for example, cannot afford the costs for getting access to these. I am sure many people will be skeptical that such testing modules as I am envisaging above can ever be created in open source, and even if that is done, that they will be viable as sure means of evaluation. I am inspired, however, by the shining examples of the creation of open source software such as LINUX and GNU, and of Wikipedia. I believe that if a large enough group of us are enrolled into thinking of this as a desirable goal, and put our creative energies to work, we can achieve it.[t]

The amount of work and the challenges involved in developing such learning content and testing modules are huge, but the payoff is enormous. Once the framework is designed and created, we will have a resource that we can continually improve and expand, based on user feedback, as well as new developments and discoveries. Somewhere along the way, when a consensus begins to emerge that the learning modules are effective in helping students to learn, and that the testing modules are indeed objective evaluators of the test taker's proficiency level in the associated content, we will be poised for a paradigm shift in the education system.

For example, we can do away with the "one size fits all" education system, with rigid schedules and time deadlines for all students to attain proficiencies in specific courses at the same pace. What is then likely to happen is that, depending on

[s]E.g., Coursera, edX, Khan academy, Tedtalks ... Many top ranking universities, such as MIT, Yale, Stanford, ..., have created and put out open courseware. California State University's MERLOT collection provides links to as many as 40000 free open course materials, 5000 free online courses and 3300 free e-textbooks.

[t]From numerous conversations I have had with colleagues in India as well as elsewhere over the years, it is my impression that while most of us who are in the teaching profession like the mentoring aspects of teaching, we find it a chore to set and grade assignments and exams. Furthermore, the evaluation is non-standard, grade inflation is pervasive, and the assigned grades are often not true measures of the proficiency attained. If we embrace the creation of open source testing modules, we can transform the process of evaluation into a collective creative enterprise, and take the chore out of teaching.

their backgrounds and abilities, and the availability of mentors, students will take different amounts of time to master a module at specific levels of proficiency. But when they do, the mastery is standardized. When the students complete the learning of a module, they will have many choices: they can put in more effort to improve their proficiency level in the same module if there is room, or go on to a higher level module in the same sub-area, or a different module in the same subject area, or a different module in a different subject area, based on the prerequisites built into the learning modules. It is likely that students will choose to attain different combinations of proficiency levels in modules and subject areas, according to their own individual preferences, abilities and career goals.

The availability of standardized and graded learning content and proficiency tests opens up the possibilities for students who have attained the required proficiency levels in any subject area to themselves act as mentors for the students who are learning appropriate modules that are a few levels lower. People who are already in the teaching profession, but have been inadequately trained, can also use the same learning modules to improve their proficiency levels and become more effective teachers. I see this as a positive feedback process that can eventually eliminate the bottleneck of the shortage of qualified and competent teachers. The opportunities that it creates for the retraining of people who are already employed as the need arises are obvious.

The above paradigm of education also helps to limit the large scale branding of students based on performances in centralized qualifying examinations that is prevalent, which is very stressful for the students. In India, coaching centers that coach students to cram for such centralized examinations are all pervasive. If the testing modules are such that it is impossible for anyone to be coached to do well in the tests without having actually mastered the learning modules, the coaching centers will be forced to turn into education centers. Employers can hire people based on the proficiency levels needed by them, with full confidence that the employees have actually learnt what they are certified to have, and without having to invest in retraining the students as they need to do at present. Students with disadvantaged backgrounds can reach the same proficiency levels as those without, simply by being provided with the required time and assistance in learning — I see this as true affirmative action.

In summary, I see the creation of publicly available, primarily web based, modular learning and evaluation content of the sort envisioned above as the key to cracking the problem of *scale and quality* in education; this especially so in developing countries like India where proprietary content is unaffordable to the majority of the students that need to be educated. The arena of teaching or learning of the publicized or equivalent content will then be open for the dance of human enterprise. Just imagine what the pace and extent of humanity's progress can be when the number of well-educated, productive and creative people is in the billions!

Acknowledgments

The seeds for the opportunities that made it possible for me to get to Cornell were sown in my upbringing, in the encouragement for learning that I received from my parents, both school teachers. Several excellent teachers in my school and undergraduate college helped as well. But the clincher, as I have stated [1], was my stint at IIT Kanpur, where H. S. Mani and T. V. Ramakrishnan, in addition to being great mentors, encouraged me to pick Cornell over Caltech. To all of these people, I owe a lot.

Much to my chagrin, my administrative commitments as the Chair of the Physics Department at the Indian Institute of Science (IISc) (from 2010–14) made it very difficult for me to keep my promise to contribute to this memorial volume within the original time schedule, and I had actually given up on the idea. However, Belal Baaquie, my classmate at Cornell, co-student of Ken, and co-editor of this memorial volume, would not give up on me. My stepping down as chair and coming away on a sabbatical leave to the University of California Santa Cruz (UCSC) opened up the possibility again. I thank Belal and K. K. Phua for their patience and encouragement, and Sriram Shastry, my host, and the Physics Department at UCSC for their enabling support. I also thank Sriram for reading through a draft of the article and making valuable suggestions for improvement.

The bouts of intense concentration required for completing such an article meant that I was not available for many things at home when I was engaged in the writing. I thank my wife, Raj, for her patience and support. She and my son Chaitanya also read through a draft, and pointed out many typos and sentences that could use improvement.

The thoughts on education I have put down here have been evolving over a period of time. They are certainly influenced by my experiences in teaching physics over the past 40 plus years, the many articles and books on education that I have read, and the many discussions I have had with colleagues in IISc and elsewhere, but in ways that are difficult for me to acknowledge specifically. I had occasion to write down some of these thoughts for a draft of an "educational technology initiative" proposal for the International Centre for Theoretical Sciences (ICTS — see http://www.icts.res.in/home/) some years ago, for which I have to thank Spenta Wadia and Avinash Dhar. My thoughts have been sharpened by my recent experiences as the Physics coordinator of the new B.S. program we started four years ago at the IISc, and as an instructor for the "Thermal and Modern Physics" course that is taught in the 3rd semester of this program. In particular, for this course we experimented with supplementing traditional lectures and lab training with computer based adaptive learning, using open course content, and an adaptive learning platform created by an educational technology startup "Lrnr Adaptive Learning Solutions (Pochys Ventures, Inc.)" (website: lrnr.us). I thank Arvind Pochiraju of lrnr for many discussions about education, and for educating me about Blooms taxonomy.

References

1. K. G. Wilson and M. E. Fisher, *Phys. Rev. Lett.* **28**, 240 (1972); K. G. Wilson, *Phys. Rev. Lett.* **28**, 548 (1972).
2. K. G. Wilson and J. Kogut, *Physics Reports*, **12C**, 74 (1974).
3. K. G. Wilson, in *Collective Properties and Physical Systems, Proceedings of Nobel Symposia — Medicine and Natural Sciences*, Vol. 24, p. 68 (1973) (Stig and Bengt Lundqvist, eds.) Academic Press, New York.
4. K. G. Wilson, *Rev. Mod. Phys.* **47**, 773 (1975).
5. H. R. Krishnamurthy, K. G. Wilson and J. W. Wilkins, *Phys. Rev. Lett.* **35**, 1101 (1975).
6. K. Wilson and B. Daviss, *Redesigning Education*, Henry Holt, Inc. (1994).
7. For example, see http://www.ey.com/Publication/vwLUAssets/EY-Government-and-Public-Sector-Reaping-Indias-demographic-dividend/$FILE/EY-Reaping-Indias-promised-demographic-dividend-industry-in-driving-seat.pdf.
8. N. M. Nilekani, *Imagining India: The Idea of a Renewed Nation.* Penguin Press HC (2009). For a brief on the ideas expounded in the book, see the "tedtalk" at the link http://www.ted.com/talks/nandan_nilekani_s_ideas_for_india_s_future?langu- age=en.
9. For example, see http://www.unicef.org/india/education_196.htm
10. E.g., see L. W. Anderson and D. R. Krathwohl, eds., *A taxonomy for learning, teaching, and assessing: A revision of Bloom's taxonomy of educational objectives*, Allyn and Bacon (2000). Other references can be found in the Wikipedia entry on Bloom's taxonomy: http://en.wikipedia.org/wiki/Bloom%27s_taxonomy.

Chapter 20

Kenneth G. Wilson: Renormalized after-dinner anecdotes*,†

Paul Ginsparg

Cornell University, 452 Physical Sciences Building, Ithaca, NY 14853, USA
ginsparg@cornell.edu

This is the transcript of the after-dinner talk I gave at the close of the 16 Nov 2013 symposium "Celebrating the Science of Kenneth Geddes Wilson" at Cornell University (see Fig. 20.1 for the poster). The video of my talk is on-line, and this transcript is more or less verbatim, with the slides used included as figures. I've also annotated it with a few clarifying footnotes, and provided references to the source materials where available.

Keywords: Kenneth G. Wilson; renormalization; after-dinner talk.

20.1. Introduction

I've been invited to give the after-dinner talk to complete today's celebration of Ken Wilson's career. It is of course a great honor to commemorate one's thesis advisor on such an occasion — and a unique opportunity. Earlier today, Ken's wife, Alison (sitting over here), asked me to be irreverent tonight. ("Don't blame it on me," she calls out. "Not hard for you to do," says someone else.) Actually I replied, "You've got the wrong person, I only do serious." Since the afternoon events ran a bit late, there wasn't quite as much time as I'd hoped to absorb the events of the day and

*This article was originally published in *J. Stat. Phys.* **157**, 610–624 (2014).

†I was unable to prepare detailed notes in advance of the talk, so instead transcribed this text more than a half year later from the video of a largely extemporaneous presentation. I retained the run-on sentences for verisimilitude, but deleted multiple instances of the word 'So...', which preceding virtually every other sentence appeared too distracting in print. The footnotes are later additions. The symposium is on line at http://www.physics.cornell.edu/events-2/ken-wilson-symposium/, and the video of my talk is at http://www.physics.cornell.edu/events-2/ken-wilson-symposium/ken-wilson-symposium-videos/paul-ginsparg-cornell-university-after-dinner-talk/.

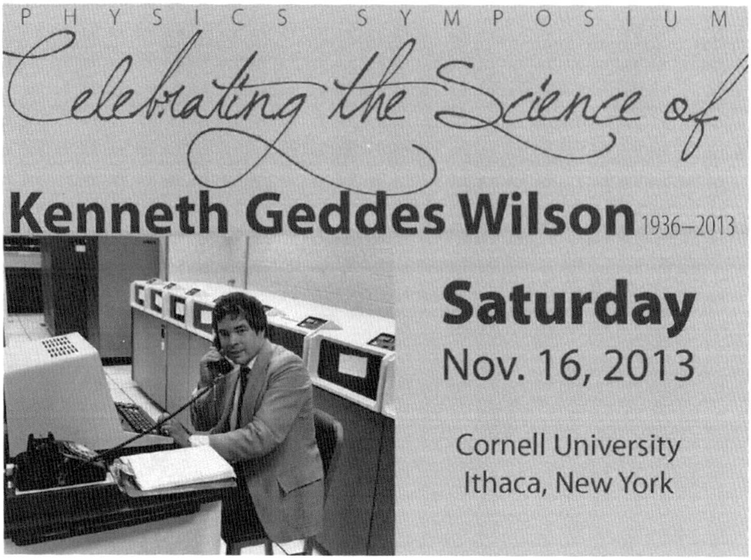

Fig. 20.1. Symposium poster.

think about what to say tonight. But by the standards of this afternoon, I hope you're all prepared to be here until midnight...

As Ken's student, I had only a brief window of interaction with him. Many of his colleagues here knew him from 1963 to 1988 at Cornell, or, like David Mermin, knew him since they were undergrads together starting in 1952. I only had two or three years of interaction with him as a graduate student,[1] after he became the chair of my special committee. But nonetheless as far as I can tell I've retained many more anecdotes about him. When I'd asked the people who knew him for decades, "Do you have any interesting stories?" They'd say simply, "No." "Well did you go hiking with him?" "Yes, I went hiking with him." "Well did he say anything then?" "No, he was very quiet"...

I suspect we all have some special place in our memory for interactions with our thesis advisor. In Ken's case, of course this was very much amplified because we were all sort of in awe of him (as Steve White also mentioned this afternoon). It's not because there was anything special about him personally, just the opposite: he looked and acted like a graduate student, had his bicycle in his office, would crack lame jokes, looked bored in seminars, etc. But we were all aware that this normal-looking guy had transformed the understanding of quantum field theory, of

[1] I was a Cornell graduate student from 1977 to 1981. Ken was chair of my committee starting in 1978. During the academic year 1979–1980, he was on sabbatical in Zurich, and I spent the year at C.E.N. Saclay, outside of Paris; but we met a few times that year. We wrote one article [1] together during my final year, related to my thesis work.

statistical mechanics, and so on. And apparently unlike his brother David,[2] we all knew this mild-mannered, unassuming figure was certain to win a Nobel prize.[3]

So in preparation for this talk, in the last couple of evenings, I've read various historical materials. I will draw from things he's written such as his Nobel lecture,[4] an oral history he did with the Dibner Institute,[5,6] and also some videos[7,8] of him that I've just watched. And I'll try to tell something of a story, putting these together, superficially discussing various periods in his career, and threading in some of the things he said directly to me for additional illustration.

To set the stage, here is a photo (Fig. 20.2) more or less along the lines of how we knew him. It is one of the Nobel promo photos. Amusingly, in so many of these photos we have from that period he's seen wearing a jacket and tie, but of course he rarely wore a jacket and tie in real life...

20.2. Caltech

I'll start with his period at Caltech, and in particular the story of why he became Gell-Mann's and not Feynman's student. To properly explain this story, I have to digress slightly and recall something that was mentioned a few times this morning, which is that Ken's father, E. Bright Wilson, Jr., was also a very prominent academic, and a pioneer in the theoretical and experimental study of the structure of molecules. Like Ken, he was a former Junior Fellow in the Society of Fellows at Harvard,[9] and in 1935 he co-authored with Linus Pauling a textbook "Introduction

[2]His brother David Wilson, a faculty member at Cornell in Biology, had commented during the afternoon session that he'd been entirely unaware that a trip to Stockholm was inevitable.

[3]I remember struggling to decide whether to say something further about the dramatic mismatch between Ken's outsized reputation within the physics community and his lack of name recognition from without. But I didn't really know how to formulate the reasons, and still don't. It could be that his work was so fundamental and far-reaching that it's impossible to encapsulate in a few short sound-bites; or that so much of his creative work was concentrated in a single decade from the mid 60s to mid 70s, and he was not as influential in his later career projects; or that he had little interest in writing popularizations (see, e.g., excerpt #13 of the video transcripts) or self-promotion. In short, he probably didn't care.

[4]`http://www.nobelprize.org/nobel_prizes/physics/laureates/1982/wilson-lecture.pdf`.

[5]Interview with Kenneth G. Wilson, 6 July 2002, conducted by Babak Ashrafi, Karl Hall, and Sam Schweber `http://authors.library.caltech.edu/5456/1/hrst.mit.edu/hrs/renormalization/Wilson/index.htm`.

[6]Ibid., `http://authors.library.caltech.edu/5456/1/hrst.mit.edu/hrs/renormalization/Wilson/Wilson3.htm`.

[7]See the ten video links at `http://www.fuw.edu.pl/~lfqcd/KGWilson-SKFiz/` (15 Nov 2010), each about 15 minutes long; See also the twenty-one video links at `http://www.fuw.edu.pl/~lfqcd/inmemoriam/` (Aug 1994).

[8]The sixteen excerpts I used are at `http://www.cs.cornell.edu/ginsparg/kgw/`.

[9]According to Diana Morse at the Harvard Society of Fellows, "They are the first and only father/son pair both being JFs. E. Bright's dates as a JF were 1934–1936; Ken's dates were 1959–1962."

Fig. 20.2. Busy at work (before the advent of the paperless office).

to Quantum Mechanics" [2], which was for many decades the standard reference text for physical chemists.

This photo (Fig. 20.3) shows Ken with his father at the Nobel prize event. I'll always remember something that Ken said to me regarding his father, the Harvard chem prof: "When I was growing up, I didn't realize that not everyone had a parent who came home every evening to dinner to fulminate against the stupidity of institutions and the people who run them."

We'll see later, in the video excerpts, many positive things Ken says about his father.[10] By a happy circumstance, I also knew his father when I was back at Harvard in the '80s, after getting my doctorate. Bright Wilson and I were both affiliates of North House,[11] so I'd introduced myself and he was congenial. It was slightly surreal I have to say — I always felt like I was talking to Ken, there was something eerily reminiscent about his speaking cadence and use of language. Here's another photo of him (Fig. 20.4), no doubt helpfully opining about the institution of monarchy.

His father had instructed Ken when he went to graduate school at Caltech to introduce himself to physicists there. As Ken wrote (see footnote 4): "When I entered graduate school, I had carried out the instructions given to me by my father and had knocked on both Murray Gell-Mann's and Feynman's doors, and asked them what they were currently doing." The sequel to this was told to me by Pierre Ramond.[12]

[10]E.g., the transcript of excerpt #4 in Sect. 20.6.
[11]An undergraduate residential house at Harvard.
[12]I spoke to Pierre about Ken in Aspen this past summer (July 2013). They met when Ken gave a talk at Syracuse University in 1967, where Pierre was a physics graduate student who followed

Fig. 20.3. Stockholm 1982: Some people really know how to wear white tie and tails.

Ken related to Pierre that he'd knocked on Feynman's closed door, and heard a gruff voice say, "What do you want?" Eventually Feynman opened the door, and Ken introduced himself, "Hi, I'm a new graduate student here, and just wanted to find out what you're working on?" And Feynman shouted, "Nothing!", and slammed the door.

So Ken went next to Gell-Mann, asked what he was working on, and as Ken wrote (see footnote 4): "Murray wrote down the partition function for the three dimensional Ising model and said it would be nice if I could solve it."[13] And that was the beginning of a relationship.

I didn't hear those stories directly from Ken, but for some reason he did once try to illustrate to me what was for him the take-away from Caltech, making it a uniquely weird place. He drew for me on the blackboard a series of concentric circles, sort of as in this photo I found (Fig. 20.5, to the left above his shoulder).

little of the talk. Coming from an undergraduate engineering background, Pierre said he also never really understood momentum space, but at least Ken was working in position space, so apparently they hit it off and kept in touch through the years.

[13]The one-dimensional Ising model was solved by Ernst Ising in 1924. The two-dimensional Ising model was solved by Lars Onsager in 1944. The three-dimensional Ising model has still not been solved exactly. Ken also refers to this problem in the transcript of video excerpt #7 in Sect. 20.6.

Fig. 20.4. So tell me, is 'king' some sort of administrative position at your institution?

Fig. 20.5. Office blackboard (handwriting to the right was later claimed by Peter Lepage).

Fig. 20.6. Caltech men's room graffiti.

This isn't the actual one he drew, unless his blackboard went four years without being cleaned. He described it as: "This is graffiti I saw in a men's room at Caltech. Underneath it, the caption said, "photon: full frontal view". That was about as risqué as it got with him (Fig. 20.6).

20.3. Cornell

Moving on to his time at Cornell, there was discussion this morning about the tenure process here for Ken, and whether there was concern over his small number of publications. Some of you here now were involved in that tenure vote in the 1965 timeframe, and everyone I've asked has said, "No, it was never an issue."[14] Following a discussion once with Vinay Ambegaokar, I tried to find the actual number of publications back then and counted *at least* three [3–5] that he had written before coming to Cornell in 1963, but there was only one article [6] that would have counted as a publication from *during* his first two years at Cornell.[15] Evidently referring to that publication, Vinay told me his comment at the meeting was something along the lines of, "Too bad he spoiled an otherwise perfect record by publishing an article." Ken himself, who would not have been present at this meeting (his tenure evaluation), related something complementary (see footnote 5): "Francis

[14]Explaining the lack of doubt on the part of Ken's colleagues about granting tenure, Vinay A. recalls: "It was obvious from spare comments in seminars and conversations that he *understood.*"
[15]From the audience, Michael Peskin pointed out that Ken had also worked on the multiperipheral model during that period, and kept returning to it at later times in his career.

Low complained that I should have made sure there was none. Just to prove that it was possible."

In Aspen this past summer, I also spoke to Steve Berry, a chemist at the University of Chicago, who happened to have been a graduate T.A. in a course Ken took as a freshman at Harvard. As a chemist there, Steve also came to know Ken's father, kept in touch with him, and during that period says Bright Wilson had expressed his concern that Ken was working on such obscure stuff that he would not be able to get tenure. So no one was worried, except his father.[16]

Ken's father's concern persisted over time. Shortly after Ken won the Nobel prize, I was at a North House function, Bright Wilson spontaneously came up to me, and without preamble said, "You know, I think Ken's going through a mid-life crisis."

20.4. Thesis advisor

Now I turn to some of my experiences as a graduate student here.

I had Ken as lecturer for a wonderful topics course in gauge theories.[17] (There's at least one other person here, Gyan Bhanot, with whom I remember taking this course; Serge Rudaz is also here but he didn't seem to go to many courses by then.) I also audited a graduate course he taught in quantum field theory, where he explained all of renormalization in terms of $\lambda\phi^4$ theory. At one point, he broke things up into momentum slices, pointed out the logarithmic divergence, and to make it vividly physical, described this as, "You see, each momentum scale contributes its penny's worth, but at the end of the day the sum is more than the U.S. national debt."

His advice to me as a graduate student, when I asked him whether I should work on a programming problem (I had some prior experience with computers, and of course he was known for his interest in them), was eminently sensible. He said, "You shouldn't choose a problem on the basis of the tool. You start by thinking about the physics problem, and the computational method should be a tool like any other. Maybe you'll solve it using computer techniques, maybe using a contour integral; but it's very important to approach it starting from the physics because otherwise you get lost in the use of the tool, and lose track of where you're trying

[16] I was told by someone at the meeting that his father later described it as "a courageus decision" on Cornell's part. Pierre Ramond told me that he saw Ken at the Aspen Center for Physics one summer, probably 1970, busily writing up eight articles (appropriately enough) in Bethe Hall. Within a couple of years, Pierre says Ken said to him, with a twinkle in his eye, "You know, I think I'm going to get more citations than my father."

[17] This would have been the spring of 1979, and the course was audited by many of the postdocs and visitors, and some faculty members as well. During the course, he calculated the QCD Beta function for n flavors of quarks, and told us that it was the first time he'd finally gotten around to doing the perturbative calculation. He'd also said that years earlier he hadn't even wanted to learn the non-abelian group theory, finding the manipulations of structure constants f_{abc} too tedious, until he understood it was essential for understanding the strong interactions.

to go." He said he'd seen lots of people get in over their heads in the computing and never get back to the physics.[18]

I took that advice to heart and didn't do anything particularly computational while I was his graduate student. But when I was reading through things that he'd written, I wasn't completely convinced he'd been true to the spirit of his own advice. When asked why he worked on the Kondo problem, his answer was (see footnote 6):

> It comes from my utter astonishment at the capabilities of the Hewlett-Packard pocket calculator, the one that does exponents and cosines. And I buy this thing and I can't take my eyes off it and I have to figure out something that I can actually do that would somehow enable me to have fun with this calculator... What happened was that I worked out a very simple version of a very compressed version of the Kondo problem, which I could run on a pocket calculator. And then I realize that this was something I could set up with a serious calculation on a big computer to be quantitatively accurate.

So we now know the origin of his inspiration for the problem.[19]

Also mentioned this morning was his 1971 article [7]. When I asked him in '79 where he'd written down the naturalness[20] idea, he replied with a smile, "Oh, that was in my paper with all possible theories of the strong interaction, ... except the correct one,"[21] and pointed me to the relevant paragraphs. The actual words in this article [7] are as incisively clear as ever:

> It is interesting to note that there are no weakly coupled scalar particles in nature; scalar particles are the only kind of free particles whose mass term does not break either an internal or a gauge symmetry.

---

[18]A decade later, I did get involved in a computer database project, and did get in over my head.

[19]In Aspen (2013), Elihu Abrahams told me, with some sheepishness, about having given a talk about the Kondo problem at Princeton in the early 70s. Afterwards, Ken said to him, "I have some ideas about this problem, would you like to work on it together?"

[20]Starting in the late 70s, there was increased discussion of the "naturalness problem": the still-unresolved question of why a light scalar particle (the Higgs) is available at "low" energies to spontaneously break the electroweak symmetry. It is one of the many theoretical constructs that crystallized from his work, and remains with us, as part of the impetus behind supersymmetric model building and recent (so far unsuccessful) searches for supersymmetric partners to known particles.

[21]In (see footnote 4), he wrote, "I should have anticipated the idea of asymptotic freedom but did not do so." In 2002 (see footnote 6), he said in addition, "I was among the last people to climb on board the quark idea ... Then I developed the presupposition ... that the Beta function would always have the wrong sign. There's a paper that I wrote in 1971 about renormalization group and field theory, and I discuss various alternatives for the Beta function, [and] omit the one case that turns out to be correct. I don't mention asymptotic freedom as a possibility because I hadn't worked on gauge theories, and I just took for granted that the Beta functions would have the other sign."

In 2010 (see footnote 7), he goes a bit further to say that he was so convinced that the gauge theory Beta function would have the other sign, he simply assumed that the deep inelastic scaling results were some intermediate phenomenon, and that some other behavior would take over at much higher energies.

In my early student days, people told me that Ken Wilson's articles were oh-so-difficult to read, and that nobody understood them at the time. The measure of how much he changed the research community was that little more than five or ten years later when I read that article, we were so thoroughly trained in the new paradigm, it read easily and beautifully: he set up the quadratic mass divergence, and then gave a blindingly clear description of the problem that we're still grappling with. And it's no longer easy to imagine what people were confused about by his articles from that period.

In person, he elaborated to me about setting the values of bare parameters (whether or not this constitutes more physical terms I'm not sure), and said, "Suppose you're throwing darts at some cosmic dartboard. You're just not going to hit with 1 part in 10^{38} precision if you're trying to get a 1 Gev particle starting from the Planck scale." This way of thinking about physics became embedded in me, and was something that I always appreciated coming directly from him.

One final comment about my graduate student experience: after Ken had looked at the first draft of my thesis, he said to me, slowly and carefully (*perhaps* not specifically referring to the thesis):

> I was once given the advice when writing that you should go back, find all of your favorite sentences, *and delete them.*

I continued to convey that to others, perhaps not necessarily regarding what they had written either, as advice given to me by my thesis advisor. It wasn't until roughly 25 years later that I mentioned the comment to Steve Strogatz, who informed me, "But that's famous literary advice from Quiller-Couch." Arthur Quiller-Couch, a well-known British author and Professor of English at Cambridge, wrote a style book called "On the Art of Writing" (1916),[22] and the actual quotation is:

> Whenever you feel an impulse to perpetrate a piece of exceptionally fine writing, obey it — wholeheartedly — and delete it before sending your manuscript to press. *Murder your darlings.*

So it was additionally fun to realize so much later that I'd long been propagating Ken's literal (i.e., physicist) transcription of this classic stylistic counsel (Fig. 20.7).

20.5. Eighties and beyond

Four years after getting my degree, I returned here in 1985 to give a seminar, and by then Ken was no longer at his office in Newman lab,[23] having moved to the nascent

[22] Sir Arthur Quiller-Couch, On the Art of Writing (1916). See section XII at `http://www.bartleby.com/190/12.html`.

[23] During my time at Cornell, he'd had an office on the 2nd floor among the experimentalists, then later moved to the 3rd floor and joined the theorists.

Fig. 20.7. Ken with Alison. This photo is here only because during dinner David Wallace asked for a photo of Ken in a kilt, so in real time I added it to the slides.

Theory Center he'd founded. I remember walking into his office, unannounced, he looked up and I couldn't tell if he was focusing on me or not. And then sort of like his father, without preamble, he said, "As you know, we've decided we don't have enough enough computational power to extract experimental numbers from lattice QCD." (I still don't know if he said that to anyone and everyone who walked unexpectedly into his office.)

During that period of the 1980s, he continued his proselytizing for the standardization of computer use (which we heard a lot about this morning). It was one of the many things that had very strongly influenced me as a graduate student: hearing him complain that physicists should be able to travel from one institution to another, then sit down and log in directly to their home institution, or anywhere else, and get immediately to work without learning a new operating system.[24] He

[24]In the pre-Apple/pre-PC days, it was typical for computer companies to introduce a new operating system for each new computer. Adopting a more open standard, such as Unix, together with compatible networking protocols, offered the possibility of a uniform interface for access to remote resources. For the Cornell high energy physics group in the 80s, Peter Lepage recalls that Ken insisted to run Unix on the DEC VAX, IBM PC's and Sun workstations they'd purchased.

very firmly had on his mind the notion of ubiquitious network connectivity, frequently spoke of the need for arrays of commodity processors to leapfrog Moore's law,[25] and also was promoting better designed programming languages for parallel computing.[26] For this reason, it came as no surprise to me when sometime in the mid '90s, George Strawn, who was among those responsible for shepherding the NSF-end implementation of the NSFnet in the mid 80s,[27] told me that on the 1982 taskforce [8] that resulted in the NSFNet recommendation, Ken was among those who argued for using the TCP/IP (Transmission Control Protocol/Internet Protocol) — the basic communication language behind the thing we now call the Internet. Apparently in the early 80s, other scientists on the committee wanted to use DECnet[28] because that's what they were familiar with, and Ken had the vision to see that we'd be better off moving in the more open Unix direction. Strawn said Ken's voice was very influential, and contributed to our having the current Internet.

George Strawn told me something else back then that I found fascinating, and will pose as a question to the audience here. He told me from his direct contemporary experience that there were two people in congress, one in the Senate, and one in the House of Representatives, who were absolutely essential to getting the Internet through congress and funded. Can anyone guess who those two people were? (Someone guesses 'Al Gore', everyone laughs...). Al Gore, Jr. is correct! He was the one who championed it in the Senate. And who in the House? (Various guesses, no one comes close.) In the House, he said it was: Newt Gingrich! Maybe it seems crazy, but then when you think about it, you remember he's always interested in off-the-wall things, and in promoting science and technology, so it actually makes sense in some respect.

Ken also shared with us his intuition for how ubiquitous computing would become. Peter Lepage mentioned this morning that as soon as he found out about the networking possibilities, he anticipated physicists using it for their social lives — though I don't know that he anticipated the rest of the planet adopting it for their social lives as well.

[25]The factor of two increase in processing speed every few years he said was not fast enough for many problems of interest. In this respect, he foresaw the development of commodity Linux clusters that became commonplace by the late 1980s.

[26]In the late 70s, he used to optimize code for lattice gauge theory simulations, running on a Floating Point Systems array processor, directly in assembly language. High quality optimizing compilers for parallelizing Fortran did not yet exist. In a physics colloquium back at Harvard in 1982, he highlighted the Gibbs project, an attempt to produce a new scientific programming language. He and Alison came to dinner with me that night at the Society of Fellows, and continued a memorable discussion.

[27]George Strawn later became NSF CIO (Chief Information Officer). When contacted in summer 2014, he added that Larry Smarr had also played an important role.

[28]DEC was the Digital Equipment Corporation, which produced the PDP-8, PDP-10, PDP-11, and later VAX lines of computers, popular among physicists in the 70s and 80s. DECnet was built into the VAX/VMS operating system, and used for both remote connectivity and e-mail by the early 80s. The company went defunct when bought in 1988 by Compaq, which then merged with Hewlett Packard in 2002.

Fig. 20.8. Toasted for 1982 Nobel prize by Hans Bethe.

And here is an email from Michael Peskin to John Cardy (apparently intercepted by the NSA) [9]:

> In 1976, he gave us a lecture about how, one day, we would all be sitting on the beach with personal computers that ran UNIX, and using them to play games that involved exploration in three dimensions. In 1976, that seemed like a dream.

Now I'm NOT trying to say that in 1976 he knew that Steve Jobs was about to found Apple computer (with Wozniak), that Jobs was going to be booted from Apple in 1985, start his own computer company NeXT that would have a new operating system based on Unix, release a Unix-based computer in 1990, in 1996 Jobs would come back to Apple, then evolve that operating system into Mac OS X which would then evolve into iOS and then be installed on the iPhone, and become the device people use on the beach when they're playing their 3d games. But I wouldn't be surprised if he...

I have a couple of other photo comments. This photo (Fig. 20.8) of a Nobel celebration in Newman Lab at Cornell was taken a year after I first left, but happily there are a few people I still recognize. There's this hippy guy standing back here in the right corner, and here in the mid-left is an upper-head I can recognize just from the eyes. I'm also very happy to see Velma Ray at the far left (loud cheers from small subset of audience) who many of us here (Serge R., Gyan B., Michael P., Steve S., ...) will never forget. Velma was Hans Bethe's secretary, an integral

Fig. 20.9. Why are these two separate devices?

part of our Newman experience, and the person who typed all of our theses in the pre-TeX era.

Finally here's a photo (Fig. 20.9) I recommended for the conference poster, basically encapsulating Ken as we knew him at that age, even though clearly posed. (It's not as though the photographer happened to be present when Ken was computing away, received a phone call, picked up the phone, and started talking.) I recall for some of you that the large device on the desk is a physical object, known as a VT100, as opposed to the name of a terminal emulator that runs in your windowing system. When I look at Ken on the phone with his hand on keyboard, in my mind he's thinking "Why are these two separate devices?"

20.6. Video transcripts

I'm going to let Ken himself have the last word tonight. For this, I have to give a special thanks to Stan Glazek (who's here). He pointed me to a number of illuminating videos (see footnote 7) from a meeting Ken had with students in Poland in 2010, and also some of him giving lectures at a school in 1994. Last night, I watched about 2.5 h of these videos (which was pure fun, and much shorter than 2.5 h because audio on computer is sped up by clipping just the pauses, so you can listen at normal pitch and not miss anything). From this exercise, I picked out about 12 min of excerpts (see footnote 8) to finish up here.[29] There are many comments

[29]It is worth watching the twelve minutes of on-line excerpts (see footnote 8) transcripted here, to hear the stories directly from Ken as the audience did.

Ken makes about his early life and graduate student experiences that few of us ever heard from him, so I'm hoping this will be novel to most of you. We're kicking ourselves here for not having invited him to come back and talk to us over the last ten years — you never realize until it's too late... But it's marvelous, after hearing so much about him all day, we can now give him the last word.

The first video excerpt is from 1994, just to illustrate to everyone his lecture style. The final clip is very short, in response to a student who asked, "When is the appropriate time to stop taking classes, and when should one start doing research?" And Ken's answer will be a perfect place for us to stop.

0. KGW: But I want to start by reminding you that in a certain sense the discovery of the formal rules of QCD represented a step backwards. (smile)
1. KGW: When I was a child, what I loved to do was mathematics and not much else, other than some climbing hills..., paddling kayak..., but I did have a wonderful project in 8th grade at school, which was making a steam engine, which is probably as close as I came to doing serious science until I came to college.
2. KGW: My earliest memories are being fascinated by the numbers on the steam engines that came every day to Woods Hole, which is near Boston but on what's called Cape Cod, and I kept records of the numbers of every steam engine coming in.
3. KGW: But by fifth grade, I was developing 'serious' mathematical skills. Namely I was taught to take cube roots of numbers by my grandfather, and I was so fascinated by the algorithm that I used to practice it in my head waiting for the bus going to school.
4. KGW: There was a national mathematics competition for students in college,[30] and I started participating in that math competition as a freshman and managed to do well enough that I got invited to a dinner by all the mathematics faculty at Harvard, as part of reward for doing well in the competition. And that was an amazing experience, since I learned that mathematics professors were nothing like my father, who was a chemistry professor.

   SG: What was the main difference?
   KGW: The mathematicians were crazy (audience laughter)... they were wonderful at mathematics but they didn't seem to me to be all that wonderful at real life, whereas my father understood real life, he understood chemistry, he understood physics. In fact he published a book which he published around 1950 which he called "An Introduction to Scientific Research".[31]
5. KGW: I knew after the 8th grade that I was going to do physics, and I knew the reason. I was going to do physics because I was certain that would give me

[30]The Putnam Competition.
[31]See [10].

more interesting problems to solve than if I'd studied mathematics, and nothing that happened in my college mathematics major[32] changed my mind.

6. KGW: I go to graduate school in physics, and I take the first course in quantum field theory, and I'm totally disgusted with the way it's related. They're discussing something called renormalization group, and it's a set of recipes, and I'm supposed to accept that these recipes work — no way. I made a resolution, I would learn to do the problems that they assigned, I would learn how to turn in answers that they would accept, holding my nose all the time, and someday I was going to understand what was really going on. And it took me ten years, but through the renormalization group[33] work I finally convinced myself that there was a reasonable explanation for what was taught in that course.
7. KGW: I met with Murray Gell-Mann to get a thesis problem, and he suggested a problem.[34] And I thought there was no way I was working on that. So he had to come up with another one, and it didn't take him very long. The thing that impresses me most about Murray Gell-Mann when I think back on it, is how quickly he got things done. I have never met anybody who could do things as fast as he could. And so it didn't take him very long to come up with another problem for me to work on. And so he asked me to think about an equation, which was called the Gell-Mann–Low equation, and it was an equation which was of interest to him for what he would call low energy phenomena. The first thing I did with that equation was study it in the limit of high energies, the exact opposite limit to anything he was interested in. As I said, it was a small problem, but it led over ten years to larger and larger problems and to some of the key publications, in fact including the publication that was cited for the Nobel prize. But I did it by working my way from small to large, not from starting at the large from the beginning.[35]
8. KGW: There was a year at Caltech when Murray Gell-Mann was off in Paris, and so Feynman was in charge of a seminar which met every week. And most of the time he would come in and say 'what should we talk about today?'

[32] During the first video from (see footnote 7), between excerpts #3 and #4 above, he explained, "I was only majoring in mathemematics because they had a thesis requirement for mathematics and I wanted to do a thesis, and physics didn't have any equivalent." Regarding its subject matter, he said, "I had done some research on propagation of sound underwater, in summer work at Woods Hole, and I believe I continued that work for my math thesis, but as far as I know I have no copy... and no memories of what was in it." As influential undergraduate experiences, he mentioned taking a sophomore year math course taught by George Mackey, and a course in American intellectual history taught by Arthur Schlesinger, Jr.

[33] In (see footnote 6), Ken said: "In retrospect, I probably made a mistake in giving it the same name. I probably should have given a name to distinguish the approach with one coupling and the approach with infinite couplings... so that... Gell-Mann–Low was renormalization group A and my work was renormalization group B. We would have gotten away from the arguments about everything reducing to perturbative renormalization group theory."

[34] The problem was to solve the 3d Ising model, see Sect. 20.2.

[35] Just as the renormalization group itself works...

And there was one seminar when he writes that question and sort of no decent conversation got started, and eventually he noticed that I was talking with my neighbor. He said, 'What are you talking about?' and I said, 'We're talking about a 16th century mathematics theorem which happens to be called "Wilson's theorem." And the next thing I knew I was up — he had me up at the blackboard explaining what this theorem was and how you proved it. And the theorem is that if n is a prime number, and only if n is a prime number, then $n! + 1$ is divisible by n.[36]

9. KGW: In a discussion with other students, we were talking with Feynman and one of the students asked Feynman, "Do you have anything that you noticed about really exceptional physicists, that characterized exceptional physicists and not normal physicists?' And he said, 'Yes, the thing that characterized them was persistence.' That they wouldn't give up, it didn't matter how long it took. And this was in the 1960's when we had this discussion.
10. KGW: When I turned in my thesis, which was based on this problem, Gell-Mann was off in Paris, so Feynman was the person who had to read my thesis. So I'll now tell you an anecdote, that after Feynman had read my thesis, it was customary to give a seminar on one's thesis. So I gave a seminar, and Feynman is there. And in the middle of it, another faculty member raises his hand and says, 'I find your discussion interesting, but what good is it?', which I had no answer for. And Feynman pipes up, and this is with an English colloquialism which I'll explain, he says, 'Don't look a gift horse in the mouth.'
11. KGW: I learned a lot that came from Gell-Mann, but not from interacting with him personally. He was the inventor of the concepts of the renormalization group in a paper that was published by him and Francis Low. But to learn about that, I didn't learn anything about that from Gell-Mann directly. I learned it from a textbook written by Bogoliubov and Shirkov, which had a chapter on the renormalization group. One of the other very important things I learned from Gell-Mann I didn't learn until very recently. I was at a celebration of his career, and he remarked that he had learned from Vicki Weiskopf that you had to ruthlessly simplify the physics of a problem that you were working or learning on.
12. KGW: Sort of the postscript to this story is thirteen years later, I found there was a problem in condensed matter physics very similar to the Gell-Mann–Low equation, but more complicated, but not so complicated that I couldn't

[36]He meant $(n-1)!+1$, and later in the video corrected that, when asked how to prove it. "Wilson's theorem" (stated by John Wilson and Edward Waring in 1770, earlier by Ibn al-Haytham c. 1000 AD, and first proved by Lagrange in 1771 (`http://en.wikipedia.org/wiki/Wilson's_theorem`)) is usually stated in the form: $(n-1)! \equiv -1 \pmod{n}$ iff n is prime. To prove it, first notice that if n is not prime, then $(n-1)! \equiv 0 \pmod{n}$, since its factors occur in the product. If n is prime, then in modern language we would note that the positive integers less than n form a group (G_n) under multiplication mod n. Each element has an inverse distinct from itself, except for $n-1$, which is its own inverse, so $(n-1)! \equiv n-1 \equiv -1 \pmod{n}$.

program it up and find worthwhile results on the computers — which were called supercomputers at that time — but which now can't even compete with an iPhone.

13. KGW:... and then figuring out what happens at about double the scale, and then figuring out what happens on double scale again. It's complicated to develop the actual procedures, but there's a Scientific American article from the late 1970s which tries to present this kind of thinking to a lay audience. I can't quote you the reference off-hand but anyone who looks up in the index of Scientific American[37] from them can find out where this article appears. And I'll say only one more thing about that article: Scientific American wanted me to write it, which is the way they usually do it, and then they assign an editor, and I told them 'nothing doing.' And finally they agreed to an arrangement where one of their editors would write the article and I would edit it. That was a very interesting experience.
14. KGW: The U.S. Congress in 1990...

    SG: You are filmed, you know it.
    KGW: What?
    SG: You are on the film, so if you criticize the U.S. Congress...
    KGW: I'm not... No, this is an amazing congressman, he was incredibly smart (audience laughter), named George Brown,[38] and he was a friend of science, and he wrote an article in which he said he had found advice from scientists, from physicists in particular, extremely valuable, because they would get interested in a topic, and they just wanted to understand it.
15. KGW: My opinion is very simple: you start research at birth, and you should never stop.

Acknowledgements

I thank the organizers of the symposium, Jeevak Parpia, Csaba Csaki, and Jim Sethna, for inviting me to give the after-dinner remarks, and Alison Brown for the photos used in Figs. 20.3, 20.4, and 20.7.

References

1. Ginsparg, P.H., Wilson, K.G.: A remnant of chiral symmetry on the lattice. Phys. Rev. D **25**, 2649 (1982). doi:10.1103/PhysRevD.25.2649

[37] See [11].
[38] George Edward Brown, Jr., Democratic congressman from California, was chairman of the Committee on Science, Space and Technology from 1991–1995, and known as a champion for science (http://en.wikipedia.org/wiki/George_Brown,_Jr).

2. Pauling, L., Wilson Jr, E.B.: Introduction to Quantum Mechanics with Applications to Chemistry. Courier Dover Publications, Mineola (1935).
3. Wilson, K.G.: Proof of a conjecture by Dyson. J. Math. Phys. **3**, 1040 (1962)
4. Amati, D., Stanghellini, A., Wilson, K.: Theory of fermion regge poles. Nuovo Cimento **28**, 639 (1963)
5. Wilson, K.G.: Regge poles and multiple production. Acta Phys. Aust. **17**, 37 (1963)
6. Wilson, K.G.: Model Hamiltonians for local quantum field theory. Phys. Rev. **140**, B445 (1965)
7. Wilson, K.G.: Renormalization group and strong interactions. Phys. Rev. D **3**, 1818 (1971). (see p, middle of left hand side column)
8. P. Lax et al., Report of the Panel on Large Scale Computing in Science and Engineering (26 Dec 1982), sponsored by DOD, NSF, DOE, and NASA, `http://www.pnl.gov/scales/docs/lax_report1982.pdf`; and later mentioned in G. Strawn, You Ain't Seen Nothin' Yet, EDUCAUSE review, Jul/Aug 2006, p. 8, at `https://net.educause.edu/ir/library/pdf/ERM0645.pdf`; Ken had earlier served on another advisory subcommittee, W. H. Press et al., Prospectus for Computational Physics. Report to the NSF Physics Advisory Committee (15 Mar 1981), `http://www.nr.com/whp/NSFCompPhys1981.pdf`
9. Cardy, J: The legacy of Ken Wilson. J. Stat. Mech. (2013) P10002. http://arxiv.org/abs/1308.1785 doi:10.1088/1742-5468/2013/10/P10002
10. Wilson Jr, E.B.: An Introduction to Scientific Research. McGraw-Hill, New York (1952). ISBN:0486665453
11. Wilson, K.G.: Problems in physics with many scales of length. Sci. Am. **241**, 158–179 (1979). doi:`10.1038/scientificamerican0879-158`

Chapter 21

My memory of Ken Wilson

A. Zee
Kavli Institute of Theoretical Physics
University of California
Santa Barbara, CA 93106, USA

In this article, I will talk about the impact Kenneth Geddes Wilson (1936–2013) had on my career.

One spring day in 1970 Roman Jackiw[1] offered to pay me to visit Aspen, which sounded to me like a really good deal. At the time I was Sidney Coleman's student at Harvard.[2] Why an MIT professor[3] would take an active interest in my career (for which I have always been grateful) is in itself a mildly interesting story. I had arrived at Harvard to discover that I was required to take a nuclear physics course which Harvard did not offer and to pass a foreign language exam.[4] I asked Sidney how I was supposed to take a non-existent course, and he said, in his characteristically biting manner,[5] that there's a technical institute down the street. So a couple times a week, a group of us would trudge down Massachusetts Avenue to attend Herman Feshbach's class. I remember Herman saying to our little group something like I know you people are not truly interested in nuclear physics but are here under duress.[6] After class, I would often hang around, and discovered, to my

[1]It is only now, when I read Jackiw's contribution to this memorial volume, that I realize that he was one of Ken's two first students.

[2]John Wheeler had practically ordered me to go to Harvard to work for Steve Weinberg, who had just arrived there from Berkeley, but Steve Adler advised me to work for the youngest guy in the group, the same advice I now give students who show up in my office.

[3]I am writing this article in Seoul, where Roman has a tremendous influence.

[4]After a few months studying German by reading dual language books in German literature, I aced the exam, which consisted of translating Newton's three laws of motion from German to English. Talk about easy!

[5]Alas, I no longer remember even an approximate version of what he said, but surely very witty and derogatory at the same time.

[6]He was rather kind to me, and told me that I could indulge my interest in symmetries by writing a report about Wigner's $SU(4)$. Perhaps this was good training for a future writer of textbooks.

utter amazement, that at MIT aspiring theoretical physicists were treated as some form of humans with the potential of joining God's chosen people. At least, we did not feel the need to pretend to be brilliant. In this more relaxed atmosphere, I got to know the junior theory faculty at MIT, in particular Roman, who at one point suggested that I look into the Deser–Gilbert–Sudarshan representation.[7]

After a few years of fooling around, which included, but not limited to, rioting and being tear-gassed by mounted police in Harvard Square, I agitated to Sidney that I wanted to be finished with graduate school and to get married. He responded by threatening to send me to a large midwestern state university as a postdoc if I didn't do something up to his standards. I said, rather defiantly, fine, I will go to the midwest. But overnight he relented, and told me the next day that he had arranged for me to go the Institute for Advanced Study.[8] It was at this point that Roman offered to pay for a month in Aspen.[9]

When I arrived in Aspen, the head secretary dismissed me with one glance and assigned me to a rather dark, almost windowless, basement apartment. I was later told that this woman treated people according to their status as perceived by her. I certainly had no complaint since I was barely out of graduate school. But apparently she didn't think much of Ken Wilson either; in the summer of 1970 Ken's days of fame and glory were still in the future. She might have taken him for an abecedarian like me. Ken had always looked boyish, and at the time he was only 34. Anyhow, I soon discovered that I was to share the basement with this rather odd Cornell professor.

Whenever I saw Ken in the apartment, he played solitaire. I believe that this was his way of relaxing himself into thinking deep thoughts. I am no longer sure — after all, it has been more than forty years — but I think that he also read pulp novels. Surely, I thought, that must be how a genius is supposed to behave. The two of us went to dinner — and it was invariably just the two of us — almost every night. I could have added to the legend by saying that we always went to the same restaurant and that Ken always ordered the same thing, but no, we would patronize two or three different restaurants and Ken did vary his selections. I know this for sure, because I was struck by how he calculated[10] to the last penny the amount we each owed when the bill came. I was, however, completely on my own when deciding how much of a tips to leave.

---

[7]Whatever that is! Never mind what it is, or was, I duly published my work: *Phys. Rev. D* **3** (1971) 3. Incidentally, in his book on the photon hadron interactions (page 185) Feynman denounced the DGS representation. Sid Drell made similarly negative comments at the time. I thank Sandip Pakvasa for telling me about these references.

[8]By the way, I did have a wedding reception on the lawns of the IAS a year later.

[9]I now realize that Jackiw was extraordinarily kind to me: not only was I not his student, I was not even a student at his school.

[10]In hindsight, it would seem that the problem could be solved by asking for separate checks, but that solution was somehow never discussed.

Since we spent a whole month together in that basement, we must have talked about physics quite a bit, although most people who knew Ken would agree that he was not one of the most talkative[11] guys in theoretical physics. I just looked up online[12] Ken's famous paper "Renormalization Group and Strong Interactions" (received 30 November 1970) and saw that in the acknowledgment he thanked me for reading the manuscript. (It also stated that the manuscript was completed at the Aspen Center for Physics.) I also see that the paper was still written in the dreadfully confusing language used by Gell-Mann and Low in their famous work. Most people breathed a sigh of relief[13] when the Callan–Symanzik equation came out in 1970. They certainly did not appreciate Ken's work at the time.[14]

In hindsight, I could only wish that I had understood everything Ken Wilson had no doubt told me as he was formulating his thoughts. Evidently, I didn't, but still, something must have seeped through to my consciousness. In the spring of 1972, I read Sidney Coleman's Erice lectures from the previous summer. After explaining the CS equation, Sidney gave a rather pessimistic assessment of the renormalization group approach, saying that this analysis, while nice, was all but worthless. How were we going to study the theory around some unknown fixed point g^*? The thought occurred[15] to me immediately: But what if g^* were zero, then we could calculate everything!

I proceeded to set up the calculation of the beta function. I thought that I would warm up with the Yukawa theory and then work my way up to the non-abelian gauge theory, which was just then rumbling on the horizon. I was also about to move to New York City.[16] In hindsight, I made the big mistake of trying to master non-abelian gauge theory by reading the formal mathematical paper by Ben Lee

[11]Once, when my then wife and I ran into Ken at the IAS, she greeted him and remarked on the nice weather. Ken looked up at the sky, thought, looked down at the ground, and then said pleasantly, "Yes, it is!" However, I do not want to give the impression that Ken is anywhere near one of the legends of profound silence in theoretical physics, talking about which I may perhaps mention that I had the uncomfortable honor of lunching once with Paul Dirac and once with John Bardeen. In contrast to these two, Ken did talk, and was sometimes even chatty, as during those dinners long long ago.

[12]Who could have imagined in 1970 what this phrase might possibly mean?

[13]I think that it is interesting to state, as a comment on how theoretical physics develops, that at the time, people thought that the CS equation had swept away Wilson's almost incomprehensible mumblings about "3.99 dimensions." Instead, the Wilsonian point of view has come to dominate physics, especially condensed matter physics, while most young people now are not even familiar with the CS equation.

[14]Decades later, when I wrote my textbook on quantum field theory, I was thoroughly struck by how profoundly Ken had turned the subject around.

[15]For some reason, I have a very vivid image of this. I was lying on the couch, with the sunlight streaming in, in my apartment on Prospect Avenue in Princeton not far from the fabled clubs.

[16]That move involves another unhappy story. Literally hours after Bram Pais had pressured me into accepting an assistant professorship at Rockefeller University, Curt Callan called to offer me an assistant professorship at Princeton University. Sam Treiman told me that I had to honor my commitment. Still, my wife and I, being young and having just spent two years in quiet dull Princeton, were excited to move to the Big Apple.

and Jean Zinn-Justin (published in 1972) rather than the more physical paper by Gerhard 't Hooft. Another factor[17] was the enormous excitement[18] at the time over the possible unification of the electromagnetic and the weak interactions.

It was David Gross who "saved" me from New York City. I ran into him at some physics event in early 1973; we chatted about what we were doing, and to my amazement and delight, he offered me an assistant professorship at Princeton University on the spot. And so in the fall of 1973 I moved back to Princeton and promptly met Frank Wilczek[19] but that's another story.

I am coming back to Ken Wilson presently. In May 1973, I was invited to a small conference[20] on the renormalization group, a subject few people had heard about at the time. I was happy to see my old apartment mate, now famous, and to meet other luminaries I knew only by name, in particular Kurt Symanzik.[21] Ken said to me, "You must be crying in your soup." Even after forty odd years I remember those as his exact words. He urged me to calculate something quick and publish.

I worked out the consequences of asymptotic freedom for electron positron annihilation fast, practically in my head. I realized that I could take the existing two-loop calculation of vacuum polarization and simply supply the appropriate group theoretic factors. When I got back to New York,[22] I immediately[23] wrote a paper.[24] I am proud to say that at the time my theoretical prediction flatly contradicted the experimental measurements. Now looking back, I realized that this work would have been child's play for Ken, and he was exceedingly generous in looking out for my best interests. Most young physicists back then were far less sophisticated than their counterparts today; I would have had no idea how to salvage the research project I have been working on and off for more than a year.

Over the years, I saw Ken less and less; neither Cornell nor Ohio State was on the circuit of the traveling circus I belonged to. I did once hear Ken lecture (perhaps in the 1980's) with the conviction of a zealot, telling the audience that if they did

---

[17] I was also busy with my work on the chiral anomaly, with Adler, Ben Lee, and Treiman, and later with Reuven Aviv. This work later evolved, in more competent hands, into the Wess–Zumino–Witten term.

[18] "Bj" Bjorken had come to the IAS to preach the gospel. At Rockefeller, Mirza Abdul Baqi Beg (who died at the tragically early age of 55) kept asking me why I was fooling with the strong interaction. He claimed that if we happened to pick the right group we could end up in a northern Europe city. This was a widely shared attitude at the time since $SU(2) \otimes U(1)$ was regarded as merely one of many possibilities.

[19] When we first met, he asked me if I were publishing under a pseudonym. I answered that it would have been cool, but I was not clever enough to have thought of that.

[20] Held at Chestnut Hill in Pennsylvania 29–31 May 1973, Temple University.

[21] Later, in Hamburg, Kurt introduced to me to his private club, which was quite an eye opener for a Chinese boy from Catholic Brazil. He passed away in 1983, at the age of 60.

[22] To this day, I have a vivid memory of looking up references in the rather ornate library at Rockefeller.

[23] Received June 25. The Chestnut Hill conference ended on May 31.

[24] Similar work was done independently by Howard Georgi and Tom Appelquist.

not learn numerical computation they would be left out of physics soon. Fortunately for people like me, that prophecy did not come to pass.

The last time I saw Wilson was at Gell-Mann's 80th birthday conference[25] held in Singapore in 2010. Ken appeared not well, and except for a short banquet speech, pretty much kept to himself. Now I regret that I did not make an effort to talk to him more. Perhaps he felt ill, or perhaps the years had put some distance between us.[26]

I thank Stanley Deser, David Gross, Joe Polchinski, and Frank Wilczek for reading this essay.

[25] Both Wilson and Coleman were Gell-Mann's students.

[26] Human interactions are often peculiar; on this occasion, I conversed at length with George Zweig, whom I had never met before, as if we were long lost friends, even though we were not in the same generation.